Acid-Base Interactions: Relevance to Adhesion Science and Technology

Frederick M. Fowkes (1915-1990)

In Honor of the 75th Birthday
of Professor Frederick M. Fowkes

Acid-Base Interactions:
Relevance to Adhesion Science and Technology

Editors: K.L. Mittal and H.R. Anderson, Jr.

Utrecht, The Netherlands
1991

VSP BV
P.O. Box 346
3700 AH Zeist
The Netherlands

First published in 1991

ISBN 90-6764-135-9

CIP-DATA KONINKLIJKE BIBLIOTHEEK, DEN HAAG

Acid-Base Interactions: Relevance to Adhesion Science and Technology:
Festschrift in Honor of the 75th Birthday of Professor F.M. Fowkes/Eds K. L. Mittal and
H.R. Anderson Jr. - Utrecht: VSP
Originally published in: *Journal of Adhesion and Technology*, ISSN 0169-4243;
Vol. 4, Nos. 4, 5 and 8, 1990; Vol. 5, No. 1, 1991
ISBN 90-6764-135-9 bound.
Subject headings: acid-base interactions / adhesion

Printed in Great Britain by Bookcraft (Bath) Ltd., Midsomer Norton

CONTENTS

Part 3: Applications of Acid-base Interactions

Acid-Base Interactions, pp. vii-viii
Eds. K.L. Mittal and II.R. Anderson, Jr.
©VSP 1991

Preface

This book chronicles the Proceedings of the Symposium on Acid–Base Interactions: Relevance to Adhesion Science and Technology held on the occasion of
the 75th birthday of Professor Frederick M. Fowkes as a part of the 64th Colloid
and Surface Science Symposium held at Lehigh University, June 18–20, 1990.
As a matter of fact the papers from this symposium were earlier published in
four issues of the *Journal of Adhesion Science and Technology* (*JAST*) as follows:
Vol. 4, No. 4 (1990); Vol. 4, No. 5 (1990); Vol. 4, No. 8 (1990); and Vol. 5, No. 1
(1991) which were dedicated to him as a Festschrift in his honor. However, a
large number of people indicated interest in acquiring these four issues
separately; concomitantly, we decided to bring out this special commemorative
volume. But as most of you may already be aware, Prof. Fowkes passed away on
October 18, 1990 and the present volume now represents a memorial to this
'giant' of adhesion science. Death came ,o him suddenly, but painlessly; and he
was quite active till the last moment.

When the idea of organizing a symposium to honor Dr. Fowkes was conceived,
the question arose as to what should be the theme of the symposium as he had
made significant contributions dealing with the many ramifications of interfacial
science and technology. However, after some deliberation, it became clear that a
most apposite topic would be 'acid–base and adhesion', as in the last several
years the leitmotif of Dr. Fowkes's research has been in the realm of acid–base
interactions. So it was decided to hold the symposium under the rubric 'Acid–
Base Interactions: Relevance to Adhesion Science and Technology'. The contributions to the Symposium were mostly by invitation only, and there was a
consensus among those invited that this event was very opportune to honor this
doyen of adhesion science.

Here, we will not expatiate upon the many and significant contributions of Dr.
Fowkes as these can be easily gathered from the following excellent write-up by
Dr. Ross. However, we would like to reinforce that Dr. Fowkes had pioneered
many ideas, and the surface and colloid community owes much gratitude and
appreciation to him for his research efforts spanning over half a century. In the
last few years, he had been a key protagonist of the concepts of acid–base interactions, had done a lot to popularize such concepts, and had shown how understanding of such interactions could be parlayed to control adhesion behavior in a
given situation. As a result of his pioneering efforts there has been a paradigm
shift in our approach to adhesion science in terms of acid–base interactions. By
the way, these days the phrase 'paradigm shift' is much in vogue in the business
and technological world, and we adhesionists have seen our own paradigm shift

(from polar–polar interactions to acid–base interactions) and much credit goes to Dr. Fowkes for this transformation.

Now turning to the present volume, here the papers have been rearranged in a more logical manner *vis-à-vis* the order in which they were published in the special issues of *JAST* (*vide supra*). The book opens with an excellent write-up summarizing the professional and personal life of Dr. Fowkes by Dr. Sydney Ross. The rest of the text containing 22 papers is divided into three parts as follows: Part I. Fundamental Aspects of Acid–Base Interactions; Part II. Characterization of the Acid–Base Properties of Materials; and Part III. Applications of Acid–Base Interactions. Among the topics covered include: Acid–base concepts: historical account, current status, and prospects for the future; quantum-mechanical approach to understanding acid–base interactions at metal–polymer interfaces; assessment of acid–base interactions at solid–liquid interfaces; quantitative characterization of the acid–base properties of solvents, polymers, and inorganic surfaces (an overview by Prof. Fowkes himself); acid–base characteristics of a variety of solid materials (clay minerals, carbon fibers, glass fibers, silicas, metals, polymers); acid–base interactions in wetting; and applications of acid–base interactions in a variety of situations, e.g. in the adhesion of polymers to metallic and ceramic substrates, mechanical properties of wood, properties of filled polymers, and behavior of fiber-reinforced polymer composites.

In essence, this volume represents the cumulative wisdom and thinking of contemporary researchers engaged actively in unravelling acid–base interactions and how these can be utilized in understanding and tailoring adhesion. Furthermore, we certainly hope this compendium will become the *vade macum* for anyone interested in acid–base interactions and adhesion. The information provided in this book should serve as a fountain of new ideas for researchers in adhesion to, hopefully, make a quantum leap in understanding the phenomenon of adhesion and concomitantly contribute substantially to the wonderful world of adhesion science.

In closing, we would like to add that this book should serve as a constant reminder of Prof. Fowkes's seminal contributions anent acid–base interactions and adhesion. Needless to say, he will be sorely missed by adhesionists world wide and particularly by those who had the good fortune to come in contact with this gentle soul and to learn from this maestro of adhesion science.

K. L. Mittal
IBM U.S. Technical Education
500 Columbus Avenue
Thornwood, NY 10594, USA
and
H. R. Anderson, Jr.
IBM Corporation
Hopewell Junction
NY 12533, USA

Acid-Base Interactions, pp. ix–xx
Eds. K.L. Mittal and H.R. Anderson, Jr.
©VSP 1991

Frederick M. Fowkes: Industrial and academic scientist

SYDNEY ROSS

Department of Chemistry, Rensselaer Polytechnic Institute, Troy, NY 12180, USA

Revised version received 20 February 1990

When Dr Mittal invited me to write a personal account of Professor Fowkes for a symposium to be held in his honor and in commemoration of his seventy-fifth birthday, I was pleased at the opportunity to present his interesting life story and the details of his productive career. I am an old friend and contemporary of the professor and yield to none in my appreciation of his contributions to science as well as in my esteem for the higher faculties of his mind and the higher qualities of his character, the former summed up as intelligence and the latter as integrity. His intelligence may be relied on for dispassionate judgments; and his integrity for a sincerity that does not always stop short of candor. (This last trait is not welcomed in all quarters.)

Frederick Mayhew Fowkes was born on January 29, 1915 in Chicago, Illinois. His father, William Herbert Fowkes, was an English-born patent lawyer who after graduating from Barnard Castle, a public school (in the English sense) in County Durham, saw action with the Northumberland Fusiliers in the Boer War, served in the Yukon with the Canadian Northwest Mounted Police and then served five years with the U.S. Army in the Philippine Insurrection before attending the Chicago–Kent College of Law. His mother, Eleanor Mayhew (Seley) Fowkes, came from a family of English heritage, which had emigrated to Minnesota and Illinois in the mid-nineteenth century. She was a graduate of the University of Chicago, with a degree in botany.

Fred Fowkes was raised in suburban Chicago and was educated in the public school system of that city. His father died in the flu epidemic of 1919–20 and afterwards his maternal grandparents lived with the family. His grandfather, Charles Arthur Seley, was an inventor and the chief mechanical engineer of the Norfolk and Western Railroad and later of the Rock Island Railroad. He loved music and engineering, and evidently became a role model to his grandson. He continued to work as a design engineer into his eighties. Fred and his brother Bob grew up in a household filled with interest in church (especially choir) music, in instrumental music, in athletics, and in science and electronics. Fred played piano, played the tuba in symphony orchestras, in marching bands, and even in dance bands. In high school he became the tuba player for the All-Chicago High School Symphony Orchestra, and actually played six concerts with the Chicago Symphony Orchestra.

He attended Morgan Park High School, which was more than fifteen miles from the center of the city in a suburban-type neighborhood, of upper-middle class background. It was a small school by Chicago standards and a large fraction of its graduates went on to college. He found the chemistry class uninteresting but he still remembers the good teaching in mathematics, biology and physics. He played on the football team every fall, failing typing as a result of breaking his right elbow in wrestling, and failing it again after tearing ligaments in his left wrist in football. He took about thirty credits more than was required to graduate.

Fowkes became an undergraduate at the University of Chicago at the time when its president Robert Maynard Hutchins (1899–1977) put into effect his exciting 'new plan', with its emphasis on a broad education, on reading the 'great books', on student contact with outstanding professors, and on much personal freedom for students. Class attendance was not required and grades were based only on final exams, consequently students were encouraged to take heavy course loads. Fowkes completed all his undergraduate course work for the bachelor's degree in Chemistry within 2 years, and started taking graduate courses and doing graduate research with William Draper Harkins (1873–1951) in 1934. In spite of his heavy work load he managed to play on the football team during his first year, to play the tuba in the University Symphony Orchestra, and to be an active member of Phi Gamma Delta fraternity, where he became a close friend of Luis W. Alvarez (1911–1988; Nobel laureate in physics, 1968).

To work in Harkins' laboratory was to be introduced to a stimulating research environment, because of the presence of many outstanding post-doctoral fellows and because of the variety of subjects under study, which included nuclear reactions. Frank Long was Fowkes' lab-mate for several years, and both of them accompanied Harkins for a semester at Cornell University during 1936–37. Long eventually became a well-known professor at Cornell.

Fowkes' first surface-chemical studies concerned two-component monolayers (fatty acids plus hydrocarbons). By means of an ultramicroscope that he built into the film balance, he observed the ejection of the oil from the mixed monolayer upon compression [1].* His Ph.D. dissertation was a study of monolayers adsorbed from aqueous solutions by various hydrophobic materials. The pressure-area relations were determined by simultaneous measurements of surface tension and contact angles [3]. This was the first of many subsequent studies of contact angles in Harkins' laboratory, and as a pioneer it had been up to Fowkes to build the equipment himself.

His first employment was a 6-month stint at the research laboratories of Nalco, while he was still a graduate student. He continued his University research during evenings and weekends. Then he was awarded a Proctor and Gamble fellowship, a rare opportunity in those years of the Great Depression. He gave up his job at Nalco, and completed all his thesis research in the next year. Meanwhile he had married Royce Elizabeth Budge, whom he had met when he was a graduate student and she was an art student at the Art Institute of Chicago. A friend in common had brought them together at a picnic meeting of Episcopalian students at the University. Their four children are Gordon Seley (born May 11, 1939), Joan Berkeley *Piper* (born December 19, 1941), Mary-Elisabeth *Tobin* (born January 18, 1950) and Virginia Mayhew *Clark* (born September 19, 1958).

---

*Numbers within brackets refer to items in the numbered List of Publications of F. M. Fowkes appended to this memoir.

Fowkes obtained the Ph.D degree in 1938 (still in the depression period) and went to work in the research laboratories of the Continental Can Company in Chicago, where he was put in charge of a group testing the interaction of polymeric can linings with all the various foods that are canned. He had asked to study the effects of foods on the iron/tin electric couple, for it was quickly evident that tin was the anode in citrus juice, but iron was the anode in canned cherries, strawberries, plums, etc.; but he was always required to put the testing of polymer coatings first, and though his group expanded to seven people, there was never time for fundamental research. So when he was called to serve in the Army in 1942, he knew he did not wish to return to his job at Continental Can.

Fowkes had been in the Reserve Officers Training Corps (ROTC) in college, and had accepted duty as a reserve officer attached to the regular army Third Field Artillery at Fort Sheridan, Illinois. However, when he was called up for active duty as a first lieutenant in February 1942, he was sent to Fort Sill, Oklahoma, together with about forty other artillery reserve officers, for a training class. Such classes started each week and continued for eight weeks. His class included about five chemists with Ph.D. degrees; it also included Temple Fielding, who later wrote the well known Fielding's Guides to Europe.

At Fort Sill Fowkes taught recruits to fire 155-mm howitzers and to drive huge Diamond T trucks. He was soon promoted to captain and put in charge of field exercises for a 10 000-man replacement training center, his task being to schedule all artillery practice and to conduct gas-mask drills. In June 1943 he was transferred to the 31st (Dixie) Infantry Division just as it was going into summer maneuvers in Louisiana, and in September he was assigned to command Battery C of the 149th Field Artillery, a 105-mm howitzer unit in support of the 124th Infantry Regiment. They trained in Virginia, where, he says, "I learned how to motivate troops. I found that all I had to do was to tell the men each morning exactly what was going on, and just what they were expected to do. Once the men knew this, they did their best. We developed high morale and efficiency." They were sent in January 1943 by ship from Virginia to the Panama Canal with a British naval convoy; from there to New Guinea they were unescorted.

After acclimatization in New Guinea, they were sent by landing craft to Aitape, New Guinea, to help defend troops there from a much larger force of attacking Japanese. They restored the broken front lines and experienced tough combat, for their immediate foe was the picked division of 6-foot Japanese who had conquered Singapore. The Americans had expected less formidable opposition and were shocked. But though their gun positions were often attacked at night, and once one of their ammunition dumps was exploded, they did a good job of defending the Drinamour River line, and turned back the attack with great losses to the enemy.

In September 1943 the 31st Division was sent by landing craft to Moratai, a small island in the Spice Islands of the Netherlands East Indies (now Indonesia). General MacArthur walked ashore with them at their landing and inspected the air field, which was soon made into a major air base for all the land-based bombing of the Philippines and even of Borneo. They, in turn, were bombed and strafed 82 nights by the Japanese.

In April 1944 Fowkes was sent to Leyte by plane, to go along with the landing of the 24th Infantry Division in Mindanao, the big southern island of the Philippines.

The 31st Division soon joined them and they were transported by boats a hundred miles up the Mindanao River, where they immediately got into heavy fighting as they drove northwards along the Sayre Highway. They were ambushed frequently and lost many men. Fowkes' assignment was as a liaison officer with an infantry battalion, which, being the most aggressive, was put into the lead and lost the most men. It was finally relieved of the lead when casualties had reduced its strength from 1000 to 245. After one tough attack in which they were subjected to a Banzai charge that splintered the battalion, Fowkes gained the Silver Star medal for Gallantry in Action: he had taken command of the remnants of five infantry companies, organized them, and led them out of danger to rejoin the rest of the battalion.

After this action the 31st went into training for the invasion of Honshu. They were to land east of Tokyo and fight their way into the city. Fortunately the atomic bomb ended the war before such a landing was necessary. By now Fowkes had the rank of major. He did not get back to the States (San Francisco) until December 19, 1945. While waiting for trains east he lived aboard ship at the Embarcadero and looked for a job in chemical research in the Bay Area. When he visited the Shell Development Company (wearing a wool uniform issued to him from a barrel on the dock), Dr Tamale, head of the Catalysis and Surface Chemistry Department, pulled out of his files a reprint of Fowkes' 1940 paper with Harkins. No further recommendation was required; he was immediately offered a job as Research Supervisor.

The Shell Development Company (of Emoryville, California; close to Berkeley) in the forties and fifties was an exciting and advanced research facility. There were about 450 scientists with Ph.D. degrees plus about another 800 people in support. It was the largest research laboratory on the west coast. There was much emphasis on buying or building the best and most advanced laboratory equipment. Faculty members from Berkeley and Stanford often came to use instruments that they yet lacked, such as an NMR spectrometer or an electron microscope. There was also much emphasis on hiring the most innovative people, who came from many countries and institutions. Fowkes' immediate boss had been the first graduate student of Jaroslav Heyrovski (1890–1967), and his boss's boss was from Holland. Fowkes was given the opportunity to build up a well instrumented surface-chemical laboratory, designing many instruments himself; and he was assigned several able researchers, such as Martin Schick and Webster Sawyer.

Fowkes' research group concentrated on the understanding of surface-active solutes, both water-soluble and oil-soluble, with emphasis on the latter, of course. They pioneered the understanding of non-aqueous micelles, of oil-soluble surface-active polymers, of electrostatic and steric stabilization in organic media, of surface-active cosolutes for aqueous systems, of the mechanism of antifoaming additives, and the concept of dispersion forces and acid–base contributions to the work of adhesion and to the surface tension of liquids and solids. The work was a fifty–fifty balance of fundamental and applied studies, in which each part supported the other. The applied problems included gasoline additives, lubricating oil additives, clay-based greases, emulsification processes for crude oils, antifoams for hydrocarbon systems, formulations of aqueous surface-active solutes, anti-corrosion formulations, studies of molecular-weight distributions of epoxies and elastomers, Ziegler catalysis within micelles for elastomer synthesis, etc. Most of this work was approved for publication, though an exception that Fowkes much regretted was the work he did in Amsterdam in 1955–56, when he studied the acidity and basicity of the polar cores of non-aqueous micelles, using indicators to determine H_0 (the Hammett acidity function) of the micellar cores. About 1960, he and his collaborators were publishing about 1% of the articles in the *Journal of Physical Chemistry*. It is clear that Fowkes owes the direction of his subsequent career to the opportunities he received at the Shell Development Company.

In 1962 the Sprague Electric Company (North Adams, Massachusetts) offered Fowkes the position of Director of Research in a new Research Center, for which he was ultimately to recruit about half of its eventual 200 people, 45 of whom had Ph.D. degrees, most of them in physics. The research was at the cutting edge of the new semiconductor and integrated-circuit business. With Fowkes came two others from Shell Development: Frank Anderson (now with IBM) and Kenneth Manchester. Manchester went on to develop the idea of ion implantation in semi-conductors, bringing it into plant operation, for the first time anywhere, in the late 1960s. Under Fowkes' direction the Sprague laboratory pioneered many other developments that are now key techniques of the semiconductor industry, such as plasma-assisted chemical deposition of silicon oxides, nitrides, and carbides. Fowkes' personal research included: (a) a study of the distribution of sodium ions or protons in the surface region of silicon dioxide or silicon nitride insulators, which proved to have a perfect Poisson–Boltzmann distribution of sodium counterions in the oxide, balancing the SiO^- surface charges [37]; and (b) a demonstration of the enhancement of mechanical properties of composite dielectrics resulting from the modification of the high dielectric-constant filler to provide the surface acidity needed to interact with a basic polymer matrix.

Fowkes was disappointed to find that Sprague Electric was too timid to intro-duce the new products and processes developed in their R&D Center. When sales declined in the late 1960s, he was ordered to lay off his fine research staff. He resigned in 1968 and accepted an offer to chair the Chemistry Department at Lehigh University. The lay-off at Sprague Electric continued; finally only about twenty people were left in the R&D Center.

As chairman of the Chemistry Department at Lehigh, Fowkes arrived in time to superintend the planning and construction of a new building, which required political scheming as well as much work with architects. By 1975 the Department moved into the $7 million S.G. Mudd Chemistry Building with attached Neville Auditorium, a modern facility with the best of laboratory furniture, hoods, and ventilation, and with a modern mirror-glass facade. Fowkes' goal as chairman was to transform a teaching faculty into a more research-oriented faculty. His style as a leader was to ask advice from professors but to make all the decisions (and to take all the blame) himself. During his 13 years as Chairman the research funding in Chemistry grew to about $2.5 million annually; it has nearly doubled since then.

During his years as Chairman his research program was small, for the job took about 60% of his time. After his retirement in 1981 from the duties of the Chair he continued to teach one course each semester, and he had much more time to write proposals and manuscripts for publication. His 'retirement' started by his becom-ing a Visiting Scientist at the Materials Laboratory of Wright–Patterson Air Force Base (WPAFB) near Dayton, Ohio. For 2 years during the Fall and Spring semesters at Lehigh he spent every Monday and Tuesday at WPAFB, driving or flying back for his Wednesday-afternoon class. In the summers he spent longer periods at WPAFB, where he had a two- to five-man research group, an FTIR spectrometer to study chemical shifts associated with polymer–solvent inter-actions, and a Cahn microbalance to measure the swelling of polymers by solvents. He had proposed to determine the acid–base characteristics of polymer–solvent interactions, which work was successfully concluded with publication of "Acid–Base complexes of polymers" [78].

During the last few years Fowkes' research group at Lehigh has averaged ten to fifteen members, with much help from Dr. T. B. Lloyd (a Research Scientist), Dr. D. W. Dwight (an Adjunct Professor), and Dr. David A. Cole (a postdoctoral fellow). He has received research grants from ONR, NSF, and DOE, but most of his support has come from industry. At present Fowkes has a DOE grant and about eight industrial grants supporting seven graduate students, four post-doctoral fellows, Lloyd, Dwight, himself and his secretary Dawne Kressler.

Fowkes came to Lehigh with 40 publications from his 27 years in industrial research. At Lehigh there have been about 65 additional publications. His early work at Lehigh was in solid-state chemistry, but soon the work reverted to colloid and surface chemistry.

On revient toujours
A ses premiers amours.

Nowadays most of his studies concern molecular interactions in solution, with emphasis on spectroscopic methods of measuring molecular interactions. His first Ph.D. student at Lehigh was Dennis Hess, who is now a professor of chemical engineering at Berkeley. Since Hess, ten other of his students have received the Ph.D. degree, and by the end of the 1989–90 academic year four others will finish, totalling fifteen. Eight other students have taken their M.S. degree with him, including Professor Cesar Silebi of Lehigh's Chemical Engineering Department.

Fowkes' other professional activities included numerous consultancies. He was a consultant to NSF on materials laboratories, to NIH on dental and pulmonary research, to NBS on polymers, to Redstone Arsenal on composite propellants, to Lawrence Livermore Laboratories and Idaho National Laboratory on floccula-tion in geothermal-energy recovery. His industrial consultancies include those with the 3M Company (12 years,) the Shell Development Company, the Union Carbon and Carbide Corporation, the Atlas Powder Company (now ICI Americas), duPont de Nemours and Company, etc. Fowkes has also been active in various scientific societies. He is a member of ACS (since 1935), the Electro-chemical Society, IEEE, and IRI (Industrial Research Institute). He was the Chair-man of the ACS Division of Colloid and Surface Chemistry in 1968, he has chaired two Gordon Research Conferences (Chemistry at Interfaces, 1971; and Science of Adhesion, 1973). The professional honors that he has received include the Hillman Award of Lehigh University for the "person who has done the most for Lehigh University", 1979. A Symposium in his honor, sponsored by the Division of Colloid and Surface Chemistry of the ACS, was held at the Colloid and Surface Science Symposium at Lehigh University in 1980. The present occasion is another such Symposium, ten years later, with a large program of invited papers under the chairmanship of Dr K. L. Mittal. One of his papers [25] earned the appellation 'Citation Classic' on May 5, 1980, having been cited over 195 times in the chemical literature. In 1989 he received the Adhesion Society Award for Excellence in Adhesion Science, sponsored by 3M.

My own association with Professor Fowkes began when he was at Sprague Electric and resided at Williamstown, Massachusetts, not far from Troy, New York. He was an adjunct professor at RPI, teaching a night course on surface and colloid chemistry. He joined me in 1967 in teaching one of the short courses of the American Chemical Society, entitled "Emulsions and Dispersions". For several years this course was given four or five times a year, but a declining economy and a growing competition have since reduced it to only once or twice a year. At present it is a 4-day course on colloid and surface chemistry applied to industrial R&D, sponsored by RPI. As teachers Fowkes and I have been joined by my former student Dr Ian D. Morrison of the Xerox Corporation. Over the 20 or so years that the course has been taught it has acquired about 2000 alumni, and has resulted in a textbook *Colloidal Systems and Interfaces*, by S. Ross and I. D. Morrison. (John Wiley and Sons, 1988.) Although Fowkes was too busy to be a co-author, this book bears the imprint of his teaching, as its authors had a unique opportunity to hear repeatedly lessons from the master.

Fowkes' experience illustrates how much benefit can be conferred on a University by taking into its faculty a mature industrial scientist of demonstrated productivity. Such a policy, like the quality of mercy, is twice blessed: "it blesseth him that gives, and him that takes". In the course of his career in industry, Fowkes discovered generalities of behavior that, later, in a University laboratory, with the aid of post-doctoral fellows and graduate students, he was able to test with purified and representative compounds. Such a professor has a research program already in mind, with more than an inkling of how it is going to come out. That knowledge is a great advantage in picking topics for Ph.D. dissertations, as he has a shrewd idea that positive results can be obtained within a limited time period. It is also useful in writing proposals for funding with confidence and with a first-hand appreciation of the ultimate significance of the work.

With so distinguished a career behind him, Professor Fowkes might well spend his remaining days in well earned leisure, but I hardly expect such an outcome, for he reverses an old Scottish proverb and believes, as well as sometimes practices, that it is better to work for nothing than to sit idle. But then there is another proverb, ascribed to St Jerome, that perhaps lies nearer the mark: Keep doing some kind of work, that the devil may not find you unemployed. We have the authority of the Hymnal that idle hands attract Satan, and for the kind of work that he then puts them to. With these awesome injunctions, which anyway he shows every sign of heeding. I wish Professor Fowkes godspeed in his undertakings, and many more years of unmischievous, even if ungainful, employment.

PUBLICATION LIST OF DR. F. M. FOWKES

Books

1. F. M. Fowkes (Ed.), *Contact Angle, Wettability and Adhesion*. ACS Adv. Chem. Ser. No. 43, American Chemical Society, Washington D.C. (1964).
2. F. M. Fowkes (Ed.), *Hydrophobic Surfaces*, Academic Press, New York (1969).
3. F. M. Fowkes, Consulting Editor, *Surfactant Science Series*, a multi-volume work published by Marcel Dekker, New York.

Research and other publications

1. F. M. Fowkes, R. J. Myers and W. D. Harkins, Ultramicroscopic examination of mixed films. *J. Am. Chem. Soc.* **59**, 593 (1937).
2. W. D. Harkins and F. M. Fowkes, Pressure-area relations of monolayers at the solid–liquid interface. *J. Am. Chem. Soc.* **60**, 1511 (1938).
3. F. M. Fowkes and W. D. Harkins, The state of monolayers adsorbed at the interface solid–aqueous solution. *J. Am. Chem. Soc.* **62**, 3377 (1940).
4. F. M. Fowkes and W. M. Sawyer, Contact angles and boundary energies of a low-energy solid. *J. Chem. Phys.* **20**, 1650 (1952).
5. F. M. Fowkes, Role of surface-active agents in wetting. *J. Phys. Chem.* **57**, 98 (1953).
6. W. M. Sawyer and F. M. Fowkes, Monolayers in equilibrium with lenses of oil on water. I. Octadecanol and tetradecanoic acid in white oil. *J. Phys. Chem.* **60**, 1235 (1956).
7. M. J. Schick and F. M. Fowkes, Foam-stabilizing additives for synthetic detergents. Interaction of additives and detergents in mixed micelles. *J. Phys. Chem.* **61**, 1062 (1957).
8. L. B. Ryland, G. S. Ronay and F. M. Fowkes, Equilibria in aqueous solutions of copper (II) chelates with α,α'-dipyridyl, *o*-phenanthroline, and ethylenediamine, *J. Phys. Chem.* **62**, 798 (1958).
9. F. M. Fowkes, G. S. Ronay and L. B. Ryland, Mechanism of reaction of diisopropyl fluorophosphonate with cupric 2,2′bipyridine chelate. *J. Phys. Chem.* **62**, 867 (1958).
10. W. M. Sawyer and F. M. Fowkes, Interaction of anionic detergents and polar aliphatic compounds in foams and micelles. *J. Phys. Chem.* **62**, 159 (1958).
11. F. M. Fowkes, G. S. Ronay and M. J. Schick, Monolayers in equilibrium with lenses of oil on water. II. Dependence of equilibrium pressures on pH and on concentration of surfactant. *J. Phys. Chem.* **63**, 1684 (1959).
12. F. M. Fowkes, M. J. Schick and A. Bondi, Oil-soluble surface-active copolymers of alkenes and polar monomers. *J. Colloid Sci.* **15**, 531 (1960).
13. F. M. Fowkes, H. A. Benesi, L. B. Ryland, W. M. Sawyer, K. D. Detling, E. S. Loeffler, F. B. Folckemer, M. R. Johnson and Y. P. Sun, Clay-catalyzed decomposition of insecticides. *J. Agr. Food Chem.* **8**, 203 (1960).
14. F. M. Fowkes, Orientation potentials of monolayers adsorbed at the metal–oil interface. *J. Phys. Chem.* **64**, 726 (1960).
15. F. M. Fowkes, Penetrated monolayers as two-dimensional solutions, in: *Proc. 3rd International Congress on Surface Activity*, Vol. 2, p. 161 (1961).
16. F. M. Fowkes, Ideal two-dimensional solutions: I. Detergent-penetrated monolayers. *J. Phys. Chem.* **65**, 355 (1961).
17. F. M. Fowkes, Ideal two-dimensional solutions. II. A new isotherm for soluble and gaseous monolayers. *J. Phys. Chem.* **66**, 385 (1962).
18. F. M. Fowkes, Ideal two-dimensional solutions. III. Penetration of hydrocarbons in monolayers. *J. Phys. Chem.* **66**, 1863 (1962).
19. F. M. Fowkes, Determination of interfacial tensions, contact angles, and dispersion forces in surfaces by assuming additivity of intermolecular interactions in interfaces. *J. Phys. Chem.* **66**, 382 (1962).
20. F. M. Fowkes, The micelle phase of calcium dinonylnaphthalene sulfonate in *n*-decane. *J. Phys. Chem.* **66**, 1843 (1962).
21. F. M. Fowkes, Rate of adsorption of calcium dinonylnaphthalene sulfonate at the oil–water interface. *J. Phys. Chem.* **67**, 1094 (1963).
22. F. M. Fowkes, Mixed monolayers of cetyl alcohol and sodium cetyl sulfate. *J. Phys. Chem.* **67**, 1982 (1963).
23. F. M. Fowkes, Additivity of intermolecular forces at interfaces. I. Determination of the contribution to surface and interfacial tensions of dispersion forces in various liquids. *J. Phys. Chem.* **67**, 2538 (1963).
24. F. M. Fowkes, Dispersion force contributions to surface and interfacial tensions, contact angles, and heats of immersion in: *Contact Angle, Wettability and Adhesion*, F. M. Fowkes (Ed.), p. 99, American Chemical Society, Washington D.C. (1964).
25. F. M. Fowkes, Attractive forces at interfaces. *Ind. Eng. Chem.* **56**, (12), 40 (1964).
26. F. M. Fowkes, Determination of Intermolecular forces by surface chemical techniques, ASTM Spec. Tech. Publ. No. **360**, 20 (1964).

27. F. M. Fowkes, Ideal two-dimensional solutions. IV. Penetration of monolayers of polymers. *J. Phys. Chem.* **68**, 3515 (1964).

28. F. M. Fowkes, Intermolecular and interatomic forces at interfaces, in: *Surface and Interfaces I. Proc. Sagamore Army Mater. Res. Conf. 13th 1966*, J. J. Burke (Ed.), p. 197, Plenum Press, New York (1967).

29. S. Patton and F. M. Fowkes, Role of plasma membrane in the segregation of milk fat. *J. Theor. Biol.* **15**, 274 (1967).

30. F. M. Fowkes, Molecular forces at interfaces, in: *Surfaces and Coatings Related to Paper and Wood*, R. H. Marchessault and C. Skaar (Eds.), p. 99. Syracuse University Press, Syracuse, New York (1967).

31. F. M. Fowkes, Attractive forces at solid–liquid interfaces, in: *Wetting*, Society of Chemical Industry (London) Monograph No. 25, p. 3 (1967).

32. F. M. Fowkes, Measurement of attractive forces at liquid-solid interfaces, in: *Proc. 4th International Congress on Surface Activity*, Vol. 2, p. 309 (1967).

33. F. M. Fowkes, Surface chemistry, in: *Treatise on Adhesion and Adhesives*, R. L. Patrick (Ed.), Vol. 1, p. 327, Marcel Dekker, New York (1967).

34. F. M. Fowkes, Comments on the calculations of cohesive and attractive energies. *J. Phys. Chem.* **72**, 3700 (1968).

35. F. M. Fowkes, Calculation of work of adhesion by pair potential summation. *J. Colloid Interface Sci.* **28**, 493 (1968).

36. T. E. Burgess, J. C. Baum, F. M. Fowkes, R. Holmstrom and G. A. Shirn, Thermal diffusion of sodium in silicon nitride-shielded silicon oxide films. *J. Electrochem. Soc.* **116**, 1005 (1969).

37. F. M. Fowkes and T. E. Burgess, Electric fields at the surface and interface of silicon dioxide films on silicon. *Surface Sci.* **13**, 184 (1969).

38. F. M. Fowkes, Calculation of work of adhesion by pair potential summation, in: *Hydrophobic Surfaces*, F. M. Fowkes (Ed.), p. 151, Academic Press, New York (1969).

39. F. M. Fowkes, F. W. Anderson and J. E. Berger, Bimetallic Coalescers: Electrophoretic coalescence of emulsions in beds of mixed metal granules, *Environ. Sci. Technol.* **4**, 510 (1970).

40. F. M. Fowkes and T. E. Burgess, Impurity concentration at clean oxide surfaces, in: *Clean Surfaces: Their Preparation, Characterization and Interfacial Studies*, G. Goldfinger (Ed.), p. 351, Marcel Dekker, New York (1970).

41. F. M. Fowkes, Calculation of Hamaker Constants from surface free energy measurements. *Croat. Chem. Acta.* **42**, 374 (1970).

42. F. M. Fowkes, R. W. Lovejoy and J. S. Chow, Anomalous water experiments with silica powders. *J. Colloid Interface Sci.* **36**, 522 (1971).

43. F. M. Fowkes, Calculation of work of adhesion. Analogy to solubility parameter, in: *Physics and Chemistry of Interfaces II*, S. Ross (Ed.), p. 153, American Chemical Society Special Publication (1971).

44. F. M. Fowkes, Donor–acceptor interactions at interfaces. *J. Adhesion* **4**, 155 (1972).

45. F. M. Fowkes, Donor–acceptor interactions at interfaces, in: *Recent Advances in Adhesion*, L. H. Lee (Ed.), p. 39, Gordon and Breach, New York (1973).

46. F. M. Fowkes and D. W. Hess, Control of fixed charge at silicon–silicon dioxide interface by oxidation reduction treatments. *Appl. Phys. Lett.* **22**, 377 (1973).

47. F. M. Fowkes and F. H. Hielscher, Spontaneous electrostatic precipitation of dust, U.S. Nat. Tech. Inform. Serv. PB Rep. No. 222850/00 (1973).

48. F. M. Fowkes and F. E. Witherell, Sodium mobility in irradiated silicon dioxide. *IEEE Trans. Nucl. Sci.* **NS-21**, 67 (1974).

49. F. M. Fowkes, W. E. Dahlke and S. R. Butler, Radiation effects of MOS gate insulators, U.S. NTIS AD Rep. No. 780186/3GA (1974).

50. F. M. Fowkes, W. E. Dahlke and S. R. Butler, Radiation effects in MOS gate insulators, U.S. NTIS, AD Rep. AD-A026017 (1975).

51. H. R. Anderson, Jr., F. M. Fowkes and F. H. Hielscher, Electron donor–acceptor properties of thin polymer films on silicon. II. Tetrafluoroethylene polymerized by RF glow discharge technique. *J. Polymer Sci., Polym. Phys. Ed.* **14**, 879 (1976).

52. M. J. Marmo, M. A. Mostafa, H. Jinnai, F. M. Fowkes and J. A. Manson, Acid-base interaction in filler–matrix systems. *Ind. Eng. Chem. Prod. R&D* **15**, 206 (1976).

53. F. M. Fowkes and D. Z. Becher, The non-aqueous micellar catalysis of the decomposition of benzyl chloroformate, in: *Colloid and Interface Science, Proc. Intl. Conf., 50th. 1976*, M. Kerker

(Ed.), Vol. 2, p. 367, Academic Press, New York (1976).

54. F. M. Fowkes, D. Z. Becher, M. Marmo, C. Silebi and C. C. Chao, Catalysis by cations in cores of non-aqueous micelles, in: *Micellization, Solubilization and Microemulsions*, K. L. Mittal (Ed.), Vo. 2, p. 665, Plenum Press, New York (1977).

55. F. M. Fowkes and S. Maruchi, Surface acidity or basicity of polymers, ACS Organic Coatings and Plastics Chemistry Preprints **37**, No. 1, 305 (1977).

56. F. M. Fowkes, Pigment-matrix interactions, in: *Corrosion Control by Coatings*, H. Leidheiser, Jr., (Ed.), pp. 483–495, Science Press, Princeton, NJ (1979).

57. F. M. Fowkes and M. A. Mostafa, Acid–base interactions in polymer adsorption. *IEC Prod. R&D* **17**, 3 (1978).

58. F. M. Fowkes, Measurement of specific and general interactions at interfaces, Proc. NSF Workshop on Interfacial Phenomena (Seattle, 1979), p. 249.

59. F. M. Fowkes, Surface effects of anisotropic London dispersion forces in *n*-alkanes. *J. Phys. Chem.* **84**, 510 (1980).

60. F. M. Fowkes, Donor-acceptor interactions at interfaces, in: *Adhesion and Adsorption of Polymers*, L.-H. Lee (Ed.), Part A, pp. 43–52, Plenum Press, New York (1980).

61. F. M. Fowkes, D. C. McCarthy and M. A. Mostafa, Contact angles and the equilibrium spreading pressures of liquids on hydrophobic solids. *J. Colloid Interface Sci.* **78**, 200 (1980).

62. F. M. Fowkes, Attractive forces at interfaces. *Current Contents*, 12 (May 5, 1980).

63. F. M. Fowkes and F. H. Hielscher, Electron injection from water into hydrocarbons and polymers. *ACS Org. Coat. Plast. Chem. Preprint* **42**, 169 (1980).

64. F. M. Fowkes, Surface tension, in: *Encyclopedia of Physics*, p. 985, Addison-Wesley (1980).

65. F. M. Fowkes, C.-Y. Sun and S. T. Joslin, Enhancing polymer adhesion to iron surfaces by acid-base interaction, in: *Corrosion Control by Organic Coatings*, H. Leidheiser Jr., (Ed.), pp. 1–3, National Association of Corrosion Engineers, Houston (1981).

66. F. M. Fowkes, Adsorption, in: *Encyclopedia of Science and Technology*, Vol. 1, p. 124, McGraw-Hill (1982).

67. Y. Kato, F. M. Fowkes and J. W. Vanderhoff, Surface energetics of the lithographic printing process. *IEC. Prod R&D* **21**, 441 (1982).

68. F. M. Fowkes, Characterization of solid surfaces by wet chemical techniques, in: *Industrial Applications of Surface Analysis*, L. A. Casper and C. J. Powell, (Eds.), ACS Symposium Series No. 199, pp. 69–88 (1982).

69. F. M. Fowkes, Surface treatments for enhanced bonding between inorganic surfaces and polymers, in: *Surface Treatments for Improved Performance and Properties*. J. B. Burke and V. Weiss (Eds.), pp. 75–84, Plenum Press, New York (1982).

70. F. M. Fowkes, B. L. Butler, P. Schissel, G. B. Butler, N. J. DeLollis, J. S. Hartman, R. W. Hoffman, O. T. Inal, W. G. Miller and H. F. Thompkins, Basic research needs and opportunities at the solid–solid interface. *Mater. Sci. Eng.* **53**, 125 (1982).

71. F. M. Fowkes, Acid-base interaction in polymer adhesion, in *Microscopic Aspects of Adhesion and Lubrication*, J. M. Georges (Ed.), pp. 119–138. Elsevier, Amsterdam (1982).

72. F. M. Fowkes, F. W. Anderson, R. J. Moore, H. Jinnai and M. A. Mostafa, Mechanisms of electric charging of particles in non-aqueous liquids, in: *Colloids and Surfaces in Reprographic Technology*, M. L. Hair and M. D. Croucher (Eds.), ACS Symp. Ser. No. 200, pp. 307–324 (1982).

73. Frederick M. Fowkes, Principles of detergency. *Chemical Times and Trends*, 31–53 (January 1983).

74. F. M. Fowkes, Acid-base interactions in polymer adhesion, in: *Physicochemical Aspects of Polymer Surfaces*, K. L. Mittal (Ed.), Vol. 2, pp. 583–603, Plenum Press, New York (1983).

75. F. N. Fowkes and J. Carnali, Oil solubilization studies, Department of Energy Report, 1983, Grant No. DE-FG19-80ET12267.

76. Robert J. Pugh, Toshiaki Matsunaga and Frederick M. Fowkes, Electrostatic and steric contributions to colloidal stability of carbon black in dodecane. *Colloids Surfaces* **7**, 183 (1983).

77. Robert J. Pugh and Frederick M. Fowkes, The dispersibility and stability of carbon black in media of low dielectric constant. 2. Sedimentation volume of concentrated dispersions, adsorption and surface calorimetry studies. *Colloids Surfaces* **9**, 33–46 (1984).

78. F. M. Fowkes, D. O. Tischler, J. A. Wolfe, L. A. Lannigan, C. M. Ademu-John and M. J. Halliwell, Acid–base complexes of polymers. *J. Polym. Sci., Polym. Chem. Ed.* **22**, 547–566 (1984).

79. F. M. Fowkes and R. J. Pugh, Steric and electrostatic contributions to the colloidal properties of non-aqueous dispersions, in: *Polymer Adsorption and Dispersion Stability*, E. D. Goddard and B.

Vincent, (Eds.), ACS Symp. Ser. No. 240, pp. 331–354 (1984).

80. F. M. Fowkes, Acid–base contributions to polymer–filler interactions. *Rubber Chem. Technol.* **57**, 328–343 (1984).

81. M. S. El-Aasser, C. D. Lack, Y. T. Choi, T. I. Min, J. W. Vanderhoff and F. M. Fowkes, Interfacial aspects of miniemulsions and miniemulsion polymers. *Colloids Surfaces* **12**, 79–97 (1984).

82. F. M. Fowkes, Interface acid–base/charge transfer properties, in: *Surface and Interfacial Aspects of Biomedical Polymers*, J. D. Andrade (Ed.), Vol. 1, Plenum Press, New York (1985).

83. F. M. Fowkes, Acid–base and electrostatic contributions to the adhesion of polymers to metal and ceramic substrates. *Materials Research Society Proceedings* **40**, 239–250 (1985).

84. C. D. Lack, M. S. El-Aasser, J. W. Vanderhoff and F. M. Fowkes, Interfacial phenomena of mini-emulsions, in *Macro- and Microemulsions: Theory and Practice*, D. O. Shah (Ed.), ACS Symp. Ser. No. 272, pp. 345–356 (1985).

85. J. O. Carnali and F. M. Fowkes, Micellar structure and equilibria in aqueous microemulsions of methyl methacrylate, *ibid*, pp. 287–302.

86. F. M. Fowkes and J. O. Carnali and J. A. Sohara, The role of the middle phase in emulsions, *ibid.*, pp. 173–183.

87. F. M. Fowkes, D. C. McCarthy and D. O. Tischler, Predicting enthalpies of interfacial bonding of polymers to reinforcing pigments, in: *Molecular Characterization of Composite Interfaces*, H. Ishida and G. Kumar, (Eds.), pp. 401–411, Plenum Press, New York (1985).

88. Donald C. McCarthy and Frederick M. Fowkes, Predicting energies of polymer–filler interactions, 40th Ann. Conf., Reinforced Plastics/Composites Institute, Soc. Plastics Industry, **17D** 1–8 (1985).

89. F. M. Fowkes, D. W. Dwight, T. B. Lloyd and D. O. Tischler, Modification of surface acidity and basicity of glass fillers, *ibid.* **17E**, 1–6 (1985).

90. Sara T. Joslin and Frederick M. Fowkes, Surface acidity of ferric oxides studied by flow micro-calorimetry. *IEC Prod R&D* **24**, 369 (1985).

91. Joseph O. Carnali and Frederick M. Fowkes, Microemulsions of methyl methacrylate in aqueous sodium lauryl sulfate: Structure and interaction energetics by SAXS, NMR spectroscopy, and G. C. headspace analysis. *Langmuir* **1**, 576–587 (1985).

92. D. L. Allara, F. M. Fowkes, J. Noolandi, G. W. Rubloff and M. V. Tirrell, Bonding and adhesion of polymer interfaces. *Mater. Sci. Eng.* **83**, 213–226 (1986).

93. Frederick M. Fowkes, Role of acid–base interactions in inorganic powder dispersion and composites, in: *Surface and Colloid Science in Computer Technology*, K. L. Mittal (Ed.), pp. 3–25, Plenum Press, New York (1987).

94. Frederick M. Fowkes, Dispersions of ceramic powders in organic media. *Adv. Ceramics* **21**, 411 (1987).

95. Frederick M. Fowkes, Role of acid–base interfacial bonding in adhesion. *J. Adhesion Sci. Technol.* **1**, 7–27 (1987).

96. F. M. Fowkes, Y. C. Huang, B. A. Shah, M. J. Kulp and T. B. Lloyd, Surface and colloid chemical studies of gamma iron oxides for magnetic memory media, *Colloids Surfaces* **29**, 243–261 (1988).

97. F. L. Riddle, Jr., and F. M. Fowkes, Spectral shifts in acid–base chemistry of polymers. *ACS Polymer Preprints* **29**(1), 188 (1988).

98. Frederick M. Fowkes, Acid–base interactions, in: *Mark–Bikales–Overberger–Menges Encylcopedia of Polymer Science and Engineering*, Supplement Volume, Second Edition, pp. 1–12, John Wiley & Sons (1989).

99. Frederick M. Fowkes, Kenneth L. Jones, Li Guozhen and Thomas B. Lloyd, Surface chemistry of coal by flow microcalorimetry, *Energy &Fuels* **3**, 97–105 (1989).

100. Frederick M. Fowkes, David W. Dwight, David A. Cole and T. Clare Huang, Acid–base properties of glass surfaces. *J. Non-Crystalline Solids*, in press.

101. Floyd, L. Riddle, Jr. and Frederick M. Fowkes, Spectral shifts in acid–base chemistry. I. Van der Waals contributions to acceptor numbers. *J. Am. Chem. Soc.* in press.

102. Frederick M. Fowkes, Floyd L. Riddle, Jr., William E. Pastore and Allen A. Weber, Interfacial interactions between self-associated polar liquids and sqalane used to test equations for solid–liquid interfacial interactions. *Colloids Surfaces* **43**, 367–387 (1990).

103. F. M. Fowkes, M. J. Kulp and J. Horothai, Adsorption of polystyrene onto noble metals from theta solvents, submitted to *Colloids Surfaces* (1990).

Ph.D. STUDENTS

David Z. Becher
Joseph Carnali
Lawrence A. Casper
Wen-Jang Chen
Dennis Hess
Kenneth J. Jones

Sara Joslin
Mary Jo Kulp
Donald McCarthy
Mohamed A. Mostafa
Joseph A. Sohara
Frederick Witherell

M.S. STUDENTS

David A. DeCrosta
Fred Hessel
Jain Horothai
Wei-Hsin Hou
Mary B. Kaczinski
John A. Kiddon

Judith Niece
Alam Shah
Prof. Cesar A. Silebi
Daniel O. Tischler
Daryl R. Williams
James E. Wolfe

RESEARCH COLLABORATORS

C. M. Ademu-John
Dr. Herbert R. Anderson, Jr. (IBM, E. Fishkill)
Dr. Christina Blom (Stockholm, Sweden)
C. C. Chao (Union Carbide, Tarrytown, NY)
David W. Dwight
Li Guozhen (Zhejiang University, China)
M. J. Halliwell (Wright-Patterson Air Force Base)
Hideo Jinnai (Saitama, Japan)
Prof. Vikram Kumar (Indian Institute of Science, Bangalore, India)

L. A. Lannigan (Abbot Co., Chicago)
Dr. Trisno Makagwinata (3M Company)
Sachio Maruchi (Tokyo, Japan)
Prof. Toshiaki Matsunaga (Akita University, Japan)
Maria Palsson (Stockholm, Sweden)
William Pastore (University of Scranton)
Dr. Robert Pugh (Stockholm, Sweden)
Prof. Howard Swain, Jr. (Wilkes College)
Patricia Webber
Allan Weber

Part 1: Fundamental Aspects of Acid-Base Interactions

Acid-Base Interactions, pp. 3-23
Eds. K.L. Mittal and H.R. Anderson, Jr.
©VSP 1991

Overview Lecture
The Lewis acid–base concepts: recent results and prospects for the future

WILLIAM B. JENSEN

Department of Chemistry, University of Cincinnati, Cincinnati, OH 45221, USA

Revised version received 6 September 1990

Abstract—Since many of the applications of the Lewis acid–base concepts to problems in the field of surface chemistry are dependent on the use of the three basic models of Lewis acid–base phenomena introduced in the 1960s (the HSAB principle, the $E \& C$ equation, and the DN–AN approach), this overview lecture briefly summarizes and comments on the published literature on these models that has appeared since they were least reviewed in 1980. It also comments on the problem of extending them to the multi-site interactions characteristic of polymers and surfaces and suggests some possible directions for future work in this area.

Keywords: Lewis acid–base; HSAB principle; $E \& C$ equation; donor–acceptor numbers.

1. INTRODUCTION

When asked to give an overview lecture for this symposium in honor of Dr. Fowkes, a number of possible themes suggested themselves. The most obvious choice would have been a review of Dr. Fowkes's own work on the application of acid–base concepts to problems in the field of surface chemistry. However, it is well to remind ourselves that, although we have selected this topic as the theme for this symposium, we are actually celebrating Dr. Fowkes's entire career, and this encompasses far more than just his work with the Lewis acid–base concepts, which, if anything, is merely the most recent entry in a long and distinguished list of contributions. To single out just this aspect would, then, be a distortion and to try and deal with the whole would be an act of hubris on my part, since I am an inorganic chemist, rather than a surface chemist, and I am really not competent to evaluate the significance of Dr. Fowkes's other activities in the field.

A second possible choice would have been to provide an overview of the relevance of the Lewis concepts to problems in the area of surface chemistry and to review recent examples of their successful application. However, given the nature of this symposium, this would be akin to carrying coals to Newcastle, as I would not only be preaching to the already converted (however pleasant that prospect), but performing a redundant service as well, since the proceedings of this symposium will in fact constitute just such a state-of-the-art review.

A third possibility would have been to provide a general introduction to the Lewis acid–base concepts which could serve as a background review for the more specialized papers which follow. Though a pedagogically laudable task, it is also one which I have performed many times in the past [1–7] and, quite frankly, I am tired of listening to myself.

As implied by my title, my final and actual choice is a bit more specialized and presumes some familiarity, not only with the original Lewis acid–base definitions, but with much of their evolution during the last 70 years. My justification for this choice is best seen by first giving a very abbreviated overview of this evolution, which I have divided into four fundamental, and largely non-overlapping, periods (Table 1).

Table 1.
The four fundamental periods in the evolution of the Lewis acid–base concepts

The classical period
- Original definitions (Lewis, 1923, 1938)
- Electrophiles and nucleophiles (Ingold, 1933)
- First monograph (Luder and Zuffanti, 1946)

The quantum-mechanical period
- Quantum-mechanical treatment of charge-transfer complexes (Mulliken, 1951)
- Frontier orbital theory of electrophilic and nucleophilic reactivity (Fukui, 1952, 1954)
- First monograph (Briegleb, 1961)

The quantitative period
- Hard–soft acid–base principle (Pearson, 1963)
- E & C equation (Drago and Wayland, 1965)
- Donor and acceptor numbers (Gutmann *et al.*, 1966, 1975)
- Perturbation theory of reactions (Hudson and Klopman, 1967)
- First comparative monograph (Jensen, 1980)

The applied period
- First applications to surface chemistry (Fowkes, 1972)
- First monograph (proceedings of this symposium)

The first of these, the classical period, corresponds to the introduction of the original definitions by Lewis in 1923 [8] (and again in 1938 [9], this time with substantially more impact) using the idiom of the localized shared electron-pair bond; the independent introduction of the same concepts in organic chemistry by Ingold in 1933 under the guise of electrophilic and nucleophilic agents [10, 11]; and the publication of the first monograph on the subject by Luder and Zuffanti in 1946 [12].

This was followed in the 1950s by the quantum-mechanical period, which saw a rephrasing of the concepts within the idioms of both valence bond and molecular orbital theory, largely as a result of Mulliken's theoretical treatment of so-called charge-transfer complexes [13] and Fukui's frontier orbital theory of nucleophilic and electrophilic reactions [14]. The first monograph dealing with Mulliken's approach was written by Briegleb in 1961 [15].

The 1960s saw the beginning of what I have called the quantitative period, in which a variety of attempts were made to quantify the Lewis concepts. Foremost among these were the hard–soft acid–base (HSAB) principle of Pearson [16], the E & C equation of Drago and Wayland [17], the donor and acceptor numbers of Gutmann and co-workers [18, 19], and the perturbation theory of Hudson and Klopman [20], which was largely an attempt to rationalize quantum-mechanically the HSAB principle. Though all of these authors published books during the 1960s and 1970s dealing with their particular approaches [21–24], the first comparative monograph was not published until 1980 [3].

The fourth and final period, that of applications, began in 1972 with Dr. Fowkes's first attempts to apply the $E \& C$ approach of Drago and co-workers to problems in surface chemistry [25] and, until recently, he has remained almost the only person to actively pursue this line of research. However, the large number of participants in this symposium is elegant testimony to the fact that this situation is rapidly changing, and there is little doubt that, when finally published in book form, the resulting symposium proceedings will constitute the first definitive monograph to deal with this aspect of Lewis acid–base chemistry.

Given that each of these periods is dependent on its predecessors and, in particular, that much of the work being done during the applied period is based on the models introduced during the quantitative period, I felt that I, as an inorganic chemist interested in the Lewis concepts, could best serve the needs of those of you who are active in applying them to the field of surface chemistry by critically reviewing some of the basic literature on the Lewis concepts that has appeared since the publication of ref. [3] in 1980, and also by reminding you (in roughly chronological order) of some of the unresolved problems still present in many of the original models. In the process, I will also comment on some potential problems in extending these models, which were largely developed for discrete molecular and ionic species, to the multi-site acid–base interactions characteristic of polymers and surfaces. Lastly, I hope to suggest some new approaches to the ongoing problem of quantifying Lewis acid–base interactions.

2. THE HSAB PRINCIPLE

The most important advance in this area since 1980 is the introduction of the concept of absolute hardness by Parr and Pearson [26–29] in 1983 using the method of density functionals. This defines the electronegativity (χ) of a species as the negative rate at which the energy (E) of the species changes with a change in its electron population (N) at constant potential (v):

$$\chi = -(\partial E/\partial N)v = (I + A)/2, \tag{1}$$

where I is the ionization potential of the species and A is its electron affinity.

The absolute hardness (η) of a species is, in turn, defined as half the negative rate at which its electronegativity changes with a change in its electron population at constant potential:

$$\eta = -(\partial \chi/\partial N)v/2 = (\partial^2 E/\partial N^2)/2 = (I - A)/2, \tag{2}$$

where the factor of 1/2 has been arbitrarily introduced in order to create a formal symmetry between the final expressions for electronegativity and absolute hardness when defined in terms of I and A.

Since most chemists are trained to think in terms of orbitals rather than density functionals, Pearson has attempted to translate these results into molecular orbital theory [30, 31]. Within this context, the absolute hardness of a closed-shell species with only doubly occupied orbitals becomes a measure of the energy separation between its frontier orbitals (i.e. its HOMO–LUMO gap)—a parameter which is large for hard species and small for soft species. This conclusion is essentially identical to the definition of softness given by Jørgensen over two decades ago [32] and also correlates with the polarizability of a species, a property which, in the case of Lewis bases at least, is commonly used as qualitative measure of softness [33]. In the case of open-shell species with a singly occupied orbital, the absolute hardness becomes a measure of intra-orbital electron repulsion. This is large for hard species with small, highly localized orbitals and small for soft species with large, diffuse orbitals.

Using the definition in equation (2), Pearson has also calculated hardness values for a large number of discrete ionic and molecular acids and bases, both in the gas phase and in aqueous solution [31, 34, 35] and later in this symposium Dr. Lee [36] will discuss its possible extension to solids in terms of the energies of their valence and conduction bands.

If one assumes that the degree of charge transfer between an interacting acid and base is a function of the difference in their electronegativities and ceases when these become equal, then the relation between hardness, as defined by equation (2), and the degree of electrostatic vs. orbital perturbation in an acid–base interaction becomes immediately apparent. The electronegativities of species with large values of η are very sensitive to changes in N and will rapidly equalize with only a small degree of charge transfer. In other words, there will be little orbital perturbation and the stability of the resulting hard–hard acid–base complex will depend, for the most part, on the initial net charges and/or dipoles present on the interacting acid and base and will be largely electrostatic in nature. For soft species, the opposite will be true. The low sensitivity of χ to changes in N will result in extensive charge transfer between the acid and base before an equalization of their electronegativities occurs and will lead to extensive orbital perturbation [37].

It should, however, be pointed out that there are many other quantitative softness scales in the literature, including several proposed since 1980 [6, 38–41], which give comparable results and that the absolute hardness concept is not without its problems. These include, in the case of anionic bases, the necessity of applying equation (2) to the neutral atom, rather than to the anion itself, in order to obtain a consistent scale; anomalous η values for Tl^+ vs. Tl^{3+}, for gas-phase vs. aqueous species, and for both H^- and H^+. In the latter case, there is no value for the ionization potential in equation (2), so η cannot be calculated, though Parr and Pearson have finessed this by arguing that its absence implies that it is really infinite in value.

Interestingly, the H^+ ion, which has been taken as the quintessential example of a hard species, is really atypical in its behavior and also has an anomalous position on many of the other softness scales that have been proposed. Thus, use of the simple criterion that the bonding in hard–hard complexes is largely ionic or polar, whereas that in soft–soft complexes is covalent, would place H^+, in contrast to the alkali metal ions, in the soft category. Likewise, the use of either gas-phase LUMO energies or orbital electronegativities ranks H^+ as the softest of all the M^+ ions [3]; the use of redox potentials places it between Cu^+ and Tl^+ [3], as does the Pearson–Mawby scale [42]; and the Ahrland scale places it between Cs^+ and Cu^+ [43]. DeKock [44] has also cited several examples of orbital-controlled electrophilic attack by H^+, behavior which is again considered to be characteristic of soft, rather than hard, species.

These placement problems reflect a long-standing ambiguity in the phraseology of both the HSAB classification and the HSAB principle. As originally derived from the Edwards equation, softness was intended to be an additional factor in the stabilization of nonprotonic acid–base complexes, as expressed by the relation

$$\log K_{AB} = S_A \cdot S_B + \sigma_A \cdot \sigma_B, \tag{3}$$

where S_A and S_B are the strength parameters of the interacting acid and base, and σ_A and σ_B are their softness parameters. In this relation, H^+ was taken as the arbitrary zero point for σ_A so that

$$\log K_{HB} = pK_a = S_A \cdot S_B. \tag{4}$$

Species which were softer than H^+ would have positive values of σ and experience additional stabilization in their complexes, whereas those which were harder would have negative values and experience destabilization. It was further assumed that the strength parameters could be calculated using an electrostatic model of some sort, whereas the softness parameters would reflect the ability of the species to undergo charge-transfer and orbital perturbations, leading to a significant covalent contribution to the bonding.

The ambiguity in the HSAB classification is the result not only of forgetting that H^+ was intended as an arbitrary zero, rather than as an absolute zero, for the softness scale, but of forgetting that the application of relation (3), in either a qualitative or a quantitative sense, requires that species be ranked according to *both* strength and softness. Indeed, it is a simple matter to show that most scales of so-called softness reproduce Pearson's original qualitative classification only if one uses the ratio σ/S rather than σ alone [3]. The simplest way to do this is to construct a 'sorting map' of species by plotting their S values vs. their σ values. If this is done for ions, using Z/r as the strength parameter, one finds that one must use diagonal lines, rather than vertical lines, to separate the species in the plot into regions corresponding to the hard and soft classification (see Figs 1 and 2). In other words, the classification is a function of σ/S (or S/σ, if you wish to maximize the ratio for hardness rather than softness) rather than σ alone. Thus, H^+ appears in the hard category, in spite of its moderate-to-large σ value on most scales, because it also has a large S value and thus gives a σ/S ratio similar to that of the alkali metal ions, which have inherently smaller values of both σ and S.

The confusion relative to the wording of the HSAB principle itself stems from the fact that relation (3) implies that any statement regarding the stability of acid–base complexes which mentions hardness and softness, but not strength, is incomplete. On the other hand, if such a statement is complete, then the current HSAB classification implies that it is only necessary to know the ratio of σ/S in order to make a prediction about stability, rather than the independent values of S and σ, as implied by relation (3).

This latter problem has been partly tested in an interesting paper published by Arbelot and Chanon in 1983 [45]. Using a purely qualitative version of the HSAB classification, in which species were placed in either the soft, hard, or

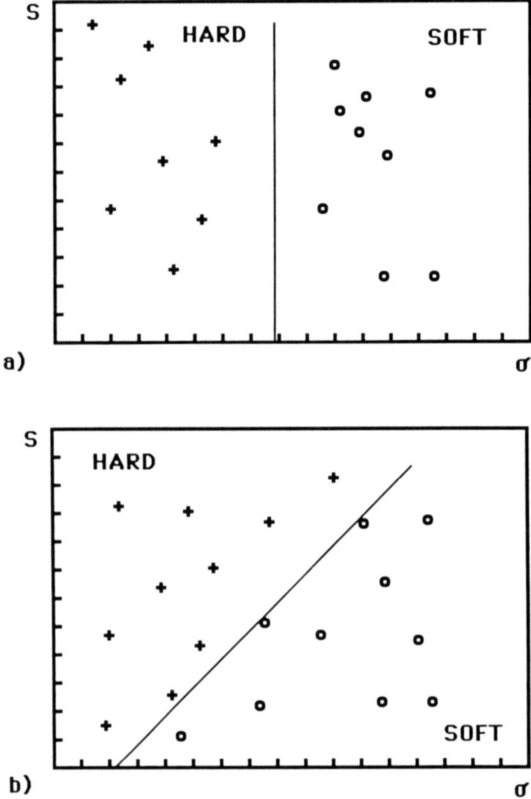

Figure 1. Two limiting cases of an idealized sorting map, based on a plot of the strength (S) vs. the so-called softness (σ) parameters for various species. Circles indicate soft species according to the HSAB classification and crosses indicate hard species. If the classes can be sorted using only vertical lines, as in (a), then σ alone reproduces the HSAB classification. If diagonal sorting lines are required, as in (b), then the classification is really based on some function of σ/S (or S/σ). For more on sorting maps and their uses, see refs [77, 78].

borderline categories, but not ranked within these categories in terms of relative softness, they first showed that such a model could be applied to only 25% of all potential acid–base interactions. That is, one could apply it only to situations in which a clear choice was presented in terms of the three categories (e.g. a hard acid selecting between a hard base and a soft base, but not between two alternative hard bases). Within this 25%, they then tested the version of the HSAB principle that ignores the operation of strength as an independent parameter on 320 example reactions involving gas-phase ionic species, gas-phase molecular species, and aqueous ionic complexes. The principle gave a correct prediction in only 60% of the cases studied, which is barely above random guess.

In order to test the version of the HSAB principle in which softness and strength operate independently, they then established a similar qualitative classification of species as strong, borderline, and weak based on the use of pK_a data. This dropped the number of applicable cases for a purely qualitative model to 9.2% of all possible acid–base interactions, though, within this more restricted set, the success rate of the extended version of the HSAB principle, based on

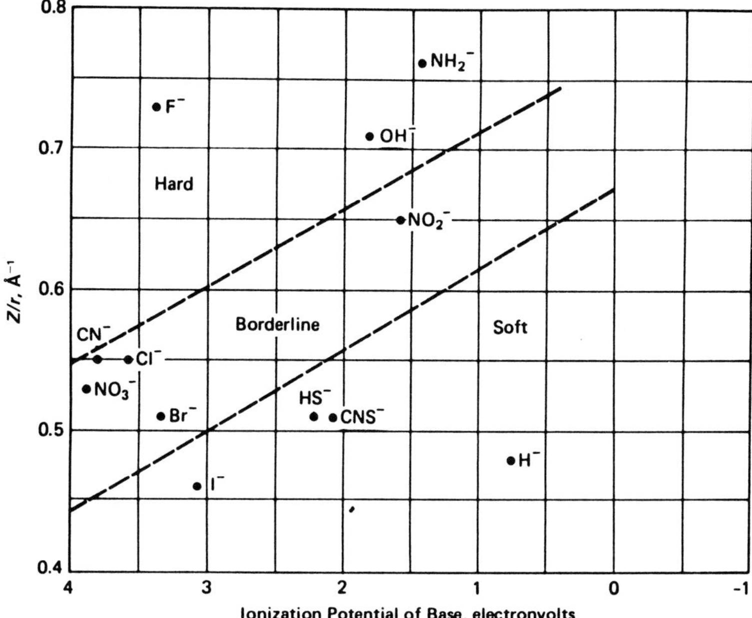

Figure 2. An example of an actual sorting map showing a plot of strength for various ionic Lewis bases, as measured by Z/r, vs. softness, as measured by the first ionization potential. Diagonal lines are required to reproduce the HSAB classification, indicating that it depends on the ratio $I \cdot r/Z$, rather than on I alone (from ref. [3]).

both strength and softness, rose to 86%. Their conclusion was that a purely qualitative classification was much too restricted in its application and that quantitative scales of both strength and softness must be used in conjunction with a relation similar to that in equation (3). In short, the purely qualitative form of the HSAB principle, which refers only to hardness and softness, is wrong.

A similar evaluation of the HSAB principle has been made by Purcell [46] within the limits of the E & C equation. He reported a success rate of 37.7% for the prediction of single displacement reactions and 71.9% for double displacement reactions.

3. THE E & C EQUATION

The E & C equation was introduced by Drago and Wayland in 1965 to predict the enthalpies of formation of one-to-one molecular adducts in the gas phase and in poorly coordinating solvents [17]:

$$-\Delta H_{AB} = E_A \cdot E_B + C_A \cdot C_B, \tag{5}$$

where the E parameters supposedly represent the electrostatic contributions of the acid and base to adduct stability and the C parameters represent the covalent contributions. The parallel between this equation and relation (3) is obvious, and it has proved highly successful for the systems for it was originally designed. The only exception appears to be the E_A and C_A values for $SbCl_5$, which Drago introduced in order to make comparisons with Gutmann's donor numbers (see

10   *William B. Jensen*

next section). As Marcus has recently noted, the predictions for the adducts of this acid agree very poorly with the experimental values reported by Gutmann [47].

It should also be noted that, in spite of the names assigned to the parameters and their implied basis in bonding theory, there is, to the best of my knowledge, no evidence that they in fact reflect the relative electrostatic and covalent contributions to the bonding in the resulting adducts. That is, they have not been shown to correlate with either a physical property (e.g. dipole moment, ionization potential, etc.) or with a quantum-mechanically calculated index thought to parallel such contributions. This observation, of course, does not damage their value as a purely empirical method of calculating enthalpies of formation for molecular adducts; it is merely intended to warn users that they should not take the implied theoretical basis too literally.

The original E & C equation was designed to deal with enthalpies of gas-phase acid–base interactions and, by implication, with bond energies. However, in the last decade, Drago and co-workers have attempted to explore its relationship to other acid–base scales, most of which are based either on spectral shifts or on free-energy data, using the modified equations [48]

$$\Delta X + W = E_A^* \cdot E_B + C_A^* \cdot C_B \tag{6}$$

and

$$\Delta X + W = E_A \cdot E_B^* + C_A \cdot C_B^*, \tag{7}$$

where ΔX is the change in the property in question (and is equal to $-\Delta H_{AB}$ in the case of the original gas-phase enthalpies), W is any constant contribution to the interaction when a given acid interacts with a series of bases [equation (6)] or a given base interacts with a series of acids [equation (7)] and is equal to zero in the case of the gas-phase enthalpy data correlated by the original equation, and the starred E and C parameters are the original unstarred values modified to incorporate the necessary conversion factors between the units of enthalpy and those of the new property, ΔX, (i.e. frequency shift, log K, etc.) being correlated. Using these relations, they have established correlations between the E and C model and such properties as the Hammett substituent coefficients [49], the Soret band [50] and IR frequency shifts which accompany acid–base adduct formation [49], the Kamlet–Taft β parameter [51], and the theoretical basis of linear free-energy correlations in general [52]. Correlations between the E and C model and both cobalt–carbon bond strengths [49] and metal–metal bond strengths [53] have also been established.

An attempt to extend the original E & C model in yet a second direction was made by Kroeger and Drago in 1981 [54]. They proposed a revised six-parameter equation:

$$-\Delta H_{AB} = e_A \cdot e_B + c_A \cdot c_B + t_A \cdot t_B, \tag{8}$$

in which the e and c parameters again represent the (redetermined) electrostatic and covalent contributions of the acid and base, and the additional t parameters reflect their ability to undergo charge transfer. It was hoped that this extended equation would allow ion–ion adducts and ion–molecule adducts, as well as the molecule–molecule adducts covered by the original E & C equation, to be incorporated within a single model. However, as far as I know, there has been no further development of this approach, though more recently a third proposal has been made which is apparently able to deal with gas-phase enthalpies for neutral molecule–ion interactions [55, 56].

The original E & C equation has been the model of choice for Dr. Fowkes in his work on the application of Lewis acid–base theory to surface chemistry, and he has attempted to determine E and C parameters for both polymers and surfaces [57]. The only potential problem with such extensions lies in the fact that the E & C equation implicitly assumes that a molecule is either a Lewis acid or a Lewis base. In reality, most molecules have both donor and acceptor sites, though usually one of these functions clearly dominates the molecule's chemistry. Nevertheless, when we are dealing with bulk liquids, in which self-association of the molecules plays an important role, this amphoteric behavior is potentially important.

4. DONOR AND ACCEPTOR NUMBERS

The donor number (DN), as a measure of Lewis basicity, was introduced in 1966 by Gutmann, Steininger, and Wychera [18], and was defined as the negative of the enthalpy of formation for the adduct formed between the base in question and the reference Lewis acid, antimony pentachloride:

$$DN_B = -\Delta H(SbCl_5 - B). \tag{9}$$

This was supplemented in 1975 by the introduction of the acceptor number (AN) as a potential measure of Lewis acidity for a species [19]. This was defined as the relative ^{31}P-NMR shift induced in triethylphosphine oxide (Et_3PO) when it was dissolved in the species in question. This was further scaled by assigning a value of 0 to the shift induced by hexane and a value of 100 to the shift produced by $SbCl_5$ upon interacting with Et_3PO in a dilute 1,2-dichloroethane solution.

Gutmann further proposed that the enthalpy of a given acid–base interaction could be approximated by a two-parameter equation of the form

$$-\Delta H_{AB} = AN_A \cdot DN_B/100, \tag{10}$$

where the factor of 100 converts the AN value from a percentage of the $SbCl_5$ value to a decimal fraction.

Some comments on both scales and on the interaction equation itself are in order. First, it is important to realize that many reported DN values are only approximate, as they have been determined indirectly by means of linear correlations with other reported measures of Lewis basicity, rather than being directly measured in the laboratory via equation (9). Thus, of the 171 values given by Marcus in 1984 [47], only 50 were determined calorimetrically. Indeed, for some of the so-called 'bulk' donicity values given in the literature, Marcus reported that he was unable to track down the origins of the original approximations. Even in the case of those values that were determined experimentally, there is evidence of errors of the order of ± 3 kcal in the reported values, due in part to the limitations of the original calorimetric procedure used [58, 59].

In 1985 Maria and Gal [59] reported a parallel measure of donor ability based on the enthalpy of interaction with the reference acid BF_3. Rather surprisingly, they found that their values correlated only moderately well with those based on the Gutmann $SbCl_5$ scale and that they gave a family of lines when correlated with the Kamlet–Taft β parameter, which measures basicity relative to the H-bonding ability of the base. They have suggested, via the classic research of Brown, that steric strain may be involved. This was also suggested earlier by Drago and Lim relative to the adducts of $SbCl_5$ [47] and would imply a further modification of equation (3), via the addition of a steric parameter (s), to give a so-called triple s correlation (strength, softness, and steric strain) for the prediction of acid–base interactions:

$$\log K_{AB} = S_A \cdot S_B + \sigma_A \cdot \sigma_B + s_A \cdot s_B. \tag{11}$$

An equation of this form has actually been developed for aqueous ionic complexes by Handcock and Marsicano [41]:

$$\log K_{AB} = E_A^{aq} \cdot E_B^{aq} + C_A^{aq} \cdot C_B^{aq} - D_A \cdot D_B, \tag{12}$$

where, in analogy with the Drago E & C equation, the E^{aq} and C^{aq} parameters reflect the electrostatic and covalent contributions to complex formation (but are numerically different from Drago's values), and the D parameters are thought to reflect desolvation phenomena, which are related, in turn, to both steric hindrance and specific solvation effects accompanying complex formation. It is also of interest to note that these authors have found, in keeping with our earlier discussion, that the HSAB classification correlates with the C^{aq}/E^{aq} ratio of the species rather than with C^{aq} alone.

More recently, Riddle and Fowkes have raised some problems relative to the AN scale [60]. They have shown that dispersion-only liquids, such as hexane, produce a significant ^{31}P shift in Et_3PO and that consequently AN values should be corrected for this dispersion effect. In many cases, this correction is quite substantial. Thus, 13.7 of the original 14.2 AN units assigned to pyridine appear to be due to dispersion rather than to specific electron-pair donor–acceptor interactions, lowering its measure of 'true' Lewis acidity from 14.2 to 0.5. These authors have also found that these dispersion-corrected AN values correlate with the enthalpies of formation of the actual adducts formed between Et_3PO and the acid in question, and have suggested that this enthalpy be taken as the true measure, AN*, of Lewis acidity for a species, thus giving both the DN and AN scales the same units. The relation between these various scales is given by the equation

$$AN^* = -\Delta H(A\text{–}Et_3PO) = 0.288(AN - AN^d), \tag{13}$$

where the AN values are the original values reported by Gutmann, and AN^d are the dispersion contributions reported by Riddle and Fowkes.

As for the two-parameter interaction equation in equation (10), this is known, on the basis of Gutmann's own work, to be incorrect. In his original work on the DN scale, he found that the enthalpies of adduct formation for a given acid linearly correlated with the DN values of the bases (Fig. 3):

$$-\Delta H_{AB} = a \cdot DN_B + b. \tag{14}$$

This implies the necessity of a three-parameter, rather than a two-parameter, equation, with one parameter characteristic of the base (DN) and two characteristic of the acid (a and b, the slope and intercept of the line, respectively).

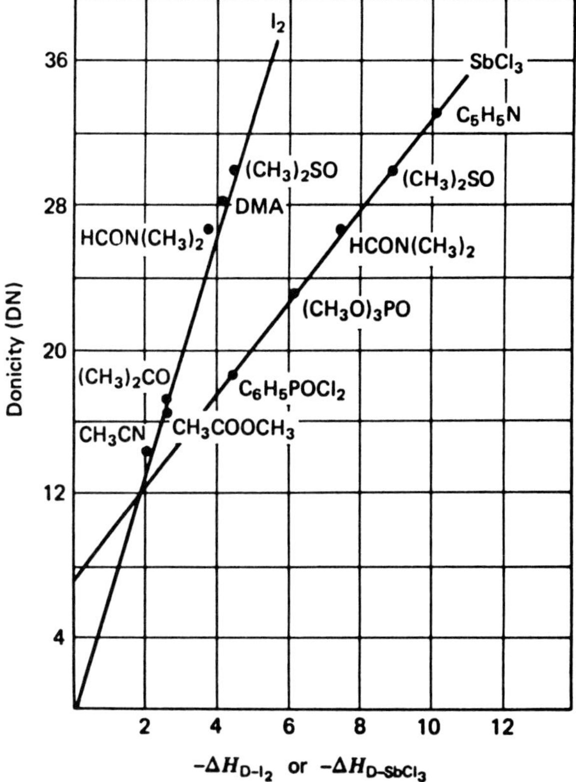

Figure 3. Two examples of a DN_B vs. $-\Delta H_{AB}$ plot. The fact that each acid gives a line with its own characteristic slope and intercept shows the necessity of using at least a three-parameter equation to correlate enthalpies of adduct formation.

Interestingly, about the same time, Satchell and Satchell [61, 62] found similar three-parameter linear correlations in their studies of adduct formation using metal halides as the Lewis acids and pK_a values as a measure of Lewis basicity:

$$\log K_{AB} = a \cdot pK_a + b. \tag{15}$$

These three-parameter correlations can be understood as an approximation to the four-parameter correlations implied by equations (3) and (5) by using the E & C equation. The bases used by both Gutmann and the Satchells are almost exclusively N and O donors, i.e. they are relatively hard, and an examination of the 39 N and O bases in Drago's list shows that the variations in their E_B values are quite small compared with the variations in their C_B values. Thus, the E_B values vary from 1.52 to 0.70, or less than one unit, whereas the C_B values vary from 13.2 to 0.53. (I have used the list of E and C parameters listed in ref. [63].) This is shown graphically in Fig. 4. Hence for this set of bases one can approximate the full E & C equation with the relation

$$-\Delta H_{AB} = 1.12\, E_A + C_A \cdot C_B, \tag{16}$$

where 1.12 is the average value of E_B. Like equations (14) and (15), this now

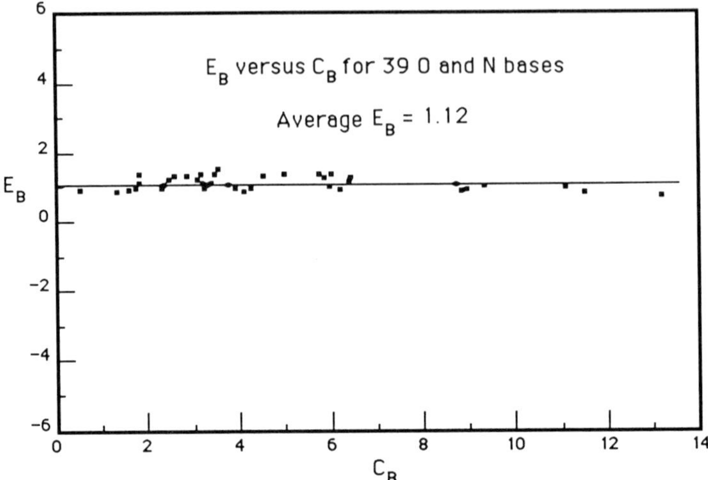

Figure 4. A plot of E_B vs. C_B for the 39 N and O donors in Drago's data set.

contains three parameters, one characteristic of the base (C_B) and two characteristic of the acid (1.12 E_A and C_A, corresponding to the intercept and slope, respectively, of a plot of $-\Delta H_{AB}$ vs. C_B). An example plot is shown in Fig. 5 for the case of iodine adducts, for which $E_A = C_A = 1.00$, and for which equation (16) becomes

$$-\Delta H_{AB} = 1.12 + 1.00\, C_B. \tag{17}$$

Obviously, the larger the E_A value of the acid, the greater the scatter will be in the corresponding graph.

Actually, the failure of the two-parameter interaction equation in equation (10) is not serious, as few authors have made use of it in their work, preferring instead

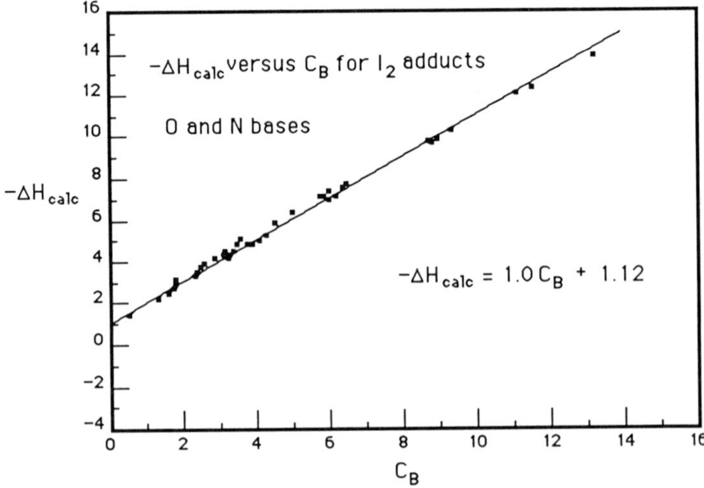

Figure 5. A plot of $-\Delta H_{AB}$ for adducts of I_2 vs. C_B of the base ($-\Delta H_{AB}$ calculated using the full *E & C* equation).

to use DN and AN values as empirical parameters within some more general linear regression correlation. For example [7],

$$P = P_0 + \alpha \cdot DN + \beta \cdot AN + \gamma \cdot \delta_d, \qquad (18)$$

where P is the property whose value is varying as a function of the solvent, P_0 is the value of the property for the reference system, α measures the sensitivity of the property to solvent basicity, β measures its sensitivity to solvent acidity, and γ measures its sensitivity to dispersion interactions as measured by the dispersion-only solubility parameter, δ_d.

One final result of interest to the donor and acceptor number approach is a paper published by Sabatino, La Manna, and Paoloni in 1980 [64] which reported linear correlations between a variety of experimental parameters as a function of change in solvent (e.g. rates and activation energies for substitution processes, free energies of solvation and transfer for ions, etc.) and the corresponding calculated *ab-initio* HOMO and LUMO energies for the solvents in question. Since similar linear correlations could be made between the properties and the DN and AN values of the solvents, the authors implied that the calculated energies of the frontier orbitals directly paralleled these parameters, though for some unknown reason they did not perform the obvious act of directly correlating them with one another. If this identification is correct, then it is of interest to note that the Parr–Pearson definitions of absolute hardness and electronegativity imply that for these solvents

$$\chi = f(DN + AN) \qquad (19)$$

and

$$\eta = f(DN - AN), \qquad (20)$$

though the exact details of the functional dependence are unknown.

5. PROSPECTS FOR THE FUTURE

As can be seen from the above comments, though there have been some interesting developments since 1980, these have not resulted in a major revision of the original quantitative models introduced in the 1960s, most of which are now more than 25 years old. This observation raises the question of whether there are any other serious alternatives, aside from basic quantum-mechanical calculations, that might be developed in the next decade. One obvious possibility is, of course, the development of more complex multi-parameter empirical equations. The six-parameter equation of Kroeger and Drago [equation (8)] has already been mentioned and the six-parameter strength, softness, steric strain relations in equations (11) and (12) are yet another possibility.

A third serious candidate is the so-called 'universal solubility equation' of Kamlet, Taft, *et al.*, which has received a good deal of publicity in recent years [65–68]. Like equation (18), this is designed to predict the way in which a change in solvent will affect a given physico-chemical property, P, the most important of which is the solubility, measured as $\log K_{sol}$:

$$P = P_0 + A\delta_1^2 \cdot V_2 + B\pi_1^* \cdot \pi_2^* + C\alpha_1 \cdot \beta_2 + D\alpha_2 \cdot \beta_1. \qquad (21)$$

The various terms in this equation are defined in Table 2. Several authors have reported good success with this relation and have indicated a preference for the Kamlet–Taft β measure of basicity over that provided by the DN scale [69]. My first reaction is that this success is hardly surprising, given that equation (21) contains at least 17 parameters, if one counts the correlation coefficients, the hidden correction parameters, and the alternative values of α and β. Indeed, one is tempted to paraphrase Cauchy's famous remark that he could graph an elephant, if given enough disposable parameters, and, if given two or more, he could make it wag its tail.

Table 2.
Definitions of the parameters in the universal solubility equation [67]

$\delta_1^2 V_2$	Product of the solubility parameter and molar volume
π^*	Polarity; may have to be modified with polarizability δ
α	H-bonding ability of the acid; may have to be modified as α_m
β	H-bonding ability of the base; may have to be modified as β_m

Actually, this is really not a fair comment, as equation (21) does not model simple acid–base adduct formation, but rather a more complex acid–base competition between solvent and solute, and a similar modeling of solubility using the conventional Lewis approach would be almost as complex. The real problem with the Kamlet–Taft approach is their denial of the usefulness of the Lewis concepts and their claim that at least three kinds of acidity exist: Lewis acidity, conventional Brønsted acidity as measured by pK_a values, and H-bonding acidity as measured by their α and β parameters [68]. This, of course, subverts the basic insight of the Lewis concepts relative to the fundamental identity of the electron-pair donor–acceptor mechanism underlying all of these phenomena.

In saying this, I do not wish to deny the uniqueness of H^+ relative to its electron-pair acceptor properties. Some of these idiosyncrasies were referred to earlier in our discussion of the HSAB classification, and they are in many ways a result of the unique position of hydrogen in the periodic table (Fig. 6). However, I do not believe that they are of such a magnitude as to necessitate a separate and unique vocabulary for describing both the reactivity and the composition of hydrogen compounds, as required by the Arrhenius and Brønsted approaches.

At this point it is perhaps worthwhile being a bit more explicit about these peculiarities and their bearing on the Lewis concepts. In contrast to other ionic Lewis acids, the H^+ ion is unique in that it is monobasic, each ion requiring only one unit of base for its neutralization:

$$H(H_2O)^+ + B^- \rightarrow HB + H_2O. \tag{22}$$

Consequently, in keeping with the classic characteristic of acids and bases, it can be titrated with solutions of monodentate bases and displays sharp end-points. Metal ions, on the other hand, typically have coordination numbers of 4, 6, or greater and require more than one unit of a monodentate base for their neutralization:

$$M(H_2O)_6^{x+} + B^- \rightarrow M(H_2O)_5 B^{z+} + H_2O \ldots \text{etc.} \tag{23}$$

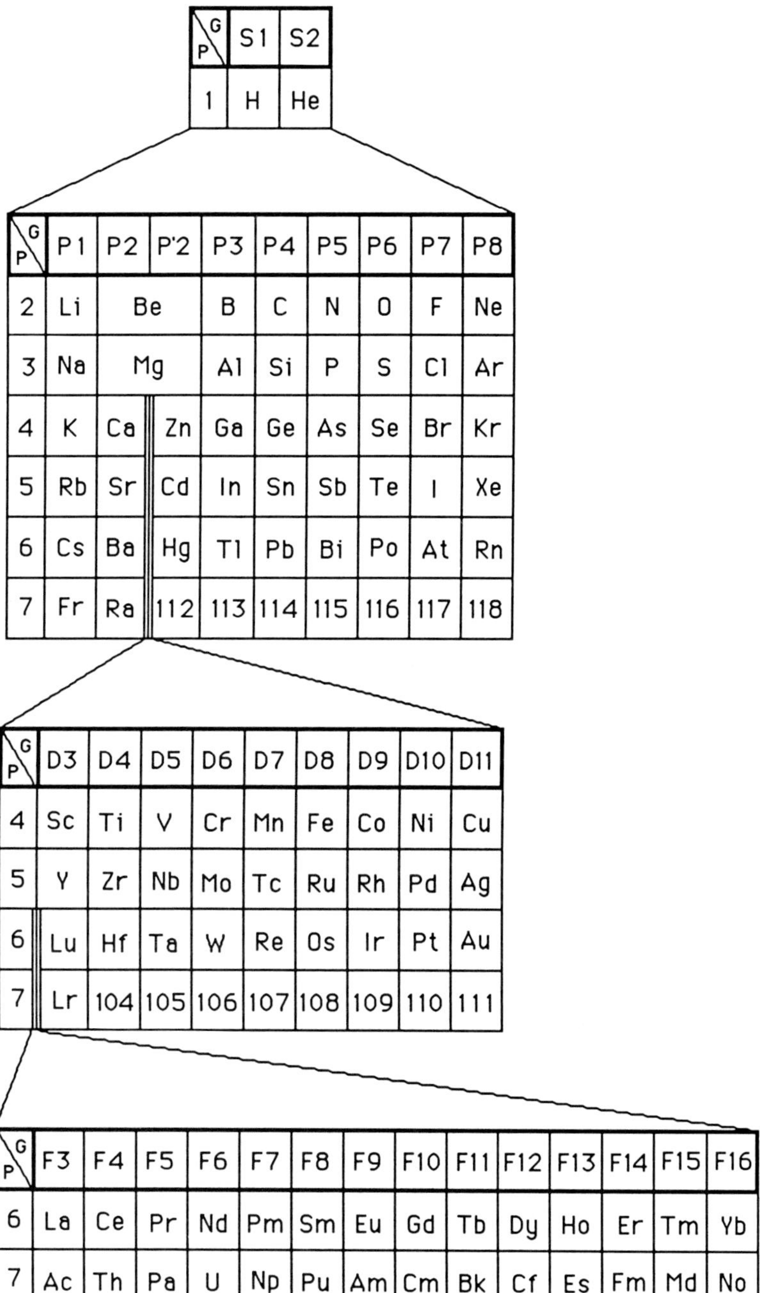

Figure 6. A 'four-block' periodic table emphasizing the unique position of hydrogen. Elements in the S-block (H and He) use only filled and/or empty *s* orbitals in their bonding. Elements in the P-block use filled and/or empty *s* and *p* orbitals; elements in the D-block use filled and/or empty *s*, *p* and *d* orbitals; and elements in the F-block use filled and/or empty *s*, *p*, *d* and *f* orbitals. This is a modification of the three-block table originally proposed by Sanderson. The smaller G/P symbols in the upper left of each block stand for group and period, respectively.

The result is a multi-step neutralization process and a smearing of the end-points, leading to an inability to determine metal ion concentrations in this manner. However, once the nature of this difference between the proton and other metal ions was understood, it became possible to design multidentate bases, such as EDTA, which could effect the neutralization of metal ions in a single step and so allow one to apply the conventional techniques of acid–base titration to their determination:

$$M(H_2O)_6^{x+} + EDTA^{4-} \rightarrow M(EDTA)^{x-4} + 6\,H_2O. \tag{24}$$

As the classic book by Tanabe shows [70], similar problems exist with the multiple donor and acceptor sites present on surfaces and polymers. Indeed, in the case of the latter especially, one would expect that the number of operative donor–acceptor sites would vary with temperature and so make the assignment of specific AN and DN values, or E and C parameters, difficult.

In contrast to metal ions, many molecular Lewis acids are, like the proton, monobasic. In this case, the difference between these species and the proton, as Brown pointed out long ago, lies in the possibility of developing steric strain in the resulting molecule–molecule adducts [71, 72]:

$$(CH_3)_3B + :N(CH_3)_3 \rightarrow (CH_3)_3B{-}N(CH_3)_3. \tag{25}$$

As long as we have a complete transfer of the proton to the base, steric strain should be at a minimum in the corresponding proton adduct, since the proton is the smallest possible acceptor site and has no attached groups to interfere with the groups attached to the donor site:

$$H^+ + :N(CH_3)_3 \rightarrow H{-}N(CH_3)_3^+. \tag{26}$$

If, however, we are looking not at a complete transfer of the proton, but at an incomplete displacement of one base by another, as in the case of hydrogen bonding, then the exact opposite may be true. Here the small size of the proton, in contrast to larger acceptor sites, will maximize the crowding of the two bases and thus maximize the potential for steric strain in the resulting H-bonded complex:

$$(CH_3)_2O: + :HN(CH_3)_3^+ \rightarrow (CH_3)_2O \cdots H \cdots N(CH_3)_3^+. \tag{27}$$

This is, at least, what Handcock and Marsicano [41] have concluded, via the use of equation (12), to describe the formation of ionic complexes in water, and it might also account for the conclusions of Kamlet, Taft *et al.* that Brønsted basicity (complete proton transfer, minimum steric strain), as measured by pK_a values, appears to be different from H-bonding basicity (incomplete proton transfer, maximum steric strain), as measured by their β parameter. Drago, on the other hand, maintains that in aqueous solution there never is complete proton transfer and that equation (22) should be explicitly written as

$$H(H_2O)_n^+ + B^- \rightarrow B{-}H(H_2O)_m + (n-m)\,H_2O. \tag{28}$$

In other words, the proton is always H-bonded and pK_a values simply measure a change in the H-bonding partners, in which case it becomes difficult to understand the basis of the distinction between β and pK_a, save, perhaps, as a function of reaction stoichiometry (i.e. the difference between a simple addition reaction and a displacement reaction).

The importance of the steric problem for surface phenomena is that the potential for steric strain (this time between similar molecules adsorbed on adjacent sites) should increase as the degree of surface coverage increases, leading to a variation in the effective acidity or basicity of the surface as a function of the degree of surface neutralization.

In closing, I would like to make one final suggestion for future work and that is the possibility of calculating Lewis acid–base parameters by means of topological indices, rather than experimentally measuring them. It is a fundamental postulate of chemistry that the specific or intrinsic properties of materials, i.e. those properties that are independent of sample size, shape, and function, and which we use to characterize specific kinds of matter or substances, are a function of molecular composition, structure, and interaction:

$$\text{intrinsic properties} = f(\text{molecular composition, structure, interaction}). \quad (29)$$

This formulation is actually slightly redundant, as the degree of molecular interaction itself also depends, in part, on molecular composition and structure, as well as on extrinsic factors, such as concentration (c) and temperature (T); so that it is perhaps more accurate to write

$$\text{intrinsic properties} = f(\text{molecular composition, structure, } c, T). \quad (30)$$

Since the advent of the electronic theory of matter, chemists have largely ignored this level of functional dependence and have instead pursued a more reduction-istic program:

$$\text{intrinsic properties} = f(\text{electronic composition and structure}). \quad (31)$$

However, the complexity of rigorous quantum-mechanical calculations has resulted in a movement to reconsider the more classical proposition in equation (30). For quantitative work, this requires the development of algorithms for reducing the classical composition and structure of a species to characteristic numerical indices which can then be used, in combination with linear regression analysis, to correlate and predict molecular properties.

Several popular reviews of this subject have appeared in recent years, the most accessible being those of Rouvray [73] and Seybold *et al.* [74], as well as a variety of topological algorithms, the most flexible of which has been the so-called Randic index, which has been extensively developed by Kier and Hall [75, 76]. In the Randic approach, each atom is assigned a characteristic 'valence' value, defined in one of two ways:*

$$\delta_i = \text{number of bonded atoms} - \text{number of bonded hydrogens} \quad (32)$$

$$\delta_i^v = \text{number of valence electrons} - \text{number of bonded hydrogens}. \quad (33)$$

These valence values are then combined in various ways to generate characteristic numerical indices that reflect various aspects of a molecule's composition

---

*The definition of δ_i^v given here is for period 2 elements. A more general definition can be found in ref. [76].

and structure. The first of these, the 0th-order $^0\chi$ and $^0\chi^v$ indices, essentially reflect only composition:

$$^0\chi = \Sigma(\delta_i)^{-0.5} \tag{34}$$

$$^0\chi^v = \Sigma(\delta_i^v)^{-0.5}. \tag{35}$$

The sums are over all atoms, except the hydrogen atoms, which serve as an arbitrary zero point for the indices, and the χ and δ's should not be confused with those used in earlier equations for electronegativity and the dispersion-only solubility parameter. The sum, $^0\chi + {}^0\chi^v$, has been shown to correlate with the molecular volume, whereas the difference, $^0\chi - {}^0\chi^v$, reflects the number of π and lone-pair electrons in the molecule and correlates with valence-state electronegativities.

First-order indices, which are summed over the bonds in the molecule, reflect information about both the composition and the molecular topology and distinguish between various topological isomers (see Fig. 7 for an example calculation for 1-propanol and 2-propanol):

$$^1\chi = \Sigma(\delta_i\delta_j)^{-0.5} \tag{36}$$

$$^1\chi^v = \Sigma(\delta_i^v\delta_j^v)^{-0.5}. \tag{37}$$

Higher-order indices can also be defined which are summed over larger fragments of the molecule:

$$^2\chi = \Sigma(\delta_i\delta_j\delta_h)^{-0.5}. \tag{38}$$

The physical interpretation of these is not always apparent, though some are thought to reflect such properties as molecular flexibility for the region of the molecule in question.

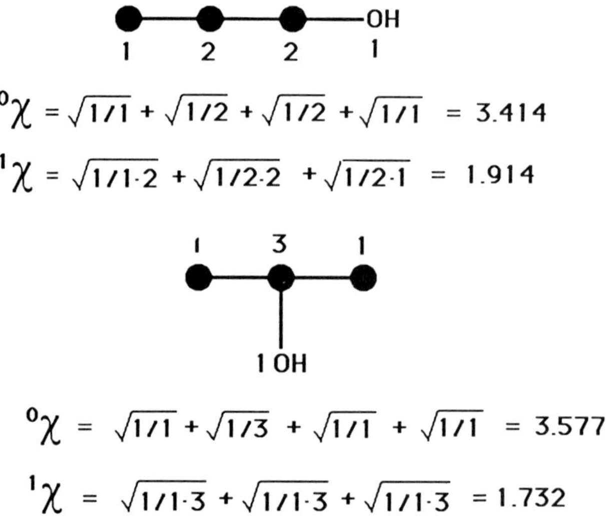

Figure 7. An example showing the calculation of the $^0\chi$ and $^1\chi$ indices for 1-propanol and 2-propanol. The H atoms are not shown, except on the OH group for ease of identification. Atoms are numbered with their δ values.

Kier and Hall, in their two monographs on this subject [75,76], have summarized an impressive array of physical and biological properties which can be correlated with these indices, some examples of which are listed in Table 3. Of particular interest to the Lewis acid–base concepts is a reported correlation between the ionization potentials, IP, of a mixed list of alcohols, amines, and ethers and the 0th- and 1st-order indices:

$$\text{IP (eV)} = 5.014 \,(^0\chi - {}^0\chi^v) + 5.166 \,(^1\chi - {}^1\chi^v) + 5.341, \tag{39}$$

where $r = 0.955$, $s = 0.299$, and $n = 24$. Likewise, the chromatographic retention times, RI, of alcohols can be correlated with $^1\chi$ and $^1\chi^v$:

$$\text{RI} = 486 \,{}^1\chi - 293 \,{}^1\chi^v + 44.9, \tag{40}$$

where $r = 0.995$, $s = 10.1$, and $n = 32$. Since ionization potentials are related to donor ability and several authors have attempted to correlate chromatographic retention times with solvent basicity and acidity, these results strongly suggest that it should also be possible to correlate such parameters as DN, AN, S, σ, s, or even $\log K_{AB}$ with such indices, thereby allowing one to calculate them solely from a knowledge of the classical structure and composition of the molecule in question.

Table 3.
Example properties successfully correlated by means of topological indices[a]

Boiling points	Solubilities
Densities	Partition coefficients
Heats of atomization	Heats of formation
Solvent polarities	Anesthetic activity
Narcotic activity	Hallucinogenic activity
Toxicity	Sweet and bitter taste

[a] Based on refs [75, 76].

In fairness, it should be pointed out that there are still a large number of unsolved problems connected with this approach. For example, there does not seem to be a systematic way of selecting ahead of time which combination of indices to use in a particular correlation. Likewise, there is at present no way of indicating higher-order geometric differences (and thus of distinguishing, for example, between *cis*- and *trans*-isomers), though it seems to me that it should be possible to develop such an algorithm based on the molecule's symmetry elements. There is also some debate as to how to extend the approach to non-molecular solids and thus, by implication, to surfaces.

REFERENCES

1. W. B. Jensen, *Chemistry* **47(3)**, 11–14; **47(4)**, 13–18; **47(5)**, 14–18 (1974).
2. W. B. Jensen, *Chem. Rev.* **78**, 1–22 (1978).
3. W. B. Jensen, *The Lewis Acid–Base Concepts: An Overview.* Wiley-Interscience, New York (1980).
4. W. B. Jensen, *Rubber Chem. Technol.* **55**, 881–901 (1982).
5. W. B. Jensen, *Chemtech* **12**, 755–764 (1982).

6. A. Beerbower and W. B. Jensen, *Inorg. Chim. Acta* **75**, 193–197 (1983).

7. W. B. Jensen, in: *Surface and Colloid Science in Computer Technology*, K. L. Mittal (Ed.), pp. 27–59. Plenum Press, New York (1987).

8. G. N. Lewis, *Valence and the Structure of Atoms and Molecules*, pp. 141–142. The Chemical Catalog Co., New York (1923).

9. G. N. Lewis, *J. Franklin Inst.* **226**, 293–313 (1938).

10. C. K. Ingold, *J. Chem. Soc.* 1120–1127 (1933). These are identical to the terms cationoid agent and anionoid agent introduced earlier by Robinson and Lapworth.

11. C. K. Ingold, *Chem. Rev.* **15**, 225–274 (1934).

12. W. F. Luder and S. Zuffanti, *The Electronic Theory of Acids and Bases*. John Wiley, New York (1946). Reprinted by Dover Books in 1961.

13. R. S. Mulliken, *J. Chem. Phys.*, **19**, 514–515 (1951). See also R. S. Mulliken and W. B. Person, *Molecular Complexes: A Lecture and Reprint Volume*. Wiley-Interscience, New York (1969).

14. K. Fukui, T. Yonezawa and H. Shingu, *J. Chem. Phys.* **20**, 722–725 (1952); K. Fukui, T. Yonezawa, C. Nagata and H. Shingu, *Ibid.* **22**, 1433–1442 (1954).

15. G. Briegleb, *Elektronen-Donator–Acceptor-Komplexe*. Springer Verlag, Berlin (1961).

16. R. G. Pearson, *J. Am. Chem. Soc.* **85**, 3533–3539 (1963).

17. R. S. Drago and B. Wayland, *J. Am. Chem. Soc.* **87**, 3571–3577 (1965).

18. V. Gutmann, A. Steininger and E. Wychera, *Monatsh. Chem.* **97**, 460–467 (1966).

19. U. Mayer, V. Gutmann and W. Gerger, *Monatsh. Chem.* **106**, 1235–1257 (1975).

20. R. F. Hudson and G. Klopman, *Tetrahedron Lett.* **12**, 1103–1108 (1967). See also R. F. Hudson and G. Klopman, *Theor. Chim. Acta* **8**, 165–174 (1967) and G. Klopman, *J. Am. Chem. Soc.* **90**, 223–234 (1968).

21. R. Pearson (Ed.), *Hard and Soft Acids and Bases*. Dowden, Hutchinson and Ross, Stroudsburg, PA (1973).

22. R. S. Drago and N. A. Matwiyoff, *Acids and Bases*. D. C. Heath, Lexington, MA (1968).

23. V. Gutmann, *Coordination Chemistry in Nonaqueous Solutions*. Springer Verlag, New York (1968); *Chemische Funktionlehre*. Springer, Vienna (1971); and *The Donor–Acceptor Approach to Molecular Interactions*. Plenum Press, New York (1978).

24. G. Klopman (Ed.), *Chemical Reactivity and Reaction Paths*. Wiley-Interscience, New York (1974).

25. F. M. Fowkes, *J. Adhesion* **5**, 155 (1972).

26. R. C. Parr and R. G. Pearson, *J. Am. Chem. Soc.* **105**, 7512–7516 (1983).

27. M. Berkowitz, S. K. Ghosh and R. G. Parr, *J. Am. Chem. Soc.* **107**, 6811–6814 (1985).

28. W. Yang, and R. G. Parr, *Proc. Natl. Acad. Sci.* **82**, 6723–6726 (1985).

29. M. Berkowitz and R. G. Parr, *J. Chem. Phys.* **88**, 2554–2557 (1988).

30. R. G. Pearson, *Proc. Natl. Acad. Sci.* **83**, 8440–8441 (1986).

31. R. G. Pearson, *Inorg. Chem.* **27**, 734–740 (1988).

32. K. Jørgensen, *Struct. Bonding (Berlin)* **1**, 234–248 (1966). See also ref. [3], pp. 264–265.

33. See ref. [3], p. 294.

34. R. G. Pearson, *J. Chem. Educ.* **64**, 561–567 (1987).

35. R. G. Pearson, *J. Am. Chem. Soc.* **108**, 6109–6114 (1986).

36. L. H. Lee, *J. Adhesion Sci. Tech.* (in press).

37. R. F. Nalewajski, *J. Am. Chem. Soc.* **106**, 944–945 (1984).

38. P. P. Singh, S. K. Srivastava and A. K. Srivastava, *J. Inorg. Nucl. Chem.* **42**, 521–532 (1980).

39. P. P. Singh, *J. Sci. Ind. Res.* **42**, 140–149 (1983).

40. Y. Zhang, *Inorg. Chem.* **21**, 3889–3893 (1982).

41. R. D. Handcock and F. Marsicano, *S. Afr. J. Chem.* **33**, 77–86 (1980).

42. R. G. Pearson and R. J. Mawby, in: *Halogen Chemistry*, V. Gutmann (Ed.), vol. 3, pp. 55–87. Academic Press, New York (1967).

43. S. Ahrland, *Chem. Phys. Lett.* **2**, 303–306 (1968).

44. R. L. DeKock, *J. Am. Chem. Soc.* **97**, 5592–5593 (1975).

45. M. Arbelot and M. Chanon, *Nouv. J. Chim.* **7**, 499–504 (1983).

46. K. F. Purcell, *Inorg. Chim. Acta* **65**, L187–L188 (1982).

47. Y. Marcus, *J. Solution Chem.* **13**, 599–624 (1984). The possible reasons for this poor agreement were discussed earlier in R. C. Drago and Y. Y. Lim, *Inorg. Chem.* **11**, 202–204 (1972).

48. R. S. Drago, *Coord. Chem. Rev.* **33**, 251–277 (1980).

49. R. S. Drago, N. Wong and G. C. Vogel, *Inorg. Chem.* **26**, 9–14 (1987).

50. R. S. Drago, M. K. Kroeger and J. R. Stahlbush, *Inorg. Chem.* **20**, 306–308 (1981).

51. P. E. Doan and R. S. Drago, *J. Am. Chem. Soc.* **104**, 4524–4529 (1982).

52. P. E. Doan and R. S. Drago, *J. Am. Chem. Soc.* **106**, 2772–2774 (1984).

53. R. S. Drago and C. J. Bilgrien, *Polyhedron* **7**, 1453–1468 (1988).

54. M. K. Kroeger and R. S. Drago, *J. Am. Chem. Soc.* **103**, 3250–3262 (1981).

55. R. S. Drago, T. R. Cundari and D. C. Ferris, *J. Org. Chem.* **54**, 1042–1047 (1989).

56. R. S. Drago, N. Wong and D. C. Ferris, *J. Am. Chem. Soc.* (in press).

57. See, for example, F. M. Fowkes, in: *Surface and Colloid Science in Computer Technology*, K. L. Mittal (Ed.), pp. 3–25. Plenum Press, New York (1987).

58. T. R. Griffiths and D. C. Pugh, *Coord. Chem. Rev.* **29**, 129–211 (1979).

59. P. C. Maria and J. F. Gal, *J. Phys. Chem.* **89**, 1296–1304 (1985).

60. F. L. Riddle and F. M. Fowkes, *J. Am. Chem. Soc.* **112**, 3259–3264 (1990).

61. D. P. N. Satchell and R. S. Satchell, *Chem. Rev.* **69**, 251–278 (1969).

62. D. P. N. Satchell and R. S. Satchell, *Q. Rev. Chem. Soc.* **25**, 171–199 (1971).

63. K. F. Purcell and J. C. Kotz, *Inorganic Chemistry*, pp. 220–221. Saunders, Philadelphia, PA (1977).

64. A. Sabatino, G. La Manna and L. Paolini, *J. Phys. Chem.* **84**, 2641–2645 (1980).

65. M. J. Kamlet, J. L. M. Abboud and R. W. Taft, *Prog. Phys. Org. Chem.* **13**, 485–630 (1981).

66. R. Rawls, *Chem. Eng. News* 20 (18 March, 1985).

67. M. J. Kamlet, R. M. Doherty, J. L. M. Abboud, M. H. Abraham and R. W. Taft, *Chemtech* **16**, 566–576 (1986).

68. M. H. Abraham, R. M. Doherty, M. J. Kamlet and R. W. Taft, *Chem. Br.* **22**, 551–554 (1986).

69. O. W. Kolling, *Anal. Chem.* **54**, 260–264 (1982).

70. K. Tanabe, *Solid Acids and Bases.* Academic Press, New York (1970).

71. H. B. Brown and S. Sujishi, *J. Am. Chem. Soc.* **70**, 2878–2881 (1948).

72. H. B. Brown, *J. Chem. Soc.* 1248–1264 (1956).

73. D. H. Rouvray, *Sci. Am.* **255**, 40–47 (1986).

74. P. G. Seybold, M. May and U. A. Bagal, *J. Chem. Educ.* **64**, 575–581 (1987).

75. L. B. Kier and L. H. Hall, *Molecular Connectivity and Drug Research.* Academic Press, New York (1976).

76. L. B. Kier and L. H. Hall, *Molecular Connectivity in Structure–Activity Analysis.* Research Studies Press, Letchworth, England (1986).

77. W. B. Jensen, *Comp. Maths. Appl.* **12B**, 487–510 (1986).

78. W. B. Jensen, *Sorting Maps and the HSAB Classification* (to be published).

Acid-Base Interactions, pp. 25-46
Eds. K.L. Mittal and H.R. Anderson, Jr.
©VSP 1991

Relevance of the density-functional theory to acid–base interactions and adhesion in solids

LIENG-HUANG LEE

Webster Research Center, Xerox Corporation, Webster, NY 14580, USA

Revised version received 10 October 1990

Abstract—In this paper, the acid–base interaction in terms of the molecular interaction is discussed from a broader background. Thus, the overall nature of the acid–base interaction consists of electrostatic, charge (or electron) transfer, exchange, polarization, and dispersion components. Among them, the electrostatic (or ionic) and charge transfer (or covalent) are the two major components. It is pointed out that one of the important criteria determining whether there is any molecular interaction is the interatomic distance that affects each of these components in different ways. The optimum distance appears to be around 2 Å.

To complement existing acid–base theories, we further demonstrate the relevance of the density-functional theory to surface interactions and solid adhesion. On the basis of the density-functional theory, two chemical parameters, i.e. chemical potential μ and absolute hardness η, will be shown to govern an acid–base interaction. From these, the number of transferred electrons ΔN can be estimated.

By extending the density-functional theory for atoms and molecules to solids, we have found two important physical properties of solids, i.e. the work function Φ and the average energy gap E_g^{Av}, which are equivalent to the above two chemical parameters. Thus, we propose to estimate the number of transferred electrons for solid interactions involving metals or polymers by the derived equation:

$$\Delta N \approx \Delta\Phi / \Sigma E_g^{Av}.$$

Basically, this equation somewhat resembles Ohm's law.

In practice, as ranked by the average energy gaps, all metals are soft for having narrow gaps; for the same reason, all semimetals and semiconductors are generally soft, and all insulators including polymers are comparatively hard for having wide gaps. Furthermore, to achieve a reasonable rate for an interfacial acid–base interaction or solid adhesion, it is advantageous to consider the HSAB principle: hard bases (or donors) prefer to interact with hard acids (or acceptors), and soft bases (or donors) with soft acids (acceptors).

Keywords: Acceptor; acid; adhesion; base; density-function; donor; interaction; solid.

NOTATION

A	Electron affinity
C	Ionic (heteropolar) component of the average energy gap
C_{El}	Coefficient of the atomic orbital of the electrophile
C_{Nu}	Coefficient of the atomic orbital of the nucleophile
dis	Dispersion
e	Electronic charge
ΔE_{int}	Interaction energy
E	Electronic energy of an atom, a molecule or an ion
E_c	Energy of the conduction band edge

E_F	Energy of the Fermi level
E_g^o	Minimum energy gap
E_g^{Av}	Average energy gap
E_h	Covalent (homopolar) component of the average energy gap
E_{HOMO}	Energy of the highest occupied molecular orbital
E_{LUMO}	Energy of the lowest unoccupied molecular orbital
E_V	Energy of the valence band edge
es	Electrostatic interaction
ex	Exchange interaction
f	Enthalpy-to-free energy correction factor
f_i	Phillips ionicity or fractional ionic character
ΔH^{a-b}	Heat of the acid–base interaction
I	Ionization potential
ΔN	Number of electrons transferred
n^{a-b}	Interface population of the acid–base pairs
po	Polarization
Q	Total electronic charge
R	Distance between an electrophile and a nucleophile
v	Potential due to nuclei, plus any external potential
V_c	Contact potential
W^{a-b}	Acid–base component of the work of adhesion
W_A	Work of adhesion

Greek letters

β	Resonance integral
ε	Permittivity
ε_∞	High-frequency dielectric constant
η	Absolute hardness
$\tilde{\eta}$	Local hardness
μ	Chemical potential
σ	Absolute softness
$\tilde{\sigma}$	Local softness
Φ	Work function
χ^M	Electronegativity (Mulliken)
Ω_p	Plasma frequency of the valence electron

1. INTRODUCTION

The donor–acceptor interactions have been studied by Gutmann [1] and
Deryagin *et al.* [2], and the acid–base interactions have been reviewed by Jensen
[3–5]. On many occasions, the two terms, though different, have been used inter-
changeably. In a broader sense, both interactions have been considered as
molecular interaction but with different emphases [6, 7]. It appears that for a
more ionic (or electrostatic) interaction, it is proper to call it the acid–base interaction; while for a more covalent (or frontier orbital) interaction, it is the donor–
acceptor interaction.

For polymers, Fowkes [8–11] and Bolger and Michaels [12] have pointed out
the important role of the acid–base interaction in the formation of an interfacial

bond. The relevance of the acid–base interaction to adhesion has further been affirmed by many authors throughout this symposium dedicated to Professor F. M. Fowkes of Lehigh University.

The purpose of this paper is to explore the scope and limitations of these interactions. Hence to apply properly the concept of the acid–base interaction to solid adhesion, we will briefly discuss (1) molecular interactions, (2) the acid–base interaction, (3) the hard–soft acid–base (HSAB) principle [13, 14], and (4) the application of the density-functional theory [15] to these interactions and adhesion of solids.

2. MOLECULAR INTERACTIONS

2.1. Perturbation theory

Hudson and Klopman [16] proposed an equation to describe the effect of orbital perturbation of two molecules on chemical reactivity. Their equation for the interaction energy ΔE_{int} can be simplified by including only two terms: the Coulombic interaction and the frontier orbital interaction between the HOMO (highest occupied molecular orbital) of a nucleophile (or base) and the LUMO (lowest unoccupied molecular orbital) of an electrophile (or acid):

$$\Delta E_{int} = -\underbrace{\frac{Q_{Nu}Q_{El}}{\varepsilon R}}_{\text{(the Coulombic term)}} + \underbrace{\frac{2(C_{Nu}C_{El}\beta)^2}{E_{HOMO} - E_{LUMO}}}_{\text{(the frontier orbital term)}}, \tag{1}$$

where Q_{Nu} and Q_{El} are the total charges of the nucleophile and electrophile, respectively; C_{Nu} and C_{El} are the coefficients of the atomic orbitals Nu and El, respectively; β is the resonance integral; ε is the permittivity; and R is the distance between Nu and El.

For the electrostatic interaction, the first term dominates, while for the electron donor–acceptor (EDA) interaction, the second term. Thus, a molecular interaction encompasses chiefly both acid–base and donor–acceptor interactions.

2.2. Decomposition of molecular interaction energies

According to Kitaura and Morokuma [17], the molecular interaction energy can be further decomposed into the following components:

$$\Delta E_{int} = \Delta E_{es} + \Delta E_{po} + \Delta E_{ex} + \Delta E_{ct} + \Delta E_{mix}, \tag{2}$$

where the subscripts are
es: electrostatic,
po: polarization,
ex: exchange,
ct: charge-transfer, and
mix: coupling terms of higher order, in some cases including the dispersion
     component of Lifshitz–van der Waals (LW) forces.

Four of the above five interactions are illustrated schematically in Fig. 1. The electrostatic interaction energy, ΔE_{es}, is the energy of interaction [18] between

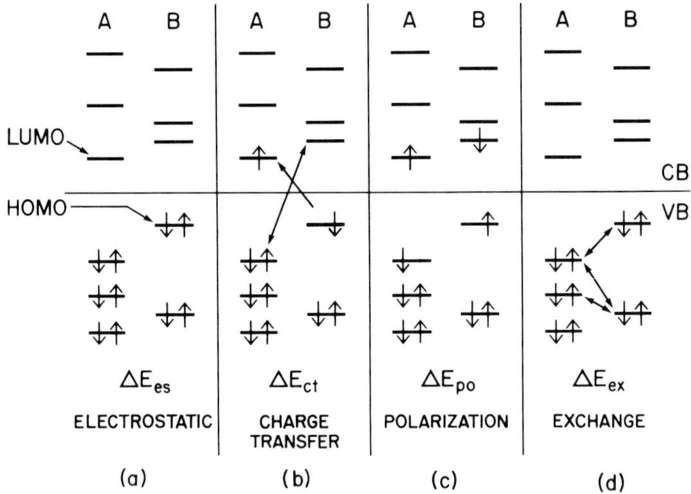

Figure 1. Molecular orbital interactions between two closed-shell molecules.

undistorted charge distributions of molecules A and B as shown in Fig. 1a. This Coulombic interaction involves permanent charges, dipoles, and higher multipoles present in both molecules; thus, the overall result could be attractive or repulsive.

The next important interaction is the charge transfer (or electron transfer). The interaction energy, ΔE_{ct}, is the contribution resulting from the mixing in of the electronic configurations representing electron transfer between an occupied orbital and an unoccupied orbital of the respective two molecules (Fig. 1b). The net result is attraction. Among them, the transfer of electrons between HOMO and LUMO frontier orbitals from each molecule is more dominant. Moreover, the HOMO–LUMO interactions are important, especially in the early stages of reaction.

Unlike the electrostatic interaction, the polarization energy, ΔE_{po}, is the energy change resulting from the distortion of the charge distribution of both molecules to give rise to a mixing in of electronic configurations as shown in Fig. 1c. In this case, the electrons from either or both molecules can be excited to occupy their own LUMOs.

The charge transfer should not be confused with the electron exchange interaction. The latter results in the loss of identity of the electrons with respect to A and B. The exchange interaction energy, ΔE_{ex}, is derived from the mixing in of configurations in which electrons from each molecule have *changed place* as in a musical chair game (Fig. 1d). Owing to the Pauli exclusion principle, a depletion of electron density in the intermolecular region results in repulsion.

Finally, the fifth term, ΔE_{mix}, is the energy involving the coupling term of higher order. In some cases, it may involve the dispersion component of Lifshitz–van der Waals interactions. Generally, for simplicity, this term may be neglected for the calculation of the molecular interaction energy.

Since both polarization and dispersion are involved in the molecular interaction, we cannot cleanly separate the acid–base interaction from the Lifshitz–van der Waals (LW) interactions, though the latter is secondary in significance.

For this reason, it is no surprise that the contribution of van der Waals inter-actions to the overall acid–base interaction has only recently been reported by Riddle and Fowkes [19].

2.3. Critical interatomic distance of molecular interactions

The effect of the interatomic distance on various energy components and the interaction energy ΔE_{int} or the binding energy ΔE_b is illustrated with the acid–base complex of BH_3 (borane) and NH_3 in Fig. 2 by Isaacs [18] on the basis of the data of Kitaura and Morokuma (KM) [17]. In this example, ΔE_{es} is the most dominant. ΔE_{ex}, which is a measure of the energy for the exchange interaction between core electrons, begins to compete strongly at a distance of less than 2 Å (or 200 pm) and rises steeply but is balanced by the growing polarization energy, ΔE_{po}, and the charge-transfer energy, ΔE_{ct}. An optimum binding energy, ΔE_{int} or ΔE_b, of -187 kcal (or -44.7 kcal mol^{-1}) is reached at 1.8 Å (or 180 pm). In general, for the borane–amine interaction, the various components decrease with increasing distance in the order $\Delta E_{es} > \Delta E_{ct}$ and ΔE_{po}.

At this point, it should be emphasized that molecular interactions including the acid–base interaction are insignificant at an interatomic distance greater than 3 Å. From Fig. 2, we can see that the optimum distance is approximately 1.8–2.0 Å. It has also been reported [20] that the optimum distance for an electrophile–nucleophile interaction is around 2.3–2.7 Å. In the case of solid interactions, this close contact between two solid surfaces is achievable more easily under high vacuum. This may be why the acid–base interaction is not as readily observed for solid-to-solid systems in spite of the fact that it is frequently reported.

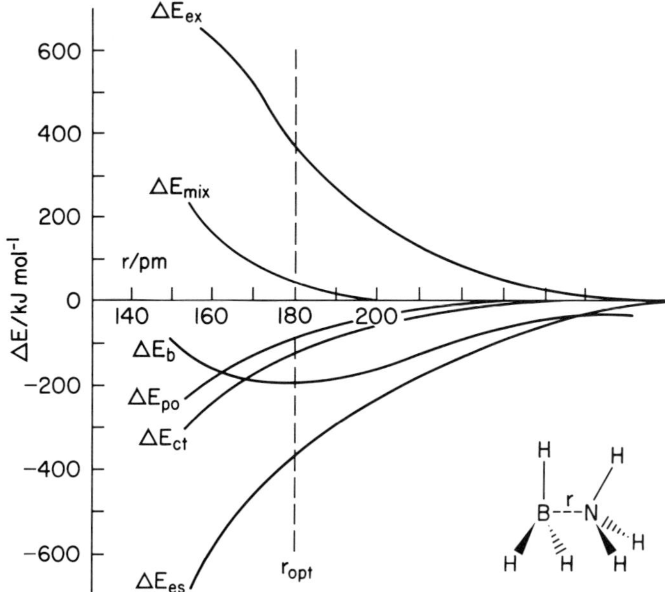

Figure 2. Interatomic distances between borane and ammonia. (After Isaacs, ref. [18], p. 59.)

Although the KM method, based on the *ab initio* self-consistent field (SCF) theory, has enjoyed a wide acceptance for many years, recently the critics [21] have claimed that the KM method has consistently led to a much smaller estimate of charge transfer. This problem has led to the development of the natural bond order (NBO) model [21, 22], which will not be elaborated in this paper.

3. ACID–BASE INTERACTION

The Lewis acid–base interactions [3–5] encompass not only hydrogen-bonding, the electron pair donor–acceptor interactions, but also electrophile–nucleophile interactions in organic chemistry [23]. The following general equation describes the Lewis acid–base interaction:

$$A \quad + \quad :B \quad \rightarrow \quad\quad A:B \tag{3}$$
$$\text{(acid)} \quad \text{(base)} \quad \text{(acid–base complex)}$$

As implied by the perturbation equation [equation (1)], both electrostatic (or ionic) and charge-transfer (or covalent) factors are involved in these interactions.

3.1. Drago's equation—electrostatic and covalent components

On the same basis of dual components, Drago *et al.* [24] introduced four parameters for the calculation of the enthalpy ΔH^{a-b} of an acid–base interaction:

$$-\Delta H^{a-b} = E_A E_B + C_A C_B, \tag{4}$$

where E_A and E_B are the susceptibilities of the acid (A) and base (B), respectively, to participate in the electrostatic interaction, while C_A and C_B are those to form covalent bonds. The inclusion of both components in the enthalpy of the acid–base interaction reflects the two primary terms in the perturbation equation [equation (1)].

3.2. Work of adhesion—the acid–base component

From the enthalpy of the acid–base interaction, ΔH^{a-b}, Fowkes [9] proposed calculating the work of adhesion, W_A^{a-b} by the following equation:

$$W_A^{a-b} = -fn^{a-b}\Delta H^{a-b}, \tag{5}$$

where f is an enthalpy-to-free energy correction factor and n^{a-b} is the interface population of the acid–base pairs in terms of the number of moles per unit area.

3.3. Role of entropy in the acid–base component of the work of adhesion

As a first approximation, Fowkes [9] assumed the f factor to be close to unity without considering the entropy effect. However, during this symposium, Vrbanac and Berg [25] have shown on the basis of the Gibbs–Helmholtz thermodynamic relation that f is substantially less than unity in most cases and increases with temperature according to the following equation:

$$f = [1 - (\mathrm{d}\ln W_A^{a-b}/\mathrm{d}\ln T)]^{-1}. \tag{6}$$

The lowering of the f value from unity, if correct as presented, can substantially affect the overall value of the acid–base component of the work of adhesion. Indeed, more work will be needed to determine f factors for other systems before we are convinced that this revision is in the proper direction.

In the following sections, in order to complement what has been published, we will discuss the importance of the hard–soft acid–base (HSAB) principle in the acid–base interaction with respect to both equilibrium and kinetics.

4. EQUILIBRIUM AND KINETICS OF THE ACID–BASE INTERACTION– HSAB PRINCIPLE

The involvement of two components, electrostatic and charge-transfer, in the acid–base interaction is illustrated best by the hard–soft acid–base (HSAB) principle proposed by Pearson [13] in 1963. This principle describes some basic rules about the kinetics and equilibrium of the acid–base interactions in solutions. In this paper, we attempt to extend the HSAB principle to solid interactions with the aid of the frontier orbital method. The HSAB principle will be described as it has evolved in recent years on the basis of the density-functional theory [15, 26–28] and the band structures of solids. After the compatibility between the HSAB principle and the band structures in the solid state is demonstrated, several examples of adhesion and surface interactions between metals and polymers will be given.

4.1. HSAB principle for inorganic interactions

Pearson [13] proposed to classify acids and bases according to hardness and softness for the inorganic reactions in solutions:
- A hard acid contains an acceptor atom of high positive charge and small size. It does not have easily excitable outer electrons and is not polarizable.
- A soft acid contains an acceptor atom of low positive charge and larger size. It has several easily excitable outer electrons and is polarizable.
- A hard base contains a donor atom of low polarizability and high electronegativity that is hard to reduce. It is associated with empty orbitals of high energy and hence is inaccessible.
- A soft base contains a donor atom of high polarizability and low electronegativity. It is easily oxidized and is associated with empty low-lying orbitals. In general, the following rules govern the acid–base interactions in solutions:
- *About equilibrium:* Hard acids prefer to coordinate with hard bases, and soft acids with soft bases.
- *About kinetics:* Hard acids react readily with hard bases, and soft acids with soft bases.

The above HSAB principle has been well tested in inorganic reactions [29–31]. However, some exceptions have been reported [5, 32].

4.2. HSAB principle for organic interactions and the frontier orbital approach

For organic reactions, the HSAB principle has also been applied for the study of kinetics and equilibrium, and the frontier orbital (Fig. 3) method has been used to illustrate the electrophilic and nucleophlic interactions [33, 34]:
- A hard electrophile (or acceptor) has a high-energy LUMO (lowest unoccupied molecular orbital) and usually has a positive charge.

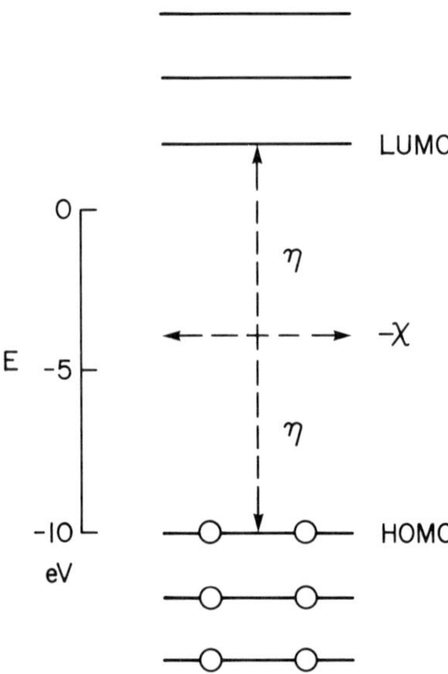

Figure 3. Orbital energy diagram for a molecule. HOMO, highest occupied MO; LUMO, lowest unoccupied MO. (After Pearson, ref. [14]; reproduced with permission.)

- A soft electrophile has a low-energy LUMO but does not necessarily have a positive charge.
- A hard nucleophile (or acceptor) has a low-energy HOMO (highest occupied molecular orbital) and usually has a negative charge.
- A soft nucleophile has a high-energy HOMO but does not necessarily have a negative charge.

For organic interactions (or reactions), the HSAB rules can be restated as follows:

- A hard–hard interaction (or reaction) is fast because of a large Coulombic attraction as described by the first term of equation (1).
- A soft–soft interaction (or reaction) is fast because of large orbital interaction between the HOMO of the nucleophile and the LUMO of the electrophile as described by the second term of equation (1).

These two rules reflect the value of the HSAB principle in establishing the importance of both electrostatic and charge-transfer contributions to the acid–base interactions.

4.3. HSAB principle and Drago's parameters

From the above discussion, although the HSAB principle and Drago's parameters differ, they share a common basis of dual components. Drago *et al.* [24] discussed the relationship between the HSAB concept and their four parameters. The ratio of C/E (covalent/electrostatic) gives a quantitative order of relative softness. They thought that the procedure used by Pearson to determine hardness and softness did not give the same result as the C/E ratio because the magnitudes of the C and E numbers, which are important in determining the magnitude of an enthalpy, are lost in the ratio. Consequently, Drago *et al.* have claimed that the original HSAB principle is rather incomplete. The major drawback is that it does not describe the strength of each component. At that time, without support from the density-functional theory, it appeared that Drago's parameters were more quantitative than the HSAB principle. However, it will become evident later in this paper that with recent theoretical support from the density-functional theory the HSAB principle has become more fundamentally significant than it first appeared.

5. DENSITY-FUNCTIONAL THEORY

5.1. *Chemical potential, electronegativity and absolute hardness*

When the HSAB principle was first introduced, the meaning of hardness was not defined theoretically. Certainly, it is not the hardness that measures the resistance against deformation. Then, what is it? It was not until the last few years that the absolute hardness received theoretical support by Parr *et al.* on the basis of the density-functional theory [15, 26–28]. However, based on this theory, absolute hardness alone is indeed incomplete in determining the number of electrons transferred during an acid–base interaction.

In the density-functional theory, two basic parameters of importance to chemistry were introduced. According to the theory, any chemical system (atom, molecule, ion, radical) can be characterized by its electronic chemical potential μ and its absolute hardness η. *Chemical potential measures the escaping tendency of an electronic cloud, while absolute hardness determines the resistance of the species to lose electrons.* It should be noted that these two parameters are molecular but not orbital properties. The exact definitions of these two quantities are

$$\mu = (\delta E/\delta N)_v, \tag{7}$$

and

$$\eta = \tfrac{1}{2}(\delta\mu/\delta N)_v = \tfrac{1}{2}(\delta^2 E/\delta N^2)_v, \tag{8}$$

where E is the electronic energy of an atom, a molecule, or an ion; N is the number of electrons; and v is the potential due to the nuclei, plus any external potential.

For atomic species, the chemical potential is the negative of electronegativity. Unfortunately, electronegativity has many definitions [36], including the newest one proposed by Allen [37]. However, this discussion will involve only the definition by Mulliken [38], i.e. electronegativity, χ^M, is the average of the energy required to remove an electron from an atom, measured as the ionization energy, I, and that released by the gain of one electron, measured as the electron affinity, A. Thus, $\chi^M = \tfrac{1}{2}(I + A)$.

Hence from equations (7) and (8), operational and approximation definitions give chemical potential μ

$$\mu \approx -\tfrac{1}{2}(I + A) \approx -\chi^M, \tag{9}$$

and absolute hardness η

$$\eta \approx \tfrac{1}{2}(I - A). \tag{10}$$

Since χ^M is the negative of chemical potential, it deserves to be called the absolute electronegativity; that is, the resistance of the chemical potential to the change in the number of electrons. It should be pointed out that both μ and η are global properties at the ground state in the sense that they characterize the species as a whole. Besides these two properties, there is also absolute softness, σ, which is the reciprocal of η.

Furthermore, according to Koopmans' theorem [35], the frontier orbital energies are given by

$$-E_{HOMO} = I; \quad -E_{LUMO} = A, \tag{11}$$

where I is the ionization potential and A is the electron affinity.

Any definition based on $E_{HOMO} - E_{LUMO}$ should be used with care because LCAO/SCF calculations place the virtual orbitals too high in energy. However, according to the frontier orbital method [33], the relationship between η and the energies of LUMO and HOMO is clearly shown in Fig. 3. Thus, from Fig. 3,

$$\eta \approx -\tfrac{1}{2}(E_{HOMO} - E_{LUMO}). \tag{12}$$

Here χ^M is the horizontal broken line in the middle of the energy gap. Hence the energy gap E_g is twice the absolute hardness η. Moreover, a hard molecule has a wide energy gap, while a soft molecule has a narrow gap. It will be demonstrated that the energy gap is an important link between chemistry and physics in the solid state.

5.2. Number of transferred electrons

The apparent success of the density-functional theory is to provide two chemical parameters from which we can calculate the number of electrons transferred resulting mainly from the charge transfer between the two molecules. Since electron (or charge) transfer is one of the major mechanisms of molecular interactions, so when two systems, A and B, are brought together, electrons will flow from that of lower χ to that of higher χ, until the chemical potential reaches an equilibrium. This follows the electronegativity equalization principle introduced by Sanderson [39]. According to this principle, when atoms (or other combining groups) of different chemical potentials unite to form a molecule with its own characteristic chemical potential, to the extent that the atoms (or groups) retain their identity, their chemical potentials must equalize.

Usually, but not always, there are electrons flowing from both directions. As a first approximation, the (fractional) number of electrons transferred, ΔN, is given by [40]

$$\Delta N \approx (\mu_B - \mu_A)/2(\eta_A + \eta_B) \approx -\Delta\mu/2\Sigma\eta, \tag{13}$$

or

$$\Delta N \approx (\chi_A - \chi_B)/2(\eta_A + \eta_B) \approx \Delta\chi/2\Sigma\eta. \tag{14}$$

Thus, the electron transfer is driven by $\Delta\chi$, but resisted by the sum of η's. Since molecular interactions involve other interactions besides electron transfer, ΔN is not the total change of electrons [41], but is still useful in determining the initial orbital interaction between A and B, and in serving as an approximation for the bond strength. Equation (14) also suggests that when $\Delta\chi \rightarrow 0$, there is no electron transfer. Since η's for metal atoms never approach zero, $\Sigma\eta \neq 0$. For the hard–hard interactions, $\Sigma\eta$ can be very large; thus, ΔN becomes too small, and the interaction will be dominated by the electrostatic interaction, instead of electron transfer. On the other hand, for the soft–soft interaction, $\Sigma\eta$ can be rather small and ΔN will be large. As expected, the interaction will be accompanied by polarization; however, it may not affect ΔN significantly.

Besides these extreme cases, generally ΔN is a fractional number of transferred electrons. This is illustrated by actual examples about the interactions between trimethylamine and boron halides [30]:

$$(CH_3)_3N + :BX_3 \rightarrow (CH_3)_3N:BX_3 \tag{15}$$

BX_3	ΔH (kcal/mol)	ΔN
BF_3	19	0.15
BCl_3	37	0.19
BBr_3	45	0.19

For the interactions between Cl_2 and several substituted benzenes [34], ΔN varied between -0.013 and 0.20. These examples indicate that the number of electrons transferred for a molecular interaction is generally fewer than that for the formation of a covalent bond. That is why the strength of molecular interaction is generally weaker than a covalent bond.

6. ACID–BASE INTERACTIONS IN SOLIDS

In the literature, there have not been many discussions about acid–base interactions in solids. Recently, Lee [42–46] has demonstrated that the extension of the HSAB principle to solid interactions is feasible in view of the electronic band structures. In the following, the physical meaning of absolute hardness will be discussed in terms of the average energy gap.

6.1. Electronic band structures of solids

Solids can be classified as metals, semimetals, intrinsic semiconductors, and insulators. The band structures of solids are illustrated in Fig. 4. Monovalent metals [47], e.g. Na, have a partially filled valence band, the lower half of which is occupied. The Fermi level is in the valence band but at the top of the occupied orbitals. Furthermore, there is still an energy gap between the valence band (occupied MO) and the conduction band (unoccupied MO). In some metals, such as the bivalent metals, the valence band is full but overlaps with a higher unoccupied conduction band. In this case, the Fermi level is between the conduction band and the overlapped valence band [48]. Thus, the electrons close to the Fermi level are still free to move as the extra bands supply the unoccupied states. In the latter case, there appears to be no minimum energy gap, E_g°, a parameter which is generally reported in the literature [e.g. *Handbook of Chemistry and Physics*, CRC Press, Florida (1989–1990)]. However, it is not obvious that there will always be an average energy gap E_g^{Av}, especially in metal atoms, which will be discussed later.

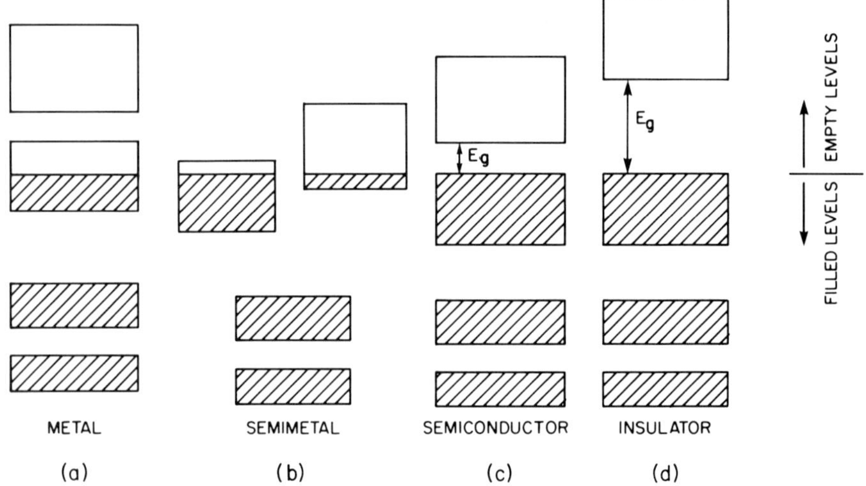

Figure 4. Band structures of solids: (a) metal; (b) semimetal; (c) semiconductor; (d) insulator. (Reproduced from ref. [48] with permission.)

In addition to metals, there are semimetals [49], such as graphite, whose valence band and conduction band can overlap. In general, their minimum energy gaps are very narrow. The third class of solids is the intrinsic semiconductor; its minimum energy gap E_g^o is generally below 3 eV. Thus, the thermal excitation alone can create an electron–hole pair to enhance conduction. The Fermi level of the intrinsic semiconductor [47] lies between the valence band (HOMO) and the conduction band (LUMO). Hence,

$$E_F = \tfrac{1}{2}(E_{HOMO} + E_{LUMO}) = \tfrac{1}{2}(E_c + E_v), \tag{16}$$

where E_F is the energy of the Fermi level. The fourth class of solids is the insulator, including most nonconducting polymers. Generally, the minimum energy gap of an insulator is above 3 eV. Therefore, thermal excitation alone cannot enhance the conduction of electricity.

In reality, the band structures in Fig. 4 are only simplistic representations of solids. The electron transfer between a donor and an acceptor is better described by the spreading distribution of energy levels between two polymers (or other solids) as shown in Fig. 5. Because of the spreading of levels influenced by the density of states (DOS), the average energy gap between LUMO and HOMO appears to be a better measure of hardness than the minimum energy gap.

6.2. Phillips' ionic and covalent components of the average energy gap

Let us now return to our familiar theme of dual components—electrostatic (ionic) and charge transfer (covalent). The average energy gap E_g^{Av} [50–52] is defined as the difference in energy between the bottom of the conduction band and the top of the valence band:

$$E_g^{Av} = -(E_{HOMO} - E_{LUMO}) = -(E_v - E_c). \tag{17}$$

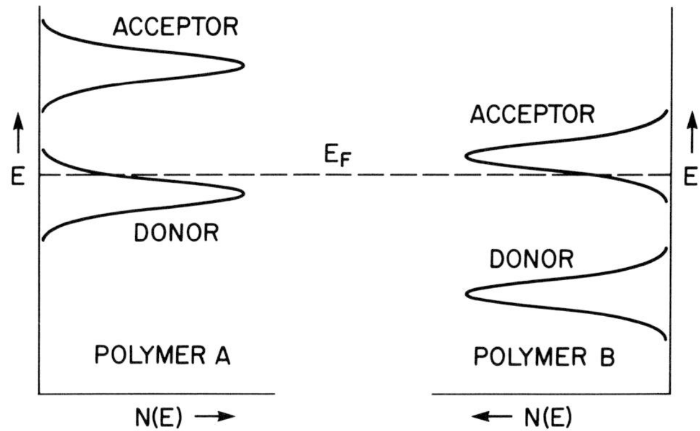

Figure 5. Electron density of state diagram for donor and acceptor levels.

Interestingly, according to Phillips [50, 51], E_g^{Av} also contains both a covalent (homopolar) component E_h and an ionic (heteropolar) component C:

$$(E_g^{Av})^2 = (E_h)^2 + (C)^2. \tag{18}$$

Equation (18) resembles the Hückel relation, and C is similar to $\Delta\chi$.

For $A^N B^{8-N}$ compounds, E_g^{Av} can be determined from the high frequency dielectric constant ε_∞ according to the following equation:

$$\varepsilon_\infty = [1 + (\hbar\Omega_p)^2/(E_g^{Av})^2], \tag{19}$$

where Ω_p is the plasma frequency of the valence electron. Thus, the average energy gap is a fundamental physical parameter. From these two components, Phillips also defines ionicity f_i (or fractional ionic character) as follows:

$$f_i = C^2/(C^2 + E_h^2). \tag{20}$$

On the basis of the density-functional theory, discussed in Section 5.2, it is insufficient to use $\Delta\chi$ (or f_i) alone to estimate the number of electrons ΔN from equation (14) involved in the electron transfer. According to the theory, another measure, the absolute hardness, has to be taken into account. Here, η is related to the average energy gap:

$$E_g^{Av} \approx 2\eta. \tag{21}$$

In Table 1, a list of C, E_h [52], E_g^{Av} [53], and η for several semiconductor compounds is shown along with their ionicities, f_i. In the case of semiconductors, f_i has been used to differentiate covalent from ionic semiconductors. For example, Si in Table 1 is covalent, while CdSe is highly ionic because it has a higher f_i.

6.3. Average energy gaps and absolute hardnesses of metals

Now let us examine again the band structures of the solids in Fig. 4. In Fig. 4a, the structure of metals with zero or low E_g signifies that all bulk metals are 'soft' as implied by $A = I$ for the finite-difference approximation. However, different metals have different softnesses, all large, but not infinite [30]; and metal atoms are as 'soft' as the low finite η values calculated from the $A-I$ data (Table 2).

Table 1.

Ionic and covalent components of $A^N B^{8-N}$ semiconductor compounds

	C (eV)[a]	E_h (eV)	E_g^{Av} (eV)[b]	η (eV) (calc.)[b]	f_i
Si	0	4.8	4.8	2.4	0
SiC	3.9	8.3	9.2	4.6	0.18
GaAs	2.9	4.3	5.2	2.6	0.31
AlSb	3.1	3.5	4.7	2.4	0.43
ZnS	6.2	4.8	7.8	3.8	0.63
CdSe	5.5	3.6	6.6	3.3	0.70

[a] C and E_h are from ref. [52].
[b] E_g^{Av} values are from ref. [53].

Indeed, metals have been classified as amphoteric materials. Most of the metals are 'soft' acids, and some of them 'soft' bases. When two metals are brought together into close contact, one of them becomes an acid and the other a base. Cain *et al.* [54] showed the reaction at the Cu–Cr interface to be an acid–base interaction. In this case, a soft base (Cu) and a soft acid (Cr) react preferentially. In fact, the Cr atom ($\eta = 3.1$ eV) is as soft as the Cu atom ($\eta = 3.3$ eV) (see Table 2).

In Table 2, a list of calculated E_g^{Av} values from η for several transition metal atoms is given. The η values falling between 3 and 4 indicate that most of these metal atoms are rather 'soft'. We will discuss the applications of the HSAB principle based on these η values.

6.4. *Average energy gaps and absolute hardnesses of semiconductor elements*

Figures 4b and 4c show that both semimetals and semiconductors should be

Table 2.

Average energy gaps and hardness values for transition metal atoms (eV)

Metal	I[a]	A[b]	χ^M	η[c]	E_g^{Av} (calc.)[d]
Ti	6.82	0.08	3.45	3.4	6.8
V	6.7	0.5	3.6	3.1	6.2
Cr	6.77	0.66	3.72	3.1	6.2
Mn	7.44	0	3.72	3.7	7.4
Fe	7.87	0.25	4.06	3.8	7.6
Co	7.8	0.7	4.3	3.6	7.2
Ni	7.64	1.15	4.4	3.3	6.6
Cu	7.73	1.23	4.48	3.3	6.6
Mo	7.10	0.75	3.9	3.2	6.4
Ru	7.40	1.5	4.5	3.0	6.0
Pd	8.34	0.56	4.45	3.9	7.8
Ag	7.58	1.30	4.44	3.1	6.2
Pt	9.0	2.1	5.6	3.5	7.0
Au	9.23	2.31	5.77	3.5	7.0

[a] From ref. [55].
[b] From ref. [56].
[c] From ref. [30].
[d] E_g^{Av} is calculated from η obtained from $A–I$ values.

rather 'soft', because of their relatively low E_g° and E_g^{Av} values (most of η's < 3.0 eV). Semiconductors generally react readily with metals even at ambient temperatures [57]. Thus, the interaction takes place as a 'soft' acid (metal) with a 'soft' base (semiconductor). Both chemical reaction and inter-diffusion [58] jointly create a diffused 'interphase' instead of an abrupt 'interface'.

For comparison, both E_g° and E_g^{Av} for several elements (diamond structure) are listed in Table 3. In general, E_g^{Av} is always larger than E_g°. From Table 2, the η^{Av} values calculated from E_g^{Av}'s are rather close to those calculated from the $A-I$ data; however, the agreement is still not good enough.

Table 3.

Minimum and average energy gaps and absolute hardness values for elements (diamond structure) (eV)

	$E_g^{\circ\ a}$	$E_g^{Av\ a}$	η^{Av} (calc.)^b	η^c
C	5.4	13.6	6.8	5.0
Si	1.1	4.8	2.4	3.4
Ge	0.7	4.3	2.2	3.4
Sn	0	3.1	1.6	3.1

^{a} E_g° and E_g^{Av} are from ref. [53].
^{b} η^{Av} is calculated from E_g^{Av}.
^{c} η is the value obtained by Pearson [30].

6.5. *Local hardness*

It is important to remember that the hardness of one material with respect to another depends on its η. The one with a lower η value is 'softer' than the one with a higher η though they both can be 'hard'. Even in the same material, one functional group can be harder than the other. In this case, a new term, local hardness, $\tilde{\eta}$ [14], has been introduced to signify the relative hardness of each functional group.

In Fig. 4d, it is noted that the band structure of an insulator contains a large gap. Thus, most insulators, regardless of whether they are organic or inorganic, are considered to be comparatively 'hard' on the basis of the energy gap alone. Generally, they are harder than metals, semimetals, and semiconductors.

7. APPLICATIONS OF THE DENSITY-FUNCTIONAL THEORY TO SOLIDS

7.1. *Number of transferred electrons, ΔN, calculated from the electronegativity and average energy gap*

On the basis of the relation between η and χ, we propose to express ΔN in terms of the average energy gap of a solid:

$$\Delta N \approx (\chi_A - \chi_B)/(E_{g_A}^{Av} + E_{g_B}^{Av}), \tag{22}$$

or

$$\Delta N \approx \Delta\chi/\Sigma E_g^{Av}. \tag{23}$$

This derived equation is very important with regard to the molecular interaction. Now if we know the difference in electronegativities of the A and B molecules and the energy gaps, we can estimate the extent of eletron transfer for the inter-action. This equation is especially useful for semiconductor compounds for which both χ and E_g^{Av} are available in the literature.

7.2. Electronegativities and work functions of solids

In a solid, the chemical potential can be represented by the negative of work function. Steiner and Gyftopoulos [59] have shown that the work function Φ for a metal surface is equal to the neutral electronegativity of the surface atoms. Thus,

$$\Phi = \chi^M. \tag{24}$$

A comparison of the work functions and χ^M of polycrystalline metals is shown in Table 4, and here again, the electronegativity of Mulliken has the best fit with the work function.

Table 4.

Mulliken's electronegativities and work functions of poly-crystalline metals (eV)

Metal	$\chi^{M\ a}$	Φ^b	Method of determination of Φ^c
Pt	5.6	5.9	P
Ni	4.4	5.15	P
Au	5.8	5.1	P
Si	4.8	4.85 (n)^d	P
Cu	4.5	4.65	CPD
Fe	4.1	4.55	P
Cr	3.7	4.50	P
Ti	3.5, 4.1	4.33	P
Al	3.2	4.28	P
Ag	4.5	4.26	P
Mg	3.8	(3.66)^e	P
K	2.4	2.3	P
Cs	2.2	2.14	P

^{a} From ref. [60].
^{b} From ref. [61].
^{c} Methods used for the determination of the work function are: P—photo-electric; CPD—contact potential difference.
^{d} n: n-type.
^{e} The number in parentheses is less certain.

Without exception, electrons tend to flow from a metal of low Φ to one of high Φ. Hence, the difference of Φ forms a contact potential V_c at the interface:

$$V_c = e(\Phi_A - \Phi_B) = e\Delta\Phi, \tag{25}$$

where e is the electronic charge.

7.3. Number of transferred electrons calculated from the work function and energy gap

Instead of chemical parameters, the number of transferred electrons can also be determined from physical properties, such as the work function and the average energy gap of a solid. Substituting equation (24) into equation (23), we obtain an important relation for solids:

$$\Delta N = \Delta\Phi/\Sigma E_{\mathrm{g}}^{\mathrm{Av}}.\tag{26}$$

This equation somewhat resembles Ohm's law, which states that the electric current is directly proportional to the voltage (or potential) and inversely proportional to the resistance. Besides metals, the work functions of polymers [62, 63] as shown in Table 5 are available; otherwise, the contact potential can be determined between a metal and another solid and substituting it into equation (26) gives

$$\Delta N = V_{\mathrm{c}}/(e\Sigma E_{\mathrm{g}}^{\mathrm{Av}}).\tag{27}$$

If $\Sigma E_{\mathrm{g}}^{\mathrm{Av}}$ approaches zero in the case of bulk metals, ΔN may go to infinity for the metal-to-metal contact, as in the case of cold-welding of metals under an ultra-high vacuum or in outer space [64]. The interesting point about equations (26) and (27) is that instead of the two basic chemical parameters, we can now use two equivalent physical properties, i.e. the work function or the contact potential, and the energy gap to estimate ΔN for any molecular interaction. These two equations signify the eventful result in meeting chemistry and physics in the solid state.

Table 5.
Work functions of polymers

	Work function Φ (eV)	
Polymer	From ref. [62]	From ref. [63]
Poly(tetrafluoroethylene)	5.75	4.26 ± 0.05
Poly(vinyl chloride)	5.13	4.85 ± 0.02
Poly(sulfone)	4.95	—
Poly(styrene)	4.90	4.22 ± 0.07
Poly(ethylene)	4.90	—
Poly(carbonate)	4.80	4.26 ± 0.13
Poly(methyl methacrylate)	4.68	—
Poly(vinyl acetate)	4.38	—
Nylon 6,6	4.30	4.08 ± 0.06
Poly(ethylene oxide)	3.95	—
Polyimide	—	4.36 ± 0.06
Polyethylene terephthlate	—	4.25 ± 0.10

7.4. HSAB principle for solids

For solids, we can now relate absolute hardness to the average energy gap that can be determined accurately through dielectric measurements. From the above discussion, we can rephrase the HSAB principle for solids in terms of the average energy gap and work function:

- A hard (more ionic) acid (or acceptor) has a high-energy LUMO and a wide average energy gap.
- A soft (more covalent) acid (or acceptor) has a low-energy LUMO and a narrow average energy gap.

- A hard base (or donor) has a low-energy HOMO, a high work function (or electronegativity), and a wide average energy gap.
- A soft base (or donor) has a high-energy HOMO, a low work function (or electronegativity), and a narrow average energy gap.

8. APPLICATIONS OF THE DENSITY-FUNCTIONAL THEORY TO ADHESION AND INTERFACIAL INTERACTIONS BETWEEN METALS AND POLYMERS

The interactions between metals and polymers are good examples of the applications of the HSAB principle to solids. When Cr is deposited on pyromellitic dianhydride–oxydianiline polyimide (PMDA–ODA PI), there appears to be some chemical bonding between the Cr atom and PI. In this case, the Cr atom is a 'soft' acid and PI is a 'hard' base. How can the bonding take place? According to Ho *et al.* [65, 66], presumably the reaction does not proceed in the manner of Cr atom attacking one of the carbonyl groups (nucleophiles). What probably happens is that the Cr atom delocalizes and forms a charge-transfer (or acid–base) complex with the PMDA unit of the PI. In the frontier orbital terminology, stabilization of the complex is achieved through the transfer of electrons from the d-states of the Cr atom (HOMO) to the LUMO of the π-system of the PMDA unit of PI. In this manner, PI acts as a 'soft' acid by accepting the electron and Cr atom ($\eta = 3.1$ eV) becomes a 'soft' base by donating the electron. However, there are other studies indicating that the metal atom, such as Cr, may indeed bond to carbonyls [67–69]. It is likely that the Cr–PI complex is a transition state which leads to the final product between the Cr atom and the carbonyls. Other transition metals, e.g. Ti ($\eta = 3.4$ eV) [70], are also very reactive with PI. The general reaction presumably follows paths similar to the acid–base interaction.

One of the exceptions is Cu [71], which does not interact with PI after the deposition and gives rise to a much weaker complex. Cain *et al.* [72, 73] used the tight binding calculations of the extended Hückel type to find the relative acid strengths of Cu compounds in the following decreasing order: $CuF_2 > CuO > CuF \approx Cu_2O > Cu$. Indeed, the Cu ions react faster than Cu with functional polymers. The Cu^{2+} ion has a Fermi level lying below the top of the Cu $3d$ band, and some of the Cu $3d$ orbitals are unoccupied and thus able to accept electron pairs by acting as an acid. In the HSAB terminology, Cu becomes a 'hard' acid in the form of Cu^{2+} ion because the hardness for Cu atom is 3.3 eV, and that for Cu^{2+}, 8.3 eV.

Generally, many transition metals react well with polymers, ceramics [74], etc., partly because of the ease of oxidation and partly because of the availability of d-orbitals. Buckley and Brainard [75] found that metals react with PTFE (Teflon) and PI during their pin-and-disc experiments under high vacuum. With PTFE, the adhesion forces are three times the applied load. For this case, these interactions may also be explained by the HSAB principle. In the literature, there are numerous examples of interfacial interactions that can be classified as the acid–base interaction. However, we are unable to discuss them all in this paper. One of the reasons is that absolute hardness values for polymers have yet to be determined indirectly. We are planning to publish the calculated values in the future.

The density-functional theory indicates that both chemical parameters are required to determine the number of electrons transferred for the interaction between A and B molecules. Thus, absolute hardness alone is insufficient to predict the extent of the interaction. It is important to point out that although absolute hardness values are rather similar for all metals, all metals have different electronegativities or work functions (see Table 4). Hence, they do not react in the same way or at the same rate with polymers. For example, the work function of Cu is 4.65 eV, and that for Cr, 4.50 eV (Table 4). In the case of polyimide ($\Phi = 4.36$ eV) (Table 5) [62], one may expect that both Cu and Cr should react in the same manner. However, the experimental results show otherwise; thus, we need to compare the Fermi levels of both metals. According to Hoffmann [76, 77], the interaction between a molecule and the surface of a metal is different from that between two different molecules, and this type of interaction depends greatly on the Fermi level of the metal involved. For example, since the Fermi level of Cr is significantly higher than that of Cu, a molecule that can readily react with Cr may not react with Cu.

Since all metals are 'soft' and most polymers are 'hard', the metal–polymer reaction or adhesion, unlike the exceptional Cr–PI reaction mentioned above, is generally difficult to achieve. To circumvent this situation, most metals have to be transformed into harder counterparts, e.g. related oxides of higher absolute hardness values. This is exemplified by the cited case about Cu^{2+} ions [72, 73], which are harder than the metal atoms, Cu. Consequently, the reactivities of the metal–polymer interactions are also determined by the ease of the formation of oxides, and that ease is, in turn, controlled by the heat of formation H_f of the oxides [78]. In theory, the more negative the H_f, the more reactive the interaction. For the same reason, Au and Ag are always nonreactive because of their high work functions (see Table 4). We have explored in detail the metal–polymer adhesion in a separate paper on the chemistry and physics of solid adhesion [79].

9. SUMMARY

In this paper, we have attempted first to discuss the acid–base interaction, from a broader perspective, in terms of the molecular interaction. Two major terms were pointed out to be involved in the interaction: the Coulombic and the frontier orbital terms. It is important to note that there is a critical interatomic distance for a molecular interaction, i.e. approximately 3 Å. However, the optimum distance is about 2.0 Å.

The focal point of this paper is the application of the density-functional theory to molecular interaction in solids. For an acid–base interaction, one needs to consider first the HSAB principle based on absolute hardness. Otherwise, without matching the hardness of both species the interaction may be too slow to be detected. Theoretically, on the basis of the density-functional theory, two basic chemical parameters should be considered simultaneously: the chemical potential and absolute hardness, from which a more fundamental parameter, ΔN, the (fractional) number of electrons transferred from molecule A to molecule B, can be estimated.

The extension of the HSAB principle and the frontier orbital concept to solid interactions has been demonstrated. For the HSAB principle, the absolute hardness has been discussed along with the average energy gap in a solid. The average energy gap also consists of two components: ionic and covalent. Through the energy gap, we found a bridge between chemistry and physics in the solid state. Furthermore, we can interpret a molecular as well as an acid–base interaction on the basis of well-defined physical properties instead of the chemical parameters derived from the density-functional theory. For example, in the case of metals or polymers, ΔN can be estimated from two physical properties, i.e. the work function and the average energy gap, instead of chemical parameters. Although we have not supplied any direct experimental evidence in this paper, we believe that our findings have a theoretical basis and should be helpful to those carrying out research on interfacial interactions and solid adhesion.

REFERENCES

1. V. Gutmann, *The Donor–Acceptor Approach to Molecular Interactions.* Plenum Press, New York (1978).
2. B. V. Deryagin, N. A. Krotova and V. P. Smilga, *Adhesion of Solids,* English Edn. Translated by R. K. Johnson. Plenum Press, New York (1978).
3. W. B. Jensen, *Chem. Rev.* **78**, 1 (1978); also W. B. Jensen, in: *Surface and Colloid Science in Computer Technology,* K. L. Mittal (Ed.), pp. 27–60. Plenum Press, New York (1987).
4. W. B. Jensen, *The Lewis Acid–Base Concepts: An Overview.* Wiley-Interscience, New York (1980).
5. W. B. Jensen, *J. Adhesion Sci. Technol.* (in press).
6. K. Morokuma, *Acc. Chem. Res.* **10**, 294 (1977).
7. K. Morokuma and K. Kitaura, in: *Molecular Interactions,* H. Ratajczak and W. J. Orville-Thomas (Eds). John Wiley, New York (1980).
8. F. M. Fowkes, *J. Adhesion* **4**, 155 (1972); also in *Recent Advances in Adhesion,* L. H. Lee (Ed.), p. 39. Gordon and Breach, New York (1973).
9. F. M. Fowkes, *J. Adhesion Sci. Technol.* **1**, 7 (1987).
10. F. M. Fowkes and M. A. Mostafa, *Ind. Eng. Chem. Prod. Res. Dev.* **17**, 3 (1978).
11. F. M. Fowkes, *J. Adhesion Sci. Technol.* (in press).
12. J. C. Bolger and A. S. Michaels, in: *Interface Conversion,* P. Weiss and D. Cheevers (Eds), Chap. 1, Elsevier, New York (1969); also J. C. Bolger, in: *Adhesion Aspects of Polymeric Coatings,* K. L. Mittal (Ed.), pp. 3–18. Plenum Press, New York (1983).
13. R. G. Pearson, *J. Am. Chem. Soc.* **85**, 3533 (1963).
14. R. G. Pearson, *J. Chem. Educ.* **64**, 563 (1987).
15. R. G. Parr and W. Yang, *Density-functional Theory of Atoms and Molecules.* Oxford University Press, New York (1989).
16. R. F. Hudson and G. Klopman, *Tetrahedron Lett.* **12**, 1103 (1967); also G. Klopman and R. F. Hudson, *Theor. Chim. Acta* **9**, 165 (1967).
17. K. Kitaura and K. Morokuma, *Int. J. Quantum Chem.* **10**, 325 (1976).
18. N. S. Isaacs, *Physical Organic Chemistry,* Longman Scientific and Technical/Wiley, New York (1987).
19. F. L. Riddle, Jr. and F. M. Fowkes, *J. Am. Chem. Soc.* **112**, 3259 (1990).
20. F. M. Menger, in: *Nucleophilicity,* J. M. Harris and S. P. McManus (eds), Chap. 14, ACS Symposium Series No. 209. American Chemical Society, Washington, DC (1987).
21. A. E. Reed, L. A. Curtiss and F. Weinhold, *Chem. Rev.* **88**, 899 (1988).
22. J. P. Foster and F. Weinhold, *J. Am. Chem. Soc.* **102**, 7211 (1980).
23. K. Fukui, *Science* **218**, 747 (1982).
24. R. S. Drago, G. C. Vogel and T. E. Needham, *J. Am. Chem. Soc.* **93**, 6014 (1971).
25. M. D. Vrbanac and J. C. Berg, *J. Adhesion Sci. Technol.* **4**, 255 (1990).
26. M. Berkowitz, S. K. Ghosh and R. G. Parr, *J. Am. Chem. Soc.* **107**, 6811 (1985).
27. W. Yang and R. G. Parr, *Proc. Natl. Acad. Sci. USA* **82**, 6723 (1985).
28. M. Berkowitz and R. G. Parr, *J. Chem. Phys.* **88**, 2554 (1988).
29. R. G. Pearson, *Hard and Soft Acids and Bases,* Dowden, Hutchinson and Ross, Stroudsburg, PA (1973).
30. R. G. Pearson, *Inorg. Chem.* **27**, 734 (1988); also *J. Am. Chem. Soc.* **110**, 7684 (1988).

31. R. G. Pearson, *Proc. Natl. Acad. Sci. USA* **83**, 8440 (1986).
32. A. D. Garnovskii, O. A. Osipov and S. B. Bulgarevich, *Russ. Chem. Rev.* **41**, 341 (1972).
33. I. Fleming, *Frontier Orbitals and Organic Chemical Reactions*. John Wiley, London (1976).
34. R. G. Pearson, *J. Org. Chem.* **54**, 1423 (1989).
35. T. Koopmans, *Physica* **1**, 104 (1933).
36. J. Mullay, in: *Electronegativity*, Structure and Bonding Series Vol. 66, p. 1. Springer, Berlin (1987).
37. L. C. Allen, *J. Am. Chem. Soc.* **111**, 9003 (1989).
38. R. S. Mulliken, *J. Chem. Phys.* **3**, 573 and 586 (1935).
39. R. T. Sanderson, *Science* **114**, 670 (1951); also R. T. Sanderson, *Chemical Bonds and Bond Energy*, 2nd edn. Academic Press, New York (1976).
40. S. Shankar and R. G. Parr, *Proc. Natl. Acad. Sci. USA* **82**, 264 (1985).
41. R. G. Pearson, *J. Am. Chem. Soc.* **107**, 6801 (1985).
42. L. H. Lee, in: *Proceedings 5th International Congress on Tribology*, K. Holmberg and I. Nieminen (Eds), Vol. 3, p. 308. Finnish Society for Tribology, Espoo, Finland (1989).
43. L. H. Lee (Ed.), *New Trends in Physics and Physical Chemistry of Polymers*, p. 185. Plenum Press, New York (1989).
44. L. H. Lee, *Prog. Colloid Polym. Sci.* **82**, 1 (1990).
45. L. H. Lee, *Polym. Mater. Sci. Eng.* **62**, 881 (1990).
46. L. H. Lee (Ed.), *Fundamentals of Adhesion*, Chap. 12. Plenum Press, New York (1990).
47. F. Gutmann and L. E. Lyons, *Organic Semiconductors*, p. 17. John Wiley, New York (1967).
48. A. J. Epstein and J. S. Miller, *Sci. Am.* **241**, 52 (Oct. 1979).
49. D. L. Carter and R. T. Bate (Eds), *The Physics of Semimetals and Narrow-Gap Semiconductors*. Pergamon Press, Oxford (1971).
50. J. C. Phillips, *Rev. Mod. Phys.* **42**, 317 (1970).
51. J. C. Phillips, *Phys. Today* **23**, 23 (1970).
52. J. A. van Vechten, *Phys. Rev.* **182**, 891 (1969).
53. E. Burstein, A. Pinczuk and R. F. Wallis, in: *The Physics of Semimetals and Narrow-Gap Semiconductors*, D. L. Carter and R. T. Bate (eds), p. 251. Pergamon Press, Oxford (1971).
54. S. R. Cain, L. J. Matienzo and F. Emmi, *J. Phys. Chem. Solids* **50**, 87 (1989).
55. C. E. Moore, *Natl. Stand. Ref. Data Ser.* NSRDS-NBS 34 (1970).
56. H. Hotop and W. C. Lineberger, *J. Phys. Chem. Ref. Data* **14**, 731 (1985).
57. L. Braicovich, in: *The Chemical Physics of Solid Surfaces and Heterogeneous Catalysts*, D. A. King and D. P. Woodruff (eds), Chap. 6. Elsevier, Amsterdam (1988).
58. L. J. Brillson, *Surf. Sci. Rep.* **2**, 123 (1982).
59. D. Steiner and E. P. Gyftopoulos, *Proc. 27th Annu. Conf. Phys. Electr.*, p. 160, Massachusetts Institute of Technology, Cambridge, MA (March 1967).
60. H. B. Michaelson, *IBM J. Res. Dev.* **22**, 72 (1978).
61. H. B. Michaelson, in: *Handbook of Chemistry and Physics*, R. C. Weast (Ed.), p. E-93. CRC Press, Florida (1989–1990).
62. D. K. Davies, in: *Advances in Static Electricity*, Vol. 1, p. 10. Proceedings of the First International Conference on Static Electricity, Vienna, Austria (4 May, 1970).
63. S. Strella, unpublished results (1974).
64. M. J. Hordon, in: *Adhesion or Cold Welding of Materials in Space Environment*, ASTM STP 431. American Society for Testing Materials, Philadelphia (1967).
65. P. S. Ho, B. D. Silverman, R. A. Haight, R. C. White, P. N. Sanda and A. R. Rossi, *IBM J. Res. Dev.* **32**, 658 (1988).
66. P. S. Ho, R. Haight, R. C. White, B. D. Silverman and F. Faupel, in: *Fundamentals of Adhesion*, L. H. Lee (Ed.), Chap. 14. Plenum Press, New York (1990).
67. J. G. Clabes, M. J. Goldberg, A. Viehbeck and C. A. Kovac, *J. Vac. Sci. Technol.* **A6**, 985 (1988).
68. M. J. Goldberg, J. G. Clabes and C. A. Kovac, *J. Vac. Sci. Technol.* **A6**, 991 (1988).
69. W. J. van Ooij, R. H. G. Brinkhuis and J. M. Park, *Surf. Interface Anal.* **12**, 505 (1988).
70. F. S. Ohuchi and S. C. Freilich, *J. Vac. Sci. Technol.* **A4**, 1039 (1986).
71. R. Haight, R. C. White, B. D. Silverman and P. S. Ho, *J. Vac. Sci. Technol.* **A6**, 2188 (1988).
72. S. R. Cain and L. J. Matienzo, *J. Adhesion. Sci. Technol.* **2**, 395 (1988).
73. S. R. Cain, L. J. Matienzo and F. Emmi, in: *Metallized Plastics 1: Fundamental and Applied Aspects*, K. L. Mittal and J. R. Susko (Eds), pp. 247–263. Plenum Press, New York (1989).

74. D. H. Buckley, *Surface Effects in Adhesion, Friction, Wear and Lubrication.* Elsevier, Amsterdam (1981).
75. D. H. Buckley and W. A. Brainard, in: *Advances in Polymer Friction and Wear*, L. H. Lee (Ed.), Vol. A, p. 315. Plenum Press, New York (1975).
76. R. Hoffmann, *Angew. Chem. Int. Ed. Engl.* **26**, 846 (1987); *Rev. Mod. Phys.* **60**, 601 (1988).
77. R. Hoffmann, *Solids and Surfaces—A Chemist's View of Bonding in Extended Structures.* VCH Publishers, Weinheim (1988).
78. D. M. Mattox, *Thin Solid Films* **18**, 173 (1973).
79. L. H. Lee (Ed.), *Fundamentals of Adhesion*, Chap. 1. Plenum Press, New York (1990).

Acid-Base Interactions, pp. 47-65
Eds. K.L. Mittal and H.R. Anderson, Jr.
©VSP 1991

Quantum-mechanical approach to understanding acid–base interactions at metal–polymer interfaces

STEPHEN R. CAIN

IBM Systems Technology Division, 1701 North Street, Endicott, NY 13760, USA

Revised version received 9 February 1990

Abstract—Molecular orbital theory is presented as a means to understanding metal–polymer interfacial chemistry. While the systems used in molecular orbital studies tend to be rather idealized, much insight may be gained from such studies. Three different investigations are cited as examples of the applicability of molecular orbital theory in describing how functional groups of an adsorbate bond to a metal surface as well as how quantum-mechanical calculations may be used to interpret experimental data, e.g. photoelectron spectra. Experimental techniques (nucleophilic displacement, isoelectric points, flow microcalorimetry, and temperature-programmed desorption) for determining acid–base interactions are also discussed briefly.

Keywords: Quantum chemistry; interfacial chemistry; metal–polymer interfaces, acid–base chemistry; surface chemistry.

1. INTRODUCTION

Interfacial chemistry is of crucial concern in a number of industries. For example, the microscopic interaction of water with inorganic oxides is of great interest in the construction industry, since this interaction in part determines the strength of concrete. Electronics is another industry concerned with chemical interactions between dissimilar materials; circuit lines must adhere to the substrate, and must not be subject to excessive corrosion. The coatings industry also takes great pains to understand interfacial chemistry, since corrosion may occur in places where the coating has been delaminated.

Acid–base chemistry plays an important role in determining interfacial interactions. For example, adhesion has been shown to be determined to a large extent by acid–base chemistry, both experimentally [1] and theoretically [2]. Recently, band calculations have been used to explain the surface acidity of copper oxides and fluorides [2a], as well as how the basic functionalities of a polymer bind to chromium, nickel, and copper (and their oxides) [2b]. Further, the reactivity of Cu_2O with a Cr_2O_3 surface has been predicted based on acid–base chemistry [2c].

Much chemical insight may be gained through molecular orbital studies used in conjunction with experimental techniques. The main utility of orbital calculations is to identify microscopic properties (e.g. ionization potential, electron count, geometry) which give rise to the macroscopic properties of interest (in this case, adhesion). Once the relationship of the microscopic and macroscopic properties has been established, predictions may be made. In this sense, quantum mechanics should not be viewed as an alternative technique, but rather as a companion technique to aid in understanding experimental data. This paper focuses on how orbital calculations may be used as a tool to interpret the available experimental data. Before proceeding to a discussion of the quantum-mechanical techniques, it is instructive to consider a few of the commonly used experimental techniques for probing interfacial chemistry.

2. SOME COMMONLY USED EXPERIMENTAL TECHNIQUES FOR DETERMINING SURFACE ACID–BASE INTERACTIONS

Perhaps the most widely used technique for determining trends in acid–base chemistry involves nucleophilic displacement, of the type given in the equation

$$\text{Nu:} + \text{AX} \xrightarrow{K_{eq}} \text{NuA} + \text{:X,} \tag{1}$$

where Nu is the nucleophile under consideration, A is the acid portion of a molecule, and X is a 'leaving group' displaced by the nucleophile. The numerical value of the equilibrium constant, K_{eq}, is a direct measure of the nucleophilicity of Nu. If Nu is much more nucleophilic than X, then K_{eq} is large ($\gg 1$); if Nu is less nucleophilic than X, then K_{eq} is small ($\ll 1$). Classification of nucleophiles from such studies [3] led Pearson to publish a landmark paper introducing the concept of hard–soft acid–base (HSAB) chemistry [4], to be discussed in the next section.

The acidity of a surface may be inferred from the isoelectric point (IEPS), discussed in ref. [5]. This technique, though applied only to aqueous dispersions, has provided valuable insight in the relative acidities of various metal and non-metal oxides. If the oxide is placed in water, the surface adsorbs water molecules which are prone to dissociate. In a formal sense, water bound to cationic sites releases protons, leaving OH^- adsorbed. Water bound to oxide ions releases OH^-, leaving H^+ adsorbed. Since the 'dry' metal or non-metal oxide is charge-neutral, this hydration/dissociation process leaves the surface charged, as illustrated in Fig. 1. If the oxide happens to be acidic, then more OH^- ions will remain bound than H^+ ions and the surface will be negatively charged. Conversely, if the oxide is basic, then more H^+ ions will be bound to the surface than OH^- ions and the surface will be positively charged. This net charging affects the 'zeta potential' of the dispersed particle, and may be measured by observing the motion of the dispersed particles in an electric field [6].

Because the relative amounts of adsorbed H^+ and OH^- ions depend on the pH of the surrounding solution, the pH can be adjusted so that the zeta potential is zero (equal amounts of adsorbed OH^- and H^+ ions). The pH required to bring the

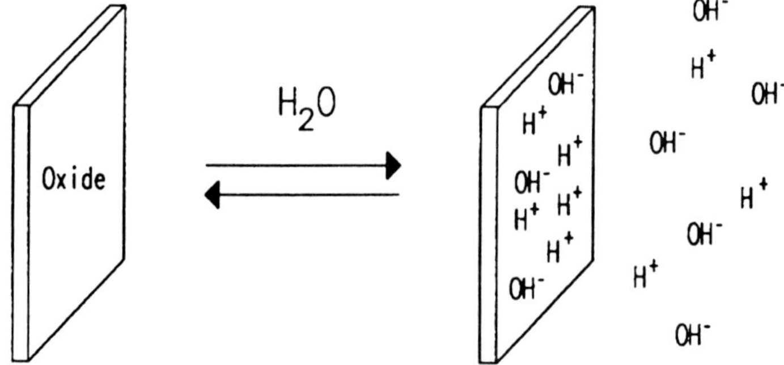

Figure 1. Water adsorbing onto a metal or non-metal oxide dissociates, releasing either a proton or an OH^- ion to the surrounding liquid. In general, the surface may be charged.

zeta potential to zero is the IEPS of the material. In this sense, it is quite proper to think of the IEPS as the pH of a surface, and the IEPS gives the same information as pH. A basic surface has a high IEPS, an acidic surface has a low IEPS, and a neutral surface has an IEPS of seven.

The IEPS has been used to establish a relationship between the surface acidity and the ratio of oxidation state to ionic radius of the cation in the oxide [7]. This relationship is roughly linear, as shown in the equation

$$\text{IEPS} \propto -Z/r, \tag{2}$$

where Z is the formal charge (oxidation state) and r is the ionic radius.

Once the acidity/basicity has been determined, polymers used for the coating may be selected on the basis of the types of functional groups. For example, one might expect polymers with amino groups to bond well to surfaces which have a low IEPS.

Another experimental technique is flow microcalorimetry, described in refs [1c] and [8]. A schematic diagram is given in Fig. 2. In flow microcalorimetry, the energy for adsorbing a material from a given solvent may be measured. A twin-flow apparatus directs either solvent or solvent plus adsorbate into the micro-calorimeter, which is equipped with two sensors. One sensor measures the temperature of the sample bed, and the other measures the concentration of adsorbate remaining in the effluent. The sample may be dried by evacuating the calorimeter for some time prior to beginning the experiment. For purposes of drying, an ultra-high vacuum system is not necessary; in one experiment, adequate drying for the hygroscopic gamma iron oxide was achieved at 1 Torr [8]. Initially, pure solvent flows over the sample. After the system has reached equilibrium, the liquid selection valve is rotated to direct the solvent/adsorbate mixture onto the sample. This sensitive technique is readily applied to non-aqueous solvent systems.

One of the problems in studying surface phenomena is the rapidity with which oxides form on the surface. Chromium, for example, oxidizes readily at 30°C in an oxygen atmosphere as low as 10^{-7} Torr [9]. Therefore, ultra-high vacuum (UHV) techniques provide a solution to the problem of controlling surface contamination. A typical UHV system, shown in Fig. 3, can maintain a vacuum of 10^{-9} Torr or better. In fact, systems which can maintain a base pressure in the 10^{-11} Torr range

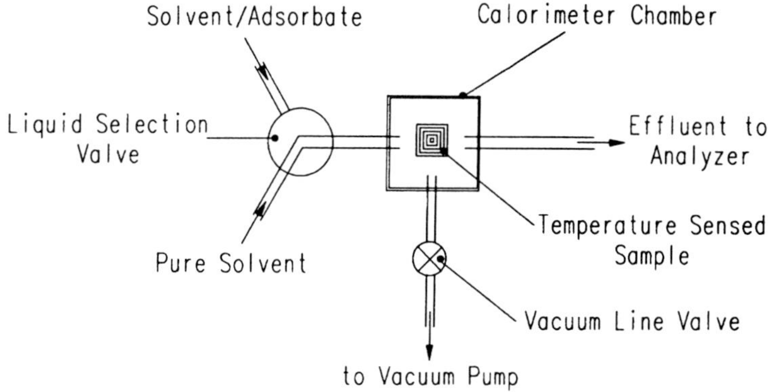

Figure 2. Flow microcalorimeter.

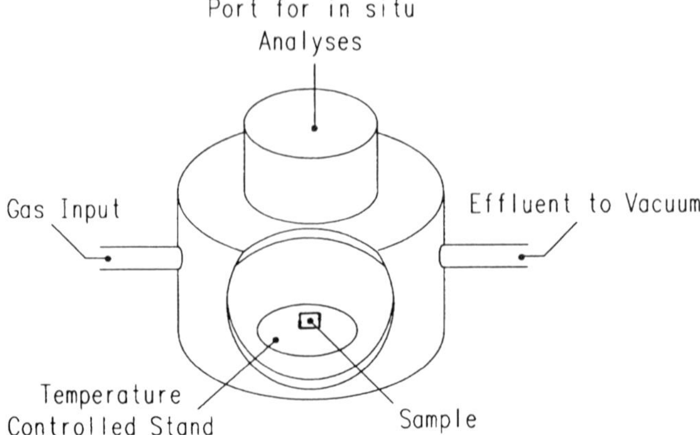

Figure 3. Typical ultra-high vacuum system. The sample may be analyzed *in situ* by a number of surface sensitive techniques.

are in use [10]. The sample sits on a temperature-controlled stage, and may be cleaned *in situ* by sputtering off the contaminated surface. Adsorbates may then be introduced into the chamber in a controlled manner through a leak valve. Analyses are performed *in situ* by a number of surface-sensitive techniques, including X-ray and ultraviolet photoelectron spectroscopy (XPS, UPS), Auger electron spectroscopy (AES), and electron energy loss spectroscopy (EELS). Further, desorption studies may be performed by analyzing the effluent, say in a mass spectrometer.

Because the strength of acid–base surface interactions governs how tightly a molecule binds to a surface, adsorption and desorption characteristics provide a useful monitor of acid–base chemistry. Many techniques are in use today, including electron impact desorption (EID), photodesorption, and field desorption (FD). These methods are discussed in ref. [11], and will not be considered here. Instead, let us focus on the widely used temperature-programmed desorption technique.

Desorption spectra are acquired by measuring (via spectrometry or pressure change) the amount of a particular material desorbing from the surface as the temperature of the sample is increased linearly throughout the experiment. It should be noted, however, that the amount of desorbing material is used as a measure of the *rate* of desorption. Thus, the desorption spectrum (frequently a plot of pressure vs. temperature) contains kinetic information which may be analyzed to give adsorption energies. The following derivation suggests how the data, specifically peaks in the spectra, may be analyzed.

For N adsorbed species, leaving a surface at a given temperature, T, the rate of change in the number of adsorbed species may be approximated by the Arrhenius equation,

$$dN/dt = -k_0 N \exp(-E_a/RT), \qquad (3)$$

where first-order kinetics have been assumed, and E_a, k_0, and R are the desorption activation energy, the pre-exponential factor, and the ideal gas constant, respectively. Since the temperature is ramped at a known rate, the time derivative may be converted to a temperature derivative using the chain rule of differential calculus:

$$dN/dT = -\frac{k_0 N}{\beta} \exp(-E_a/RT), \qquad (4)$$

where β is the ramping rate, dT/dt. The peak positions may be found by maximizing equation (4):

$$d(\text{rate})/dT = d^2N/dT^2 = -[dN/dT + NE_a/RT^2]\frac{k_0}{\beta}\exp(-E_a/RT). \qquad (5)$$

The final steps in this brief analysis are to use equation (4) for dN/dT in equation (5) and set equation (5) equal to zero.

$$\frac{k_0}{\beta} = \frac{E_a}{RT_p^2}\exp(E_a/RT_p), \qquad (6)$$

where T_p is the temperature at which the peak in the desorption spectrum occurs. The activation energy and the pre-exponential factor may be estimated by running the experiments at different heating rates. In systems where k_0 is reasonably constant, an increase in T_p also implies an increase in E_a. Interpretation may be complicated by surface reactions (e.g. dissociation) as well as inter-adsorbate interactions (e.g. hydrogen bonding). More details are given in ref. [11].

As an example of a temperature-programmed desorption study, consider the work by Netzer and Madey involving ammonia adsorbing onto nickel [12]. Figure 4 gives a representation of the spectra that they reported. The peaks at low temperature (120 K) result from loosely bound second-layer adsorption. The broad bands at higher temperature result primarily from molecularly bound NH_3. When Ni was predosed with oxygen, the broad peaks were shifted to higher desorption temperature, implying a stronger binding energy.

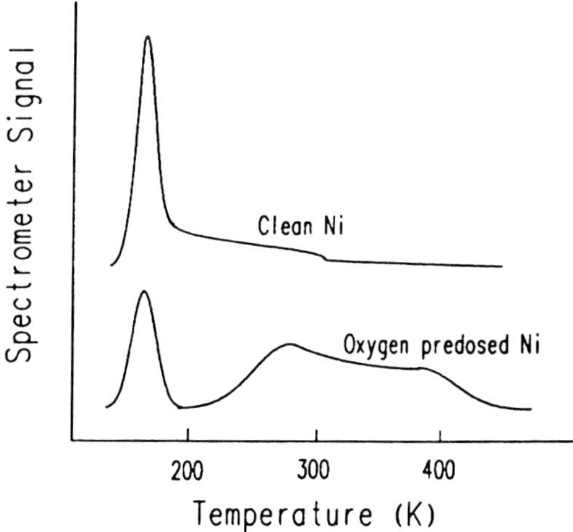

Figure 4. Temperature-programmed desorption spectrum, after Netzer and Madey [12]. Species were analyzed using a mass spectrometer (mass 17 peak). The top curve shows desorption of NH_3 from a clean Ni surface, while the bottom curve shows desorption of NH_3 from Ni predosed with enough oxygen to give NiO on the surface.

3. HARD–SOFT ACID–BASE THEORY

In 1963, Pearson published a landmark paper [4] which introduced the concept of acids and bases being divided into two classes of behavior. Based on nucleophilic substitution reactions of the type in equation (1), Pearson observed that the substitution for certain acid–base systems is governed by the basicity of the nucleophile, while for other acid–base systems the substitution is governed by the polarizability of the nucleophile. Acid–base systems in which basicity dominates were classified as 'class a' or 'hard'. Conversely, those systems in which polariziability dominates were classified as 'class b' or 'soft'.

Pearson found that these classifications assisted in understanding and predicting nucleophilic displacements in that hard acids prefer to bond to hard bases, while soft acids prefer to bond to soft bases. This preference has been related to fundamental properties such as charge, ionization potential, and size [13]. Hard acids and bases are characterized by high electronegativity and small size, thus increasing positive charge tends to make the species harder. Soft acids and bases are characterized by high polarizability and relatively large size. Soft acid–base interactions appear to be described by covalent bonding, and include such important concepts as π back-bonding and orbital energy match.

The preference of hard acids for hard bases has been related to the match between orbital sizes, which assumes that orbitals of comparable size overlap better than do orbitals which are not well matched in size. Though plausible, this explanation is not necessarily realistic, as shown by the case of H^+ approaching NH_3 or PH_3. Contour plots of the H^+ $1s$ orbital (contracted because of the positive charge), the NH_3 lone-pair orbital, and the PH_3 lone-pair orbital are shown in the left-hand portion of Fig. 5. From the plots, it is evident that the H^+ orbital is better matched in size to the NH_3 lone-pair orbital than to the PH_3 lone-pair orbital. Yet the $1s$–lone-pair overlap is greater for the PH_3–H^+ system than for the NH_3–H^+ system for all reasonable base–H^+ separations, as shown in the right-hand portion of Fig. 5. As pointed out by Pearson, the preference of hard acids for hard bases may be due primarily to ionic interactions, even though there may be substantial covalent character in the bond.

4. MOLECULAR ORBITAL THEORY

4.1. Overview of the molecular orbital approach

One of the major successes of quantum mechanics is the explanation and prediction of chemical reactions. The chemical behavior of a species is determined by the motion of electrons about the nuclei. In the case of atoms, the problem only involves a cetral potential. However, in the case of molecules, the description is more complicated. Mathematical representation of the orbitals and the orbital energies may be found, in principle, by solving the time-independent Schrödinger equation. However, because electron–electron repulsions appear explicitly in the Schrödinger equation, no analytical solution exists; numerical methods must be employed. One technique in use today is the $X\alpha$ method, but in order to keep the discussion focused, the $X\alpha$ approach will not be considered here. The interested reader is referred to a review by Slater [14]. Another commonly used approach is the so-called linear combination of atomic orbitals (LCAO) and this will be

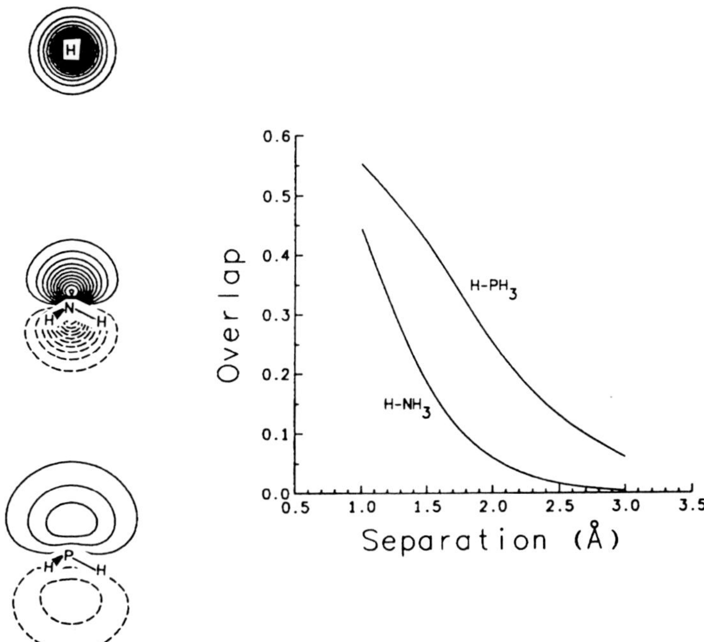

Figure 5. Contour plots for the H^+ $1s$ orbital (top left), the NH_3 lone-pair orbital (middle left), and the PH_3 lone-pair orbital (bottom left) and the overlap between the contracted H^+ $1s$ orbital and the lone-pair orbital of NH_3 or PH_3 (right).

discussed exclusively in the remaining portion of this paper. Molecular orbitals are assumed to have the form

$$\Phi_j = \sum_i C_{ij}\varphi_i, \qquad (7)$$

where Φ_j is the jth molecular orbital, φ_i is the ith atomic orbital, and C_{ij} is a coefficient in the summation. Even with the LCAO assumption, the problem of finding the coefficients and the molecular orbital energies is formidable. The major difficulty lies in properly accounting for the electron–electron repulsions; the electron motion in one orbital affects all of the other orbitals. To circumvent this problem, the self-consistent field (SCF) approach is used. An initial guess as to the coefficients [equation (7)] is made. From this initial guess, the electron–electron interactions are calculated and used in the Schrödinger equation. A new and presumably better set of coefficients is found, from which the electron–electron interactions are again calculated. This procedure is continued until the change in the set of coefficients is within a predetermined range. Sometimes, however, instead of converging the iteration actually diverges.

There are several SCF approaches in use. In *ab initio* calculations, all electron–electron interactions as well as the overlap integrals are calculated explicitly [15]. For computational economy, it is convenient to treat the electrons in the outermost shell (the valence electrons) only. This approximation is valid for describing chemical phenomena because the chemistry is governed by the valence electrons. SCF methods, which employ the valence shell approximation, include complete neglect of differential overlap (CNDO) [16], intermediate neglect of differential overlap (INDO) [17], and modified intermediate neglect of differential overlap (MINDO) [18]. These techniques not only treat the valence electrons only, but also neglect and parameterize the certain electron–electron interactions. The CNDO, INDO, and MINDO methods also neglect the overlap integrals. A complete discussion is beyond the scope of this work, but may be found in ref. [19]. The wave functions and energies calculated from all of the methods mentioned above are reasonably accurate.

Though the SCF methods account for electron–electron repulsions, they over-estimate the magnitude of these interactions. Basically, this is because the electron–electron interactions in the SCF approach are represented by un-weighted averages over all electron–electron separations. Thus, the energy is too high because the SCF procedure does not properly take into account the fact that the electrons will tend to avoid each other. To compensate for this deficiency, the SCF wave functions of higher energy configurations (electrons occupying certain orbitals not found in the ground state) are mixed into the ground state configuration. This approach is the basis of the so-called configuration interaction (CI) as well as the self-consistent field multiconfiguration (SCFMC) schemes. More details on the topic of electron correlation may be found in ref [15b].

More severe approximations than those discussed above are made in the extended Hückel (EH) method [20]. In EH calculations, all overlap integrals are calculated, but none of the electron–electron repulsions is explicitly calculated. Extended Hückel calculations are very approximate, but with careful application they can yield useful insights into the electronic structure of a molecule. Typically, only the valence atomic orbitals are considered in EH calculations.

At this point, a few comments concerning the analysis of the wave functions are in order. With a wave function of the form in equation (7), charge distribution, as well as the degree of bonding between two atoms, may be calculated [21]. These represent the 'Mulliken population analyses'. In general, the charge distribution is found according to the formula

$$Q_i = \sum_j \sum_k \sum_l n_j C_{kj} C_{lj} S_{kl}, \tag{8}$$

where Q_i is the number of electrons 'localized' on the ith atom, n_j is the occupation number of the jth molecular orbital (either 0, 1, or 2), and S_{kl} is the overlap integral between the kth and lth atomic orbitals. The 'l' summation runs over all atomic orbitals, the 'j' summation runs over all molecular orbitals, and the 'k' summation runs over all atomic orbitals centered on the ith atom. A useful measure of the degree of bonding between two atoms, i and j, is the overlap population,

$$P_{ij} = \sum_k \sum_l \sum_m n_k C_{lk} C_{mk} S_{lm}, \tag{9}$$

where n, C, and S have the same definitions as in equations (7) and (8). Equation (9) looks very similar to equation (8), save for the index labels. However, the 'l' and 'm' summations run over only those atomic orbitals centered on atoms i and j, respectively. The 'k' summation runs over all molecular orbitals. In the case of zero differential overlap calculations (CNDO, INDO, MINDO), the S_{lm} factor is omitted, and equation (9) becomes

$$P_{ij} = \sum_k \sum_l \sum_m n_k C_{lk} C_{mk}, \qquad (10)$$

where P_{ij} in equation (10) is a 'bond order' instead of an 'overlap population', as in equation (9). A large overlap population tends to imply strong covalent bonding. However, overlap population analyses, while useful in determining bonding trends among compounds with similar structure, may not be used to compare bond strengths in dissimilar systems.

In the molecular orbital approach, acids and bases are treated in the Lewis sense, i.e. as electron-pair accepters and donors. For example, a simple interaction diagram for the classic Lewis acid–base pair, NH_3 and BH_3, is shown in Fig. 6. Three features in the figure are worth noting: (1) the base orbital acquires some characteristics of the acid orbital and vice versa; (2) lobes of the acid and base orbitals are in phase in the attractive or 'bonding' component of the interaction, but out of phase in the repulsive or 'anti-bonding' component; and (3) the anti-bonding orbital, if occupied, would destabilize the system more than the bonding orbital stabilized it, a consequence of properly accounting for orbital overlap.

The strength of the interaction is related to the overlap, as well as the energy match between the empty acid orbital and the filled base orbital. From second-order perturbation theory, the interaction energy (decrease in the energy of the lower orbital) is given by

$$E_2 = \frac{|\langle \varphi_a / H / \varphi_b \rangle|^2}{E_b - E_a}. \qquad (11)$$

Here, E_a, E_b, φ_a, and φ_b are the unperturbed energies and wave functions of the acid and base, and H is the Hamiltonian operator. As can be seen from equation (11), the interaction can be strong if $E_b \approx E_a$, and if $\langle \varphi_a / H / \varphi_b \rangle$ is large. Usually $\langle \varphi_a / H / \varphi_b \rangle$ will be significant if the overlap between the acid and base orbitals is significant. Essentially, the interaction is strong if the acid and base orbitals are of comparable energy, and overlap each other well.

To deal with solids, 'band' calculations are employed. Although different techniques exist for deriving band structures, the following discussion focuses on the 'tight binding' approach, which has been deferred until now so that the concepts of bonding and anti-bonding orbitals, population analyses, and orbital mixing could be discussed in a simpler context. Tight binding band calculations appear to differ from the techniques discussed above, but in reality the approach is the same. A tight binding calculation may be thought of as a molecular orbital calculation for a very large molecule (the entire solid crystal). Rather than considering every atom in the crystal, only a small repeating piece of the crystal, called the 'unit cell', is used. Since the entire crystal can be generated by translations of the unit cell, the entire wave function can also be generated by translations of the unit cell wave function, weighted by a plane wave:

$$\Phi_i(\mathbf{r},\mathbf{k}) = \sum_j \varphi_i(\mathbf{r} - \mathbf{r}_j) \exp\left[i\mathbf{k} \cdot (\mathbf{r} - \mathbf{r}_j)\right]. \qquad (12)$$

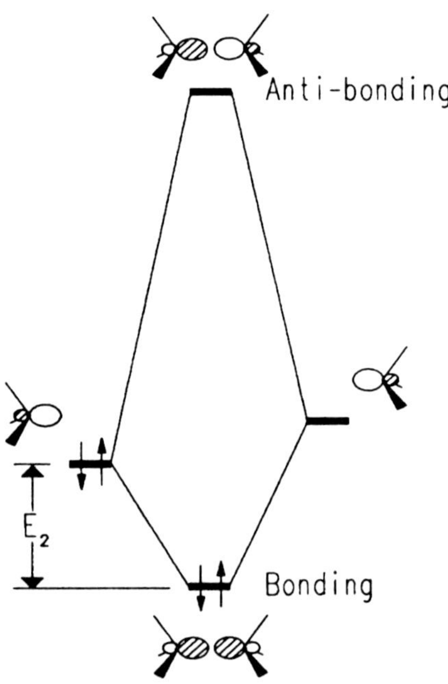

Figure 6. Diagram of NH_3 interacting with BH_3. Only the NH_3 lone-pair and empty BH_3 orbitals are shown.

In equation (12), $\Phi_i(\mathbf{r},\mathbf{k})$ is the ith crystal wave function derived from the ith unit cell orbital, φ_i. The factor $\exp[i\mathbf{k}\cdot(\mathbf{r}-\mathbf{r}_j)]$ introduces the appropriate phase relationship between unit cell orbitals located at difference lattice positions, $\mathbf{r}_j$, in the crystal. A 'band' is the collection of crystal orbitals (with common index i) over the allowed range of the wave vector, $\mathbf{k}$. For more detailed discussions, consult ref. [22].

Because of the periodicity of the unit cell, the problem is reduced to solving Schrödinger's equation for the unit cell plus a few near-neighbors for various $\mathbf{k}$ points. For a discussion of how to select a set of $\mathbf{k}$ points to use in the band calculations, see ref. [23]. Tight binding calculations may be based on the *ab initio*, CNDO, INDO, MINDO, or EH treatments. Orbital mixing and population analyses from a tight binding calculation are completely analogous to those from a molecular orbital calculation. One may even construct an orbital interaction diagram by plotting the density of states (DOS) as a function of the energy. The density of states is the number of crystal orbitals in the energy range E to $E+dE$, and may be thought of as a degeneracy factor. A quantity that appears frequently in the literature is the 'Fermi level' or 'Fermi energy', henceforth denoted as ε. Crystal orbitals above the Fermi level are unoccupied, while those below the Fermi level are occupied. In the context of solid-state chemistry, the Fermi level in a band calculation is analogous to the energy of the highest occupied molecular orbital (HOMO) in a molecular orbital calculation.

To illustrate the similarities between the molecular orbital and tight binding approaches, consider the results of calculations performed on hydrogen, illustrated in Fig. 7. A molecular orbital description of the H_2 molecule consists of two H $1s$ atomic orbitals mixing in phase (bonding) and out of phase (anti-bonding).

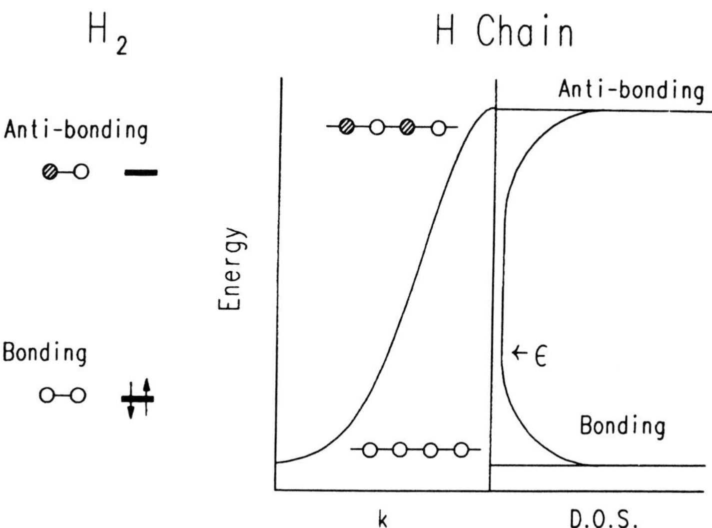

Figure 7. Molecular orbitals for H_2 (left) and the band and density of states plots for a linear chain of hydrogen atoms (right).

The lower, bonding orbital is filled and the upper, anti-bonding orbital is empty, resulting in a bond between the two hydrogen atoms. On the other hand, the tight binding approach describes the hydrogen chain as a unit cell containing a single H $1s$ atomic orbital. By translating the H $1s$ function along the chain, the wave function for the entire chain may be generated. The band energy is plotted as a function of $\mathbf{k}$; each point on the E vs. $\mathbf{k}$ curve represents a crystal orbital. In the case of the hydrogen chain, the lower portion of the band is comprised of crystal orbitals with the H $1s$ functions in phase with each other. The upper portion of the band is comprised of crystal orbitals with the H $1s$ functions out of phase with each other. Information contained in the band plot may be condensed into a density of states plot, shown to the right of the band plot. Note that there are two peaks in the DOS plot. The lower peak arises from bonding interactions, while the upper peak arises from anti-bonding interactions. Further, the Fermi level falls at the top of the lower peak; the bonding portion of the band is occupied, while the anti-bonding portion of the band is unoccupied. Because this discussion has been brief, the reader interested in more detail should consult ref. [24].

Now consider the following examples from the literature as applications of quantum mechanics to the field of interfacial chemistry.

4.2. Cluster calculations for bases on a copper surface

Application of quantum mechanics to solid surfaces is somewhat tricky because, on a molecular scale, the solid is very large. One way to circumvent the problem is to assume that the interactions are fairly local. Thus, an entire surface may be satisfactorily modeled by a small piece of the surface, usually a cluster of several atoms. The most thorough studies probe the effect cluster size has on the properties being calculated [25]. A more complete discussion of the cluster technique is given in ref. [26].

As an exampleof the cluster approach, consider a report by Bagus *et al.* [27] which describes how various bases bond to a copper cluster. In this study, the *ab initio* approach was used. A pyramidal cluster of five Cu atoms represented the surface. The geometry is shown in the left-hand portion of Fig. 8. Three bases were studied: CO, NH_3, and PF_3. From their calculations, the authors concluded that bonding to the cluster in a σ fashion was not particularly important, owing to the net cancellation of bonding and anti-bonding components. Because copper has 11 valence electrons ($3d^{10}4s^1$), σ anti-bonding orbitals are occupied. However, CO and PF_3 have empty orbitals at fairly low energy which can accept electron pairs from the Cu cluster. Bagus *et al.* pointed out that the resulting π-type back-donation accounts for significant covalent bonding between the cluster and CO or PF_3.

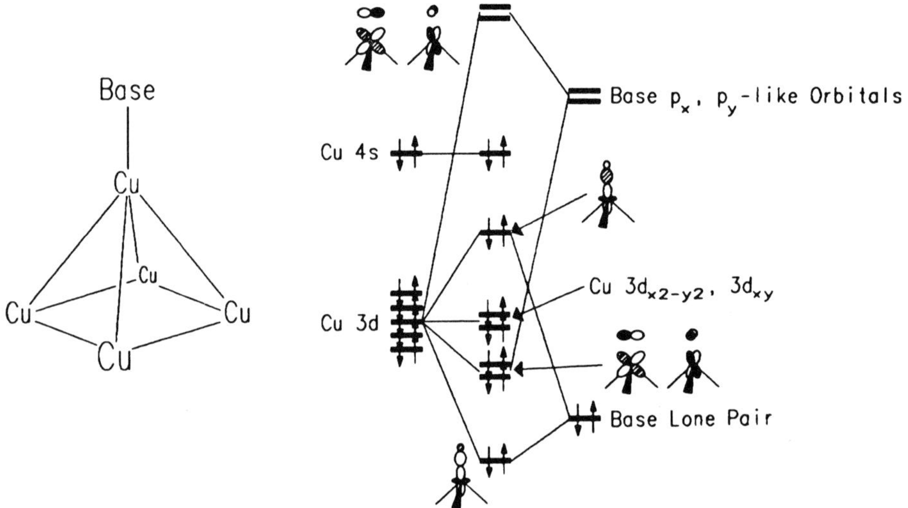

Figure 8. Geometry of a base attached to the Cu cluster (left) and the interaction diagram (right). Only those Cu_5 orbitals with a significant contribution from the apical Cu atom are shown.

The results of Bagus *et al.*'s study might be summarized in part by the simplified interaction diagram given in the right-hand portion of Fig. 8. The results of EH calculations show that the molecular orbitals derived from the Cu $3d$ states fall within a 0.7 eV range. Because of this narrow energy range, the cluster orbitals with a significant contribution from the apical Cu $3d$ states are represented in Fig. 8 simply as the five degenerate Cu $3d$ orbitals. Cluster orbitals derived mainly from the basal Cu atomic orbitals are not shown, since they do not overlap well with the orbital of the base. As indicated in the diagram, the σ-type interaction involves two orbitals (Cu $3d_{z^2}$ and the 'lone pair' of the base) and four electrons, leading to no net bonding. The π-type interactions each involve two orbitals (Cu $3d_{xz}$ or $3d_{yz}$ and higher energy virtual orbitals of the base) and two electrons, resulting in net bond formation. In the case of NH_3, the virtual orbitals are too high in energy to participate in the bonding scheme; the difference in energy between the NH_3 virtual orbitals and the Cu $3d$ orbitals is large [cf. equation (11)]. On the other hand, CO and PF_3 have virtual orbitals at lower energy; the π back-bonding component is significant, as reported by Bagus *et al.*

At this point, it is appropriate to consider how well the cluster represents the entire surface or even bulk copper. This is done conveniently by constructing a density of states histogram. The results for EH calculations by Cain (not previously reported) are summarized in Fig. 9, which shows the DOS histograms for five-atom and nine-atom clusters, as well as the DOS from a band calculation on 'bulk' copper. The histograms from the five- and nine-atom clusters both resemble (at least approximately) the bulk DOS. Further, the Fermi energy obtained from the cluster calculations is nearly the same as that obtained from the band calculation. These results indicate that even a small cluster of five atoms may accurately reflect the chemistry of the bulk (or entire surface). Bagus *et al.* also reached this conclusion [27]. Band dispersion (energy range of the band), however, is not well described by a small cluster. From the calculations, the $3d$ states are dispersed 0.7 eV in Cu_5, 0.9 eV in Cu_9, and 1.7 eV in the bulk.

Although these results apply specifically to molecular adsorbates bonding to a metal surface, the concepts may be applied readily to the case of polymers bonding to a metal. The molecular adsorbates represent functional groups in the polymer. In the above example, PF_3 and NH_3 might represent terminal or secondary phosphine or amine groups in a polymer. Such an application disregards the conformational degrees of freedom experienced by the unbound portions of the polymer, but gives useful insights into the interfacial chemistry.

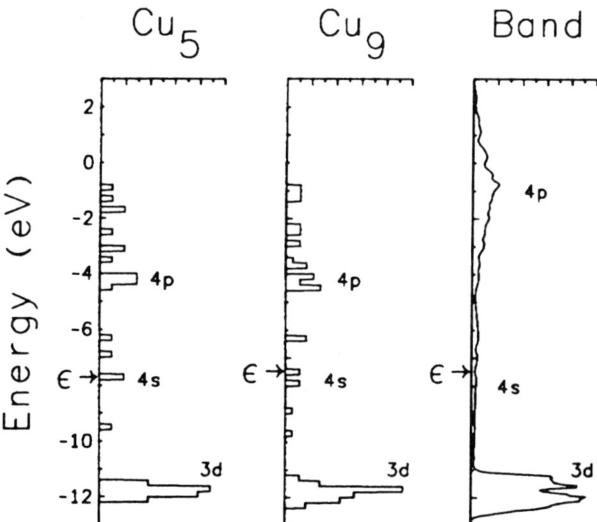

Figure 9. DOS histograms for Cu_5 (left) and Cu_9 (center) and the DOS plot from a band calculation on bulk copper (right).

4.3. Photoelectron spectra of chromium on polyimide

Quantum mechanics may also be used to interpret spectra. Consider the case of photoelectron spectroscopy, in which the sample is irradiated with either ultra-violet light (UPS) or X-rays (XPS). Irradiation excites electrons from bound states (orbitals) into the unbound continuum. In the continuum, electrons have only kinetic energy, which is easily measured by the detector. Essentially, the binding energy of the ejected electron is found by taking the difference between its kinetic energy and the energy of the incident photon. X-Rays are considerably more energetic than ultraviolet light: ultraviolet photons ionize only the valence electrons; X-rays ionize both core and valence electrons. Thus, UPS is sensitive to the valence orbitals, while XPS is sensitive primarily to the core orbitals. See ref. [28] for a more thorough, yet elementary, discussion. Molecular orbital energies are approximately the same as the binding energies [29], which is an application of Koopman's theorem. This allows quantum-chemical calculations to be used to interpret XPS and UPS data. In fact, with a proper choice of ionization cross-sections, simulated spectra may be derived from MO wave functions and energies [30]. Such a spectrum may be calculated according to the formula

$$I_i = \sum_j \sum_k \sum_l \sigma_j C_{ji} C_{ki} S_{jk}, \tag{13}$$

where I_i is the probability of ionizing an electron from the ith molecular orbital, σ_j is the cross-section for the jth atomic orbital (relative values for some elements are tabulated in ref. [30]), and C and S have the same definitions as in equations (7) and (8). The total intensity is calculated by summing the contributions from all of the molecular orbitals within a given energy range. Frequently, peaks are represented by Gaussian functions centered at the energy of the corresponding molecular orbital. The height of the Gaussian may be taken to be I_i [equation (13)], and the width is a resolution factor characteristic of the spectrometer. It should be noted, however, that treatments based on the approach given above are only approximate. In the first place, the ionization energy may differ from the orbital energy because the final state of the species is an ion, with orbitals which are different from those in the neutral species. Also, the spectrum may be complicated by 'shake-up' peaks which result form secondary electronic transitions. A good discussion of these effects may be found in ref. [29b].

An example of this application in the field of metal–polymer interfaces has been discussed by Ho *et al.* [31]. Although their conclusions are still being debated, it is instructive to consider their approach, particularly as it applies directly to metal–polymer adhesion. In their studies, *ab initio* calculations were performed on various complexes of chromium and polyimide, shown in Fig. 10. For computational economy, they treated the pyromellitic dianhydride (PMDA) and oxydianiline (ODA) portions separately. The relative stability of the complexes appears to follow the trend, a > b > c. Submonolayer coverage by Cr was accompanied by a substantial decrease in the carbonyl carbon $1s$ peak in the XPS. This decrease was attributed to the formation of the Cr–arene charge-transfer complex, a. At higher coverages, Cr was assumed to occupy sites b, then c. To check their model, Ho *et al.* used the core, or subvalence, orbitals from the MO calculations to generate a series of simulated XPS spectra for different Cr coverages. These spectra agreed quite well with their experimental data.

This same group has also studied Cu–polyimide interfaces [32]. The results of the calculations indicated that the 'b'-type arene complex (cf. Fig. 10) is the most stable for Cu. Again, the model was checked by using the orbitals to generate simulated photoelectron spectra. This time the valence orbitals were considered; the experimental UPS spectra compared well with the simulated spectra, giving credence to the model.

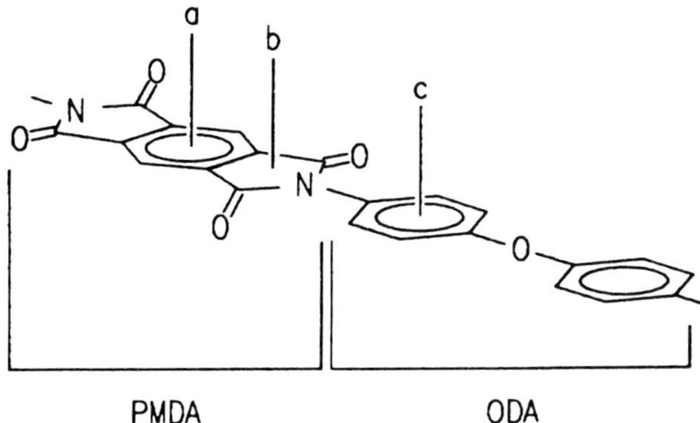

Figure 10. Polyimide structure, with a, b, and c indicating the various Cr– and Cu–arene complexes which can be formed.

4.4. *Chemical reactions at the chromium–polyimide interface*

In this example, tight binding calculations at the EH level of appxoimation were used [33]. Boundary conditions for a surface are not properly addressed by a three-dimensional band calculation. This problem is sidestepped by treating the surface as a two-dimensional slab. As with cluster calculations, the question of how well the model represents a surface must be considered. Obviously, an infinite number of layers stacked in one direction is an exact representation, but is not computationally manageable. To address this question, DOS analyses for slabs with different numbers of layers may be performed; the results are shown in Fig. 11. The DOS plots for the three- and six-layered slabs are very similar to that of bulk Cr, suggesting that only three or four layers are adequate to model the surface. Hoffmann and co-workers have also come to this conclusion [34].

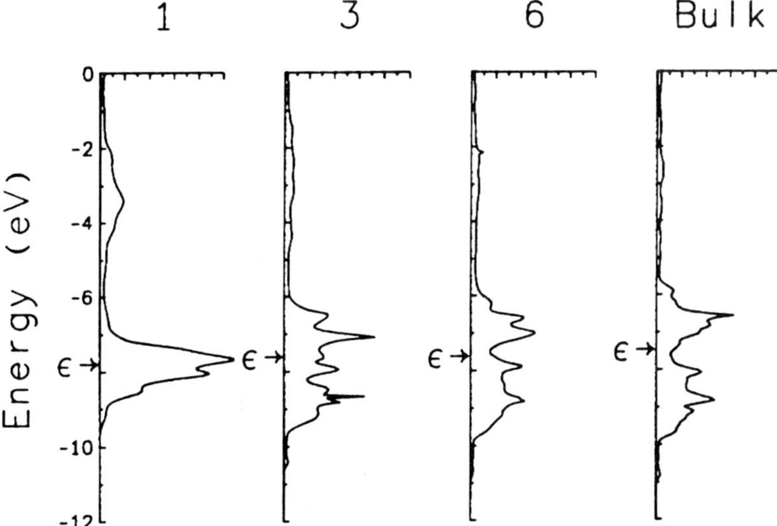

Figure 11. Total DOS plots from a two-dimensional calculation on Cr slabs of one layer (far left), three layers (middle left), and six layers (middle right), and a three-dimensional calculation on bulk Cr (far right).

In the study on reactivity, the interaction of Cr with the polyimide carbonyls was considered, since an earlier report [35] had suggested that interfacial bonding involves a Cr–carbonyl redox reaction. Surfaces of Cr and Cr_2O_3 were modeled by three-layered slabs of material, the unit cells of which are shown in Fig. 12. Chromium oxide was studied because the surface is oxidized in all commercial applications where UHV systems are not used. For computational economy, formaldehyde was used as a model for the carbonyls.

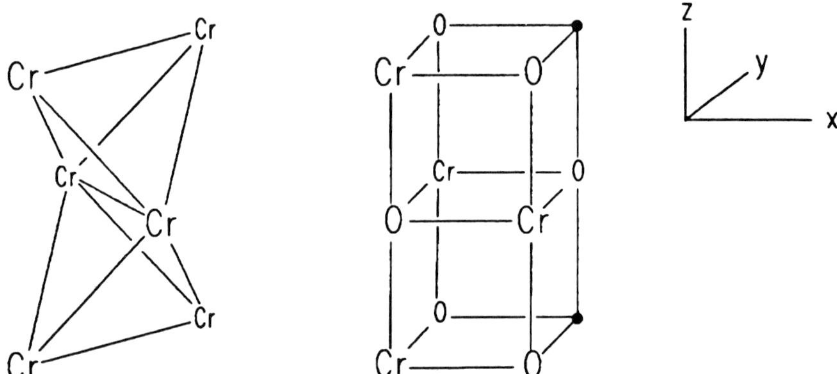

Figure 12. Geometries of the Cr and Cr_2O_3 unit cells. The entire crystal is generated by translations in the x–y plane.

The investigators [33] found two major components of the interaction: (1) 'donation' of a formaldehyde pair of electrons to the empty Cr $3d_{z^2}$ orbital ('forward bonding'); and (2) 'back-donation' of Cr electrons from the $3d_{xz}$ orbital into the empty π^* orbital of formaldehyde. These interactions operate in the same manner as those discussed in the section on cluster calculations. Further, the Cr $3d$ orbitals and the Fermi level of Cr_2O_3 were found to be about 3 eV lower than in metallic Cr, resulting in a difference in the energy match between the Cr and formaldehyde orbitals [cf. equation (11)]. As a result, they concluded that the forward-bonding component dominates (almost to the exclusion of the back-bonding component) in the formaldehyde–Cr_2O_3 system, and that the back-bonding component is very strong (maybe even stronger than the forward-bonding component) in the formaldehyde–Cr system. Because the forward-bonding component is maximized by end-on type bonding, they suggested that formaldehyde tends to bond end-on to a Cr_2O_3 surface, but in a π manner to a Cr surface, as shown in Fig. 13.

Since formaldehyde is prone to decompose on a clean metal surface, usually to CO and H_2 [36], reactivity was also probed. Adsorbed formyl radical, shown in Fig. 13, is a plausible intermediate. Reductive elimination involves breaking C—H bonds; hence the effect of stretching one of the C—H bonds was investigated, and may be summarized by the orbital correlation diagram in Fig. 14. When the C—H anti-bonding orbital crosses the Fermi level, electrons are transferred from the substrate to that orbital, after which the bond is broken easily. The only cost in energy comes from stretching the C—H bond early in the reaction. As shown in Fig. 14, less energy is spent in stretching the C—H bond if the electron transfer occurs early in the reaction than if it occurs late in the reaction. Since the Fermi

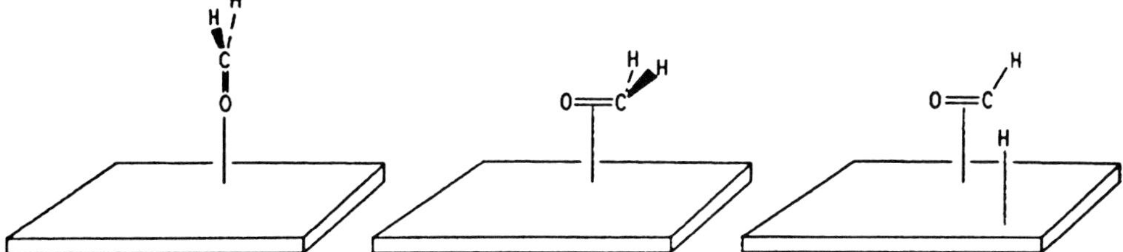

Figure 13. Geometries of formaldehyde binding end-on to Cr_2O_3 (left) and in a π manner to Cr (middle). Bound formaldehyde may dissociate to hydrogen and formyl (right).

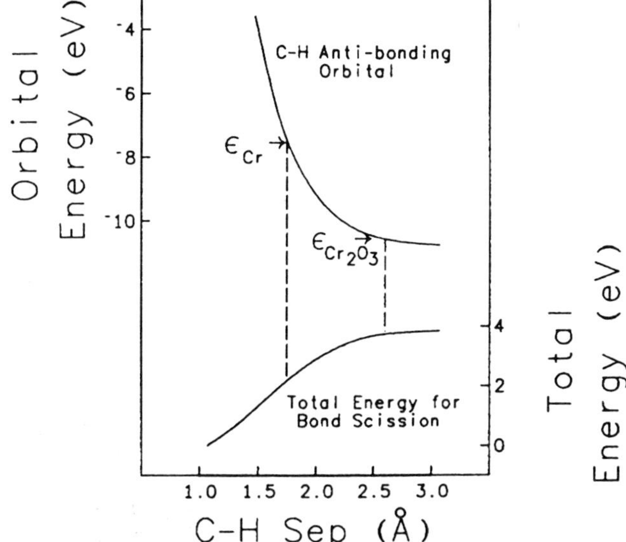

Figure 14. Correlation diagram for the C—H anti-bonding orbital (top curve) and total energy for C—H bond scission (bottom curve) in the free formaldehyde molecule. The Fermi levels for Cr and Cr_2O_3 are indicated by arrows.

level is higher in Cr than in Cr_2O_3, the electron transfer takes place earlier in the formaldehyde–Cr system than in the formaldehyde–Cr_2O_3 system. The reaction barrier would be considerably smaller if the substrate were Cr rather than Cr_2O_3. Hence formaldehyde is more likely to decompose on Cr than on Cr_2O_3. Further, as depicted in Fig. 13, the preferred bonding geometry in the Cr–formaldehyde system is quite similar to that for the Cr–formyl intermediate, which also explains why formaldehyde is more prone to decompose on a clean metal surface than on a metal oxide surface.

From their studies, the investigators concluded that polyimide is very apt to decompose at the Cr–polyimide interface, which may or may not be detrimental to adhesion depending on how completely the polymer reacts. On the other hand, the polyimide functional groups should retain their integrity at a polymer–Cr_2O_3 interface.

5. SUMMARY

With proper application of molecular orbital techniques, as well as prudent choices of model compounds on which to perform calculations, much insight may be gained into the nature of metal–polymer interfacial chemistry. The examples discussed above have illustrated three types of applications. Orbital and band calculations are useful for explicitly predicting material behavior, but in a deeper sense find utility in extracting the important properties of a compound (e.g. charge, electronic structure, ionization potential) which govern chemical interactions and reactivity. The strength of orbital calculations is not in making predictions, but rather in understanding the factors which govern basic chemical phenomena.

Finally, even the most sophisticated theory is only a model. All predictions ultimately must be verified in the laboratory.

REFERENCES

1. (a) J. C. Bolger and A. S. Michaels, in: *Interface Conversion*, P. Weiss and D. Cheevers (Eds), Ch. 1. Elsevier, New York (1969). (b) F. M. Fowkes, in: *Physicochemical Aspects of Polymer Surfaces*, K. L. Mittal (Ed.), vol. 2, p. 583. Plenum Press, New York (1983). (c) F. M. Fowkes, *J. Adhesion Sci. Technol.* **1**, 7 (1987).
2. (a) S. R. Cain and L. J. Matienzo, *J. Adhesion Sci. Technol.* **2**, 395 (1988). (b) S. R. Cain, L. J. Matienzo and F. Emmi, in: *Metallized Plastics 1: Fundamental and Applied Aspects*, K. L. Mittal and J. R. Susko (Eds), pp. 247–264. Plenum Press, New York (1989). (c) S. R. Cain, L. J. Matienzo and F. Emmi, *J. Phys. Chem. Solids* **50**, 87 (1989).
3. (a) S. Ahrland, J. Chatt and N. R. Davies, *Q. Rev. London* **12**, 265 (1958). (b) J. O. Edwards and R. G. Pearson, *J. Am. Chem. Soc.* **84**, 16 (1962).
4. R. G. Pearson, *J. Am. Chem. Soc.* **85**, 3533 (1963).
5. J. C. Bolger, in: *Adhesion Aspects of Polymeric Coatings*, K. L. Mittal (Ed.), p. 3. Plenum Press, New York (1983).
6. The reader is referred to an appropriate physical chemistry text, e.g. A. W. Adamson, *A Textbook of Physical Chemistry*, p. 1015 ff. Academic Press, New York (1973).
7. G. A. Parks, *Chem. Rev.* **65**, 177 (1965).
8. F. M. Fowkes, Y. C. Huang, B. A. Shah, M. J. Kulp and T. B. Lloyd, *Colloids Surf.* **29**, 243 (1988).
9. J. S. Arlow, D. F. Mitchell and M. J. Graham, *J. Vac. Sci. Technol. A* **5**, 573 (1987).
10. See, for example, (a) T. Miyano, K. Kamei, Y. Sakisaka and M. Onchi, *Surf. Sci.* **148**, L645 (1984). (b) N. D. Shinn and T. E. Madey, *Surf. Sci.* **173**, 379 (1986).
11. D. Menzel, in: *Topics in Applied Physics: Interactions on Metal Surfaces*, R. Gomer (Ed.), vol. 4, p. 101 ff. Springer, New York (1975).
12. F. P. Netzer and T. E. Madey, *Surf. Sci.* **119**, 422 (1982).
13. For textbook discussions, see (a) G. C. Demitras, C. R. Russ, J. F. Salmon, J. H. Weber and G. S. Weiss, *Inorganic Chemistry*, p. 268 ff. Prentice-Hall, Englewood Cliffs, NJ (1972). (b) J. E. Huheey, *Inorganic Chemistry: Principles of Structure and Reactivity*, SI Units edn, p. 225 ff. Harper and Row, New York (1975).
14. J. C. Slater, *Adv. Quantum Chem.* **6**, 1 (1972).
15. (a) For a thorough discussion of the Hartree–Fock method, see A. C. Hurley, *Introduction to the Electron Theory of Small Molecules*. Academic Press, London (1976). (b) A. C. Hurley, *Electron Correlation in Small Molecules*. Academic Press, New York (1976).
16. J. A. Pople, D. P. Santry and G. A. Segal, *J. Chem. Phys.* **43**, S129 (1965).
17. J. A. Pople, D. L. Beveridge and P. A. Dobosh, *J. Chem. Phys.* **47**, 2026 (1967).
18. N. C. Baird and M. J. S. Dewar, *J. Chem. Phys.* **50**, 1262 (1969).
19. J. A. Pople and D. L. Beveridge, *Approximate Molecular Orbital Theory*. McGraw-Hill, New York (1970).
20. R. Hoffmann and W. N. Liopscomb, *J. Chem. Phys.* **36**, 2179 (1962).
21. R. S. Mulliken, *J. Chem. Phys.* **23**, 1833, 1841, 2388, 2343 (1955).

22. (a) C. Kittel, *Introduction to Solid State Physics*, 5th edn, pp. 185 ff, 260 ff. John Wiley, New York (1976). (b) M.-H. Whangbo and R. Hoffmann, *J. Am. Chem. Soc.* **100**, 6093 (1978).
23. See, for example, (a) D. J. Chadi and M. L. Cohen, *Phys. Rev. B* **8**, 5747 (1973). (b) H. J. Monkhorst and J. D. Pack, *ibid.* **13**, 5188 (1976).
24. R. Hoffmann, *Solids and Surfaces: A Chemist's View of Bonding in Extended Structures.* VCH, New York (1988).
25. See, for example, D. D. Coolbaugh, Ph.D. thesis, State University of New York, Binghamton (1987).
26. R. P. Messmer, in: *The Nature of the Surface Chemical Bond*, T. N. Rhodin and G. Ertl (Eds), p. 53, North-Holland, Amsterdam (1979).
27. P. S. Bagus, K. Hermann and C. W. Bauschlicher, *J. Chem. Phys.* **81**, 1966 (1984).
28. L. J. Matienzo, F. Emmi and R. W. Johnson, in: *Principles of Electronic Packaging*, D. P. Seraphim, R. Lasky and C.-Y. Li (Eds), p. 723. McGraw-Hill, New York (1989).
29. (a) A. D. Baker and C. R. Brundle, in: *Electron Spectroscopy: Theory, Techniques, and Applications*, C. R. Brundle and A. D. Baker (Eds), vol. 1, p. 1. Academic Press, New York (1977). (b) R. L. Martin and D. A. Shirley, *ibid.*, p. 75.
30. U. Gelius and K. Siegbahn, *Faraday Discuss. Chem. Soc.* **54**, 257 (1972).
31. (a) P. S. Ho, R. Haight, R. C. White and B. D. Silverman, *J. Phys. Colloq.* **49**, 49 (1988). (b) P. S. Ho, B. D. Silverman, R. A. Haight, R. C. White, P. N. Sanda and A. R. Rossi, *IBM J. Res. Dev.* **32**, 658 (1988).
32. P. S. Ho, P. O. Hahn, J. W. Bartha, G. W. Rubloff, F. K. LeGoues and B. D. Silverman, *J. Vac. Sci. Technol. A* **3**, 739 (1985).
33. S. R. Cain and F. Emmi, *Surf. Sci.* (in press).
34. J.-Y. Saillard and R. Hoffmann, *J. Am. Chem. Soc.* **106**, 2006 (1984).
35. C. A. Kovac, J. L. Jordan-Sweet, M. J. Goldberg, J. G. Clabes, A. Viehbeck and R. A. Pollak, *IBM J. Res. Dev.* **32**, 603 (1988).
36. See, for example, (a) T. E. Madey, J. T. Yates, Jr. and M. J. Dresser, *J. Catal.* **30**, 260 (1973). (b) J. T. Dickinson and R. J. Madix, *Int. J. Chem. Kinet.* **10**, 871 (1978). (c) J. B. Benziger and R. J. Madix, *J. Catal.* **74**, 55 (1982).

Acid-Base Interactions, pp. 67-78
Eds. K.L. Mittal and H.R. Anderson, Jr.
©VSP 1991

The use of wetting measurements in the assessment of acid–base interactions at solid–liquid interfaces

MICHAEL D. VRBANAC* and JOHN C. BERG†

Department of Chemical Engineering BF-10, University of Washington, Seattle, WA 98195, USA

Revised version received 12 January 1990

Abstract—It is now generally recognized that the principal forces contributing to the work of adhesion between two phases, W_A, are the Lifshitz–van der Waals forces (which include a small contribution from permanent and induced dipoles) and acid–base interactions, taken in the most general 'Lewis' sense. One may thus write

$$W_A = W_A^{LW} + W_A^{ab} = 2\sqrt{\sigma_S^{LW}\sigma_L^{LW}} + fN(-\Delta H^{ab}),$$

where W_A^{LW} and W_A^{ab} are the Lifshitz–van der Waals and acid–base contributions to the work of adhesion, and σ_S^{LW} and σ_L^{LW} are the Lifshitz–van der Waals contributions to the surface free energies of the solid and the liquid, respectively; ΔH^{ab} is the enthalpy (per mol) of the acid–base adduct formation between the acid or base functional groups on the adherend and in the adhesive; N is the number (moles) of accessible functional groups per unit area of the adherend; and f is an enthalpy-to-free energy correction factor (which has normally been assumed to be ≈ 1). The present work seeks to evaluate W_A^{ab} for several systems using wetting measurements and, for at least one system, to obtain a quantitative check of the above equation using independently measured values of f, N, and $(-\Delta H^{ab})$. The total work of adhesion is determined from the measured surface tension of the liquid, σ_L, and its contact angle, θ, against the solid: $W_A = \sigma_L(1 + \cos\theta)$. σ_S^{LW} and σ_L^{LW} are determined using probe liquids, N is determined from conductometric titrations of the solid in finely divided form, and ΔH^{ab} is determined by flow microcalorimetry. f is determined from a Gibbs–Helmholtz analysis of surface tension and contact angle data obtained over a range of temperatures. Conclusions reached are that the f factor is significantly below unity in most cases and that even including this effect, the above equation is still not verified quantitatively when the terms are measured independently.

Keywords: Acid–base interactions; wetting measurements; adhesion; temperature effects.

1. INTRODUCTION

Starting with the work of Professor Fowkes in 1978 [1], recognition of the central role played by acid–base interactions in adhesion has emerged. Fowkes has continued to take the lead in the development of these ideas, which he comprehensively reviewed in the lead-off paper of this journal in 1987 [2].

The point of departure for all work relating surface chemistry or molecular interactions across interfaces to adhesion is the assumption that the strength of an adhesive bond is at least proportional to the thermodynamic work of adhesion, W_A

---

*Presently with the Weyerhaeuser Company, Weyerhaeuser Technology Center, Tacoma, WA 98477, USA.

†To whom correspondence should be addressed.

[2]. The latter is defined without ambiguity as the reversible work (per unit area) of disjoining the adhesive–adherend interface:

$$W_{\mathrm{A}} = \sigma_{\mathrm{S}} + \sigma_{\mathrm{L}} - \sigma_{\mathrm{SL}}, \tag{1}$$

where σ_{S} and σ_{L} are the surface free energies of the solid (adherend) and liquid (adhesive) surfaces, respectively, assuming them to be in contact with air, and σ_{SL} is the interfacial free energy of the solid–liquid interface. σ_{L} is readily measurable as the surface tension of the liquid, but σ_{S} and σ_{SL} are not generally accessible. For liquids yielding finite contact angles with the solid, however, Young's equation in the form

$$\cos\theta = \frac{(\sigma_{\mathrm{S}} - \pi_{\mathrm{e}}) - \sigma_{\mathrm{SL}}}{\sigma_{\mathrm{L}}} \tag{2}$$

may be used to eliminate the difference $(\sigma_{\mathrm{S}} - \sigma_{\mathrm{SL}})$ from equation (1), so that

$$W_{\mathrm{A}} = \sigma_{\mathrm{L}}(1 + \cos\theta) + \pi_{\mathrm{e}}, \tag{3}$$

where θ is the contact angle of the liquid against the solid, and π_{e} is the equilibrium spreading pressure of the liquid's vapor upon the solid. The contact angle is readily measured using optical goniometry or from the wetting force on the solid dipping into the liquid (Wilhelmy method). The spreading pressure, π_{e}, may be determined from the measured adsorption isotherm of the vapor on the solid using

$$\pi_{\mathrm{e}} = RT \int_0^1 \Gamma \, \mathrm{d} \ln \frac{p}{p^{\mathrm{s}}}, \tag{4}$$

where p is the partial pressure of the liquid's vapor in the gas phase, p^{s} is its vapor pressure, RT is the gas constant times absolute temperature, and Γ is the adsorption. For many low-energy solids, π_{e} is effectively zero [3]. The thermodynamic work of adhesion is thus a quantity measurable in the laboratory for cases in which a finite contact angle between the liquid and the solid is formed. It cannot, of course, be set equal to the mechanical energy associated with the destruction of the adhesive bond because it contains no information on the intimacy of molecular contact across the interface in the actual bond nor any accounting for the rheological effects associated with the disjoining event. Nonetheless, it has been convincingly demonstrated [4] that the measured mechanical adhesion is directly proportional to the work of adhesion expressed as $\sigma_{\mathrm{L}}(1 + \cos\theta)$. It may thus be asserted that good thermodynamic adhesion is a *necessary*, if not sufficient, condition for good practical adhesion.

Early modelling of the physical interactions between molecules across interfaces [5–9] led to expressions of the type

$$W_{\mathrm{A}} = 2\sqrt{\sigma_{\mathrm{S}}^{\mathrm{d}} \sigma_{\mathrm{L}}^{\mathrm{d}}} \tag{5}$$

or

$$W_{\mathrm{A}} = 2\sqrt{\sigma_{\mathrm{S}}^{\mathrm{d}} \sigma_{\mathrm{L}}^{\mathrm{d}}} + 2\sqrt{\sigma_{\mathrm{S}}^{\mathrm{p}} \sigma_{\mathrm{L}}^{\mathrm{p}}}. \tag{6}$$

The assumption underlying equation (5), due to Fowkes, was that only dispersion forces were operative across the interface, while equation (6) (or its equivalents [10]) took note of the fact that in most cases involving other than completely apolar materials, the dispersion forces alone were unable to account for the magnitudes measured for W_A [11]. This excess contribution was arbitrarily assigned to 'polar effects', even though its magnitude seldom correlated with the dipole moments of the materials.

As indicated above, it was Fowkes who first pointed out that this excess contribution could be identified with acid–base interactions, considered in their most general 'Lewis' sense. Such interactions embrace not only hydrogen bonding, but also any in which the sharing of an electron pair may be identified. The near universality of such interactions follows from the observation that so many materials are capable of acting as either acids *or* bases. Examples of such materials include water, alcohols, carboxylic acids, nitriles, and amides. They are thus self-associated in the bulk and have been referred to as 'bipolar' [12]. Fowkes's original contention that purely physical interactions across interfaces are limited to dispersion forces only has proved to be essentially correct: the small extent to which Keesom forces play any role can be combined with the 'dispersion' force contribution to the interfacial free energy, now more properly designated as Lifshitz–van der Waals interactions [13]. In view of the above, the work of adhesion has been expressed in the form

$$W_A = 2\sqrt{\sigma_S^{LW}\sigma_L^{LW}} + W_A^{ab}, \tag{7}$$

where W_A^{ab} refers to the contribution of acid–base interactions. Surface free energies (surface tensions) may be decomposed into terms arising from Lifshitz–van der Waals interactions between the molecules and those attributable to self-association. Thus, for example,

$$\sigma_L = \sigma_L^{LW} + \sigma_L^{SA}. \tag{8}$$

For materials incapable of self-association,

$$\sigma_L = \sigma_L^{LW}, \tag{9}$$

regardless of their possession of permanent dipoles.

Fowkes has suggested further [1] that the acid–base contribution be expressed as

$$W_A^{ab} = fN(-\Delta H^{ab}), \tag{10}$$

where $(-\Delta H^{ab})$ is the (exothermic) enthalpy per mol of acid–base adduct formation at the interface, N is the number of moles of accessible acid or base functional groups per unit area on the solid surface, and f is a factor converting the enthalpic quantity $N(-\Delta H^{ab})$ to a free energy. Fowkes has assumed that f may be set equal to unity. In such circumstances, a quantitative evaluation of W_A^{ab} should, in principle, be possible. N should be accessible through appropriate titrations of the solid material, and $(-\Delta H^{ab})$ may be measured directly by microcalorimetry or indirectly by measuring the shift in stretching frequency of the relevant bond in the functional group on the solid surface upon formation of the acid–base adduct. Greatly increasing the practical utility of equation (10) is the likelihood of being able to predict the value of $(-\Delta H^{ab})$ in terms of tabulated parameters which characterize the acid or base nature of the functional groups involved. Fowkes has suggested using the Drago C and E constants [14, 15], which characterize the covalent and electrostatic contributions, respectively. In terms of these constants,

$$(-\Delta H^{ab}) = C_1 C_2 + E_1 E_2, \tag{11}$$

where the subscripts 1 and 2 refer to the acid and base components, respectively. Direct measurements of $(-\Delta H^{ab})$ and N for solid–liquid interactions should permit evaluation of the Drago constants for solids.

Fowkes and others have provided numerous qualitative examples of the importance of acid–base interactions, including their effect on the adsorption of dissolved polymers onto solid surfaces [1], the peelability of polymeric films cast onto glass [16], and the toughness of filled polymeric materials [17]. Most convincing of all was a set of wetting measurements yielding W_A^{ab} for a series of Drago-characterized acidic or basic liquids on polymers of varying acidity or basicity [18]. While the solids themselves were not quantitatively characterized (N, C, and E values were not determined independently), the results showed appreciable values for W_A^{ab} for acidic liquids on basic solids and basic liquids on acidic solids, but no measurable acid–base contribution when both materials were either acidic or basic.

The promise of using equations (9)–(11) to predict the work of adhesion when acid–base interactions are important is in view, but its implementation awaits the accumulation of a larger database of the type described above, most importantly one in which W_A^{ab} values obtained by wetting measurements can be *quantitatively* compared with computed values based on independent measurements of the quantities in equation (10). The present work addresses itself to this task, focusing particular attention on the value of the f factor. Setting it equal to unity, as has been done to this point, is tantamount to ignoring any entropy change associated with the formation of acid–base adducts at the solid–liquid interface, an assumption which has not yet been justified.

2. EXPERIMENTAL APPROACH AND ANALYSIS

The acid–base contribution to the work of adhesion, W_A^{ab}, may be obtained entirely by wetting measurements (for liquids which form a finite contact angle against the solid and whose vapors do not adsorb measurably on the solid) in accord with

$$W_A^{ab} = \sigma_L(1 + \cos\theta) - 2\sqrt{\sigma_S^{LW}\sigma_L^{LW}}. \tag{12}$$

The surface tension of the liquid and its contact angle may be measured directly, while the Lifshitz–van der Waals contribution to the solid surface energy is obtained using a non-self-associating probe liquid with no acidic or basic character (i.e. a 'neutral' liquid). σ_S^{LW} is then obtained with such a liquid using the appropriate forms of equations (2) and (8):

$$\sigma_S^{LW} = \frac{\sigma_{probe}}{4}(1 + \cos\theta_{probe})^2, \tag{13}$$

where σ_{probe} is the surface tension of the probe liquid and θ_{probe} is the contact angle it makes against the solid. A probe liquid of sufficiently high surface tension must be chosen such that $\theta_{probe} > 0°$.

For test liquids which are non-self-associating acids or bases, σ_L^{LW} is just equal to the surface tension itself. For self-associating test liquids, σ_L^{LW} is obtained from interfacial tension measurements of the test liquid against an immiscible neutral probe liquid.

Values of W_A^{ab} obtained by equation (12) through measurements of the above type may be compared with values obtained from equation (10), using independent measurements of N and $(-\Delta H^{ab})$. The latter are obtained by conductometric titrations of the test solid in finely-divided form and flow microcalorimetry, respectively, as mentioned above. The main obstacle to the final comparison is the factor f. Nonetheless, a comparison can be made using the assumption that $f = 1$.

The factor f itself is examined in this work using a Gibbs–Helmholtz analysis [19] of measurements of surface tension and contact angle over a range of temperatures:

$$(-N \cdot \Delta H^{ab}) = T^2 \left[\frac{\mathrm{d}\,\dfrac{W_A^{ab}}{T}}{\mathrm{d}T} \right] = W_A^{ab} - T\frac{\mathrm{d}\,W_A^{ab}}{\mathrm{d}t} = \frac{1}{f}\,W_A^{ab}. \qquad (14)$$

This leads directly to the expression for f:

$$f = \left[1 - \frac{\mathrm{d}\ln W_A^{ab}}{\mathrm{d}\ln T} \right]^{-1}. \qquad (15)$$

W_A^{ab} is obtained as a function of temperature from measurements of σ_L, θ, and the surface tension and contact angles of the probe liquid (for evaluation of σ_S^{LW}) as functions of temperature.

Finally, a comparison of the W_A^{ab} values obtained from wetting measurements alone is made with $fN(-\Delta H^{ab})$, including the f factor obtained from separate wetting measurements taken over a range of temperatures.

3. MATERIALS AND METHODS

Measurements were performed first with a variety of liquids against a poly(ethylene) solid, which was expected to yield no acid–base interactions. The liquids included two non-self-associating acids (bromoform and pyrrole), one non-self-associating base (dimethyl sulfoxide), one self-associating liquid (2-iodoethanol), and five aromatic liquids (quinoline, benzaldehyde, 2-iodothiophene, benzyl benzoate, and 1-iodonaphthalene), all of which were expected to be non-self-associating bases. The neutral probe liquids used were di-iodomethane and n-octane. The test solids used were poly(ethylene) (PE) (neutral), poly(vinyl chloride) (PVC) (acid), poly(ethylene) of varying extents of chlorination (varying acidity), poly(methylmethacrylate) (PMMA) (base), copolymers of poly(ethylene) and poly(vinyl acetate) (PVAc) (varying basicity), and copolymers of poly-(ethylene) and poly(acrylic acid) (PAA) (acidic or basic). Both the liquids and solids used are summerized in Table 1. The liquids were reagent grade. All materials were used as received, i.e. without further purification. The purity of the poly(ethylene) ($<1\%$ oxygen) was verified by XPS. All of the test solids had a relatively low surface energy (<40 erg/cm^2), so that finite spreading pressures of the test liquids were not anticipated.

The solids were received in powdered form, and for purposes of contact angle measurement, were cast as thin films from solution onto glass slides, from which after drying they were peeled using a razor blade. PE, PE/PAA copolymers,

Table 1.
Test materials used

	Type	Source
Liquids		
Bromoform	Acidic	Kodak
Pyrrole	Acidic	Baker
Dimethyl sulfoxide	Basic	Baker
Quinoline	Basic	Aldrich
Benzaldehyde	Basic	Baker
2-Iodothiophene	Basic	Aldrich
Benzyl benzoate	Bi-functional	Fluka
2-Iodoethanol	Bi-functional	Aldrich
Di-iodomethane	Neutral	Baker
n-Octane	Neutral	Aldrich
Solids		
Poly(vinyl chloride)	Acidic	Scientific Polymer Products
Chlorinated poly(ethylene) (25%, 36%, 48%)	Acidic	Scientific Polymer Products
Poly(methyl methacrylate)	Basic	Scientific Polymer Products
Poly(ethylene)/poly(vinyl acetate) (50%, 100%)	Basic	Scientific Polymer Products
Poly(ethylene)/poly(acrylic acid) (5%, 20%)	Bi-functional	Scientific Polymer Products
Poly(ethylene)	Neutral	Scientific Polymer Products

chlorinated PE, and PMMA were cast from hot xylene solutions, while the PVC and the PE/PVAc copolymers were cast from warm solutions of tetrahydrofuran.

Surface tension and contact angle measurements were made using the Wilhelmy method, as described in detail elsewhere [20]. The microtensiometer used was equipped with a heater and could be temperature-controlled, permitting measurement of the above properties as functions of temperature at 5°C intervals between 20 and 60°C for bromoform against PMMA and for dimethyl sulfoxide against PVC, PE/PAA (5%), and PE/PAA (20%). Surface tensions were obtained using a platinized platinum plate, and the contact angles reported are the static advanced values (corresponding to the limiting wetting force as the interline velocity was reduced).

The two non-self-associating acids (bromoform and pyrrole), one non-self-associating base (dimethyl sulfoxide), and the one self-associating liquid (2-iodoethanol) were tested against the monofunctional acidic (PVC) and basic (PMMA) solids to yield the acid–base contribution to the work of adhesion. The model acid (bromoform) and base (dimethyl sulfoxide) were tested against the solids of varying degrees of acidity, basicity, or bi-functionality.

For one system, dimethyl sulfoxide against PE/PAA (5%) copolymer, independent measurements of N and $(-\Delta H^{ab})$ were made. The moles of acid–base pairs per unit weight was obtained by conductometric titration of the powdered solid using a Radiometer RTS822 titrator system with a CDM 83 conductivity meter under a nitrogen blanket using CO_2-free standard NaOH. A small amount of Triton X-100 nonionic surfactant was necessary to force the polymer powder into suspension, because in its absence the powder was non-wetting and floated on the surface. Single-point BET measurements using a Micromeritics Flowsorb II 2300 surface area analyzer yielded the surface area of the powder so that the moles of acid–base pairs could be put on a per unit area basis.

The enthalpy of acid–base interaction $(-\Delta H^{ab})$ was obtained using a flow microcalorimeter (Microscal Ltd, Mark IV), as described elsewhere [21].

4. RESULTS AND DISCUSSION

Figure 1 shows the measured W_A^{ab} as computed by equation (12) for all of the acidic and basic test liquids against the neutral solid poly(ethylene). Surprisingly, the five aromatic compounds all showed significant apparent acid–base interactions with the poly(ethylene). The reasons for this behavior must be investigated further because such results suggest that the present formulation of equation (7) for the work of adhesion may be incomplete. At any rate, these aromatic liquids were eliminated from further investigation in the present study. The remaining liquids displayed the expected negligible acid–base interaction with PE.

Figures 2 and 3 show W_A^{ab} for the remaining test liquids against the monofunctional acidic and basic solids, respectively. The results shown are of the type anticipated, i.e. significant interactions between the acidic liquids and the basic solid, between the basic liquid and the acidic solid, and between the bi-functional liquid and both solids. The W_A^{ab} values between the acidic liquids and the acidic solid, and between the basic liquid and the basic solid were small to negligible.

Figures 4–6 show the interaction of both the acidic liquid, bromoform, and the basic liquid, dimethyl sulfoxide, against the test solids of varying acidity, varying basicity, and varying functionality of either acidic or basic character. Again, the anticipated results were obtained. The interaction of the acidic liquid and the basic liquid increased with increasing basicity and acidity of the solids, respectively, and the acidic liquid against the acidic solid, and basic liquid against basic solid

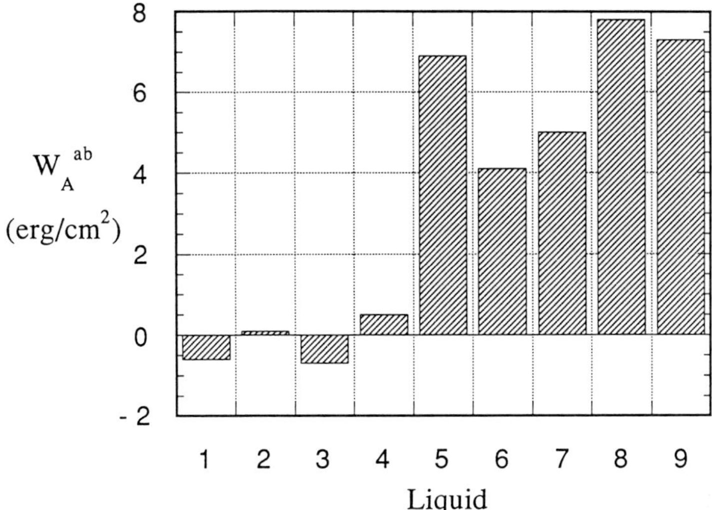

Figure 1. Apparent acid–base contribution to the work of adhesion of various test liquids against poly(ethylene): (1) bromoform; (2) pyrrole; (3) dimethyl sulfoxide; (4) 2-iodoethanol; (5) quinoline; (6) benzaldehyde; (7) 2-iodothiophene; (8) benzyl benzoate; and (9) 1-iodonaphthalene.

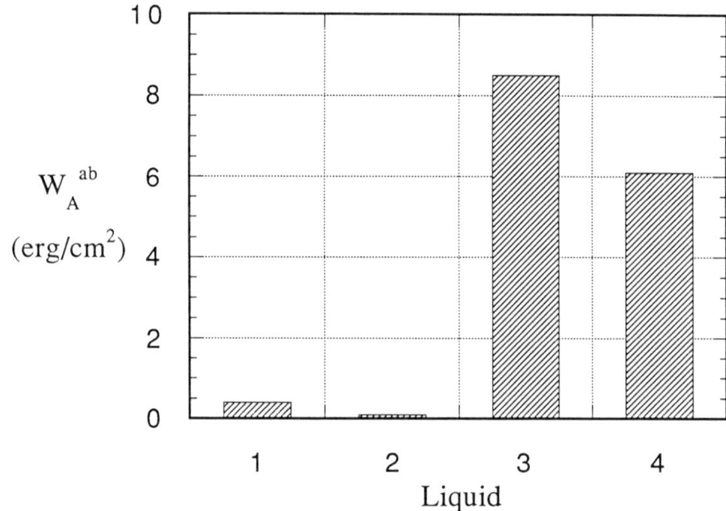

Figure 2. Acid–base contribution to the work of adhesion of selected test liquids against poly(vinyl chloride): (1) bromoform; (2) pyrrole; (3) dimethyl sulfoxide; and (4) 2-iodoethanol.

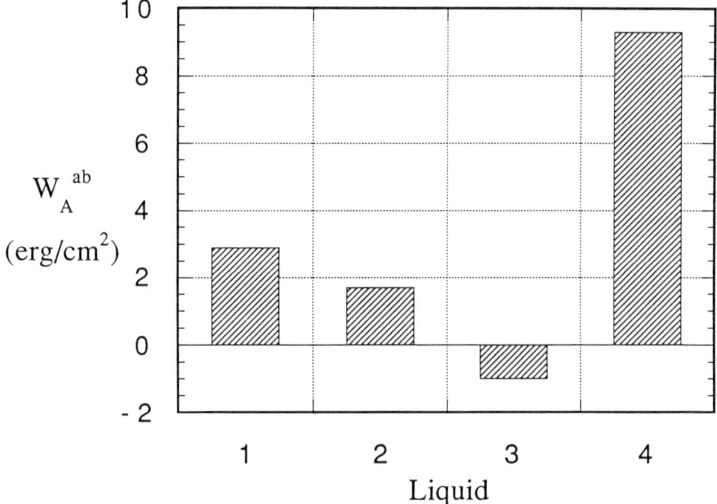

Figure 3. Acid–base contribution to the work of adhesion of selected test liquids against poly(methyl methacrylate): (1) bromoform; (2) pyrrole; (3) dimethyl sulfoxide; and (4) 2-iodoethanol.

showed negligible acid–base interaction at all levels. The bi-functional solid showed acid–base interaction with both test liquids.

Results for the temperature dependence of W_A^{ab} for the four systems for which such data were taken are shown in Fig. 7. These results required the measurement of σ_L, cos θ, σ_{probe}, and cos θ_{probe} all as functions of temperature, and their assembly in accord with equations (12) and (13). Differentiation in accord with equation (15) yields the enthalpy-to-free energy correction factor, f, as a function of temperature, as shown in Fig. 8. These results show, at least for the systems

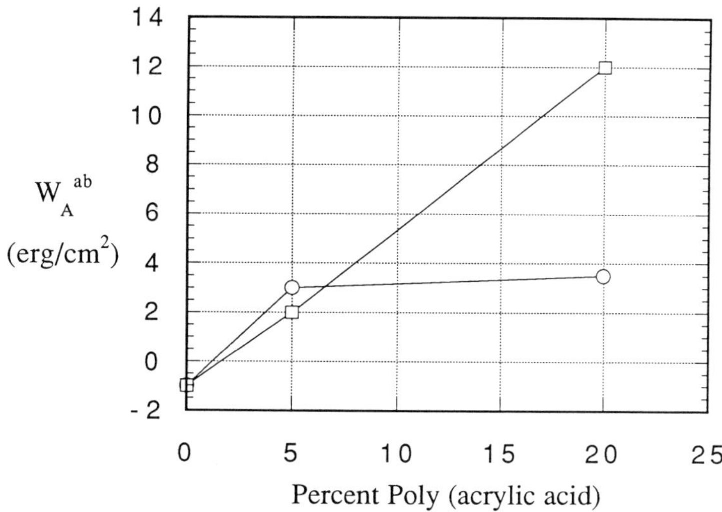

Figure 4. Acid–base contribution to the work of adhesion of (—O—) bromoform and (—□—) dimethyl sulfoxide against copolymers of poly(ethylene) and poly(vinyl acetate) of varying composition.

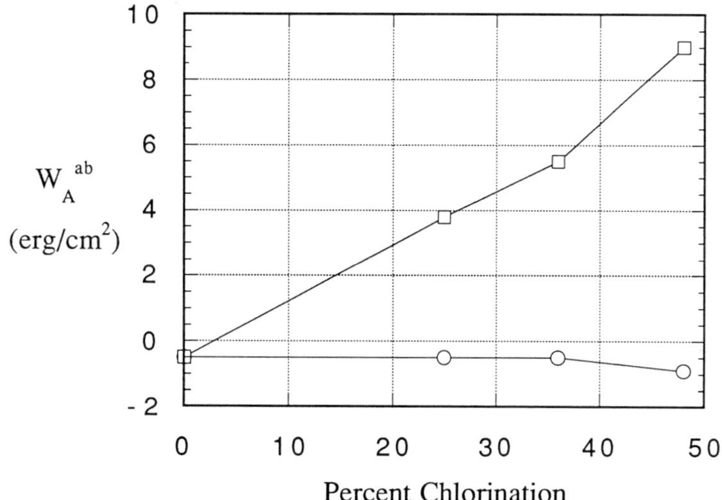

Figure 5. Acid–base contribution to the work of adhesion of (—O—) bromoform and (—□—) dimethyl sulfoxide against chlorinated poly(ethylene) of varying extent of chlorination.

studied, that the f factor is substantially less than unity in most cases and increases with temperature. It may also suggest that adhesive bonds formed at higher temperature may take better advantage of acid-base interactions.

Finally, a comparison between W_A^{ab} as determined solely by wetting measurements against those obtained by equation (10) and employing independent measurements of N and $(-\Delta H^{ab})$ is shown in Table 2. General conclusions cannot be drawn from such results for only a single system, but they do suggest that agreement is not yet complete. Even the rather large uncertainty which exists with

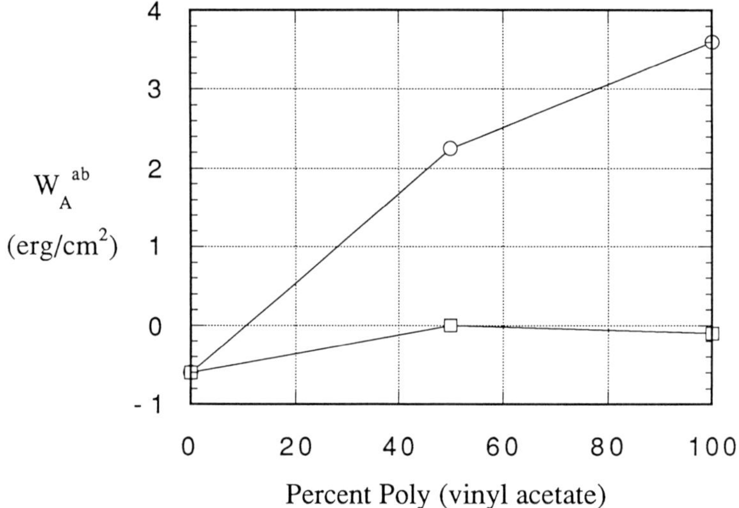

Figure 6. Acid–base contribution to the work of adhesion of (—◯—) bromoform and (—◻—) dimethyl sulfoxide against copolymers of poly(ethylene) and poly(acrylic acid) of varying composition.

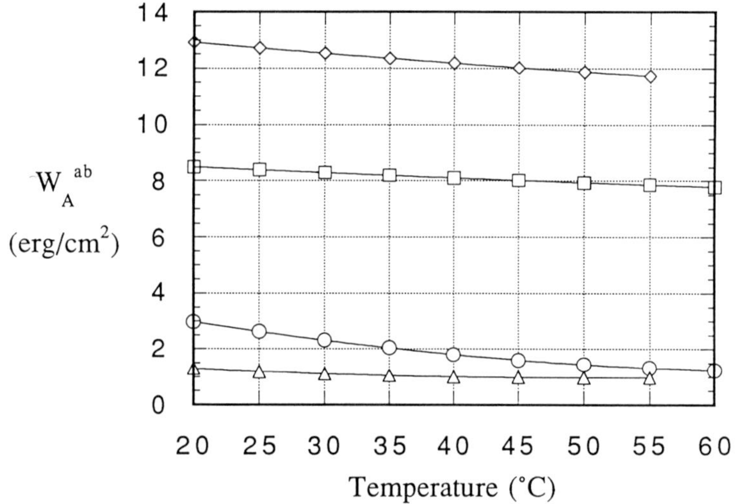

Figure 7. Temperature dependence of the acid–base contribution to the work of adhesion in four systems: (—◯—) bromoform against poly(methyl methacrylate); (—◻—) dimethyl sulfoxide against poly(vinyl chloride); (—△—) dimethyl sulfoxide against copolymer of poly(ethylene)/poly(acrylic acid) (5%); and (—◇—) dimethyl sulfoxide against copolymer of poly(ethylene)/poly(acrylic acid) (20%).

respect to ($-\Delta H^{ab}$), viz. ± 2 kcal/mol, does not embrace the disparity observed. One factor that is clear, however, is that setting the f factor equal to unity would have made the disparity substantially greater. Further conclusions require data of the type given in Table 2 for several different test systems.

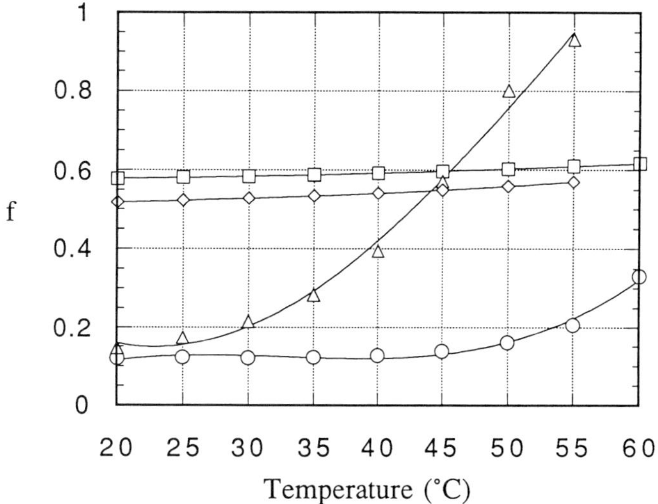

Figure 8. Enthalpy-to-free energy correction factor, *f*, as a function of temperature for the four systems of Fig. 7: (—O—) bromoform against poly(methyl methacrylate); (—□—) dimethyl sulfoxide against poly(vinyl chloride); (—△—) dimethyl sulfoxide against copolymer of poly(ethylene)/poly(acrylic acid) (5%); and (—◇—) dimethyl sulfoxide against copolymer of poly(ethylene)/poly(acrylic acid) (20%).

Table 2.

Comparison of the acid–base contribution to the work of adhesion as computed from wetting measurements and from independent measurement of the effective enthalpy of interaction for the system dimethyl sulfoxide against copolymer of poly(ethylene)/poly(acrylic acid) (5%)

Wetting measurements

$$W_A \quad = \sigma_L(1 + \cos\theta) \quad = 69.4 \text{ erg/cm}^2$$
$$W_A^{LW} = 2(\sigma_S^{LW}\sigma_L^{LW})^{1/2} = 68.1 \text{ erg/cm}^2$$
$$W_A^{ab} = W_A - W_A^{LW} \quad = 1.3 \text{ erg/cm}^2$$

Independent measurements

$$f = 0.15$$
$$\Sigma = 0.30 \text{ m}^2/\text{g (specific area of powdered solid)}$$
$$N = 6.1 \times 10^{-11} \text{ mol/cm}^2$$
$$-\Delta H^{ab} = 8 \text{ kcal/mol}$$
$$W_A^{ab} = fN(-\Delta H^{ab}) \quad = 3.0 \text{ erg/cm}^2$$

---

5. SUMMARY

(1) The splitting of the thermodynamic work of adhesion between a liquid and a solid into terms accounting for the Lifshitz–van der Waals interactions and acid–base interactions, respectively, allows both terms to be evaluated by wetting measurements. Such measurements are found to be in qualitative agreement with theory, except for an apparent acid–base contribution to the interaction between several aromatic liquids and poly(ethylene).

(2) A Gibbs–Helmoltz analysis of wetting measurements obtained as a function of temperature reveals that the enthalpy-to-free energy correction factor

cannot be set equal to unity, but instead has values which are temperature-dependent and substantially less than one over most of the temperature range studied for most of the systems investigated.

(3) Quantitative agreement between the acid–base contribution to the work of adhesion as derived solely from wetting measurements and that obtained from independent measurement of the enthalpy of adduct formation and the number of interacting functional groups per unit area of the solid surface cannot be claimed or conclusively denied for the system investigated.

Acknowledgements

This work was supported by the 3M Company, St. Paul, MN, and the University of Washington Center for Surfaces, Polymers and Colloids.

REFERENCES

1. F. M. Fowkes and M. A. Mostafa, *Ind. Eng. Chem. Prod. Res. Dev.* **17**, 3 (1978).
2. F. M. Fowkes, *J. Adhesion Sci. Technol.* **1**, 7 (1987).
3. F. M. Fowkes, D. C. McCarthy and M. A. Mostafa, *J. Colloid Interface Sci.* **78**, 200 (1980).
4. K. L. Mittal, in: *Adhesion Science and Technology, Part A*, L.-H. Lee (Ed.), p. 129. Plenum Press, New York (1975).
5. F. M. Fowkes, *J. Phys. Chem.* **66**, 382 (1962).
6. F. M. Fowkes, *Ind. Eng. Chem.* **12**, 40 (1964).
7. L. A. Girifalco and R. J. Good, *J. Phys. Chem.* **61**, 904 (1957).
8. D. H. Kaelble, *Physical Chemistry of Adhesion*. Wiley-Interscience, New York (1971).
9. D. K. Owens and R. C. Wendt, *J. Appl. Polym. Sci.* **13**, 1741 (1969).
10. S. Wu, *Polymer Interface and Adhesion*. Marcel Dekker, New York (1982).
11. J. R. Dann, *J. Colloid Interface Sci.* **32**, 302, 321 (1970).
12. C. J. van Oss, M. K. Chaudhury and R. J. Good, *Adv. Colloid Interface Sci.* **28**, 35 (1987).
13. C. J. van Oss and R. J. Good, *Colloids Surf.* **8**, 373 (1984).
14. R. S. Drago, G. C. Vogel and T. E. Needham, *J. Am. Chem. Soc.* **93**, 6014 (1971).
15. R. S. Drago, L. B. Parr and C. S. Chamberlin, *J. Am. Chem. Soc.* **99**, 3203 (1977).
16. F. M. Fowkes, D. O. Tischler, J. A. Wolfe, L. A. Lannigan, C. M. Ademu-John and M. J. Halliwell, *J. Polym. Sci.* **22**, 547 (1984).
17. M. Marmo, H. Jinnai, M. A. Mostafa, F. M. Fowkes and J. A. Manson, *Ind. Eng. Chem. Prod. Res. Dev.* **15**, 206 (1976).
18. F. M. Fowkes, in: *Adhesion and Adsorption of Polymers, Part A*, L.-H. Lee (Ed.), p. 43. Plenum Press, New York (1980).
19. G. N. Lewis, M. Randall, K. S. Pitzer and L. Brewer, *Thermodynamics*, 2nd edn, p. 165. McGraw-Hill, New York (1961).
20. J. C. Berg, in: *Composite Systems from Natural and Synthetic Polymers*, L. Salmén, A. DeRuvo, J. C. Seferis and E. B. Stark (Eds), p. 23. Elsevier, Amsterdam (1986).
21. M. D. Vrbanac, Ph.D. dissertation, University of Washington, Seattle (1989).

Acid-Base Interactions, pp. 79-89
Eds. K.L. Mittal and H.R. Anderson, Jr.
© VSP 1991

Theory of the acid–base hydrogen bonding interactions, contact angles, and the hysteresis of wetting: application to coal and graphite surfaces

R. J. GOOD,[1,*] N. R. SRIVATSA,[1,†] M. ISLAM,[1] H. T. L. HUANG[1] and C. J. VAN OSS[1,2]

Department of [1] Chemical Engineering and [2] Microbiology, SUNY at Buffalo, Buffalo, NY 14260, USA

Revised version received 27 March 1990

Abstract—An understanding of the acid–base behavior of coal surfaces is needed, if the wetting of coal by water and other liquids, and the separation processes such as floatation and oil agglomeration of coal, are to be exploited. The modern theory of the acidic and basic parameters of surface free energy, $\gamma^\oplus$ and $\gamma^\ominus$, as well as the apolar component γ^{LW}, is reviewed, together with the equations for contact angles in terms of these properties.

Contact angles of water, glycerol and methylene iodide on 14 different coals, and on graphite, have been measured. The advancing angle data and the retreating angle data lead to two decidedly different sets of the parameters, γ^{LW}, $\gamma^\oplus$ and $\gamma^\ominus$. γ_r^{LW}, based on retreating angles, is 10 mJ/m^2 higher than γ_a^{LW}, based on advancing angles; and both are independent of % carbon and % oxygen, i.e. of coal rank. $\gamma_r^\oplus$ lies between 1.2 and 2.4 mJ/m^2, and $\gamma_a^\oplus$ is close to 0.2 mJ/m^2, independent of composition. $\gamma_r^\ominus$ and $\gamma_a^\ominus$ exhibit downtrends with increasing carbon content. For each of the three parameters, the data for graphite fit the trends found with the various coals. The surface structures that would account for these trends are briefly discussed.

The data point to a two-phase surface structure for coal and for graphite.

Keywords: Graphite; coal; hydrogen bond; wetting; hysteresis; contact angle; surface free energy; acids and bases.

1. INTRODUCTION

In the important contribution that Fowkes *et al.* [1–3] have made to the theory of interfacial interactions, surface free energy was found to have two components, one that is apolar, and the other that involves acid–base interactions. For a solid, s,

$$\gamma_s = \gamma_s^d + \gamma_s^{AB} \tag{1}$$

where d refers to the dispersion force and AB to acid–base interactions. The critical surface tension for wetting, γ_c [4, 5] gives a satisfactory approximation to γ_s^d, if the liquids employed in determining γ_c are chosen correctly.

We will, in this paper, show how the acidic and basic parameters of γ_s^{AB}, as well as the apolar component of γ_s, can be determined for coal. These properties are needed for the understanding of the beneficiation processes of floatation [6, 7] and oil agglomeration [8–10], by which hydrophillic impurities can be eliminated from coal.

---

*To whom correspondence should be addressed.

†Present Address: Corporate Research Center, International Paper Co., Tuxedo, NY 10987, USA.

There is, currently, a considerable need to remove SO_2-forming impurities such as iron pyrite (FeS_2) as well as other ash-forming components such as clay, shale, limestone, silica, etc., from coal. The two beneficiation processes depend upon the surface properties of the solid, for their functioning.

It has been found by Aplan *et al.* [11, 12] that the critical surface tension for coal, γ_c, is very nearly the same for a wide variety of coals, independent of rank. This was true of coals ranging from lignite to anthracite; the coals that they studied had oxygen content ranging from 20% down to 1.9%, and carbon content from 46 to 95% daf (dry, ash free). Graphite was found to have the same γ_c as the coals.

This being the case, the well known, wide variation of behavior of coal with rank, in froth floatation and in oil agglomeration (e.g. that lower-rank coals do not respond as well as higher-rank coals) must be due to variations in the acid–base component of γ_s. We will, below, discuss how the experimental determination of γ_s^{AB} may be carried out. We will show that contact angle measurements on coal constitute a very interesting system for applying the recently-proposed [13–18] method of determining the components of γ_s^{AB}.

Our direct interest in the surface chemistry of coal arose as a major part of an investigation of oil agglomeration. This technique has recently been developed into a very economical and effective process [8, 9]. In it, coal (which always contains mineral matter) is wet-ground to a particle size sufficiently fine that the mineral matter is released; and an oil is added. Agitation at high shear rates leads to the formation of aggregates in which coal particles are enveloped in oil, with water droplets as liquid bridges between coal particles; the liquid bridges provide cohesive mechanical strength to the aggregates (see Fig. 1). Hydrophobic materials such as coal are held in the agglomerate; and hydrophilic materials are rejected. Iron pyrite, when cleaved under water, is hydrophilic. In oil agglomeration by the Otisca T process [8, 9], which uses pentane as the oil, FeS_2 particles remain in the bulk aqueous phase. The coal–oil–water aggregates, which resemble black cottage cheese, are separated from the water by screening; and the FeS_2 and other (hydrophilic) ash-forming minerals are carried through with the aqueous phase. The pentane is recovered by evaporation under vacuum. The coal may be

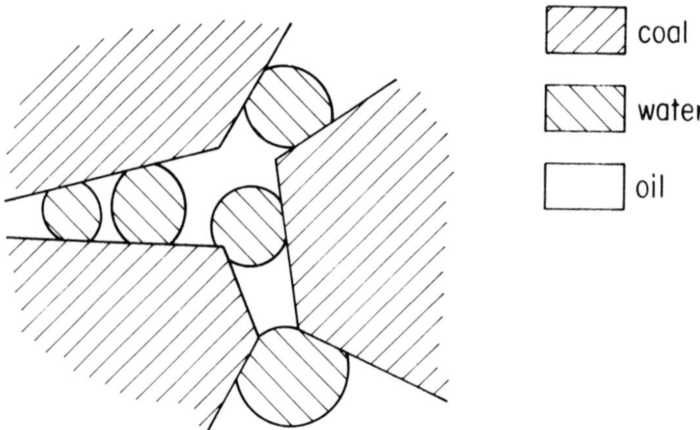

Figure 1. Schematic drawing of the internal structure of a coal–oil–water agglomerate. Coal particles are held together by aqueous liquid bridges.

dried and burned, or it may be used in the form of a water slurry. The inorganic sulfur content of a typical coal can be reduced from, say, 3%, to below 1%, in a single cycle of this process.

The rank of coal, from lignite through various grades of bituminous and anthracite (with graphite as the terminal member of the series) is principally controlled by the composition.* Low rank coals have carbon content ranging from as low as about 40% daf (dry, ash-free basis); and anthracite is about 95% carbon. Oxygen content decreases with rank, from about 20% in lignite, down to about 2% in anthracite. The trend of hydrogen content is parallel to that of the oxygen content. The chemical composition and structure of surface species, on coal, and their variation from point to point over a cleavage surface, are not yet fully understood. We can contribute to elucidating this structural problem, if we can discover how the acidic and basic character, and their contribution to hydrogen bonding, vary from one coal to another.

2. THEORY

γ_i^{AB} is, by its nature, a property that depends on the mutual interaction of *two* unlike species, i.e. an acid and a base. (In this, it differs from the *apolar* component of surface free energy, which is a property *of a single type of substance*. That is to say, two groups or molecules, that have the *same type* of force field, interact.) It has been pointed out [13–18] that a pure substance, i, will have a non-zero γ_i^{AB} only if i has both acidic and basic character—as is the case with water. (Chloroform is weakly acidic, and, when pure, exhibits negligible basic character. Acetone is basic and, when pure, exhibits no acidic character. They are referred to as monopolar. Their acidic and basic properties are manifested in their interaction with compounds or groups that have opposite acid–base polarity, or with substances such as water. By contrast, water may be referred to as bipolar.)

We have introduced [13–18] two new surface parameters of condensed-phase substances, which are independent of the physical presence of another substance:

$$\gamma^\oplus \equiv \text{Lewis (or Brönsted) acid component of surface free energy}$$
$$\gamma^\ominus \equiv \text{Lewis (or Brönsted) base component of surface free energy.}$$

These, together, yield the acid–base component of surface free energy, γ^{AB}. Thus, for substance i,

$$\gamma_i^{AB} = 2\sqrt{\gamma_i^\oplus \gamma_i^\ominus} \quad \text{(substance } i \text{ bipolar)}$$
$$= 0 \qquad \text{(substance } i \text{ monopolar).}$$

$$(2)$$

For a monopolar substance, either $\gamma_i^\oplus = 0$ or $\gamma_i^\ominus = 0$, and hence $\gamma_i^{AB} = 0$.

For an interface between immiscible phases, substances i and j, it has been shown [14, 15] that

$$\gamma_{ij}^{AB} = 2(\sqrt{\gamma_i^\oplus} - \sqrt{\gamma_j^\oplus})(\sqrt{\gamma_i^\ominus} - \sqrt{\gamma_j^\ominus}). \tag{3}$$

We have shown, elsewhere, (see, e.g. refs [15] and [22]) that it is very desirable to replace the Fowkes γ^d, in equation (1), with γ^{LW}, given by

---

*Composition, however, is not the primary variable by which rank is established. See Carpenter's detailed review, *Coal Classification* [19], and the discussion of the relation between compositional variables, by Neavel *et al.* [20]. See also the general discussion in ref. [21].

$$\gamma^{\text{LW}} = \gamma^{\text{d}} + \gamma^{\text{i}} + \gamma^{\mu} \tag{4}$$

where γ^{i} is the induction (Debye) component and γ^{μ} is the dipole–dipole (Keesom) component of surface free energy. (The superscript, LW, was chosen to denote 'Lifshitz–van der Waals'.) Both γ^{i} and γ^{μ} are generally small (though not zero) in comparison to γ^{d}. When, in place of equation (1), we write,

$$\gamma_i = \gamma_i^{\text{LW}} + \gamma_i^{\text{AB}} \tag{5}$$

we specify that γ^{AB} does *not* contain the components of surface free energy, γ^{i} and γ^{μ}, which do not make any contribution to acid–base interactions. See refs [15, 18, 22] for extensive discussion of these points regarding equations (4) and (5).

In addition to equation (5), we may write, for an *i–j* interface,

$$\gamma_{ij} = \gamma_{ij}^{\text{LW}} + \gamma_{ij}^{\text{AB}} \tag{6}$$

The apolar component of γ_{ij} is related to the apolar components for *i* and for *j*, by the well-known equation

$$\gamma_{ij}^{\text{LW}} = (\sqrt{\gamma_i^{\text{LW}}} - \sqrt{\gamma_j^{\text{LW}}})^2 \tag{7}$$

which is a rearranged version of the 'geometric mean' equation [23],

$$\gamma_{ij}^{\text{LW}} = \gamma_i^{\text{LW}} + \gamma_j^{\text{LW}} - 2\sqrt{\gamma_i^{\text{LW}}\gamma_j^{\text{LW}}} \tag{8}$$

in which the interaction parameter, Φ, is unity. Contact angles of liquids on solids are given by the familiar Young equation,

$$\gamma_1 \cos\theta = \gamma_{\text{s}} - \gamma_{\text{sl}}. \tag{9}$$

(In this form of Young's equation, we neglect the spreading pressure, π_{e}.) The Good–Girifalco equation [5] (see also ref. [1]) in the form that applies to the contact angle of an *apolar* liquid on a solid (i.e. for the case in which $\Phi = 1$) is

$$\gamma_{\text{s}}^{\text{LW}} = \frac{\gamma_1^{\text{LW}}(1 + \cos\theta)^2}{4} \tag{10a}$$

or,
$$\gamma_1^{\text{LW}}(1 + \cos\theta) = 2\sqrt{\gamma_{\text{s}}^{\text{LW}}\gamma_1^{\text{LW}}}. \tag{10b}$$

For a bipolar liquid, with surface tension γ_1, and acidic and basic surface parameters $\gamma_1^{\oplus}$ and $\gamma_1^{\ominus}$, respectively, and apolar surface component γ_1^{LW}, the equation corresponding to (10b) is

$$\gamma_1(1 + \cos\theta_1) = 2(\sqrt{\gamma_1^{\text{LW}}\gamma_{\text{s}}^{\text{LW}}} + \sqrt{\gamma_1^{\oplus}\gamma_{\text{s}}^{\ominus}} + \sqrt{\gamma_1^{\ominus}\gamma_{\text{s}}^{\oplus}}) \tag{11a}$$

For a second polar liquid, with surface parameters γ_2^{LW}, $\gamma_2^{\oplus}$ and $\gamma_2^{\ominus}$, the corresponding equation is,

$$\gamma_2(1 + \cos\theta_2) = 2(\sqrt{\gamma_2^{\text{LW}}\gamma_{\text{s}}^{\text{LW}}} + \sqrt{\gamma_2^{\oplus}\gamma_{\text{s}}^{\ominus}} + \sqrt{\gamma_2^{\ominus}\gamma_{\text{s}}^{\oplus}}) \tag{11b}$$

Equations (10b), (11a) and (11b) constitute a set of three simultaneous equations, in terms of the parameters of the two liquids γ_i^{LW}, $\gamma_i^{\oplus}$, $\gamma_i^{\ominus}$ ($i = 1$ or 2) and the three contact angles that are measured on the solid: θ_{LW}, θ_1 and θ_2. These can be solved for $\gamma_{\text{s}}^{\text{LW}}$, $\gamma_{\text{s}}^{\oplus}$ and $\gamma_{\text{s}}^{\ominus}$. Three appropriate liquids for use on coal are di-iodomethane (DIM), glycerol (GL) and water (W). The parameters for these liquids have been determined [16–18] and are shown in Table 1.

Table 1.
Surface parameters for test liquids (mJ/m^2)

	γ_l	γ_l^{LW}	γ_l^{AB}	$\gamma_l^{\oplus}$	$\gamma_l^{\ominus}$
Water	72.8	21.8	51.0	25.5[a]	25.5[a]
Glycerol	64	34	30	3.92	54.7
Diiodomethane	50.8	50.8	0.0	0.0	0.0

[a] $\gamma^{\oplus}$ and $\gamma^{\ominus}$ for water are reference values; and all reported values of $\gamma^{\oplus}$ and $\gamma^{\ominus}$ are relative to these values, with the convention that $\gamma_W^{\oplus} = \gamma_W^{\ominus}$.

3. EXPERIMENTAL

3.1. Equipment, procedure and materials

A modified Ramé Hart contact angle instrument with a controlled-atmosphere chamber was employed. It was modified by the insertion of a zoom lens in the optical train, which enables the operator to change magnification with almost no loss of focus. An image analyzer system, consisting of a COHU solid state camera and a Javelin TV screen and printer, increases the ease of operation. Each sample was sawed, and ground on an abrasive paper of various grit sizes, down to #600, and then polished, in anerobic conditions in a glove bag, with a Buhler Minimet polisher, using 0.05 μm alumina.

The Ramé Hart chamber was placed in the glove bag before the polishing step was started. The sample was transferred to the chamber; and the chamber was closed and taken from the glove bag, and mounted on the Ramé Hart sample table, where it was connected to the nitrogen gas supply. Liquid, for contact angle measurement, was introduced via a hollow needle on a micrometer pipet, through a rubber septum. The captive drops method was employed; and both advancing and retreating angles were always measured.

Samples were also prepared by polishing in air, and these were measured in air. The precautions such as employment of a glove bag and controlled-atmosphere chamber were, of course, omitted in these measurements.

The liquids employed were double distilled water, Fisher certified ACS grade glycerol, purified by means of an activated alumina column; and Fisher reagent grade diiodomethane, purified by means of an activated charcoal column.

Samples of coal from 14 seams were furnished by Dr. D. V. Keller, Jr., of Otisca Industries, Ltd., Syracuse, N.Y. Graphite from Sri Lanka was obtained from Ward's Scientific Establishment, Rochester, N.Y.

The value of γ^{LW} for a coal was determined using CH_2I_2 data, and employing equation (10a). Then, contact angles of water and of glycerol were determined. When the measuring liquid was changed, the coal sample was taken out and re-polished before a new liquid was used. About 30 contact angle measurements were made, for each reported value of θ.

4. RESULTS

Typical contact angle data are given in Table 2, obtained with Upper Freeport (Mv bituminous) coal. This coal contains 6.2% oxygen, 5.6% hydrogen and 85.95% carbon (daf).

Table 2.
Contact angles (in nitrogen) of liquids on Upper
Freeport coal (85.95% C, daf), and calculated surface
free energy components

	Contact angles in nitrogen	
	θ_a, deg	θ_r, deg
Water	87.2 ± 1.7	28.3 ± 5.8
Glycerol	72.9 ± 1.9	12.9 ± 2.8
Diiodomethane	39.7 ± 1.9	11.5 ± 1.3
	Computed surface free energy parameters (mJ/m^2)	
γ^{LW}	39.8 ± 1.0	49.8 ± 0.2
$\gamma^{\oplus}$	0.0	2.0
$\gamma^{\ominus}$	2.5	31.7
γ^{AB}	0	15.9

The presence of hysteresis is notable, in the angles reported in Table 2. The hysteresis in θ translates into the existence of two distinct sets of free energy parameters, γ_s^{LW}, $\gamma_s^{\oplus}$ and $\gamma_s^{\ominus}$ e.g., as shown in Table 2. (The standard deviation is given only for γ_s^{LW}, in Table 2, because for $\gamma^{\oplus}$ and $\gamma^{\ominus}$, the computation of standard deviations using equations (9b), (10a) and (10b) is complex.) The effective standard deviation is of the order of ± 0.1 in $\gamma^{\oplus}$ and in $\gamma_a^{\ominus}$, and 1.0 in $\gamma_r^{\ominus}$.

This behavior, of having two sets of surface free energy parameters, was found with all the coals that were studied. See Fig. 2, for γ_s^{LW}, and Figs 3–5 for $\gamma_s^{\oplus}$ and $\gamma_s^{\ominus}$. To a first approximation, within the experimental error, (e.g. ± 1 mJ/m^2 in single γ^{LW} values) all the values of γ_r^{LW} (based on retreating angles) can be considered as fitting a horizontal line at 49.6 mJ/m^2; and all the values of γ_a^{LW} (from advancing angles) fit a line at 39.5 mJ/m^2. The small displacement of the values for samples handled in nitrogen, below the corresponding values for samples handled in air, may be considered a second-order effect, and the same is true of the minor downtrend of values of γ_r^{LW} (air) with coal rank.

Our γ^{LW} results are quite compatible with the earlier results of Aplan *et al.* [11, 12], whose estimate of the critical surface tension for wetting of coal led to a γ_c of 46.5 mJ/m^2, independent of coal rank. Aplan *et al.* did not make an explicit effort to measure both advancing and retreating angles.

It has been suggested [24] that a major cause of contact angle hysteresis is the presence of heterogeneities in a solid surface. On the basis of well known microscopic studies of coal, it would be surprising not to find that distinct patches or bands existed, with different surface energies. So we may interpret the value of γ_a^{LW}, of about 39.5 mJ/m^2, as probably representing the surface free energy of a carbonaceous component of coal (possibly, CH_2 groups) independent of rank. The value of γ_r^{LW}, of about 49.6 mJ/m^2, corresponds to the surface free energy of a higher-energy carbonaceous component (possibly, aromatic groups). This, too, is independent of rank.

Figure 3 shows the calculated values of $\gamma^{\oplus}$ as a function of % carbon (daf). (Note the expanded energy scale, compared to that in Fig. 2. The absolute scatter of data in Fig. 3 is actually less than that in Fig. 2.) We may first discuss the results

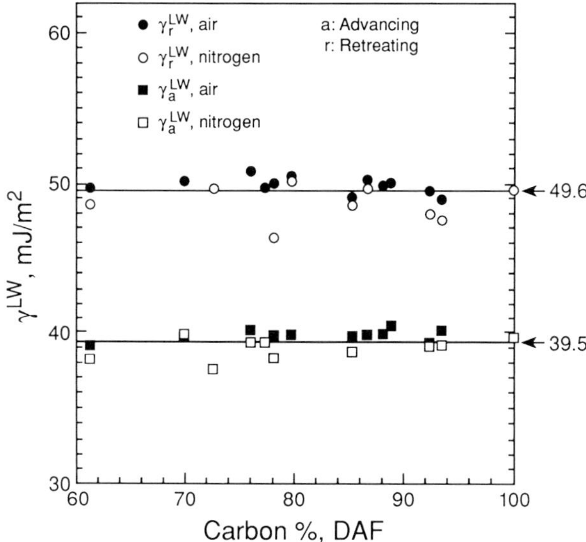

Figure 2. Apolar component of surface free energy of coal and graphite, γ^{LW}, based on advancing and retreating contact angles, as a function of % carbon.

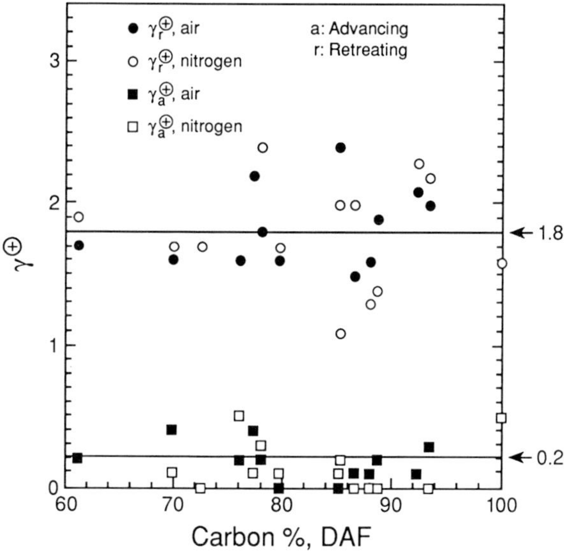

Figure 3. Lewis acid component of surface free energy of coal and graphite, $\gamma^{\oplus}$, based on advancing and retreating contact angles, as a function of % carbon.

for $\gamma_a^{\oplus}$, i.e. the values from advancing angles. These all lie in the range, zero to 0.55 mJ/m^2; and there is no significant trend with % carbon. The mean is 0.2 mJ/m^2; and the difference between this mean and zero is barely significant.

The interpretation of this result is that the areas that are responsible for the advancing contact angles contain no (or practically no) acidic groups, or groups that can contribute a hydrogen atom to a hydrogen bond. This is exactly what would be expected if these areas consist of aliphatic or naphthenic groups, or aromatic groups, and no oxygen.

The $\gamma_r^\oplus$ values are appreciably different from zero, and all the $\gamma_r^\oplus$ values lie between 1.1 and 2.5 mJ/m^2. The mean for all the coal samples, both handled in nitrogen and handled in air, is 1.8 mJ/m^2. The relative scatter is considerable for both techniques of handling the specimens, so that the two means do not differ significantly. The interpretation of these observations is that the areas that correspond to the retreating angles contain acidic hydrogens, probably as hydroxyls; and that the oxidation that occurs during polishing (and measuring) in air does not lead to an important increase in the concentration or the acidity of the hydroxyls.

Figures 4 and 5 show the values of $\gamma^\ominus$. It appears to be reasonable to extrapolate the coal data to the graphite points at 100% carbon. It is clear that the basic parameter is strongly dependent on % carbon and on % oxygen. We can attribute the basic behavior to oxygen atoms in the surface. These are, probably, ether groups and carbonyl groups [25], and possibly carboxyl groups. Hydroxyl groups may also contribute basic behavior, by way of the unshared pair electrons of the oxygens.

In Fig. 4, the moderate downtrend trend, i.e. a decrease in both $\gamma_r^\ominus$ and $\gamma_a^\ominus$ with rank and with increasing % carbon (decreasing % oxygen) is evident. The same trend was found with samples polished and measured in air (see Fig. 5). The values of both $\gamma_a^\ominus$ and $\gamma_r^\ominus$ for low-rank coals are nearly the same in Figs 4 and 5. For high-rank coals, both curves for the samples handled in air were above the nitrogen curves. It is not possible, at this time, to separate out the contribution to $\gamma_r^\ominus$ due to aromatic π electrons from that due to basic oxygens.

The fact that $\gamma_a^\ominus$ based on the advancing angle, in Figs 4 and 5, is considerably smaller than $\gamma_r^\ominus$, but is quite significantly larger than zero (except for graphite) indicates that the lower-energy areas as well as the higher-energy areas contain significant concentrations of basic oxygen.

The relative magnitudes of $\gamma^\ominus$ and $\gamma^\oplus$, with the former being as much as an

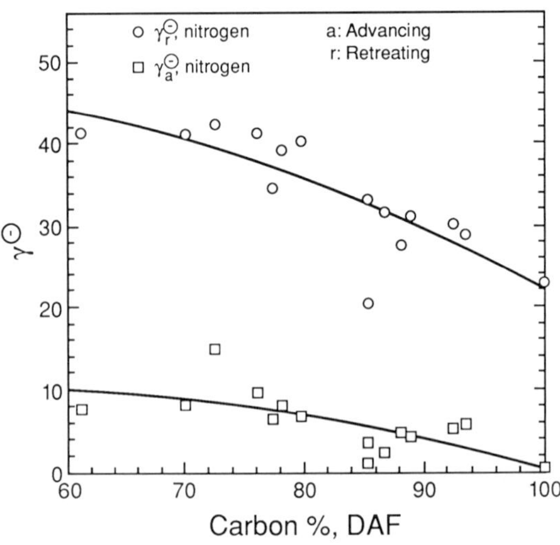

Figure 4. Lewis base component of surface free energy of coal and graphite, $\gamma^\ominus$, based on advancing and retreating contact angles, as a function of % carbon. Samples polished and measured in nitrogen.

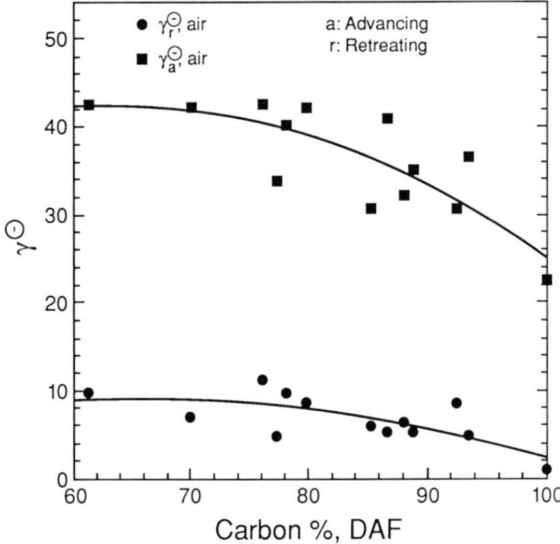

Figure 5. Lewis base component of surface free energy of coal and graphite, $\gamma^\ominus$, based on advancing and retreating angles, as a function of % carbon. Samples polished and measured in air.

order of magnitude larger, are notable. The same order of difference has been observed on polymeric surfaces such as flame-treated polyethylene [26]. See also, ref. [18].

The values of $\gamma^\ominus$ and $\gamma^\oplus$ for graphite are given in Table 3.

While further investigation of graphite is needed, it seems that the graphite surface has considerable resemblance to coal. Areas exist (quite possibly, as patches) that have higher and lower surface energy. The higher energy patches (which have nearly the same $\gamma^\ominus$, i.e. close to 23 mJ/m^2 for samples handled in N$_2$ and for those handled in air) are quite appreciably basic. The lower energy patches have very nearly zero basic character. The acidic character of both the higher and lower energy patches resembled that of coal.

Table 3.

Surface acidic and basic parameters of graphite, for two different methods of preparation (mJ/m^2)

		$\gamma^\ominus$	$\gamma^\oplus$
Solid handled in N$_2$	advancing	0.5	0.5
	retreating	23.0	1.6
Solid handled in air	advancing	1.2	0.6
	retreating	22.7	1.9

5. CONCLUSIONS

As indicated above, this study is part of an ongoing investigation of the surface chemistry of coal, in aid of applications to separation processes. It is not clear, at present, how many different coals, from different sources, will have to be examined before we can generalize confidently about 'all coals'. But the following, preliminary generalizations appear likely to hold up.

(1) The model, in which patches or bands of surface that have different surface free energy, and (no doubt) different chemical composition, will account for the large contact angle hysteresis.

(2) As far as the apolar component of surface free energy is concerned, the value of γ_a^{LW} is about 39.5 mJ/m^2 for all the coals studied, and the value of γ_r^{LW} is about 49.6 mJ/m^2, independent of rank.

(3) The suggestion can be made, that a two-phase structure exists, in the carbonaceous component (the macerals) of coal. The same appears to be true of the surface structure of graphite.

6. GENERAL COMMENT

This study has some methodological implications with respect to the study of contact angles themselves. First, there is likely to be chemical information residing in retreating contact angle data, that is different from the information in advancing contact angles for the same solid. *Therefore, retreating angles as well as advancing angles should be measured wherever possible.*

Second, the analysis of contact angle and interfacial tension data of polar liquids and solids, as acid–base phenomena, has proved to be very fruitful. For example, there was no way that anybody could have anticipated that the acidic parameter of interfacial free energy, $\gamma_s^{\oplus}$, would be nearly independent of coal rank, while the basic parameter, $\gamma_s^{\ominus}$, would show the decreasing trend that we have reported, above.

As a final generalization, we may point out that the one-parameter interpretation of contact angle data known as the 'equation of state' theory [27] is incapable of yielding any conclusions at all, in the areas where we have found results that cast considerable light on the chemistry and physics of coal surfaces. A minimum of three surface energy parameters for a polar solid are needed; and two of the parameters must be complementary to each other, as acidic and basic functions are complementary.

Acknowledgement

The authors thank the Pittsburgh Energy Technology Center, of the US Department of Energy, for support of this work.

REFERENCES

1. F. M. Fowkes, *J. Phys. Chem.* **66**, 682 (1962); *Advan. Chem. Ser.* **43**, 99 (1964).
2. F. M. Fowkes and M. A. Mostafa, *I & EC Prod. Res. Dev.* **17**, 3 (1978).
3. F. M. Fowkes, in: *Physicochemical Aspects of Polymer Surfaces*, K. L. Mittal (ed.), vol. 2, p. 583, Plenum Press, New York (1983).
4. H. W. Fox and W. A. Zisman, *J. Colloid Interface Sci.* **5**, 514 (1950).
5. R. J. Good and L. A. Girifalco, *J. Phys. Chem.* **64**, 561 (1960).
6. J. Leja, *Surface Chemistry of Froth Floatation.* Plenum Press, New York (1982).
7. D. W. Fuerstenau, G. C. C. Yang and J. S. Laskowski, *Colloids Surf.* **8**, 153 (1983).
8. D. W. Keller, Jr., *Colloids Surf.* **22**, 21 (1987).
9. D. W. Keller, Jr. and W. Bury, *Colloids Surf.* **22**, 37 (1987).
10. C. E. Capes, *Can. J. Chem. Eng.* **54**, 3 (1976).
11. B. K. Parekh and F. A. Aplan, in: *Recent Developments in Separation Science*, N. N. Li, R. B. Long, S. A. Stern and P. Somasundaran (Eds), vol. 4, pp. 107–113. CRC Press, Cleveland, OH (1978).
12. J. A. Gutierrez-Rodriguez and F. A. Aplan, *Colloids Surf.* **12**, 27 (1984).

13. C. J. van Oss, R. J. Good and M. K. Chaudhury, *J. Chromatogr.* **376**, 111 (1986).
14. C. J. van Oss, M. K. Chaudhury and R. J. Good, *Adv. Colloid Interface Sci.* **28**, 35 (1987).
15. C. J. van Oss, M. K. Chaudhury and R. J. Good, *Chem. Rev.* **88**, 927 (1988).
16. C. J. van Oss, M. K. Chaudhury and R. J. Good, *Separation Sci. Technol.* **24**, 13 (1989).
17. C. J. van Oss, R. J. Good and H. J. Busscher, *Dispersion Sci. Technol.* **11**, 77–81 (1990).
18. R. J. Good, M. K. Chaudhury and C. J. van Oss, in: *Fundamentals of Adhesion*, L. H. Lee (Ed.). Plenum Press, New York (in press).
19. A. M. Carpenter, *Coal Classification*, IEACR No. 12, IEA Coal Research, [International Energy Agency] Gemini House, 10–18 Putney Hill, London SW15 6AA, England.
20. R. C. Neavel, S. E. Smith, E. J. Hippo and R. N. Miller, *Fuel*, **65**, 312 (March 1986).
21. M. A. Elliot (Ed.), *Chemistry of Coal Utilization*, 2nd Supplementary Volume. John Wiley and Sons, New York (1981).
22. R. J. Good and M. K. Chaudhury, in: *Fundamentals of Adhesion*, L. H. Lee (Ed.), Plenum Press, New York (in press).
23. L. A. Girifalco and R. J. Good, *J. Phys. Chem.* **61**, 904 (1957).
24. A. W. Neumann and R. J. Good, in: *Colloid and Surface Science*, R. J. Good and R. R. Stromberg (Eds), vol. 11. Plenum Press, New York (1979).
25. L. M. Stock, in: *New Trends in Coal Science*, Y. Yürüm (Ed.), NATO ASI Series No. 244. Kluwer Academic Publishers, Dordrecht (1987).
26. L. K. Shu, unpublished work in this Laboratory, 1989–90.
27. J. K. Spelt, D. R. Absolom and A. W. Neumann, *Langmuir* **2**, 620 (1986).

Part 2: Characterization of the Acid-Base Properties of Materials

Acid-Base Interactions, pp. 93-115
Eds. K.L. Mittal and H.R. Anderson, Jr.
©VSP 1991

Quantitative characterization of the acid–base properties of solvents, polymers, and inorganic surfaces

FREDERICK M. FOWKES

Department of Chemistry, Lehigh University, Bethlehem, PA 18015, USA

Revised version received 16 June 1990

Abstract—The growing realization of the importance of intermolecular acid–base interactions in promoting the solubility, adsorption, and adhesion of polymers to other materials has caused a demand for the quantitative characterization of the acid–base properties of the commonly used solvents, polymers, and inorganic fillers and substrates.

There have been several recent advances in the measurement techniques for such determinations, especially in the fields of inverse gas chromatography, microcalorimetry, ellipsometry, FTIR, NMR, and XPS spectroscopy, all leading to the capability of determining the Drago E and C constants or the Gutmann acceptor numbers (AN) or donor numbers (DN) for the acidic or basic sites of solvents, polymers, or inorganic surfaces. In the last year, new studies have also allowed the characterization of the specific acid–base cohesive interactions in solvents and polymers, and the determination, from contact angle measurements on polymers, of the surface concentration and strength of acidic and basic surface sites. All of these techniques are discussed in this paper and it is expected that they will soon become standard laboratory practices.

Keywords: Acid–base interactions; acid–base spectroscopy; acid–base calorimetry; donor–acceptor interactions; molecular interactions; polymer solubility; polymer adhesion.

1. INTRODUCTION

The quantitative understanding of the intermolecular forces in liquids and at interfaces has been developing over the past century in the fields of solution and surface chemistry, chromatography, and spectroscopy. In liquids and at interfaces the intermolecular forces are of a somewhat different nature than the van der Waals forces in gases, which were so well characterized by Keesom, Debye, and London in the 1920s as depending on dipole interactions and on the quantum-mechanical charge-density fluctuations of the London dispersion forces. In liquids and at interfaces the contribution of dipole interactions to molecular interactions is much smaller than in gases; for example, in water vapor 76% of the cohesive energy comes from dipole interactions, but in liquid water only 2.4% of the cohesive energy depends in any way on dipoles. In liquids and at interfaces the dispersion force contribution to the van der Waals intermolecular interactions is always of appreciable magnitude. In some cases (such as water), there is also a small contribution related to the static dielectric constant [1], but in 'polar' liquids such as water the main contribution is from hydrogen bonding, an acid–base interaction completely unrelated to dipole moments [2].

Acid–base interactions (such as hydrogen bonding) provide an appreciable fraction of the cohesive energy in 'polar' liquids. These interactions require both acidic (electron-accepting) sites and basic (electron-donating) sites. Liquids which are basic but have negligible acidity (or which are acidic but have negligible basicity) have no specific self-association, and their cohesive energy and surface tension are a function of only their dispersion force interactions [3]. Van Oss *et al.* have invented the term 'monopolar' to describe materials which have only acidic or only basic sites, and the term 'bipolar' to describe materials which have both types of sites [4].

Experimental tests to distinguish bipolar from monopolar liquids include the solubility in saturated liquid hydrocarbons, for saturated hydrocarbons have neither acidic nor basic sites, and tend to dissolve only very weakly self-associated organic liquids [5]. For instance, dimethyl sulfoxide (DMSO) has both acidic and basic sites, as evidenced by appreciable heats of acid–base interaction with either acids or bases (-29.8 kcal/mol with Gutmann's test acid, $SbCl_5$ in 1,2-dichloroethane; and -3.1 kcal/mol with Gutmann's test base, triethylphosphine oxide) [6, 7]. The self-association of DMSO has been discussed in some detail by Gutmann [8], so it is not surprising that DMSO is insoluble in saturated hydrocarbons (hexane, cyclohexane, decahydronaphthalene, and squalane). In Drago's studies of DMSO, attention was centered on only its basicity, just as in his studies of hydrogen-bonded liquids, where only the acidic character was assessed [9]. Some investigators have assumed (incorrectly) that if only the basicity of a compound was assessed in Drago's studies the compound therefore has no acidic sites. In the van Oss *et al.*'s contact angle studies of acid–base interactions, DMSO was considered (incorrectly) to be a monopolar base [4]. Such errors by very competent investigators indicate that laboratory tests are urgently needed for determining whether a compound is acidic, basic, 'monopolar', or 'bipolar'. Such definitions are especially important for the acidic or basic test liquids which are used to determine the acid–base character of other materials, a need addressed in this paper.

2. THE DRAGO AND GUTMANN DETERMINATIONS OF LEWIS ACIDITY AND BASICITY

Lewis originally defined acids as electron acceptors and bases as electron donors [10]; more recently, Mulliken received the Nobel prize for his characterization of acid–base interactions as charge-transfer complexes in which there are two contributions to the energy of interaction: electrostatic and covalent [11]. The understanding of the importance of the electrostatic and covalent contributions led to the hard–soft acid–base (HSAB) concepts of Pearson [12], and the four-constant 'E and C' equation of Drago, where there are terms for the electrostatic (E) and covalent (C) contributions to the heats of complexation of acid A with base B [9]:

$$-\Delta H = C_A C_B + E_A E_B. \tag{1}$$

Pearson's studies concerned the equilibrium constants and the rates of complexation of aqueous anions, cations, and ligands, while Drago's studies concerned the heats of acid–base complexation in organic liquids. In the latter studies, the E and C constants were determined for about 40 acids and 40 bases, but, as already mentioned, Drago chose to ignore the basic sites of hydrogen-bonding liquids such as water and alcohols, and to ignore the acidic sites of amines, amides, ketones, and DMSO.

Mayer and Gutmann [6] were more concerned with the acid–base self-association of organic liquids, and sought to characterize both functionalities of organic liquids, for they taught that all organic liquids have some degree of specific self-association. Basic strengths were characterized by the calorimetric heats of the acid–base interaction of organic compounds with antimony pentachloride in dilute solutions in 1,2-dichloroethane, and these were reported as the donor number (DN) in kcal/mol. Acceptor numbers (AN) were determined from the chemical shifts of ^{31}P-NMR spectra of a basic probe molecule (triethylphosphine oxide, Et_3PO) in the tested liquid. The NMR shift observed for Et_3PO in *n*-hexane was assigned an AN of 0, and the shift observed for Et_3PO in a dilute solution of $SbCl_5$ in 1,2-dichloroethane was assigned an AN of 100.

The calorimetric heats of acid–base complexation used in the determination of the Drago *E* and *C* constants, and in the determination of the donor numbers of Mayer and Gutmann, were all measured in dilute solution in a non-interacting solvent (cyclohexane and CCl_4 in Drago's work, and 1,2-dichloroethane in the work of Mayer and Gutmann). In all solutions the solute molecules have van der Waals interactions with neighbor molecules, with appreciable heats and spectral shifts due to the van der Waals interactions. However, when acid–base interactions are measured in dilute solution in a non-interacting solvent, the van der Waals interactions of the tested molecules and of the test molecules are confined almost entirely to the solvent, so the measured heats of interaction have no contribution owing to changes in the van der Waals interactions with neighbor molecules. However, in the determination of acceptor numbers by Mayer and Gutmann the ^{31}P-NMR spectral shifts were not determined in solutions of the tested liquids, but in the liquids themselves, so that the spectral shifts resulted from van der Waals interactions as well as from acid–base interactions. The van der Waals contributions to AN values (AN^d) have been calculated from surface tension measurements, and AN values have been corrected for the van der Waals contributions ($AN - AN^d$), together with AN^* values (proportional to $AN - AN^d$) which give the heat of acid–base complexation of the tested liquid with Et_3PO in kcal/mol [7].

It should be noted that the Mayer and Gutmann donor-number and acceptor-number approach gives single-valued scales of acidity and basicity, and ignores the concepts of Mulliken, Pearson, and Drago concerning the importance of both the covalent and the electrostatic contributions to acidity and basicity. The donor number determined by the heats of interaction of tested liquids with $SbCl_5$, a very soft acid, gives an order of basic strengths in which the basicity of softer bases (nitrogen and sulfur bases) is much greater than if a hard test acid such as phenol had been used. Similarly, the corrected acceptor number, AN^*, determined by heats of interaction of tested liquids with Et_3PO, a hard base, will give an order of acid strength which favors hard (hydrogen) acids much more than if a soft base (such as Et_3PS) had been used as the test base.

3. HEATS OF ACID-BASE SELF-ASSOCIATION OF ACIDIC AND BASIC SOLVENTS

A most important aspect of acid–base studies is to be able to determine whether a solvent, polymer, or other material is neutral, acidic, basic, or amphipathic (both acidic and basic), termed by van Oss, Good, and Chaudhury as 'bipolar'.

Investigators rushing into acid–base studies have usually been concerned only with the acidity of their model acids and with the basicity of their model bases. This section reports current studies which show that all solvents (except saturated hydrocarbons) have some amphipathic character, and that the degree of amphipathic behavior can be quantitatively characterized by calorimetric determination of the acid–base and van der Waals contributions to heats of vaporization, $\Delta H_{\text{vap}}^{\text{ab}}$ and $\Delta H_{\text{vap}}^{\text{d}}$, respectively.

Gutmann contended that all organic solvents have some degree of self-association resulting from the presence of both acidity and basicity (DN and AN numbers). Such a contention can now be tested using calorimetric findings of heats of vaporization of a liquid to be tested, and of heats of mixing of such a tested liquid into a neutral hydrocarbon liquid such as n-hexane [13]. The energy of vaporization of a liquid has two contributions: a van der Waals contribution ($\Delta U_{\text{vap}}^{\text{d}}$; where nearly all of this contribution is due to dispersion force interactions) and an acid–base contribution ($\Delta H_{\text{vap}}^{\text{ab}}$), resulting from the dissociation of the acid–base complexes upon evaporation. Self-associated liquids (1) have appreciable $\Delta U_{\text{vap}}^{\text{ab}}$ values, and when these liquids are mixed into hexane or cyclohexane, the limiting heat of mixing at zero concentration of such a tested liquid ($\Delta H_{1,0}^{\text{M}}$) is proposed to be equal, but of opposite sign, to the energy of dissociation of such complexes upon evaporation:

$$\Delta U_{\text{vap}}^{\text{ab}} = -\Delta H_{1,0}^{\text{M}}. \tag{2}$$

Solubility parameters ($\delta_i = [\Delta U_i^{\text{vap}}/\overline{V}_i] = [\{\Delta H_i^{\text{vap}} - RT\}/\overline{V}_i]^{1/2}$), where $\overline{V}_i$ is the partial molar volume, are useful for predicing heats of mixing of liquids 1 and 2 when only London dispersion force interactions are involved, because then the ΔU of mixing (usually referred to as the 'heat of mixing') can be predicted by Hildebrand's equations [14]:

$$\Delta U_{12}^{\text{M}} = \overline{V}_1 \overline{V}_2 \phi_1 \phi_2 (\delta_1 - \delta_2)^2, \tag{3}$$

where ϕ_i stands for the volume fraction of component i, and the partial molar energies of mixing are given by

$$\Delta \overline{U}_1^{\text{M}} = \overline{V}_1 \phi_2^2 (\delta_1 - \delta_2)^2, \tag{4}$$

and its limiting value as $x_1 \to 0$ is

$$\Delta \overline{U}_{1,0}^{\text{M}} = V_1 (\delta_1 - \delta_2)^2. \tag{5}$$

The acid–base self-association of acetone is a good example. The limiting value at $x_1 = 0$ for the partial molar heat of mixing of acetone in n-hexane is predicted by equation (5) to be $+1.70$ kJ/mol, but calorimetry (Fig. 1) shows that the experimental value of the limiting slope at $x_1 = 0$ is five times greater ($+9.02$ kJ/mol). This discrepancy results from the specific acid–base self-association of acetone, which raises the energy of vaporization well above that resulting from van der Waals interactions.

The energy of acid–base self-association (ΔU_1^{ab}) of acetone (1) can be calculated from the heat of mixing into a non-associated liquid such as hexane (2), and from the molar volumes and energies of vaporization ($V_i \delta_i^2$) of the two liquids:

$$\Delta U_{1,0}^{\text{M}} = \Delta U_1^{\text{ab}} + \Delta U_1^{\text{d}} = (\Delta U_1^{\text{vap}} - \Delta U_1^{\text{d,vap}}) + V_1 (\delta_1^{\text{d}} - \delta_2^{\text{d}})^2, \tag{6}$$

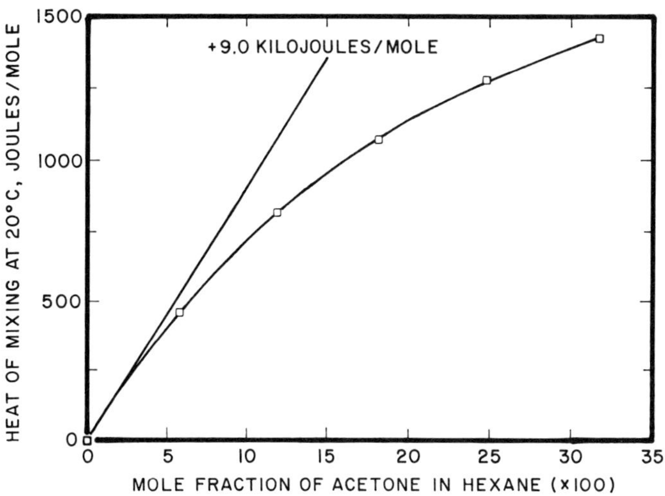

Figure 1. Heat of mixing of acetone into *n*-hexane [13].

where

$$\delta_1^d = (\Delta U_1^{d,vap}/V_1)^{1/2}. \tag{7}$$

Therefore

$$\Delta U_1^{d,vap} = (V_1\delta_2^2 - \Delta H_{1,0}^M + \Delta U_1^{vap})^2/4V_1\delta_2^2 \tag{8}$$

and

$$\delta_1^d = (\delta_2 + \delta_1^2/\delta_2 - \Delta H_{1,0}^M/V_1\delta_2)/2. \tag{9}$$

For the case of acetone (1) added to *n*-hexane (2), $\Delta U_{1,0}^M = +9.02$ kJ/mol, $\Delta U_1^{vap} = 28.7$ kJ/mol, and $V_1 = 74$ ml. Equation (6) gives $\Delta U_1^{d,vap} = 19.9$ kJ/mol, and $\Delta U_1^{ab} = 28.7 - 19.9 = 8.8$ kJ/mol, the energy of intermolecular specific (acid–base) interaction in acetone. The corrected solubility parameter δ_1^d is $(19\,900/74)^{1/2} = 16.4$, as compared with the 19.7 $(J/ml)^{1/2}$ for the total solubility parameter (δ_1^t) shown in Table 1. This value of δ_1^d is very precise, and is not related in any way to the values of δ^d found in tables of 'three-dimensional solubility parameters' [15].

In the same way, we can determine the values of ΔU_{vap}^d, ΔU_{vap}^{ab}, and δ_1^d for other liquids for which we have the limiting heats of mixing at infinite dilution in *n*-hexane or cyclohexane $(\Delta U_{1,0}^M)$, and for which we know the energy of vaporization (ΔU_1^{vap}) and molar volume (V_1), as shown in Table 1. Most of the data needed for these calculations are available from tabulations of the measurements of other investigators [16], but the calculations in Table 1 are novel in this report.

In Table 1 the acidic or basic liquids are listed in decreasing degree of acid–base self-association, expressed as the fraction of the heat of vaporization at 25°C required to dissociate the acid–base bonds. It is seen that *n*-butanol has the highest degree of self-association (46%) in this list, but it may come as a surprise that acetone has 31%, THF has 27%, *n*-butylamine 22%, ethyl acetate 18%, and even benzene 8%. The acetone results were expected, for it had been observed

Table 1.
Contributions of van der Waals and of acid–base interactions to energies of vaporization of organic liquids in kJ/mol, and solubility parameters in $(J/ml)^{1/2}$. Listed in order of decreasing self-association (%SA)

Liquid	ΔU_{vap}	ΔU_{vap}^{d}	ΔU_{vap}^{ab}	V (ml)	δ^{t}	δ^{d}	%SA
n-Butanol	51.68	27.7	24.0	92.8	23.7	17.4	46
Acetone	28.72	19.9	8.8	74.0	19.7	16.4	31
Tetrahydrofuran	28.37	20.8	7.6	81.8	18.6	17.6	27
n-Butylamine	22.90	17.8	5.1	73.1	17.7	15.6	22
Ethyl acetate	32.46	26.8	5.7	98.5	18.2	16.5	18
Pyridine	38.85	32.8	6.1	80.9	21.9	20.4	16
Benzene	31.23	28.8	2.4	89.3	18.7	18.0	8
Ethyl ether	24.90	23.3	1.6	104.8	15.4	14.9	6
Chloroform	28.32	27.9	0.46	80.7	18.7	18.55	1.6
CCl_4	29.89	29.4	0.46	96.5	17.6	17.46	1.6
Triethylamine	32.35	32.3	0.08	140.0	15.2	15.18	0.1

that the carbonyl stretching frequency of acetone shifted from 1720 to 1728 cm^{-1} upon dilution to about 100 ppm in cyclohexane [17], and this corresponded to a heat of acid–base dissociation of $+8$ kJ/mol. The results for benzene were also expected, for in recent studies by Jorgensen at Purdue University [18], he and others showed that aromatics self-associate in a T-shaped 'hydrogen-bonded' complex.

Table 1 will eventually be extended to include many candidate acid–base probes, for liquids chosen to test the acidity or basicity of other materials should be 'monopolar' (able to interact either as acids or as bases, but not both). Thus, triethylamine is seen to have very little acid–base self-association and is a good base for assessing the acidity of other materials; but acetone, THF, and ethyl acetate are seen to be to some degree amphipathic ('bipolar'), having the ability to interact by either functionality.

4. ACID–BASE PROPERTIES OF POLYMER SOLUTIONS

All polymers, except saturated hydrocarbons such as polyethylene and polypropylene, have acidic or basic functional sites. Just as ester solvents are predominantly basic but have sufficient acidity to result in measurable acid–base self-association as shown for ethyl acetate in Table 1, so polyesters such as polymethylmethacrylate (PMMA) are predominantly basic but also have sufficiently acidic sites so that some degree of self-association is observed. Similarly, polyvinyl chloride has the predominant acidity of methylene chloride, but also some degree of self-association due to the weak, but undeniable basicity of the fluorine atoms. The acid–base properties of polymers can be determined on solid samples, but there are several advantages in making such measurements in solution.

4.1. Calorimetry

The calorimetric titration of the acidic or basic sites of polymers can be done with polymers that are sufficiently soluble in fairly neutral solvents such as xylene, carbon tetrachloride, or tetrahydronaphthalene. In order to eliminate van der Waals contributions to such heats of mixing, both the polymer and the acidic or basic titrant should be in as inert a solvent as can be used. The polymer solution should be titrated step-wise into a solution of a test acid or base in the same solvent, so that the limiting partial molar heats of mixing (per mol of repeat units) can be determined at infinite dilution, as was done in Fig. 1 for acetone in *n*-hexane. Swain has done some measurements of this kind with polypropylene oxide dissolved in benzene [19], enough to show that such measurements are quite possible.

4.2. FTIR and NMR spectral shifts

Spectral shift techniques appear to have several advantages for determining heats of acid–base interaction of polymers with test acids or bases, from which the Drago E and C constants or the Gutmann donor and acceptor numbers can be determined. The infra-red shifts of carbonyl groups ($\Delta v_{C=O}$) can be used for measuring the acid–base complexation of the ester groups of polyesters and polyacrylates, of the carbonate group of polycarbonates, of the amide groups of polyamides, and of the amide, ester, carbonate, and ketone groups of test solvents. The initial studies were carried out with ethyl acetate, and it was found that the carbonyl stretching frequency was shifted from the 1764 cm^{-1} observed in ethyl acetate vapor as a result of intermolecular interactions with solvent molecules, which included both a van der Waals contribution and an acid–base contribution. The van der Waals (dispersion force) contribution (Δv^{d}) is proportional to the van der Waals contribution to surface tension or surface free energy (γ^{d}), and the acid–base contribution (Δv^{ab}) is proportional to the heat of acid–base interaction (ΔH^{ab}) of ethyl acetate with the tested acid [17]:

$$v_{C=O} = 1764 \text{ cm}^{-1} - 0.714 \, \gamma^{d} \text{ cm}^{-1}/\text{mJ/m}^2 + 0.99 \, \Delta H^{ab} \text{ cm}^{-1}/\text{kJ/mol}. \qquad (10)$$

The proportionality to γ^{d} was established with neutral or non-acidic liquids, and the proportionality to ΔH^{ab} was established with a series of acids or acidic solutions of known calorimetric heats of acid–base complexation with ethyl acetate, and having nearly the same γ^{d} (26 ± 1 mJ/m^2). Figure 2 illustrates the acid–base shift of the carbonyl stretching frequency of ethyl acetate and of polymethylmethacrylate (PMMA), where it is seen that the acid–base shifts from PMMA complexation with each acid tested are directly proportional to the shifts observed with ethyl acetate. The heats of acid–base interaction of PMMA with various test acids, determined from the infra-red shifts of PMMA complexes in solution, were used to calculate the Drago constants for PMMA: $C_B = 0.96 \pm 0.07$, $E_B = 0.68 \pm 0.01$. Such Drago E and C constants for polymers are needed, but few are available as yet.

Infra-red shifts upon complexation of polymers with test acids or bases have been helpful in determining the relative acidity or basicity of polymers or of solvents. The spectral shifts of acidic polymers upon complexation with ethyl acetate indicate that acidic strength decreases in the order polyvinyl fluoride > polyvinylidene fluoride > polyvinyl butyral > polyvinyl chloride [17].

The calorimetric calibration of the carbonyl shift in acid–base complexation (4.24 cm^{-1}/kcal/mol or 0.99 cm^{-1}/kJ/mol) was determined originally with ethyl acetate, and since all acid–base carbonyl shifts for acid complexation of PMMA

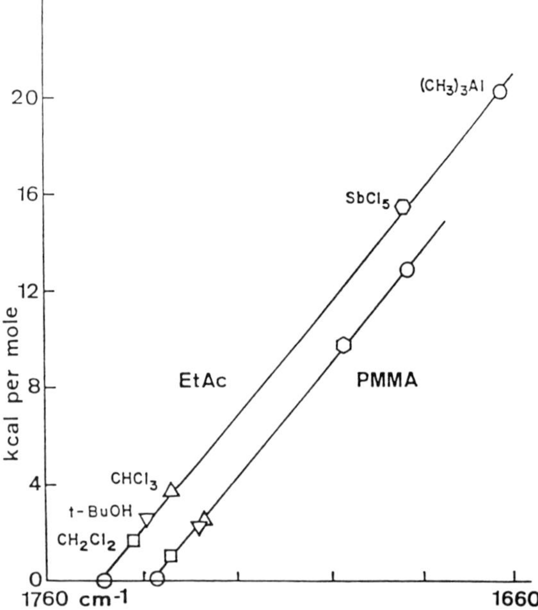

Figure 2. Heats of acid–base interaction of ethyl acetate and of PMMA with various test acids vs. their carbonyl stretching frequencies. Published with permission of John Wiley & Sons, copyright owners [17].

were proportional to the ethyl acetate shifts with the same acids, it was assumed that the same constant applied to all carbonyl groups [17]. This assumption was checked by Kwei *et al.* [20] and found to be acceptable. In that study, infra-red shifts of OH groups and of C=O groups were determined for an acidic co-polymer of styrene and vinylphenyl hexafluorodimethylcarbinol forming acid–base (hydrogen-bonded) complexes with a wide variety of basic polymers (various polyesters, two polyethers, a polynitrile, and a polycarbonate). Of special interest is that polyvinylmethyl ether was the strongest basic polymer ($\Delta H^{ab} = 8.0$ kcal/mol), a much stronger base than the aromatic ether poly-phenylene oxide ($\Delta H^{ab} = 5.3$ kcal/mol).

Infra-red spectral shifts for the measurement of the heats of acid–base inter-action of test acid–base probes with solid polymer powders have also been determined with photoacoustic FTIR techniques [21]. In these studies the infra-red spectrum of a polymer powder was determined by the photoacoustic technique; the powder was then exposed in a closed desiccator to test acidic or basic vapors for 4 h (which penetrated the surface region and complexed with the functional sites of the polymer); and then new photoacoustic scans were made. Mirror velocities were set to obtain spectra from polymer functional sites within about 7 μm from the surface. The observed infra-red shifts allowed calcu-lation of the heats of acid–base complexation, from which Drago E and C constants could be calculated. Table 2 shows results with three polymers: a phenoxy resin from Union Carbide, polyethylene oxide (PEO), and PMMA. The downfield shifts of the acidic OH stretching frequency for the phenoxy resin were measured upon complexation with pyridine, dimethyl formamide, and

Table 2.
Photoacoustic measurement of Drago E and C constants of polymers [21]

Polymer	Test probe	IR shift (cm^{-1})	ΔH^{ab} (kcal/mol)	E and C values (kcal/mol)$^{1/2}$
Polyethylene oxide	t-Butanol	-19	-3.28	$E_B = 0.77 \pm 0.03$
	CDCl$_3$	-32	-3.40	$C_B = 5.6 \pm 0.3$
	Phenol	-274	-5.90	
	t-Butylphenol	-212	-5.26	
PMMA	Chloroform	-8.5	-2.0	$E_B = 0.45 \pm 0.1$
	Iodine	-7	-1.65	$C_B = 1.35 \pm 0.1$
	SbCl$_5$	-48	-11.3	
	Phenol	-10.5	-2.1	
Phenoxy resin	Pyridine	-228	-5.4	$E_A = 1.53 \pm 0.05$
	Dimethyl formamide	-193	-5.07	$C_A = 0.24 \pm 0.01$
	Trimethylamine	-158	-4.7	

trimethylamine. The enthalpy change of PEO upon complexation with acidic probes (t-butanol, phenol, t-butylphenol, and CDCl$_3$) was measured from the downfield shifts of the OH and CD stretching frequencies for these acidic probes, and the enthalpy change of PMMA upon complexation with chloroform, phenol, iodine, and antimony pentachloride was measured by the downfield shifts of the carbonyl group of PMMA upon complexation with these acids.

The basic probe of Gutmann's NMR studies, Et$_3$PO, has also been calibrated for its infra-red shifts upon forming acid–base complexes in liquids of surface tension γ_L^d [7, 22]:

$$\nu_{P=O} = 1210\ \text{cm}^{-1} - 1.06\ (\text{cm}^{-1}/\text{mJ/m}^2)\gamma_L^d + 0.646\ (\text{cm}^{-1}/\text{kJ/mol})\Delta H^{ab}. \quad (11)$$

The dispersive shift was calibrated with a series of neutral or non-acidic liquids having a wide range of surface tensions, and the acid–base shift was calibrated from the shift of the OH or NH stretching frequencies of acid–base complexes with Et$_3$PO with a series of hydrogen acids in carbon tetrachloride solutions.

Quite similar results have been obtained for the ^{31}P-NMR shifts of Et$_3$PO which measure electron density changes in the phosphorus atom of the P=O group, in which the oxygen is a basic site which complexes strongly with acids. The electron density of the ^{31}P decreases upon complexation with acids, but it also has an appreciable decrease due to van der Waals interactions with solvents. The NMR peak position (relative to the ^{31}P-NMR peak of an external diphenyl-phosphinic chloride standard) is [7, 22]

$$\delta(\text{Et}_3\text{PO}) = -7.37\ \text{ppm} + 0.312\ (\text{ppm/mJ/m}^2)\gamma_L^d - 0.145\ (\text{ppm/kJ/mol})\Delta H^{ab}. \quad (12)$$

The dispersion force shift and the acid–base shift were calibrated with the same solutions as those used for calibration of the infra-red shifts of Et$_3$PO. Figure 3 illustrates measurement with the ^{31}P-NMR shift method of the acid strength of two polymers: an epoxy resin and a phenolic resin. The NMR shifts are extrapolated towards infinite dilution of Et$_3$PO, and this figure is taken from the most dilute solution, where the signal-to-noise ratio is rather low. The phenolic resin was supposedly allyl-capped, but the acid strength determined (in

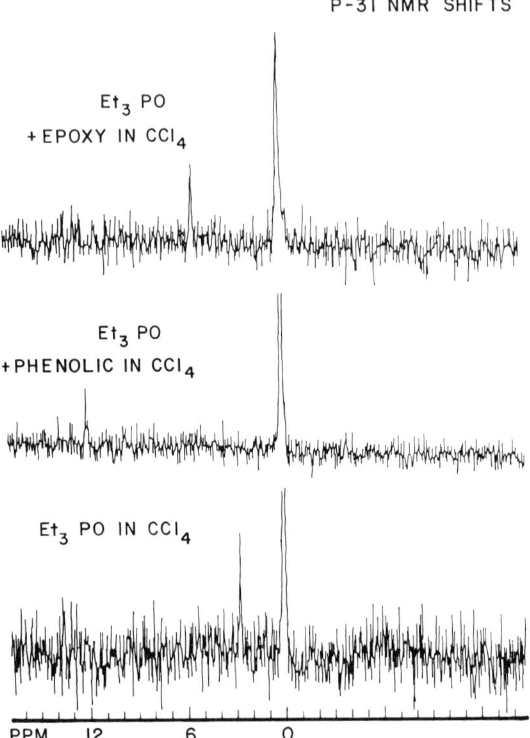

Figure 3. ^{31}P-NMR shifts of Et$_3$PO upon interaction with the acidic sites of an epoxy resin and a phenolic resin in CCl$_4$ solutions [22].

CCl$_4$ solution) was as strong as that of phenol. The secondary alcohol groups of the epoxy resin Epon 2004 are seen to be appreciably weaker acid sites, but nevertheless of definite acidity. From the measured ^{31}P-NMR spectral shift of Et$_3$PO and of tricyclohexylphosphine sulfide in carbon tetrachloride solutions of Epon 2004, the Drago E_A and C_A constants were determined to be 1.72 and 0.29 (kcal/mol)$^{1/2}$ [22].

A series of NMR test probes are under development for determining the hard–soft acid–base (HSAB) properties of solvents, polymers, and inorganic solids. These include a hard and soft base [Et$_3$PO and (C$_6$H$_{12}$)$_3$PS], and a hard and a soft acid (triphenyl silanol and diphenyl mercury). With these probes the E and C constants are now being determined for ten polymers [22].

5. ACID–BASE PROPERTIES OF FLAT SURFACES OF POLYMERS AND INORGANIC SOLIDS

There is much need of methods for the direct determination of the acid–base properties of polymer films or of other flat surfaces. In the case of hydrophobic surfaces, as with polymers and hydrophobic minerals (graphite, talc, metal sulfides), contact angle measurements have been the traditional method for surface study, but recently ellipsometric methods have been developed for measurement of the adsorption of test acids or bases out of neutral solvents onto any flat reflective surfaces, not just hydrophobic surfaces. Either technique provides determination of surface free energy changes resulting from acid–base interaction, determination of surface concentrations of acidic or basic sites, and determination of molar heats of acid–base interaction of surface sites with test acids or bases.

5.1. Contact angle studies with pure liquids

The use of contact angles (θ) on a solid S of liquid L with surface tension γ_L to calculate the work of adhesion W_{SL} is the starting point of contact angle measurements, and this work of adhesion is now known to have a van der Waals and an acid–base component [3]:

$$W_{SL} = \gamma_L(1 + \cos \theta) = W_{SL}^d + W_{SL}^{ab} = 2(\gamma_S^d \gamma_L^d)^{1/2} + W_{SL}^{ab}, \qquad (13)$$

in which the dispersive (van der Waals) component of the work of adhesion is determined from the geometric mean of the dispersive components of the surface energies of the solid S and liquid L [3]. The dispersive component of the work of adhesion has been calculated so successfully from the geometric mean of the dispersive components of the liquid and solid surface energies that several investigators tried to calculate the acid–base ('polar') contribution with such an equation. For instance, water has a total surface free energy of 72.8 mJ/m^2 at 20°C, of which $\gamma_L^d = 22.0$ mJ/m^2, and the balance (50.8 mJ/m^2) is the acid–base contribution, γ_L^{ab}, which has often been referred to as the polar contribution γ_L^p. The values for γ_L^d can be determined with equation (13) when the solid surface has only dispersive interactions (i.e. $W_{SL}^{ab} = 0$), but they are more easily and more accurately determined from interfacial tensions of a polar liquid (1) against a neutral hydrocarbon (2) [3, 5, 7]:

$$W_{12} = \gamma_1 + \gamma_2 - \gamma_{12} = W_{12}^d + W_{12}^{ab} = 2(\gamma_1^d \gamma_2^d)^{1/2} + W_{12}^{ab}. \qquad (14)$$

By first determining the γ_1^d value for polar liquids from interfacial tensions against a neutral liquid such as squalane, one can then determine W_{12}^{ab}, provided the two polar liquids are immiscible [5].

The common error in analysis of 'wetting energetics' is to assume that the geometric mean expression can be used for polar (or acid–base) interactions, as in $W_{12}^p = 2(\gamma_1^p \gamma_2^p)^{1/2}$. This equation (unfortunately called the 'extended Fowkes equation') by Owens and Wendt [23] and used extensively by Kaelble [24] is quite incorrect because the 'polar' (or acid–base) contribution to the surface energy γ^p is usually a very inadequate measure of 'polarity' or hydrophilicity. Striking examples are acetone and ethanol, which have γ^p values of only 0.2 and 1.1 mJ/m^2, respectively. Despite these very low values of γ_p, acetone and ethanol are very hydrophilic and interact so strongly with water that they are miscible in all proportions. If we were to use the geometric mean expression for estimating the work of adhesion and interfacial tension between water (1) and acetone (2), $W_{12}^p = 2(\gamma_1^p \gamma_2^p)^{1/2} = 2(0.2 \times 50.8)^{1/2} = 6.4$ mJ/m^2, and substitute this value into equation (14) where $W_{12}^d = 2(\gamma_1^d \gamma_2^d)^{1/2} = 2(22.0 \times 23.6)^{1/2} = 45.6$ mJ/m^2, these two liquids would be predicted to be completely immiscible with an interfacial tension γ_{12} of 44.6 mJ/m^2, a completely ridiculous prediction. The γ_2^p of 0.2 for acetone is quite correct, but not at all predictive of the extent of acid–base self-association in acetone shown in Table 1. Obviously, γ_2^p is quite useless for predicting the strength of hydrogen bonding of acetone to water. No investigators are using $W_{12}^p = 2(\gamma_1^p \gamma_2^p)^{1/2}$ for liquid–liquid systems because measured interfacial tensions show such an equation to be ridiculously incorrect, but hundreds of investigators are using the same equation for estimating the values of W_{SL}^p for solid/liquid interfaces, largely because solid/liquid interfacial tensions γ_{SL} cannot be measured experimentally and the errors of these very incorrect predictions are never detected.

It is very important to be able to predict values of W_{SL}^{ab} at solid/liquid or solid/solid interfaces, and two methods may be recommended: the 1978 method of Fowkes and Mostafa [25], and the 1988 method of van Oss *et al.* [4]. The former method makes use of the molar heats of acid–base interaction between the two materials in contact, ΔH_{SL}^{ab}, and the interfacial concentration, n_{ab}, of interfacial acid–base bonds (in mol/m^2):

$$W_{SL}^{ab} = -f n_{ab} \Delta H^{ab}, \tag{15}$$

where f is a constant close to unity for converting heats of interfacial acid–base interaction into free energies of interfacial acid–base interaction. This method was tested with the benzene/water interface, using Drago's E_A and C_A constants for water [5.01 and 0.67 (kJ/mol)$^{1/2}$, respectively] and the E_B and C_B constants for benzene [0.75 and 1.8 (kJ/mol)$^{1/2}$, respectively]; these predict a ΔH^{ab} of -5.0 kJ/mol. If each interfacial benzene molecule lies flat at the interface, occupying 0.50 nm^2 each, $n_{ab} = 3.3$ μmol/m^2, and $W_{12}^{ab}/f = 16.5$ mJ/m^2, which is very close to the value measured at 20°C:

$$W_{12}^{ab} = \gamma_1 + \gamma_2 - \gamma_{12} - 2(\gamma_1^d \gamma_2^d)^{1/2}$$
$$= 72.8 + 28.9 - 35.0 - 2(22 \times 28.9)^{1/2} = 16.3 \text{ mJ/m}^2.$$

The method of van Oss *et al.* also recognizes that interfacial acid–base interaction requires that acidic sites of one phase interact with basic sites of the other, so that if either phase is neutral or if both phases have only basic or only acidic sites, there can be no acid–base interaction. In this approach the form of the geometric mean equation remains, but this is not a geometric mean of the 'polar' (acid–base) contribution to the surface free energy:

$$W_{12}^{ab} = 2(\gamma_1^{\ominus} \gamma_2^{\oplus})^{1/2} + 2(\gamma_1^{\oplus} \gamma_2^{\ominus})^{1/2}, \tag{16}$$

where $\gamma_1^{\ominus}$ is a measure of the surface basicity of phase 1, $\gamma_1^{\oplus}$ is a measure of the surface acidity of phase 1, etc. The units of these functions are the same as those for the surface energy (mJ/m^2), and the values of these functions can be determined from measured values of W_{12}^{ab} on the assumption that for water

$$\gamma_1^{\ominus} = \gamma_1^{\oplus} = \gamma_1^{ab}/2 = 25.4 \text{ mJ/m}^2 \text{ at } 20°C. \tag{17}$$

While this approach avoids the difficulties of using the geometric mean of the actual polar components of the surface free energy as in the 'extended Fowkes equation', accurate values of $\gamma^{\ominus}$ and $\gamma^{\oplus}$ are not easy to determine with accuracy, for the degree of acidity or basicity of the test liquids is still under investigation (as in Table 1), and has not yet been established. Furthermore, the use of a single parameter to characterize acidity or basicity ignores the hard and soft character of acids and bases, as with Gutmann's acceptor numbers or donor numbers.

5.2. Contact angle studies with solutions of acids or bases

Contact angle studies of surface acidity and basicity can also be done on hydrophobic surfaces with solutions of test acids or test bases in a neutral liquid of high surface tension, such as methylene iodide (CH_2I_2), which has a surface free energy at 20°C of 49.6 mJ/m^2 and which is a good solvent for test acids or bases. By determining the dependence of the adhesion tension ($\gamma_L \cos \theta$) on the concentration c_2 of the test acid or base ($d\gamma_L \cos \theta/d\ln c_2$), the Gibbs adsorption equation may be used to determine the adsorption isotherms of the test acid or base (2) [26]:

$$\Gamma_2 = -(1/RT)(d\gamma_{SL}/d\ln c_2) = (d\gamma_L \cos \theta/d\ln c_2)/RT, \qquad (18)$$

where Γ_2 is the interfacial concentration. By determining several such points of an adsorption isotherm the straight-line Langmuir plot may be determined, from which the surface concentration of acidic or basic sites (Γ_m) can be calculated and the equilibrium constant K_{eq} for adsorption determined [27]:

$$c_2/\Gamma_2 = 1/\Gamma_m K_{eq} + c_2/\Gamma_m. \qquad (19)$$

If isotherms are determined at two or more temperatures, the molar heat of adsorption of the test acid or base can be determined from the van't Hoff relation [25]:

$$\Delta H^{ab} = [RT_1 T_2/(T_2 - T_1)] \ln[(K_{eq} \text{ at } T_1)/(K_{eq} \text{ at } T_2)]. \qquad (20)$$

If one thus determines the molar heats of adsorption of hard and soft test acids or bases, the Drago E and C constants for the surface sites may be calculated.

In initial studies [28], the basic surface sites of PMMA films were assessed by contact angle studies of solutions of phenol in CH_2I_2; the addition of phenol did not decrease the surface tension measurably. At 23°C the contact angle decreased from 45° to 36.8° as the phenol concentration was increased to 20 mmol/l, which corresponded to a decrease in the interfacial free energy between PMMA and CH_2I_2 of 4.7 mJ/m^2. Treatment of the data with equations (18) and (19) determined a surface concentration of basic ester sites of only 0.55 μmol/m^2, indicating that the ester groups tend to self-associate and are largely buried below the surface [29]. Equation (20) was used with contact angle data for phenol solutions in CH_2I_2 at 10 and 23°C, and the heat of adsorption was -20 kJ/mol, in agreement with the literature [20]. Iodine has proven suitable as the soft acid for adsorption from CH_2I_2 onto PMMA for the determination of its Drago C_B and E_B constants. This method should prove useful for determining E and C constants for polymers and for determining the chemical nature of surface modifications used to enhance the wettability and adhesion of polymers.

5.3. Ellipsometric studies of acid–base adsorption on flat surfaces

Flat surfaces of polymers, metals, or oxides (whether hydrophobic or hydrophilic) may be similarly characterized with adsorption isotherms determined by ellipsometry, provided the surface is smooth enough to be sufficiently reflective. The surface is immersed in a neutral solvent as indicated in Fig. 4, and adsorption isotherms such as those for pyridine adsorbing from octane onto silica can

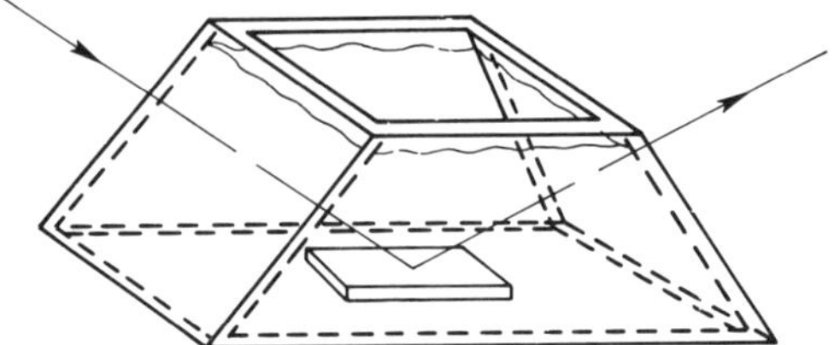

Figure 4. Ellipsometric cell for measuring test acid or base adsorption from a neutral solvent onto flat reflective specimens [30].

be determined with solutions of test acids or bases [30]. Similar studies have been carried out recently with polymer films [31]. Such adsorption isotherms for acidic or basic test molecules adsorbing from neutral solvents may be analyzed with equation (19) to determine the surface concentration of acidic or basic sites, and if the adsorption isotherms are obtained at different temperatures, equation (20) may be used to determine the heats of adsorption of these probes, from which the Drago E and C constants for the surface sites may be determined.

6. MEASUREMENT OF THE ACID–BASE SURFACE PROPERTIES OF POWDERS AND FIBERS

Powders and fibers usually have large enough specific surface areas that there is sufficient adsorption to observe color changes of adsorbed dyes, to observe infra-red or NMR spectral shifts, or to observe heats of adsorption with calorimeters.

6.1. Indicator dye techniques

The surface acidity of inorganic powders such as cracking catalysts [32] or clays [33] has traditionally been determined by using basic indicator dyes of various pK_a values (in water). These dyes (in dry benzene or toluene) are used to determine end-points in butylamine titrations of the acidic surface sites [32–34]. Basic surface sites can also be titrated in dry toluene with an acidic indicator dye such as Bromphenol Magenta E and a titrant such as trifluoroacetic acid [35]. Although such non-aqueous titrations provide quantitative surface concentrations of acidic or basic sites, the estimates of acid or base strength are only approximate, because the relative basic strength of the indicator dyes is estimated from the pH of the color change in water, a rather inexact standard for base strength in non-aqueous solvents.

The acidity of titanium dioxide powders and of the oxide surface of a titanium alloy (before and after pretreatments for enhanced adhesive bonding) has also been investigated by the color changes of adsorbed indicator dyes [36].

6.2. Calorimetric techniques

Calorimetric methods can provide quantitative thermodynamic measurements of acidic or basic strength, and can be used to determine the acid–base surface properties of inorganic or organic powders or fibers. Molar heats of adsorption of test acids or bases interacting with the surface sites of such powders can be determined with a precision which depends to some degree on the specific surface area of the powder. If the adsorption is determined from neutral solvents at moderate concentrations so that the adsorbed films are less than half-saturated, the van der Waals interactions of the adsorbed molecules are still controlled by the solvent, so that the heats of adsorption are entirely the result of acid–base interactions with the acidic or basic surface sites.

6.2.1. Batch calorimetry methods. Calorimetric titrations of the acidic or basic surface sites of a powder may often be done with stirred suspensions of powders in a batch calorimeter. Such titrations require that the particles do not flocculate appreciably nor settle out in the stirred calorimeter vessel, and also require specific surface areas of about 20 m²/g in order to have enough sensitivity. For example, we have had no success with batch calorimetric titrations with magnetic iron oxides because of the fast flocculation resulting from magnetic attraction, but silica and rutile dispersions in cyclohexane behaved quite well, as indicated in Fig. 5, an E and C plot for the determination of the E_A and C_A values for silica [37]. These data were obtained at test base concentrations which provided about 40% of a saturated monolayer, and two silicas (a high surface area silica made by pyrolysis by Degussa, and a lower surface area silica made by precipitation by PPG) were used with identical results. The silica was dried and equilibrated with

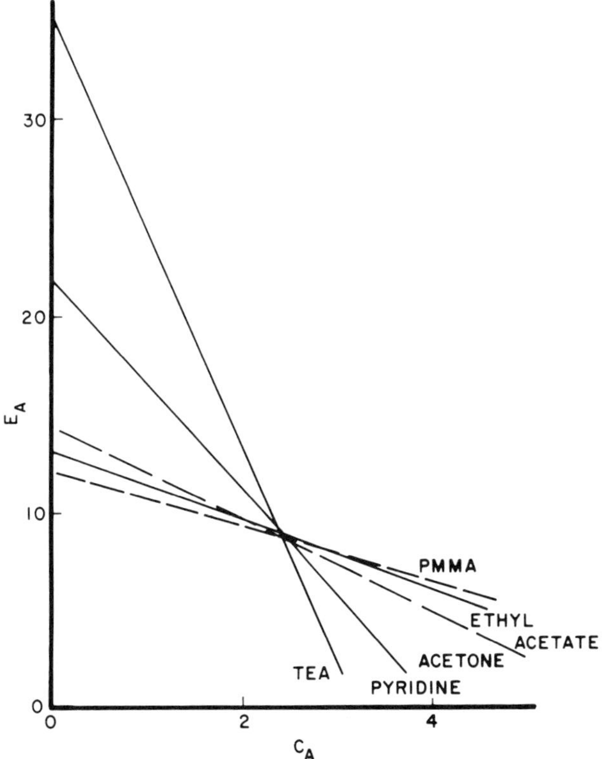

Figure 5. Drago E and C plot for the determination of the E_A and C_A constants of the acidic surface sites of silica, using heats of adsorption determined by batch calorimetry (solid lines) and by FTIR spectral shifts (dashed lines) [37].

cyclohexane dried to about 10 ppm of water; then about 0.5 g was stirred in
25 ml of the cyclohexane in a Tronac isothermal calorimeter, the base (triethylamine, pyridine, or ethyl acetate) was titrated into the stirred suspension, and the
heat of adsorption was monitored for several hours. The concentration of
unadsorbed base was determined by UV spectroscopy of the supernatant, which
allowed calculation of the molar heat of adsorption, ΔH^{ads}. Drago E_A and C_A
constants were evaluated with the plot shown in Fig. 5, based on an equation derived from equation (1):

$$E_A = -\Delta H^{ads}/E_B - C_A(C_B/E_B),\tag{21}$$

in which the intercept at $C_A=0$ is $-\Delta H^{ads}/E_B$ and the slope is C_B/E_B. By
choosing hard and soft bases with appreciably different ratios of C_B/E_B, statisti-
cally significant intersections are obtained. For silica the intersection occurs at
$E_A = 8.93 \pm 0.02$ (kJ/mol)$^{1/2}$, and $C_A = 2.32 \pm 0.02$ (kJ/mol)$^{1/2}$ [31].

Batch titrations with phenol and with pyridine have also been made with the
Tronac calorimeter to assess the surface acidity and basicity of pyrene derivative
organic pigments suspended in cyclohexane [38].

6.2.2. Flow microcalorimetry. Microscal flow calorimeters have been used
with downstream concentration-detectors (UV-absorbance detectors or refrac-
tive-index detectors) to titrate powders in the adsorption bed of the calorimeter
and with a laboratory computer to record twice a second both the heat of
adsorption detected with thermistor sensors and the concentration of the acidic
or basic probe flowing from the adsorption bed [27]. The sensitivity to the heat
of adsorption is at the microcalorie level, so that this technique can be used for
powders or fibers of rather limited specific surface area, as illustrated in Fig. 6
for glass fibers of 14 μm in diameter [39]. Figure 6 shows the titration of the
acidity with cyclohexane solutions of pyridine and of the basicity with cyclohexane solutions of phenol. The bare glass is seen to have very weak acidity and
weak basicity, but after treatment with silane coupling agents A-174 and A-1100
(γ-methacryloxypropyl- and γ-aminopropyl-triethoxysilanes) the acidity increased

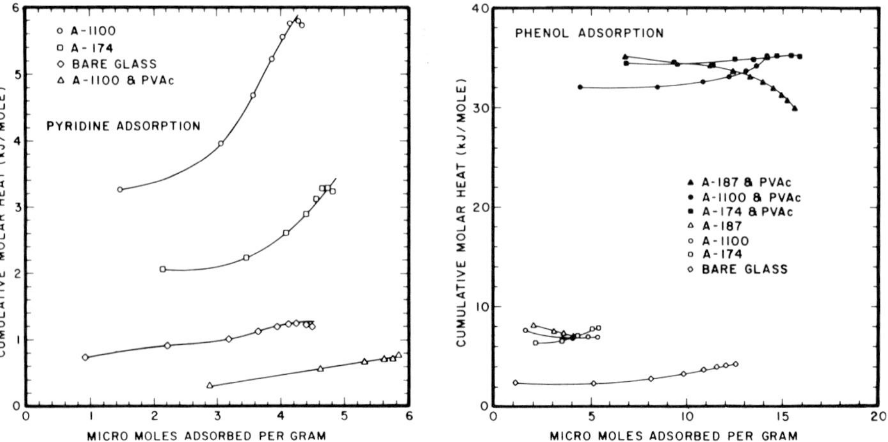

Figure 6. Flow microcalorimetric titration of the acidic and basic sites of glass fibers, untreated and
treated with silane coupling agents (A-1100, A-174, and A-187) and/or with polyvinyl acetate [39].

three-fold and five-fold, respectively, while the basicity doubled in strength, but decreased at least two-fold in surface concentration. The greatest changes occurred when the glass fibers were coated with polyvinyl acetate, for this basic polymer apparently covered up most of the acidic sites and increased the basicity four- to five-fold and the surface concentration of basic sites three-fold. These findings appear to confirm the electron micrographs which suggest that the coupling agents cover only a fraction of the glass surface, and that the polyvinyl acetate covers the fibers much more completely.

Flow microcalorimetry has been used successfully with powdered coal having specific surface areas of only about 1–4 m^2/g; with coal powders the acid–base probes (pyridine and phenol) were partly dissolved into the coal, but *t*-butylphenol and *t*-butylpyridine were not, for with these probes the concentrations and heats of desorption observed in the desorption cycles matched the concentrations and heats of adsorption observed in the adsorption cycles [40].

Flow microcalorimetry has also been used most successfully for titration of the acidic surface sites of magnetic iron oxides, which cannot be titrated in suspension because the stong magnetic attraction between particles causes such suspensions to flocculate [41].

6.3. Spectral shift measurements of the surface acidity or basicity of powders

The infra-red spectral frequency of carbonyl groups has been shown to shift to lower frequencies upon complexation with acids, with a shift of 0.99 cm^{-1} per kJ/mol heat of acid–base complexation, as shown in equation (10) and Fig. 2. The same shift is observed for the adsorption of carbonyl groups onto the acidic surface sites of inorganic solids, and consequently such infra-red shifts can be used for determination of the heats of adsorption of carbonyl-containing bases. Figure 7 shows the infra-red spectral shift observed with acetone adsorbed onto silica from mineral oil; the heat of adsorption calculated from this shift was used to determine the dashed line for acetone in the E and C plot of Fig. 5 from which the E_A and C_A constants for the surface SiOH sites of silica were determined. It is seen that the dashed lines determined by infra-red shifts (one for acetone and one for PMMA) intersect very well with the solid lines determined by batch calorimetry. Such infra-red measurements usually take less effort than batch calorimetry, especially when the surface sites tend to be uniform in acid or base strength.

Solid-state NMR spectral shifts of appropriate NMR probes resulting from adsorption by acid–base complexation show promise. In catalyst studies, adsorbed basic probes are found to have chemical shifts indicative of the acid strength of surface sites [42], and in our NMR studies the four acid–base probes (hard and soft acids and bases) which have been calibrated for solution studies are being studied by solid-state NMR spectroscopy when adsorbed on the acidic or basic surface sites of fine powders, especially coal powders [22].

6.4. Inverse gas chromatography

Gas chromatographic studies are used to determine the nature of the surface sites of powders to be tested, which are packed into columns through which elution volumes (V_N) are determined with test vapors of volatile probe molecules

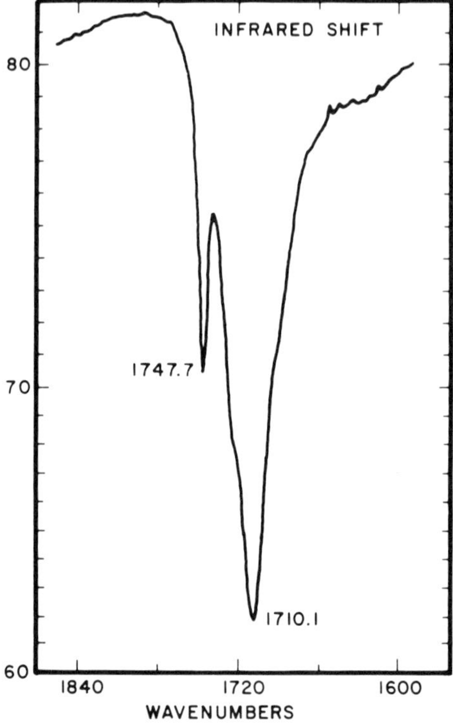

Figure 7. Infra-red spectrum for the carbonyl stretching frequency of acetone in a dry mineral oil mull of Degussa 380 silica gel. The peak with stronger absorption is for acetone carbonyls bonded to the acidic SiOH surface sites of silica; the smaller peak is for acetone in solution in the oil [37].

which adsorb on the surface of the sample in the column. By using a homologous series of saturated hydrocarbons of known γ_L^d the elution volumes at infinite dilution can be used to determine the γ_S^d value of the powder, based on the principle that $\Delta G°$ of adsorption ($RT \ln V_N$) equals the product of the work of adhesion W_{SL} and the area per adsorbed mole (Na, where N is Avogadro's number and a is the area per adsorbed molecule). For saturated hydrocarbons W_{SL} equals $2(\gamma_S^d \gamma_L^d)^{1/2}$ [43–46], so the slope of the straight-line plot of $RT \ln V_N$ vs. $a(\gamma_L^d)^{1/2}$ equals $2(\gamma_S^d)^{1/2}$, from which γ_S^d can be determined. This line represents the van der Waals (mainly dispersive) contribution to the molecular free energy of adsorption, so that when acid–base contributions enhance the free energy of adsorption by aW_{SL}^{ab}, values of $RT \ln V_N$ plotted against $a(\gamma_L^d)^{1/2}$ fall above the line and allow the determination of aW_{SL}^{ab}. Thus, inverse gas chromatography (IGC) can be used to determine both van der Waals and acid–base contributions to interfacial interactions. Since most laboratories use gas chromatography, this method may become the most widely used to determine γ_S^d and the acidity or basicity of polymers and of inorganic solids.

IGC has been used to measure the acidity of polyvinyl chloride and the acidity or basicity of plasma-treated calcium carbonate [47]. It has been used for the study of chemical modification of the acid–base surface properties of silica particles [45], and to study the acid–base character of carbon fibers and the acid–base effects of surface treatments of carbon fibers for enhanced adhesion in epoxies [46, 48], and in similar studies with glass fibers [49].

In future IGC studies of the acid–base surface properties of powders or fibers of polymers or inorganic materials, it is recommended that the acidic or basic probes be chosen from those which have minimal acid–base self-association, as illustrated in Table 1. For instance, a frequently used acidic probe is *n*-butanol, but as shown in Table 1 this alcohol has enough basicity that its acid–base self-association accounts for 46% of its heat of vaporization. Chloroform may be more acceptable, because the acid–base self-association of this acid probe provides less than 2% of its heat of vaporization. Triethylamine or ethyl ether may be more satisfactory basic probes than the frequently used *n*-butylamine or THF. Acid–base self-association provides an appreciable fraction of the heat of vaporization at 25°C in *n*-butylamine (22%) and in THF (27%), but only 6% in ethyl ether and less than 1% in triethylamine.

6.5. X-ray photoelectron spectroscopy (XPS)

XPS spectra also have chemical shifts related to acid–base interaction, especially the basic oxygens or nitrogens of surface sites, and high resolution XPS spectra are most helpful in identifying exactly which atoms in the surface of polymers, metal oxides, or other inorganic surfaces are involved in acid–base interactions. The electron binding energies of inner core electrons of atoms such as carbon, oxygen, or nitrogen vary over a range of about 10 eV, with lower electron binding energies for basic (electron-donating) atoms and higher energies for acidic (electron-accepting) atoms. For example, in the case of the nitrogen 1*s* electrons, the binding energies are low for the amine form and high for the ammonium form. This point is illustrated in the right-hand diagram of Fig. 8, which shows multipeak resolution of an XPS spectrum for an adsorbed film of A-1100 (γ-aminopropyl triethoxysilane) on glass fibers [39]. The larger peak centered at about 399 eV is for the amine form, while the smaller peak centered at about 401.4 eV is for the ammonium form.

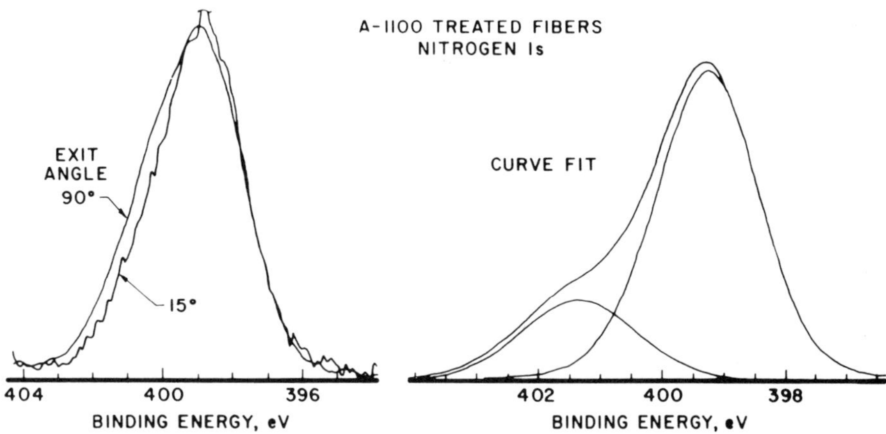

Figure 8. Nitrogen 1*s* XPS spectra for E-glass fibers treated with an aminosilane coupling agent (A-1100) [39].

Angle-dependent XPS spectra allow the determination of the orientation of silane coupling agents adsorbed on glasses or other inorganic surfaces [37], a finding of importance for the understanding of how silane coupling agents promote adhesion. Figure 8 illustrates the shift of the high resolution nitrogen $1s$ binding energy spectrum observed at exit angles of 15° and 90° for glass fibers treated with silane coupling agent A-1100, the same sample as that shown in Fig. 6. Electrons exiting at 15° come from nitrogens within less than 1 nm from the outer surface, while those exiting at 90° can come from much deeper. The left-hand diagram shows that at a 90° exit angle the intensity at higher electron binding energies (where ammonium groups are found) is appreciably greater than at a 15° exit angle; this shows that the ammonium groups are at the glass interface with the coupling agent, where acid–base interactions occur between the amine groups and the acidic sites of the glass. On the other hand, the proportion of amine in the ammonium form is only about 10–20%, as can be estimated from the areas of the two peaks in the right-hand diagram of Fig. 8. Thus, we conclude that the outer surface of the glass fibers treated with A-1100 still has an appreciable proportion of amine groups, which is also evidenced by the increased basicity of the treated fibers in the flow microcalorimetric findings of Fig. 6.

7. CHARGE TRANSFER BETWEEN ELECTRON-DONATING AND ELECTRON-ACCEPTING MATERIALS

Electrons are usually transferred when unlike materials are brought into contact. When unlike metals are brought into contact, electrons are transferred from the less noble to the more noble metal, and at equilibrium a potential difference is established between the two metals in contact. When insulators with acidic sites (electron traps) are brought into contact with the less noble metals, electrons tend to be donated to the electron traps of the insulator from the Fermi level of the metals. Similarly, when insulators with electron-donor (basic) sites are brought into contact with the more noble metals, electrons are transferred to the metals.

Charge transfer between materials in contact can also result from proton transfer from acidic to basic materials, especially when one of the materials is a liquid.

7.1. Zeta potentials

Dispersions in a liquid of either immiscible droplets or solid particles are normally electrically charged, and the charge is usually easily related to differences in electron donor–acceptor properties [50]. Acidic (electron-accepting) particles such as silica are negatively charged in organic bases such as ethers, esters, ketones, aromatics, and amines; they are also negatively charged in aqueous media when the pH is above 2. Basic (electron-donating) particles such as hydrated alumina are positively charged in acidic liquids such as chlorinated solvents, phenols, alcohols, and nitro-compounds; they are also positively charged in aqueous solutions when the pH is below 9.

Since water has both acidic and basic characteristics, it is not surprising that it can charge particles negatively at higher pH and positively at lower pH. For each dispersed material there is a dividing pH value (the isoelectric point, IEP) above which this material is charged negatively and below which it is charged positively. Acidic materials have low isoelectric points and basic materials have high isoelectric points [51]. Silica is acidic and has an isoelectric pH of about 2, while alumina is basic and has an isoelectric pH of about 9. However, some materials have both acidic and basic surface sites, such as iron oxides or titanium dioxide, and these have isoelectric pHs near 7. Such neutral isoelectric pHs do not mean that these oxides are neutral; actually, iron oxides are about as strongly acidic as silica and about as strongly basic as hydrated alumina.

Acetonitrile is also both acidic and basic, and particles of acidic materials in acetonitrile are charged negatively, while particles of basic materials in acetonitrile are charged positively. Measurements of the zeta potential of dispersed particles in acetonitrile or in other liquids with both acidic and basic character are a useful tool for observing the acidity of basicity of particle surfaces. The effectiveness of silane coupling agents for modifying the surface acidity or basicity of glass powders was determined by zeta-potential measurements in a mixture of methanol and methylisobutylketone (MIBK). The untreated glass powders had zeta potentials of less than 5 mV; powders treated with amino-silanes became quite basic and had zeta potentials of about −60 mV, while those treated with methacryloxysilane became quite acidic and had zeta potentials of about +50 mV [52].

7.2. *Charge injection from conductors into polymers*

Polymers have trapping sites for both negative and positive charges (electrons and holes), and upon contact with metals, electrons tend to be injected into the polymer from the Fermi level of active metals of low work function (alkali and alkaline earth metals), while electrons are removed from polymers in contact with noble metals of high work function such as gold [53]. The work function of a metal is the energy required to emit electrons from the Fermi level into the vacuum level, and it varies from about 2 eV for active metals to 5.5 eV for noble metals. PMMA, for instance, accepts electrons from metals of work function less than 4 eV, but gives up electrons to metals of work function greater than 4 eV [54]. Thus, the metal work function for no charge injection into PMMA is about 4 eV, as opposed to about 4.8 eV for the less basic polystyrene and about 3.5 eV for the more basic polyvinylpyridine [54].

Charge injection into polymers from silicon is easily measured with metal/insulator/silicon (MIS) structures with thin polymer layers as the insulator, using capacitance changes vs. applied potential bias to determine charge injection [55]. It is found that only positive charges are injected from silicon into some polymers (such as polystyrene), only negative charges are injected into polytetrafluoroethylene [56], and with many other polymers (depending on the sign and magnitude of the bias) either negative or positive charges are injected from silicon [55].

8. SUMMARY

Determination of the acidic and basic strengths of the functional sites of solvents, polymers, and inorganic solids allows a quantitative measure of the 'polar' (acid–base) interactions between materials which are so important to solubility, wettability, adsorption, and adhesion. Many quantitative laboratory techniques may now be used to determine the Drago E and C constants of polar groups, from which the energies of interaction may be calculated.

Acknowledgements

Help in the preparation of this manuscript by D. A. Cole, T. B. Lloyd, D. W. Dwight, and K. L. Mittal is gratefully acknowledged.

REFERENCES

1. J. N. Israelachvili, *Intermolecular and Surface Forces*, Chap. 11. Academic Press, New York (1985).
2. G. C. Pimentel and A. L. McClellan, *The Hydrogen Bond*. W. H. Freeman, San Francisco (1960).
3. F. M. Fowkes, *Ind. Eng. Chem.* **56** (12), 40–52 (1964).
4. C. J. van Oss, R. J. Good and M. K. Chaudhury, *Langmuir* **4**, 884–891 (1988).
5. F. M. Fowkes, F. L. Riddle, Jr., W. E. Pastore and A. A. Weber, *Colloids Surf.* **43**, 367–387 (1990).
6. V. Gutmann, *The Donor–Acceptor Approach to Molecular Interactions*. Plenum Press, New York (1978).
7. F. L. Riddle, Jr. and F. M. Fowkes, *J. Am. Chem. Soc.* **112**, 3259–3264 (1990).
8. V. Gutmann, *The Donor–Acceptor Approach to Molecular Interactions*, pp. 134–136. Plenum Press, New York (1978).
9. R. S. Drago, G. C. Vogel and T. E. Needham, *J. Am. Chem. Soc.* **93**, 6014–6026 (1971).
10. G. N. Lewis, *J. Franklin Inst.* **226**, 293 (1938).
11. R. S. Mulliken, *J. Phys. Chem.* **56**, 801 (1952).
12. R. G. Pearson, *Hard and Soft Acids and Bases*. Dowden, Hutchinson and Ross, Stroudsburg, PA (1973).
13. F. M. Fowkes and D. A. Cole, manuscript in preparation.
14. J. H. Hildebrand and R. L. Scott, *Solubility of Non-Electrolytes*, 3rd edn. Reinhold, New York (1950).
15. F. M. Barton, *Handbook of Solubility Parameters and Other Cohesion Parameters*. CRC Press, Boca Raton, FL (1983).
16. J. J. Christensen, R. W. Hanks and R. M. Izatt, *Handbook of Heats of Mixing*. John Wiley, New York (1982).
17. F. M. Fowkes, D. O. Tischler, J. A. Wolfe, L. A. Lannigan, C. M. Ademu-John and M. J. Halliwell, *J. Polym. Sci., Polym. Chem. Ed.* **22**, 547–566 (1984).
18. W. L. Jorgensen, J. D. Madura and C. J. Swenson, *J. Am. Chem. Soc.* **106**, 6638 (1984).
19. H. Swain, private communication (1984).
20. T. K. Kwei, E. M. Pearce, F. Ren and J. P. Chen, *J. Polym. Sci., Polym. Phys. Ed.* **24**, 1597 (1986).
21. D. Valia, Ph.D. Thesis, Lehigh University (1988).
22. F. L. Riddle, Jr. and F. M. Fowkes, manuscript in preparation.
23. D. K. Owens and R. C. Wendt, *J. Appl. Polym. Sci.* **13**, 1741 (1969).
24. D. H. Kaelble, *J. Adhesion* **2**, 66 (1970).
25. F. M. Fowkes and M. A. Mostafa, *IEC Prod. Res. Dev.* **17**, 3 (1978).
26. F. M. Fowkes and W. D. Harkins, *J. Am. Chem. Soc.* **62**, 3377 (1940).
27. S. T. Joslin and F. M. Fowkes, *IEC Prod. Res. Dev.* **24**, 369 (1985).
28. F. M. Fowkes, M. B. Kaczinski and P. M. Kelly, manuscript submitted to *Langmuir*.
29. F. M. Fowkes, *J. Adhesion* **4**, 155 (1972).
30. L. A. Casper, Ph.D. Thesis, Lehigh University (1985).
31. K. Backstrom, B. Lindman and S. Engstrom, *Langmuir* **4**, 872 (1988).
32. O. Johnson, *J. Phys. Chem.* **59**, 827 (1955).
33. F. M. Fowkes, H. A. Benesi, L. B. Ryland, W. M. Sawyer, K. D. Detling, E. S. Loeffler, F. B. Folckemer, M. R. Johnson and Y. P. Sun, *J. Agric. Food Chem.* **8**, 203–210 (1960).

34. D. Atkinson and G. Curthoys, *Chem. Soc. Rev.* **8**, 475–497 (1979).
35. F. M. Fowkes, in: *Industrial Applications of Surface Analysis*, L. A. Casper and C. J. Powell (Eds), ACS Symposium Series 199, pp. 69–88. Am. Chem. Soc., Washington, DC (1982).
36. J. G. Mason, R. Siriwardane and J. P. Wightman, *J. Adhesion* **11**, 315–328 (1981).
37. F. M. Fowkes, D. C. McCarthy and D. O. Tischler, in: *Molecular Characterization of Composite Interfaces*, H. Ishida and G. Kumar (Eds), pp. 401–411. Plenum Press, New York (1985).
38. G. W. Heebner, Ph.D. Thesis, Lehigh University (1990).
39. F. M. Fowkes, D. W. Dwight, D. A. Cole and T. C. Huang, *J. Non-Cryst. Solids* **120**, 47–60 (1990).
40. F. M. Fowkes, K. L. Jones, G. Li and T. B. Lloyd, *Energy Fuels* **3**, 97–105 (1989).
41. F. M. Fowkes, Y. C. Huang, B. A. Shah, M. J. Kulp and T. B. Lloyd, *Colloids Surf.* **29**, 243–261 (1988).
42. L. Baltusis, J. S. Frye and G. E. Maciel, *J. Am. Chem. Soc.* **108**, 7119 (1986).
43. G. M. Dorris and D. K. Gray, *J. Colloid Interface Sci.* **77**, 353 (1980).
44. S. Katz and D. G. Gray, *J. Colloid Interface Sci.* **82**, 318–351 (1981).
45. E. Papirer, H. Balard and A. Vidal, *Eur. Polym. J.* **24**, 783–790 (1988).
46. J. Schultz and Lavielle, in: *Inverse Gas Chromatography*, D. R. Lloyd, T. C. Ward and H. P. Schreiber (Eds), ACS Symposium Series 391, p. 185. Am. Chem. Soc., Washington, DC (1989).
47. A. E. Bolvari and T. C. Ward, in: *Inverse Gas Chromatography*, D. R. Lloyd, T. C. Ward and H. P. Schreiber (Eds), ACS Symposium Series 391, pp. 217–229. Am. Chem. Soc., Washington, DC (1989).
48. H. P. Schreiber, C. Richard and M. R. Wertheimer, in: *Physicochemical Aspects of Polymer Surfaces*, K. L. Mittal (Ed.), vol. 2, pp. 739–748. Plenum Press, New York (1983).
49. E. Osmont and H. P. Schreiber, in: *Inverse Gas Chromatography*, D. R. Lloyd, T. C. Ward and H. P. Schreiber (Eds), ACS Symposium Series 391, pp. 230–247. Am. Chem. Soc., Washington, DC (1989).
50. F. M. Fowkes, H. Jinnai, M. A. Mostafa, F. W. Anderson and R. J. Moore, in: *Colloids and Surfaces in Reprographic Technology*, ACS Symposium Series 200, M. Hair and M. D. Croucher (Eds), pp. 307–324. Am. Chem. Soc., Washington, DC (1982).
51. J. C. Bolger and A. S. Michaels, in: *Interface Conversion and Polymer Coatings*, P. Weiss (Ed.), pp. 3–60. Elsevier, New York (1968).
52. F. M. Fowkes, D. W. Dwight, J. A. Manson, T. B. Lloyd, D. O. Tischler and B. A. Shah, *Mater. Res. Soc. Symp. Proc.* **119**, 223–234 (1988).
53. D. K. Davies, *Br. J. Appl. Phys.* **2**, 1533 (1969).
54. C. B. Duke and T. J. Fabish, *J. Appl. Phys.* **49**, 315 (1978).
55. F. M. Fowkes, in: *Surface and Interfacial Aspects of Biomedical Polymers*, J. D. Andrade (Ed.), vol. 1, pp. 337–372. Plenum Press, New York (1985).
56. H. R. Anderson, F. M. Fowkes and F. H. Hielscher, *J. Polym. Sci., Polym. Phys. Ed.* **14**, 879 (1976).

Acid-Base Interactions, pp. 117-134
Eds. K.L. Mittal and H.R. Anderson, Jr.
©VSP 1991

Characterization of the acid–base nature of metal oxides by adsorption of TCNQ

KENJIRO MEGURO* and KUNIO ESUMI

*Department of Applied Chemistry, Institute of Colloid and Interface Science,
Science University of Tokyo, Kagurazaka, Shinjuku-ku, Tokyo 162, Japan*

Revised version received 26 February 1990

Abstract—This paper reviews the acid–base (electron donor–acceptor) interaction of 7,7,8,8-tetra-cyanoquinodimethane (TCNQ) with metal oxides. When TCNQ is adsorbed onto metal oxides, TCNQ anion radicals are formed as a result of electron transfer from the metal oxides to TCNQ. The measurement of the TCNQ radical concentration allows the determination of the electron donicity of single and two-component metal oxides. TCNQ adsorption is also affected significantly by the acid–base interaction between the metal oxide and solvent, the metal oxide and TCNQ, and TCNQ and the solvent. In addition, the strength and distribution of electron donor sites and enhancement of the electron donor properties of metal oxides are discussed.

Keywords: Metal oxide; acid–base; TCNQ; electron transfer; TCNQ radical; electron donicity.

1. INTRODUCTION

Recently, acid–base interactions at interfaces have been studied extensively in colloidal systems [1, 2]. Fowkes *et al.* [3–5], in particular, have studied the adsorption of acidic and basic molecules from neutral solvents on inorganic powders such as iron oxides, silica, and titania. They found that the calorimetric heats of adsorption are actually the heats of acid–base interactions, governed by the Drago equation [6], and that the Drago equation constants can be accurately determined for the surface sites of these inorganic solids. Furthermore, Fowkes [7] has extended this acid–base interaction theory to polymer–powder interfaces.

On the other hand, it is known [8–11] that when strong electron acceptors (Lewis acids) or donors (Lewis bases) are adsorbed on metal oxides, the corresponding radicals are formed as a result of electron transfer between the adsorbate and the metal oxide surface. Flockhart *et al.* [12] attempted the adsorption of tetracyanoethylene (TCNE) for the estimation of the electron donor properties of alumina surface. In this respect, they associated the electron donor sites with the unsolvated hydroxyl ions and the defect centers involving oxide ions. Che *et al.* [13] carried out a systematic study of the adsorption of TCNE on the surface of titania and magnesia.

Such electron donor–acceptor interactions at interfaces become important in elucidating the adhesion forces at these interfaces.

In this paper, the results obtained in our laboratory are summarized briefly. We

---

*To whom correspondence should be addressed.

first describe the electron donor properties of metal oxides using the adsorption of 7,7,8,8-tetracyanoquinodimethane (TCNQ). Secondly, the strength and distribution of the electron donor sites of the metal oxides are discussed. Finally, the solvent effect on the electron donor–acceptor interaction and the enhancement of the electron donor–acceptor interaction by plasma treatment of the metal oxide are described.

2. ELECTRON DONOR PROPERTIES OF ONE-COMPONENT METAL OXIDES

The electron donor properties of MgO, Al_2O_3, SiO_2, TiO_2, ZnO, and NiO were studied by means of TCNQ adsorption [11].

When TCNQ was adsorbed on the surfaces of these solids from its acetonitrile solution at 25°C, their surfaces developed a remarkable coloration: MgO, blue-green; Al_2O_3, blue-green; SiO_2, yellow; TiO_2, violet; ZnO, greyish green. The same colors were observed when the metal oxides were ground with added TCNQ crystals. The coloration of the metal oxide surfaces suggests that new adsorbed species are formed on the surfaces.

This suggestion was confirmed by ESR spectra indicating the presence of free radical species. The first derivative ESR spectra showed only one broad absorption with $g = 2.011$ (Fig. 1); this is due to the hindered rotational freedom of the adsorbed species, which obscures the hyperfine structure of the spectra.

The electronic state of the adsorbed species was studied by UV spectroscopy in addition to ESR spectroscopy.

The electronic spectra of the colored samples are also given in Fig. 1. The bands appearing below 400 nm are common to all the oxides and may be assumed to correspond to the physically adsorbed state of neutral TCNQ, which has an absorption band at 395 nm. In the case of TiO_2 and ZnO, this assignment does not hold completely, because TiO_2 and ZnO have characteristic bands in the same region. The bands near 600 nm are probably related to the dimeric TCNQ anion radical, which absorbs light at 643 nm [14]. This tentative attribution is supported by the characteristic features that neutral TCNQ absorbs only at 395 nm, that TCNQ has a high electron affinity, and that the TCNQ anion radical derivatives are stable even at room temperature [15–18]. Therefore, the ESR and electronic spectra provide evidence that the TCNQ anion radicals are formed as a result of electron transfer from the metal oxide surface to adsorbed TCNQ.

The electron donicities of the metal oxide surfaces, estimated by means of the radical concentration ratios for unit surface area (spin number/m^2), are as follows: MgO, $100 > ZnO$, $47 > Al_2O_3$, $38 > TiO_2$, $1.6 > SiO_2$, $0.11 > NiO$, 0.00. Thus, this TCNQ adsorption can be applied to characterize the electron donor properties of metal oxides.

The nature of the sites responsible for the electron-transfer process is not well understood. However, it may be suggested that two types of donor site exist on the surfaces. The first type consists of an anion deficiency (n-type semiconductor), and the second is a hydroxyl ion on the oxide surface.

Semiconductors [19] which have an excess of conductive electrons are normally called 'n-type', because their conductivity stems from negative charge carriers. Those with an excess of holes are called 'p-type', as the carriers are positive. n-Type semiconductors have anion deficiencies (or cation excesses), while p-type

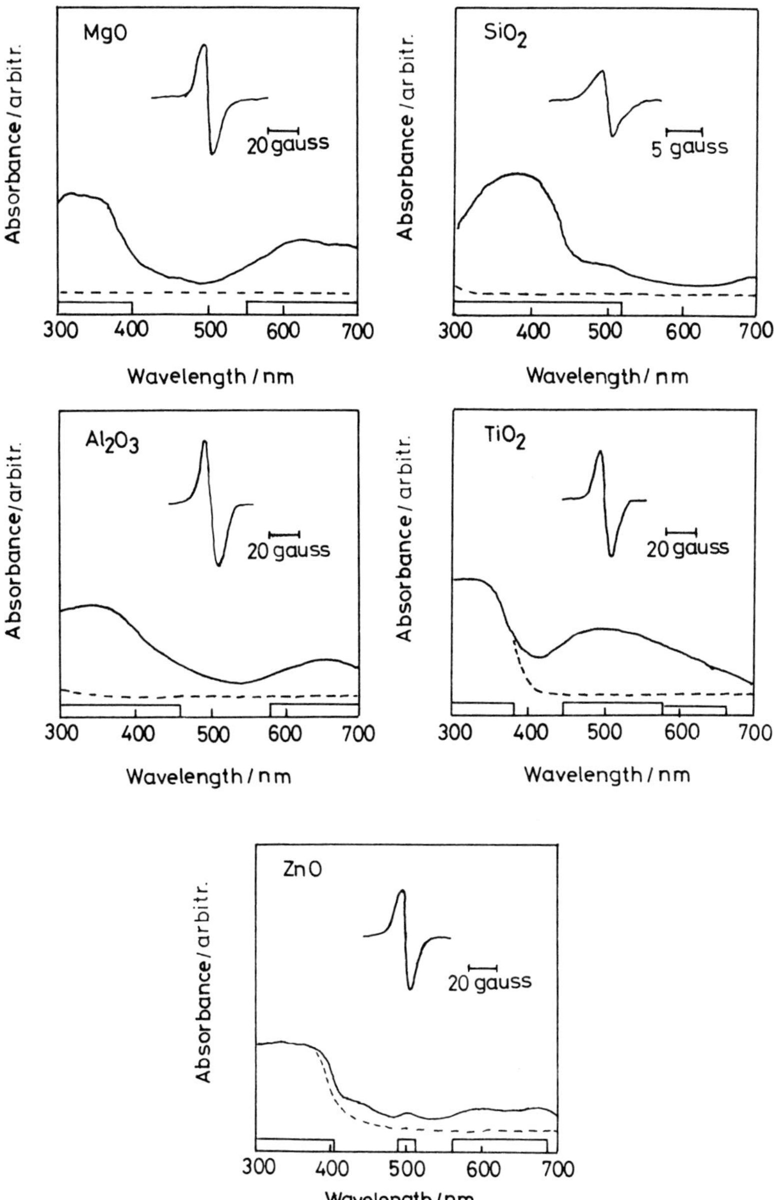

Figure 1. ESR and electronic spectra from TCNQ adsorbed on metal oxides. Dotted line: metal oxides prior to adsorption.

semiconductors have anion excesses (or cation deficiencies). From this fact it is reasonable that NiO has no electron donor property. If TCNQ anion radicals are formed by the transfer of free electrons from intrinsic or extrinsic defects which are present on the surface or in the bulk of the solid, MgO, Al_2O_3, and SiO_2, which are insulators, would normally not be expected to participate in a charge-transfer adsorption process because of the lack of free electrons.

The second type of donor site may be a surface hydroxyl ion. Surface hydroxyl groups can be expected on all metal oxides. The ionization potential of hydroxyl ions is comparatively small (2.6 eV in the gas phase [20]); therefore, the possibility of its participation in oxidation–reduction processes of the type

$$OH^- + A \rightarrow OH + A^-,$$

where A is an electron acceptor, cannot be excluded. Indeed, Fomin *et al.* [21] have shown that electron transfer from the hydroxyl ion can and does occur in certain solvent systems provided a suitable acceptor molecule is present. As Al_2O_3, MgO and SiO_2 surfaces are usually covered with hydroxyl groups [22] which have ionic and covalent characters depending on the nature of the metal, the charge-transfer adsorption of TCNQ on the oxides may result from electron transfer from the hydroxyls. Actually, Stober [23] found that, even after thorough outgassing at 100°C, one molecule of physically adsorbed water is retained for every two silanol groups on the surface of amorphous silicon dioxide. This surprising result was confirmed by Boehm *et al.* [24], using chemical reactions as well as deuterium exchange. Lee and Weller [25] confirmed the existence of surface hydroxyl groups on Al_2O_3 which had been dehydrated at 500°C. Similarly, the existence of surface hydroxyl groups on MgO has been observed [26]. It is reasonable to assume that the electrons transfer to TCNQ from hydroxyl ions on the metal oxides which have been calcined at a high temperature such as 500°C. Surface hydroxyls may exist on all the samples investigated. However, surface hydroxyls on metal oxides are shown to differ in chemical properties. Surface silanol groups are much more stable than Al–OH groups [22]. Differences in acidity between the hydroxyl groups on several oxide surfaces have been reported [27]. These results suggest that the hydroxyl ions of the metal oxide surfaces have different electron donor properties.

3. THE ELECTRON DONOR PROPERTIES OF TWO-COMPONENT METAL OXIDE SYSTEMS

It is well known that two-component metal oxide systems exhibit characteristic surface properties which are not qualitatively predictable from consideration of the independent properties of the parent oxides. The following section gives the electron donor properties of SiO_2–Al_2O_3, SiO_2–TiO_2, Al_2O_3–TiO_2 [28], and ZrO_2–TiO_2 [29] systems as studied by means of TCNQ adsorption.

The abbreviations used are as follows: S, SiO_2; A, Al_2O_3; T, TiO_2; Z, ZrO_2; 4SA, SA, and S4A, SiO_2–Al_2O_3 containing 35, 64, and 81% Al_2O_3; 4ST, ST, and S4T, SiO_2–TiO_2 containing 21, 55, and 83% TiO_2; 4AT, AT, and A4T, Al_2O_3–TiO_2 containing 28, 62, and 83% TiO_2; 4Zt, ZT, and Z4T, ZrO_2–TiO_2 containing 13, 38, and 80% TiO_2, respectively. The metal oxide systems, except the ZrO_2–TiO_2 system, were prepared by co-hydrolysis of the mixtures of the metal alkoxides, while the ZrO_2–TiO_2 system was prepared from mixtures of the metal chlorides.

When TCNQ was adsorbed on the surfaces of SiO_2–Al_2O_3, SiO_2–TiO_2, Al_2O_3–TiO_2, and ZrO_2–TiO_2 from its acetonitrile solution, the surfaces of the metal oxides showed the characteristic coloration of each system. From the ESR and electronic spectra, similar to those of the parent oxide, it was confirmed that the

TCNQ anion radicals are formed as a result of electron transfer from the metal oxide surfaces to TCNQ.

The TCNQ anion radical concentration of the metal oxide surfaces is shown in Fig. 2. The electron donicity is reflected by the saturation of the color of TCNQ on the metal oxide surfaces and the increase in absorbance at the charge-transfer band. The order of electron donicity was found to be: S4A > SA > 4SA; 4ST > S4T > ST; 4AT > A4T > AT; Z4T > ZT > 4ZT. The SiO_2–Al_2O_3, SiO_2–TiO_2, Al_2O_3–TiO_2, and ZrO_2–TiO_2 systems showed much less electron donicity than the parent oxides—SiO_2, Al_2O_3, TiO_2, and ZrO_2. Thus, these systems exhibit electron donor properties which would not be qualitatively predicted from consideration of the independent properties of the parent oxides [30, 31].

It is well known that since the —Si—O—Al— structure forms a strong acid site

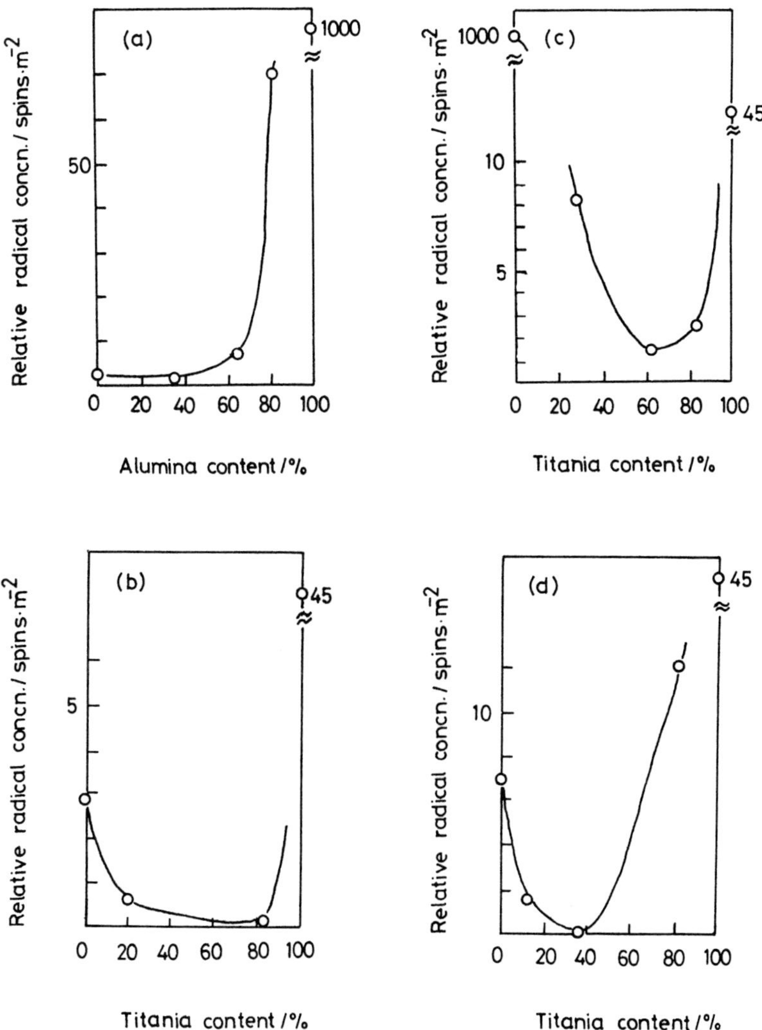

Figure 2. The change in TCNQ anion-radical forming activity of two-component metal oxides with the composition: (a) SiO_2–Al_2O_3; (b) SiO_2–TiO_2; (c) Al_2O_3–TiO_2; (d) ZrO_2–TiO_2.

[25, 32–42], the formation of an electron donor site cannot be expected. In the present work, all the SiO_2–Al_2O_3 systems showed appreciable electron donicity, but had much less electron donicity than Al_2O_3. The steep decrease in the electron donicity from the SiO_2–Al_2O_3 system to Al_2O_3 suggests that the pure Al_2O_3 phase is present only in small amounts on the SiO_2–Al_2O_3 surfaces. The existence of the pure Al_2O_3 phase was confirmed by the appearance of a weak band at about 650 nm in S4A corresponding to the charge-transfer band of TCNQ adsorbed on Al_2O_3.

The SiO_2–TiO_2 systems had negligible electron donicity, and the charge-transfer band scarcely appeared at all, suggesting that the pure TiO_2 phase is present only in small amounts on the SiO_2–TiO_2 surfaces. It is noteworthy that a weak anatase line in the X-ray diffraction pattern of S4T suggests the presence, in the conglomerate, of a pure TiO_2 phase, but S4T has negligible electron donicity.

The Al_2O_3–TiO_2 systems showed much less electron donicity than the parent oxides. The electron donicity decreased steeply from Al_2O_3 to 4AT, decreased gradually from 4AT to AT, increased from AT to A4T, and increased markedly from A4T to TiO_2. AT showed the most characteristic surface properties among the Al_2O_3–TiO_2 systems [30].

The ZrO_2–TiO_2 systems had a much lower electron donicity than TiO_2. The same tendency was observed in the SiO_2–Al_2O_3, SiO_2–TiO_2, and Al_2O_3–TiO_2 systems.

4. STRENGTH AND DISTRIBUTION OF ELECTRON DONOR SITES ON METAL OXIDE SURFACES

This section describes an investigation of the strength and distribution of electron donor sites on Al_2O_3, TiO_2, and ZrO_2–TiO_2 by adsorption of some electron acceptors measured by means of an ESR spectrometer [43].

The electron donor strength on a metal oxide can be defined as the conversion ratio of an electron acceptor adsorbed on the surface into its anion radical. If a strong electron acceptor is adsorbed on a metal oxide, its anion radical is formed at every donor site present on the metal oxide surface. On the other hand, if a weak electron acceptor is adsorbed, the formation of an anion radical will be expected only at the strong donor sites. Finally, in the case of very weak electron acceptor adsorption, its anion radical will not be formed even at the strongest donor sites. Therefore, the electron donor strength on a metal oxide can be expressed as the limiting electron affinity value at which free anion radical formation is not observed at the metal oxide surface.

The following electron acceptors with different electron affinities were used: TCNQ, 2,5-dichloro-*p*-benzoquinone (DCQ), *p*-dinitrobenzene (PDNB), and *m*-dinitrobenzene (MDNB).

When these electron acceptors were adsorbed on the metal oxides from their acetonitrile solutions, the surface of the metal oxides developed a characteristic coloration with each electron acceptor. The colors of the metal oxide surfaces after the adsorption are given in Table 1. The colored samples of Al_2O_3 and TiO_2 obtained by the adsorption of TCNQ, DCQ, and PDNB showed ESR absorption with a g value of 2.003 for TCNQ, 2.005 for DCQ, and 2.004 for PDNB owing to the anion radicals of the respective acceptors [44], but the sample in the case of

Table 1.
Coloration of the metal oxide surfaces and the electron affinity values of electron acceptors

Electron acceptor	Electron affinity (eV)	Metal oxide		
		Al_2O_3	TiO_2	ZrO_2–TiO_2
TCNQ	2.84	Blue-green	Violet	Violet
DCQ	2.30	Yellow	Brown	Brown
PDNB	1.77[a]	Pale orange	Brown	Pale brown
MDNB	1.26	Colorless	Colorless	Colorless

[a] The value given by Briegleb [46] plus 1.07 eV.

MDNB did not show any absorption. On the other hand, the colored samples of ZrO_2–TiO_2 showed ESR absorption, except in the case of PDNB and MDNB. By comparing these results with the electron affinity values [45, 46] of the electron acceptors, the limiting value of the affinity of an electron acceptor for electron transfer on Al_2O_3 and TiO_2 ranges between 1.77 and 1.26 eV, but that on ZrO_2–TiO_2 ranges between 2.30 and 1.77 eV. These results indicate that the electron donor strengths of Al_2O_3 and TiO_2 are similar and stronger than that of ZrO_2–TiO_2.

When the anion radical concentrations per m^2 formed on the metal oxides are plotted against the equilibrium concentration of electron acceptors, the isotherms are of the Langmuir type. The saturated anion radical concentrations per m^2 formed on the metal oxides, estimated from Langmuir plots, are plotted as a function of the electron affinity value of the electron acceptor in Fig. 3. The plots of the log of the anion radical concentration per m^2 vs. the electron affinity value are linear, i.e. $\log(c_R) = AX + B$, where c_R is the anion radical concentration per m^2,

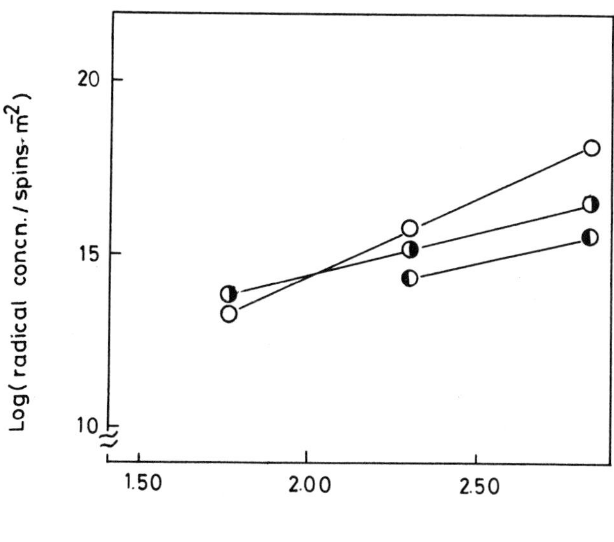

Figure 3. The change in the log of the saturation anion radical concentration as a function of the electron affinity of the electron acceptor: (O) Al_2O_3; (◑) TiO_2; (◐) ZrO_2–TiO_2.

A is the slope, X is the electron affinity value of the electron acceptor, and B is the intercept. For the interpretation of this linearity, we assume that there exists a distribution of strengths among the electron donor sites on the metal oxides. On this assumption, only the strong donor sites can interact with the weak electron acceptors, but both the strong and the weak ones can react with the strong electron acceptors. The anion radical concentration per m^2 formed on the metal oxide should decrease with decreasing electron affinity value of the electron acceptor. Thus, the distribution of electron donor sites with different strengths can be evaluated from the slope of the straight line. The gradient of the straight line becomes larger with an increase in the number of active donor sites. The values of the slopes A follow the order $Al_2O_3 > TiO_2$, ZrO_2-TiO_2. This means that the distribution of electron donor sites having different strengths on Al_2O_3 is larger than that on TiO_2 and $ZrO-TiO_2$. It is clear that there are no significant differences between the slopes obtained for TiO_2 and ZrO_2-TiO_2. It is noteworthy that the weak anatase pattern of ZrO_2-TiO_2 suggests the presence, in the conglomerate, of a pure TiO_2 phase. Therefore, it may be supposed that the amount of pure TiO_2 phase present on the surface of ZrO_2-TiO_2 plays an important role in the slope obtained for ZrO_2-TiO_2.

The electron donor sites of the metal oxides have already been discussed in Section 2. Under the experimental conditions, where the samples were pretreated at 120°C at 10^{-5} Torr, it is considered that surface hydroxyl ions on the metal oxides operate mainly as electron donor sites. Indeed, it has been reported that the surfaces of metal oxides contain various types of hydroxyl groups [22, 47–50]. Therefore, it should be anticipated that the various surface hydroxyl ions might have an energy distribution corresponding to those found for the electron donor sites.

5. THE SOLVENT EFFECT ON THE ACID–BASE INTERACTION OF TCNQ WITH METAL OXIDES

The work reported here was undertaken to study the solvent effect on the acid–base interaction of electron acceptors with metal oxides (Al_2O_3 and TiO_2) [51, 52].

The amount of TCNQ adsorbed from solutions on the surfaces of Al_2O_3 and TiO_2 was determined using three basic solvents (acetonitrile, ethyl acetate, and 1,4-dioxane) and two acidic solvents (dichloromethane and chloroform). The adsorption isotherms of TCNQ from these basic and acidic solvents were measured. In Fig. 4 the adsorption isotherms of TCNQ from acetonitrile on Al_2O_3 and TiO_2 are given. It can be seen from Fig. 4 that the isotherms are of the Langmuir type for both metal oxides. The other isotherms were almost of the Langmuir type. From the Langmuir plots of these isotherms, the saturated amounts of TCNQ adsorbed were obtained; their values are listed in Table 2. It can be seen that the saturated amount of TCNQ adsorbed decreases considerably with increasing basicity of the solvent or acidity of the solvent for both metal oxides. To interpret these results in terms of the acid–base theory, the heats of mixing of substances with acid–base interactions are expressed by the Drago equation [6]:

$$-\Delta H^{ab} = C_A C_B + E_A E_B,$$

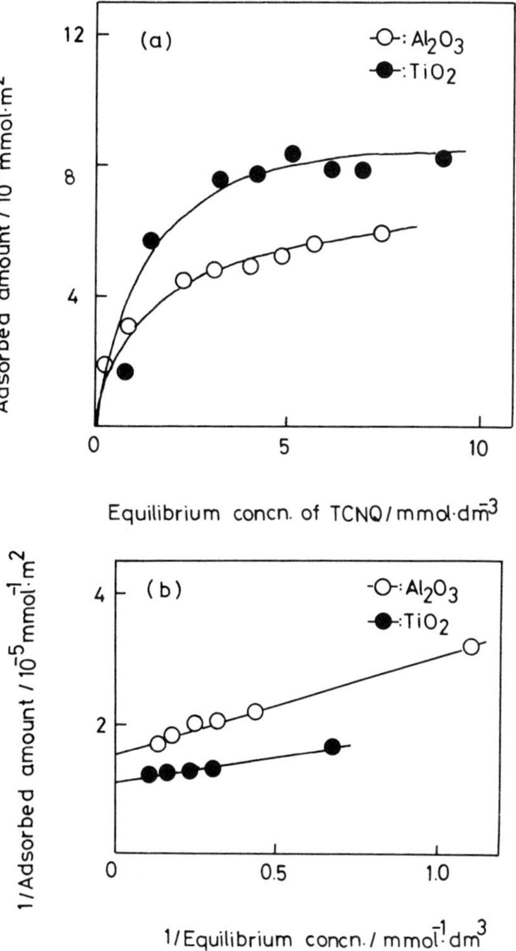

Figure 4. (a) Adsorption isotherms of TCNQ from acetonitrile on Al$_2$O$_3$ and TiO$_2$. (b) Langmuir plots.

Table 2.
Saturated amounts of TCNQ adsorbed on metal oxides from various solvents

Substrate	Saturation amount adsorbed (10^7 mol m^{-2})				
	Acetonitrile	Ethyl acetate	1,4-Dioxane	CH$_2$Cl$_2$	CHCl$_3$
Al$_2$O$_3$	6.2	4.4	2.9	7.4	5.4
TiO$_2$	7.7	5.7	1.9	8.2	3.7

where E and C are the Drago constants for the acidic compound (A) and the basic compound (B). Drago *et al.* [6, 53] determined many E and C values for organic compounds from measurements of the heat of acid–base interactions made by his group and by others using calorimetric and spectroscopic methods. Their studies demonstrated that the Drago equation usually predicted ΔH^{ab} values within 3%.

Accordingly, a very useful approach for relating the interfacial interactions quantitatively has been provided by the Drago equation of enthalpy changes in acid–base complexation. When the Drago equation is applied to this work, the E and C values for five solvents are required and these values are listed in Table 3.

The saturated amount of TCNQ adsorbed onto Al_2O_3 and TiO_2 is plotted as a function of the acidity or basicity of the solvent in Fig. 5. The basicity of the solvents was estimated by the heat of interaction with a given acid (TCNQ); as the values of E and C for TCNQ are not available, the E and C values of TCNE, which is similar to TCNQ, were employed. The acidity of the solvents is illustrated by their heat of interaction with the electron donor sites of the metal oxide surfaces, where for convenience we chose acetone as the base because no suitable compound can be found in the literature [6, 53] to serve as a model of electron donor sites.

Table 3.

Enthalpy changes in acid–base complexation calculated by the Drago equation [6, 53]

Solvent	C_B	E_B	C_A	E_B	$-\Delta H^{ab}$ with TCNE[a] or acetone (kcal mol^{-1})
Acetonitrile	1.34	0.886	—	—	3.51
Ethyl acetate	1.74	0.975	—	—	4.27
1,4-Dioxane	2.38	1.09	—	—	5.23
Dichloromethane	—	—	0.01	1.66	1.66
Chloroform	—	—	0.150	3.31	3.62

[a]TCNE ($C_A = 1.51$, $E_A = 1.68$), acetone ($C_B = 2.33$, $E_B = 0.987$).

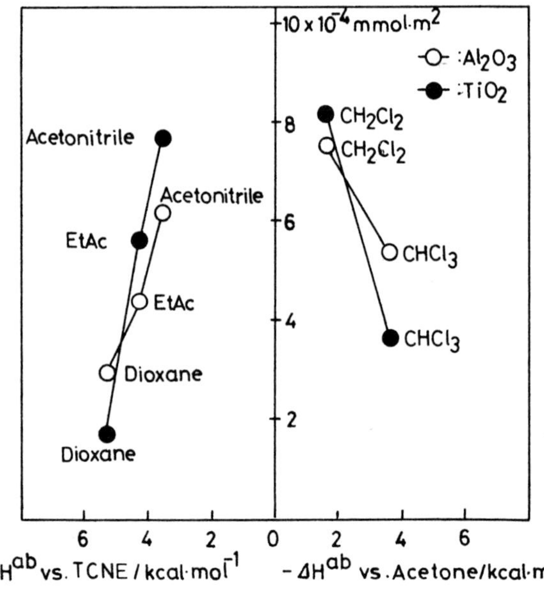

Figure 5. The change in the saturated amount of TCNQ adsorbed on Al_2O_3 and TiO_2 as a function of the acidity or basicity of solvent ($-\Delta H^{ab}$).

In Fig. 5 the adsorption of TCNQ onto the metal oxides is shown to decrease with more acidic or more basic solvents. Apparently, the adsorption of TCNQ is dominated by acid–base interactions. On the left-hand side of Fig. 5, the decrease in adsorption of TCNQ with increasing basicity of the solvents shows the competition between the basic solvents and basic sites (electron donor sites) of the metal oxides for TCNQ. Similarly, on the right-hand side of Fig. 5, the decrease in adsorption of TCNQ with increasing acidity of the solvents illustrates the competition between the acidic solvents and TCNQ for the basic sites of the metal oxides.

The samples colored by the TCNQ adsorption gave unresolved ESR spectra with a g value of 2.003. These spectra have been identified as being those of TCNQ anion radicals [11, 44].

The TCNQ-radical concentrations formed from acetonitrile on the surfaces of the metal oxides are plotted against the equilibrium concentration of TCNQ in Fig. 6. In the other systems, similar trends in the TCNQ-radical concentrations were observed to those in the TCNQ–acetonitrile system. For the three basic solvents, the TCNQ-radical concentration for both metal oxides decreased in the following order: acetonitrile > ethyl acetate > dioxane, i.e. decreasing with increasing basicity of the solvents. On the other hand, for the two acidic solvents, the TCNQ-radical concentration for Al_2O_3 was altered, but for TiO_2 no difference in the TCNQ-radical concentration was observed. Furthermore, in order to compare the TCNQ-radical concentration as a function of the basicity or acidity of the solvents, the TCNQ-radical concentration at equilibrium concentration corresponding to half the value of the saturated amount of TCNQ adsorbed was adopted because the isotherms of TCNQ radicals did not obey the Langmuir equation; the results are illustrated in Fig. 7. The TCNQ-radical concentration on Al_2O_3 decreased gradually in the weakly basic solvent and steeply in the moderately basic solvent. On the right-hand side of Fig. 7, the TCNQ-radical

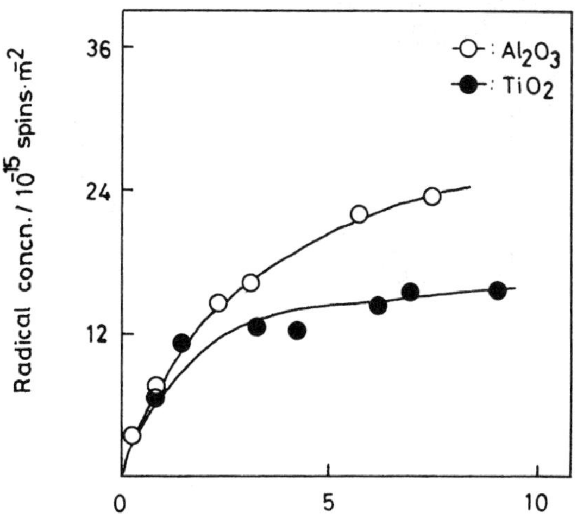

Figure 6. The TCNQ-radical concentrations on the surfaces of the metal oxides obtained by adsorption of TCNQ from acetonitrile vs. the equilibrium concentration of TCNQ.

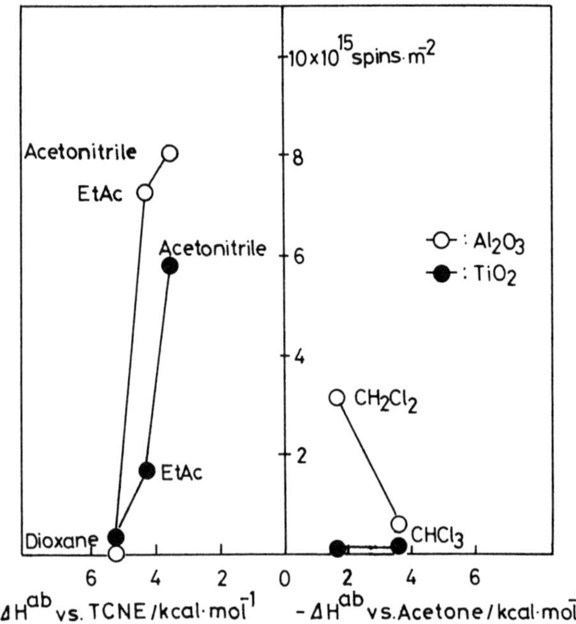

Figure 7. The change in TCNQ-radical concentration on the metal oxides as a function of the basicity or acidity of the solvent ($-\Delta H^{ab}$).

concentration on Al_2O_3 decreased with increasing acidity of the solvent, while that on TiO_2 was almost the same even for different acidities of the solvent. It should be noted that less than 3% of the TCNQ adsorbed on the metal oxide is converted into TCNQ radicals. Further, Fig. 8 shows that the zeta potential of Al_2O_3 decreases with increasing concentration of TCNQ in acetonitrile and ethyl acetate, indicating that TCNQ anion radicals formed on Al_2O_3 contribute to the decrement in the zeta potential [54].

Comparing the TCNQ-radical concentration formed by the adsorption from the basic and acidic solvents, the TCNQ-radical concentration was greater from the basic solvents than from the acidic solvents. This result may suggest that the acidic solvent is adsorbed preferentially on the electron donor sites, preventing the formation of TCNQ radicals, rather than the competition between solvents and TCNQ for the metal oxide. This effect is particularly observed for the adsorption of TCNQ on TiO_2 from an acidic solvent. Similarly, the acid–base interaction at the solid–liquid interface has also been confirmed to be important for the adsorption of tetrachloro-*p*-benzoquinone from various solvents [55]. We have also studied [56] the solvent effect of several aromatic solvents on the charge-transfer adsorption of TCNQ onto metal oxides and found that the TCNQ-radical concentration depends on the ionization potential of the solvent.

Thus, TCNQ adsorpton on metal oxides is strongly influenced by the interaction between basic solvents and TCNQ, or between acidic solvents and donor sites of the metal oxides, and these interactions can be estimated from the acid–base enthalpy predicted by the Drago equation.

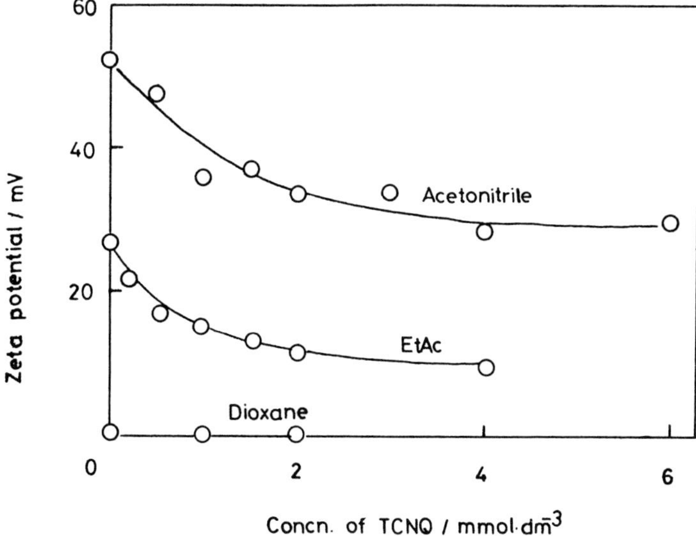

Figure 8. The change in zeta potential of Al_2O_3 as a function of concentration of TCNQ in various solvents.

6. ENHANCEMENT OF ELECTRON DONICITY OF METAL OXIDES BY PLASMA TREATMENT

Recently, plasma treatment has become attractive [57] as a method for surface treatment, probably because it is a dry process at low temperatures with a relatively low pressure gas. We have studied the surface modification of meso-carbon microbeads [58–60] by various plasma treatments so far, and found that oxygen plasma treatment renders the surface more acidic owing to the formation of carboxyl groups, whereas nitrogen or ammonia plasma treatment renders the surface more basic owing to the formation of amino groups. Taking into consideration these plasma treatments for carbon, there is a possibility of modifying the electron donor properties of metal oxides by plasma treatment.

The interaction of plasma-treated metal oxides with TCNQ in acetonitrile solution was studied by measuring the absorption intensity of TCNQ in acetonitrile solution [61]. A typical spectrum of TCNQ in acetonitrile is given in Fig. 9. The band at 395 nm has been assigned to neutral TCNQ, while the bands at 420 and 842 nm have been attributed to the monomer TCNQ radical and those at 372 and 643 nm to the dimer TCNQ radical [14]. The effect of plasma treatment on the acid–base interaction was studied using a UV spectrophotometer. The acetonitrile solution of TCNQ which was in contact with the plasma-treated TiO_2 showed an increase in the absorbances at 420 and 842 nm assigned to the monomer TCNQ radical. This result is explained by assuming that monomer TCNQ radicals are formed as a result of electron transfer from the electron donor sites at TiO_2 to TCNQ and desorption of the radicals from the TiO_2 surface. Thus, from the magnitude of the absorbance at 420 or 842 nm, the change in electron donicity of TiO_2 produced by the plasma treatment can be estimated. Subsequently, in this section, the electron donicity of TiO_2 is expressed by the concen-

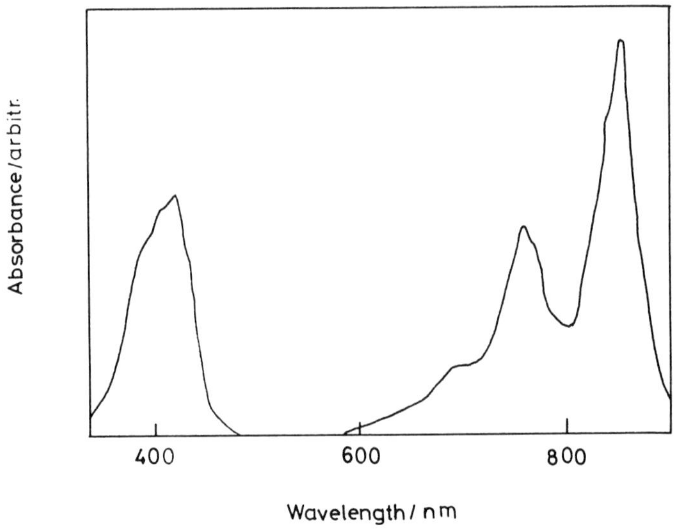

Figure 9. Typical spectrum of TCNQ in acetonitrile.

tration of monomer TCNQ radical formed from the TiO_2 surface in acetonitrile solution. Figure 10 shows the concentration of monomer TCNQ radical formed with TiO_2 by various plasmas. Here, about 3 g of TiO_2 was charged into a reactive vessel, followed by degassing to 1.5 Torr. Each gas was then passed at a rate of 100 ml min^{-1} into the vessel, while evacuated TiO_2 samples were treated using glow discharges generated by a radio frequency of 13.57 MHz under a high-frequency power of 50 W. The pressure during the treatment was kept at 3 Torr. It is apparent that the concentration of monomer TCNQ radical is remarkably increased as a function of the treatment time by the ammonia and nitrogen plasmas; the concentration produced by the ammonia plasma treatment is almost three times greater than that produced by the nitrogen plasma, whereas the concentration is slightly decreased by the oxygen plasma. Furthermore, if the concentration of monomer TCNQ radical formed is expressed per unit surface area, the increment of electron donicity caused by the ammonia and nitrogen plasmas will be much emphasized, compared with that without plasma treatment. These results imply that the electron donicity of TiO_2 is affected by the plasma treatment, and in particular the ammonia plasma enhances very effectively the electron donicity of TiO_2.

Similar experiments were performed on Al_2O_3 under a 100 ml min^{-1} flow rate and high-frequency power, 50 W conditions. Figure 11 demonstrates that the change in concentration of monomer TCNQ radical produced by various plasmas is very similar to that for TiO_2, except that the magnitude of the concentration is much smaller for Al_2O_3 than for TiO_2; the ammonia plasma also induces a greater TCNQ radical concentration, compared with the nitrogen and oxygen plasmas. It was found that the surface of TiO_2 is greatly affected by the plasma, and that the concentration of electron donor sites formed on TiO_2 is greater than that on Al_2O_3.

In order to elucidate the changes in the surfaces of TiO_2 and Al_2O_3 caused by

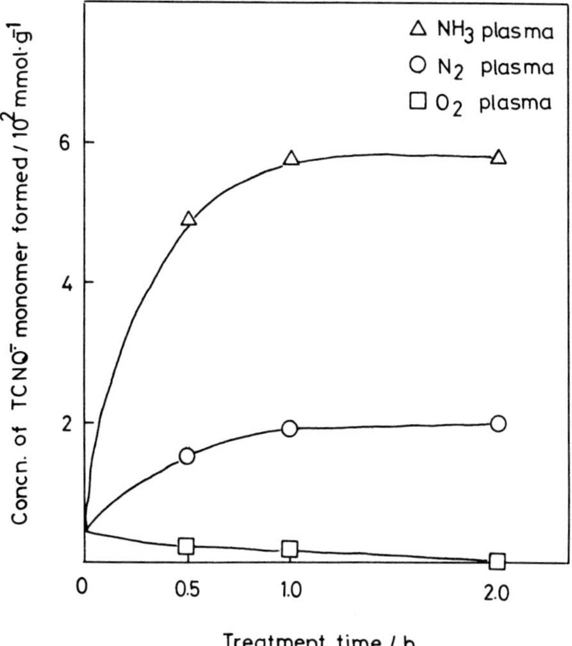

Figure 10. Concentration of TCNQ⁻ monomer formed with TiO_2 treated by various plasmas.

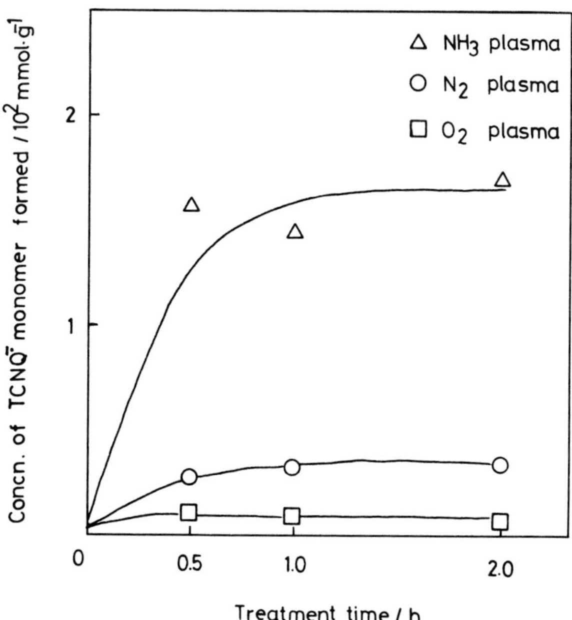

Figure 11. Concentration of TCNQ⁻ monomer formed with Al_2O_3 treated by various plasmas.

the plasmas, the ESCA spectra for TiO_2 and Al_2O_3 treated with various plasmas were obtained. The oxygen and nitrogen plasmas do not provide any new species on TiO_2, but monitoring of the N_{1s} core level at 398.9 eV by the ammonia plasma provides direct confirmation of NH_2 groups at the surface (Fig. 12). Accordingly, it is shown that nitrogen incorporation occurs at the surface of TiO_2 treated with ammonia plasma. These groups are expected to enhance the electron donicity of TiO_2. On the other hand, on Al_2O_3 treated with various plasmas, no new species were observed, although they are expected to be formed, especially with the ammonia plasma. Although this does not exclude the possibility that the properties of surface hydroxyl ions as electron donor sites are affected by the plasma treatment, the electron donor sites for TiO_2 induced by the ammonia plasma appear to be NH_2 groups, as evidenced by ESCA.

7. SUMMARY

The electron-donor properties of metal oxides have been estimated by the adsorption of TCNQ. From the measurement of the TCNQ anion radical concentration formed on the metal oxides by the adsorption of TCNQ, the order of electron donicity is as follows: $MgO > ZnO > Al_2O_3 > TiO_2 > SiO_2 > NiO$. The electron donicity of two-component metal oxide systems exhibits a characteristic tendency, depending on the ratio of the two components. By the adsorption of various electron acceptors with different electron affinity values, the strength and distribution of electron donor sites on metal oxides have been evaluated.

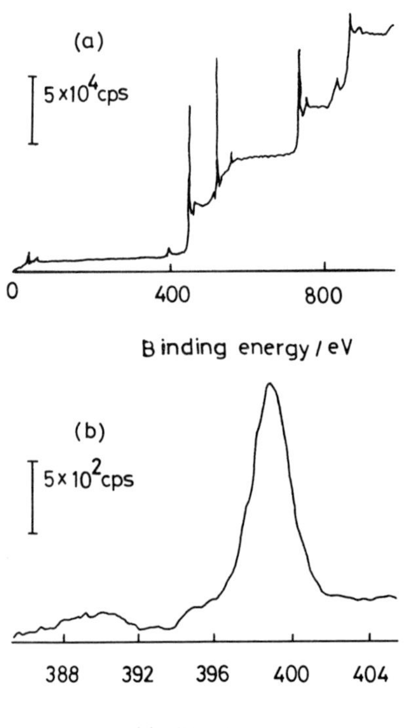

Figure 12. ESCA spectra of TiO_2 treated by ammonia plasma: (a) wide spectrum; (b) N_{1s} peak.

The adsorption of TCNQ on metal oxides is affected by the solvent: the adsorption behavior is correlated with the acid–base interaction estimated by the Drago equation.

Furthermore, in order to enhance the electron donicity of metal oxides, a low-temperature plasma treatment was performed: the electron donicity is increased by the ammonia and nitrogen plasma treatments.

REFERENCES

1. W. B. Jensen, *The Lewis Acid–Base Concepts*, Wiley, New York (1980).
2. P. Sorensen, *J. Paint Tech.* **47**, 31 (1975).
3. S. T. Joslin and F. M. Fowkes, *Ind. Eng. Chem. Prod. Res. Dev.* **24**, 369 (1985).
4. F. M. Fowkes, D. C. McCarthy and J. A. Wolfe, *J. Polym. Sci., Polym. Chem. Ed.* **22**, 547 (1984).
5. F. M. Fowkes, Y. C. Huang, B. A. Shah, M. J. Kulp and T. B. Lloyd, *Colloids Surf.* **29**, 243 (1988).
6. R. S. Drago, L. B. Parr and C. S. Chamberlain, *J. Am. Chem. Soc.* **99**, 3203 (1977).
7. F. M. Fowkes, in: *Physicochemical Aspects of Polymer Surfaces*, K. L. Mittal (Ed.), vol. 2, p. 583. Plenum Press, New York (1983).
8. J. J. Rooney and R. C. Pink, *Trans. Faraday Soc.* **58**, 1632 (1962).
9. B. D. Flockhart, J. A. N. Scott and R. C. Pink, *Trans. Faraday Soc.* **62**, 730 (1966).
10. A. J. Tench and R. L. Nelson, *Trans. Faraday Soc.* **63**, 2254 (1967).
11. H. Hosaka, T. Fujiwara and K. Meguro, *Bull. Chem. Soc. Jpn.* **44**, 2616 (1971).
12. B. D. Flockhart, I. R. Leith and R. C. Pink, *Trans. Faraday Soc.* **65**, 542 (1969).
13. M. Che, C. Naccache and B. Imelik, *J. Catal.* **24**, 328 (1972).
14. R. H. Boyd and W. D. Phillips, *J. Chem. Phys.* **43**, 2927 (1965).
15. D. S. Acker, R. J. Harder, W. R. Hertler, W. Mahler, L. R. Melby, R. E. Benson and W. E. Mochel, *J. Am. Chem. Soc.* **82**, 6408 (1960).
16. R. G. Kepler, P. E. Bierstedt and R. E. Merrifield, *Phys. Rev. Lett.* **5**, 503 (1960).
17. D. B. Chesnut, H. Foster and W. D. Phillips, *J. Chem. Phys.* **34**, 684 (1961).
18. L. R. Melby, R. J. Harder, W. R. Hertler, W. Mahler, R. E. Benson and W. E. Mochel, *J. Am. Chem. Soc.* **84**, 3374 (1962).
19. N. B. Hannay, in: *Semiconductors* N. B. Hannay (Ed.), p. 20. Reinheld, New York (1959).
20. V. M. Vedeneev, L. V. Guvvich, *et al.*, *Cleavage Energies of Chemical Bonds. Ionization Potentials and Electron Affinity Handbook*. Izd. AN SSSR (1960) (in Russian).
21. G. V. Fomin, L. A. Blyumenfel'd and V. I. Sukhorukov, *Proc. Acad. Sci.* **157**, 819 (1964).
22. H. P. Boehm, *Adv. Catal.* **16**, 179 (1966).
23. W. Stober, *Kolloid-Z. Z. Polym.* **145**, 17 (1956).
24. H. P. Boehm, M. Schneider and F. Arendt, *Z. Anorg. Allg. Chem.* **320**, 43 (1963).
25. J. K. Lee and S. W. Weller, *Anal. Chem.* **30**, 1057 (1958).
26. P. J. Anderson, R. F. Horlock and P. J. Oliver, *Trans Faraday Soci.* **61**, 2754 (1965).
27. M. L. Hair and W. Hertl, *J. Phys. Chem.* **74**, 91 (1970).
28. H. Hosaka, N. Kawashima and K. Meguro, *Bull. Chem. Soc. Jpn.* **45**, 3371 (1972).
29. K. Esumi, H. Shimada and K. Meguro, *Bull. Chem. Soc. Jpn.* **50**, 2795 (1977).
30. H. Murayama, K. Kobayashi, M. Koishi and K. Meguro, *J. Colloid Interface Sci.* **32**, 470 (1970).
31. H. Murayama and K. Meguro, *Bull. Chem. Soc. Jpn.* **43**, 2386 (1970).
32. C. J. Plank, *J. Colloid Interface Sci.* **2**, 413 (1947).
33. M. Yamaguchi and S. Tsutsumi, *Nippon Kagaku Zasshi* **69**, 6 (1948).
34. C. L. Thomas, *Ind. Eng. Chem.* **41**, 2564 (1949).
35. B. S. Greesfelder, H. H. Voge and G. M. Good, *Ind. Eng. Chem.* **41**, 2573 (1949).
36. T. H. Millikan, Jr., G. H. Mills and A. G. Oblad, *Discuss. Faraday Soc.* **8**, 279 (1950).
37. G. A. Mills and S. G. Hindin, *J. Am. Chem. Soc.* **72**, 5549 (1950).
38. R. C. Hansford, *Adv. Catal.* **4**, 17 (1952).
39. A. G. Oblad, S. G. Hindin and G. A. Mills, *J. Am. Chem. Soc.* **75**, 4096 (1953).
40. J. D. Danforth, *J. Phys. Chem.* **59**, 564 (1955).
41. R. G. Haldeman and P. H. Emmett, *J. Am. Chem. Soc.* **78**, 2917 (1956).
42. M. W. Tamele, *Discuss. Faraday Soc.* **8**, 270 (1960).
43. K. Esumi and K. Meguro, *J. Colloid Interface Sci.* **66**, 192 (1978).

44. K. Meguro and K. Esumi, *J. Colloid Interface Sci.* **59**, 93 (1977).
45. E. C. M. Chen and W. E. Wentworth, *J. Chem. Phys.* **63**, 3183 (1975).
46. G. Briegleb, *Angew. Chem.* **76**, 326 (1964).
47. J. B. Peri, *J. Phys. Chem.* **69**, 211, 220, 231 (1965).
48. P. A. Agron, E. L. Fuller, Jr. and H. F. Holmes, *J. Colloid Interface Sci.* **52**, 553 (1975).
49. H. F. Holmes, E. L. Fuller, Jr. and R. A. Beh. *J. Colloid Interface Sci.* **47**, 365 (1974).
50. M. Primet, P. Pichat and M. V. Mathieu, *J. Phys. Chem.* **75**, 1216 (1971).
51. K. Esumi, K. Miyata and K. Meguro, *Bull. Chem. Soc. Jpn.* **58**, 3524 (1985).
52. K. Esumi, K. Miyata, F. Waki and K. Meguro, *Colloids Surf.* **20**, 81 (1986).
53. R. S. Drago, G. C. Vogel and T. E. Needham, *J. Am. Chem. Soc.* **93**, 6014 (1971).
54. K. Esumi, K. Magara and K. Meguro, Unpublished work.
55. K. Esumi, K. Miyata, F. Waki and K. Meguro, *Bull. Chem. Soc. Jpn.* **59**, 3363 (1986).
56. H. Hosaka and K. Meguro, *Colloid Polym. Sci.* **252**, 322 (1974).
57. J. R. Hollahan and A. T. Bell, *Techniques and Applications of Plasma Chemistry.* Wiley, New York (1974).
58. M. Sugiura, K. Esumi, K. Meguro and H. Honda, *Bull. Chem. Soc. Jpn.* **58**, 2638 (1985).
59. K. Esumi, M. Sugiura, T. Mori, K. Meguro and H. Honda, *Colloids Surf.* **19**, 331 (1986).
60. K. Esumi, S. Nishina, S. Sakurada, K. Meguro and H. Honda, *Carbon* **25**, 821 (1987).
61. K. Esumi, N. Nishiuchi and K. Meguro, *J. Surf. Sci. Technol.* **4**, 207 (1988).

Acid-Base Interactions, pp. 135-143
Eds. K.L. Mittal and H.R. Anderson, Jr.
©VSP 1991

Determination of the acid–base characteristics of clay mineral surfaces by contact angle measurements— implications for the adsorption of organic solutes from aqueous media

P. M. COSTANZO,[1,*] R. F. GIESE[1] and C. J. VAN OSS[2,3,†]

[1]*Department of Geology,* [2]*Department of Microbiology, and* [3]*Department of Chemical Engineering, State University of New York at Buffalo, Buffalo, NY 14120, USA*

Revised version received 10 January 1990

Abstract—The apolar and the polar (electron-acceptor and electron-donor, or Lewis acid–base) surface tension components and parameters of solid surfaces can be determined by contact angle measurements using at least three different liquids, of which two must be polar. With swelling clay minerals (e.g. smectite clay minerals), smooth contiguous membranes can be fabricated, upon which contact angles can be measured directly. With non-swelling clay minerals (e.g. talc), contact angles can be determined by wicking, i.e. by the measurement of the rate of capillary rise of the liquids in question through thin layers of clay powder adhering to glass plates. The apolar and polar (acid–base) surface tension components and parameters thus found for various untreated and quaternary ammonium base-treated clays allowed the determination of the net interfacial free energy of adhesion of human serum albumin onto the various clay particle surfaces immersed in water. The free energies of adhesion, thus found, correlate well with the experimentally observed degree of adsorption of human serum albumin.

Keywords: Author; to supply; key; words; eight; words; maximum.

1. INTRODUCTION

Fowkes [1] pioneered the consistent practice of distinguishing between the apolar and polar contributions to surface and interfacial tension. Fowkes [2] was also among the first to recognize that the polar contributions are essentially congruent with Lewis acid–base interactions. As far as apolar interactions are concerned, Fowkes [1] identified these with van der Waals–London, or dispersion, forces. In aqueous media, this is certainly correct for more than 90% of the total apolar interaction energy. More recently, however, it could be shown (using Lifshitz's approach [3]) that, in condensed media, apolar macroscopic interactions comprise all three van der Waals forces, i.e. van der Waals–Keesom or orientation forces, van der Waals–Debye or induction forces, and van der Waals–London or dispersion forces, all of which obey the same combining rules and thus should be treated together [4]. These three (apolar) electrodynamic forces were together designated as Lifshitz–van der Waals (or LW) interactions [5, 6]. Chaudhury's

---

*Present address: Unilever Research US, Inc., 45 River Road, Edgewater, NY 17020, USA.
†To whom correspondence should be addressed.

analysis [4] proved it to be legitimate to study the polar aspects of surface inter-actions completely separate from the apolar interactions, as it had become clear that the van der Waals–Keesom and van der Waals–Debye forces had only been considered 'polar' in contrast with 'apolar' through a semantic confusion. (Dipole moments are, of course, involved in the latter two electrodynamic interactions, but at a totally different level from polar interactions of the Lewis variety.)

The way thus became clear for the development of a treatment by the Lewis acid–base, or electron-acceptor/electron-donor, approach of the polar contribution to interfacial interactions. These polar, or acid–base, interactions are designated by the superscript AB.* It soon became obvious that, following the classical Lewis acid–base treatment, AB interactions obey rules which differ drastically from those followed by LW forces [6–8]. These differences in behavior also made it essential to modify the approach by which contact angle analysis is used for the measurement of interfacial interaction energies [6–8]. This new analytical approach makes it possible to obtain reliable values for the Lifshitz–van der Waals components of the surface tension of various materials (i): γ_i^{LW}, as well as of their electron-aceptor and electron-donor parameters, designated respectively as $\gamma_i^{\oplus}$ and $\gamma_i^{\ominus}$. In view of the restricted combining rules governing the interactions involving the electron-acceptor and electron-donor parameters, it was found to be more prudent consistently to use the $\oplus$ and $\ominus$ superscripts for these symbols rather than alphabetical characters such as a and b, or A and B [9]. The application of this approach to clay minerals and other similar fine-grained inorganic materials or surfaces has opened the way to the discovery of a whole new array of materials with often widely differing surface properties, which can be put to a variety of novel uses through the proper utilization of their different adhesive or adsorbing properties.

2. THEORY

The theory of apolar and acid–base (polar) interfacial interactions has been extensively treated elsewhere [6–8]; therefore, only the essentials are given here. The total surface tension of a given (non-metallic) material (i) is the sum of its apolar and polar components:

$$\gamma_i = \gamma_i^{\mathrm{LW}} + \gamma_i^{\mathrm{AB}}. \tag{1}$$

The total acid–base free energy of interaction between two polar materials i and j involves two independent interactions and hence must be expressed as (see, for example, ref. [11])

$$\Delta G_{ij}^{\mathrm{AB}} = -2(\sqrt{\gamma_i^{\oplus}\gamma_j^{\ominus}} + \sqrt{\gamma_i^{\ominus}\gamma_j^{\oplus}}). \tag{2}$$

---

*In 1983, Fowkes [2] first used the connotation ab (lower case) for these interactions but for two reasons it appears preferable to use instead upper case AB. One reason is that LW and AB should be noted in the same manner, and especially in typescripts 'lw' (lower case) leads to errors and, in general, to confusion. The other motive is that for some reason some authors are more readily tempted to split the superscript 'a' from the 'b', and then indulge in non-permissible combinations [9, 10], than would probably be the case with the superscript AB.

As we may state that

$$\Delta G_{ij}^{AB} = -2\gamma_{ij}^{AB}, \tag{3}$$

it follows from equations (2) and (3) that

$$\gamma_{ij}^{AB} = 2\sqrt{\gamma_i^{\oplus}\gamma_i^{\ominus}}, \tag{4}$$

Using the Young–Dupré equation in the form of

$$(1 + \cos\theta)\gamma_i = -\Delta G_{ij} \tag{5}$$

and re-expressing equation (1) [cf. equation (3)] as

$$\Delta G_{ii}^{TOT} = \Delta G_{ii}^{LW} + \Delta G_{ii}^{AB} \tag{6a}$$

and therefore also as

$$\Delta G_{ij}^{TOT} = \Delta G_{ij}^{LW} + \Delta G_{ij}^{AB}, \tag{6b}$$

we may, by combining equations (1), (2), (4), and (6b), express the Young–Dupré equation (5) as

$$(1 + \cos\theta)\gamma_i = 2(\sqrt{\gamma_i^{LW}\gamma_j^{LW}} + \sqrt{\gamma_i^{\oplus}\gamma_j^{\ominus}} + \sqrt{\gamma_i^{\ominus}\gamma_j^{\oplus}}). \tag{7}$$

In this equation it should be noted that

$$\gamma_{ij}^{LW} = (\gamma_i^{LW} - \gamma_j^{LW})^2 \tag{8}$$

and

$$\Delta G_{ij}^{LW} = \gamma_{ij}^{LW} - \gamma_i^{LW} - \gamma_j^{LW}, \tag{9}$$

so that

$$\Delta G_{ij}^{LW} = -2\sqrt{\gamma_i^{LW}\gamma_j^{LW}}. \tag{10}$$

In all the above expressions of the Young equation, i stands for solid and j for liquid.

The values for γ_i^{LW}, $\gamma_j^{\oplus}$, and $\gamma_j^{\ominus}$ can be determined by contact angle (θ) measurement using equation (7). Clearly, to solve for γ_i^{LW}, $\gamma_j^{\oplus}$, and $\gamma_j^{\ominus}$, contact angle measurements must be done with three different, completely characterized liquids (i), of which two must be polar.

Once $\gamma_i^{\oplus}$ and $\gamma_i^{\ominus}$, as well as γ_j^{LW}, $\gamma_j^{\oplus}$, and $\gamma_j^{\ominus}$, are determined for two different materials, i and j, their interfacial tension γ_{ij} follows from the Dupré equation (9) and equations (1), (2), and (4):

$$\gamma_{ij} = (\gamma_i^{LW} - \gamma_j^{LW})^2 + 2(\sqrt{\gamma_i^{\oplus}\gamma_i^{\ominus}} + \sqrt{\gamma_j^{\oplus}\gamma_j^{\ominus}} - \sqrt{\gamma_i^{\oplus}\gamma_j^{\ominus}} - \sqrt{\gamma_i^{\ominus}\gamma_j^{\oplus}}). \tag{11}$$

It should be noted that the polar part of equation (11) corresponds closely to the expression found earlier by Small for hydrogen-bonding interactions [12]. From equations (2), (3), (6b), and (10) various free energies of interaction can be obtained:

$$\Delta G_{iji}^{TOT} = -2\gamma_{ij} \tag{12}$$

(for the interaction between two particles or molecules, i, immersed in a liquid, j);

$$\Delta G_{ij}^{TOT} = -2\sqrt{\gamma_i^{LW}\gamma_j^{LW}} - 2(\sqrt{\gamma_i^{\oplus}\gamma_j^{\ominus}} + \sqrt{\gamma_i^{\ominus}\gamma_j^{\oplus}}) \tag{13}$$

(for the interaction between particles or molecules, i and j, *in vacuo*);

$$\Delta G_{ikj} = \gamma_{ij}^{\mathrm{LW}} - \gamma_{ik}^{\mathrm{LW}} - \gamma_{jk}^{\mathrm{LW}} + 2[\sqrt{\gamma_k^{\oplus}}(\sqrt{\gamma_i^{\ominus}} + \sqrt{\gamma_j^{\ominus}} - \sqrt{\gamma_k^{\ominus}})$$
$$+ \sqrt{\gamma_k^{\ominus}}(\sqrt{\gamma_i^{\oplus}} + \sqrt{\gamma_j^{\oplus}} - \sqrt{\gamma_k^{\oplus}}) - \sqrt{\gamma_i^{\oplus}\gamma_j^{\ominus}} - \sqrt{\gamma_i^{\ominus}\gamma_j^{\oplus}}] \qquad (14)$$

(for the interaction between particles or molecules, i and j, immersed in a liquid, k).

The total interfacial free energy of interaction in general terms is expressed in equations (6a) and (6b), where ΔG^{TOT} can be found for particular cases using equations (12), (13), or (14), as the situation warrants. If, in addition, an electrostatic interaction occurs, one can obtain the total free energy of interaction by adding the electrostatic (EL) energy term:

$$\Delta G = \Delta G^{\mathrm{TOT}} + \Delta G^{\mathrm{EL}}, \qquad (15)$$

where ΔG^{EL} may be derived from the surface, or ψ_0 potential, of the material in question, immersed in a given liquid medium. The ψ_0 potential is obtainable from the measured potential at the slipping plane (i.e. the ζ potential), through an electrokinetic determination [13, 14]. In most of the cases treated below, ΔG^{EL} does not play a preponderant role, and will not be taken into account.

3. METHODOLOGY

3.1. Contact angles

Advancing contact angles (θ) were measured with a number of appropriate liquids on the surfaces of membranes formed with swelling clay such as the smectites of which the Wyoming montmorillonite (SWy-1 of the Clay Minerals Repository), used here, is an example [15]. With non-swelling clays, such as talc (Fisher Scientific), $\cos\theta$ was determined by thin-layer wicking [16], making use of the Washburn equation:

$$l^2 = (tR\gamma_{\mathrm{L}} \cos\theta)/4\eta, \qquad (16)$$

where l is the length of the capillary rise in time, t; R is the average equivalent interstitial pore radius of the spaces between the particles; and η is the viscosity of the wicking liquid. At least three liquids were used (two of which were polar) for determining the various γ_i components and parameters [equation (7)]. For direct contact angle measurement on (swelling) clay membranes, di-iodomethane, α-bromonaphthalene, water, glycerol, and formamide were used; and for the thin-layer wicking, the same liquids were used except for glycerol (which is too viscous for wicking), which was replaced by ethylene glycol. In additon, decane was used for wicking, as a spreading liquid (for which $\cos\theta = 1$), to obtain the value of R [equation (16)]. In Table 1 the relevant properties are given for these contact angle liquids. For the γ^{LW} values, either di-iodomethane or α-bromonaphthalene may be used alone; however, we prefer to use them both, to serve, by comparison, as an indication of the accuracy of the observations. In the same manner, for the γ^{AB} values, either water and glycerol, or water and formamide can be used as the polar liquids. However, again, for the sake of internal control, all three liquids were used in most cases. In wicking, ethylene glycol replaces glycerol, but the surface tension

Table 1.
Selected values of the surface tension components and parameters for the liquids used in this study (in mJ/m^2, at 20°C)

Liquid	γ	γ^{LW}	$\gamma^{\oplus}$	$\gamma^{\ominus}$	η^a
Di-iodomethane	50.8	50.8	0	0	0.028
α-Bromonaphthalene	44.4	44.4	0	0	0.0489
Water	72.8	21.8	25.5	25.5	0.01
Glycerol	64	34	3.92	57.4	14.90
Formamide	58	39	2.28	39.6	0.0455
Ethylene glycol	48	29	3.0	30.1	0.199
Decane	23.8	23.83	0	0	0.0092

^{a} The viscosity is given in poises, at 20°C.

Table 2.
Contact angles (θ) for the solid substrates described in the text. The measurements were derived either from direct contact angle measurements on dry films (SWy-1, SWy-1:HDTMA, and HSA) or from the determinations of cos θ values from wicking measurements with thin layers of powder (talc)

Liquid	SWy-1	SWy-1:HDTMA	Talc	HSA
α-Bromonaphthalene	27.4	27.5	43.7	23.2
Di-iodomethane	35.5	34.3	57.2	37
Water	42.5	67.8	80.4	63.5
Glycerol	40.9	60.0	—	59.5
Formamide	16.4	47.5	47.3	—
Ethylene glycol	—	—	48.8	—
Decane	—	—	0	—

components and parameters of ethylene glycol have not yet been as rigorously established as they have been for the other liquids listed in Table 1. In Table 2, the measured contact angles with the above liquids are given for the materials studied.

3.2. Clays

The clay minerals used were the fine fractions (≤ 2 μm) of the SWy-1 smectite (Clay Minerals Repository) and of talc (Fisher Scientific) for the purpose of having a rather hydrophilic material (SWy-1) and a much more hydrophobic one (talc). Also, SWy-1 was given a hydrophobic surface treatment with hexadecyltri-methyl ammonium bromide (HDTMA) [17]. To that effect 300 mg of SWy-1 was dispersed in 25 ml of distilled water, to which was added a solution of its approximate charge equivalent of 110 mg of HDTMA dissolved in 25 ml of water, and left overnight at room temperature. The next day the SWy-1:HDTMA com-plex was subjected to several centrifugal washes, first with water and subsequently with methanol, to remove any residual unbound HDTMA. SWy-1:HDTMA flocculates in water, and membranes for contact angle determination could only be made by settling from a suspension of the organo-clay in methanol. The relevant surface properties of SWy-1, SWy-1:HDTMA, and talc are given in Table 3.

Table 3.
Values of the components and parameters of the surface tension of the materials used in this study. The values are in mJ/m^2

Mineral	γ^{LW}	$\gamma^{\oplus}$	$\gamma^{\ominus}$
SWy-1	40.7	1.6	28.9
SWy-1:HDTMA	41.0	0.3	12.3
Talc	31.5	2.4	2.7
HSA	41.0	0.002	20.0

3.3. Albumin adsorption

A 0.3% (w/v) human serum albumin (HSA) (Calbiochem, La Jolla, CA) solution in distilled water was used for the adsorption assays. HSA concentrations were determined by UV absorption at 280 nm wavelength, using 0.3% HSA as the standard. For adsorption purposes, the surface properties of dried HSA [18] were used (see also Table 3). The interfacial free energy of adsorption (ΔG_{132}) of HSA onto the various clays was calculated from the data given in Table 4, using equation (15); for the surface properties of the solvent, water, see Table 1. The degree of adsorption, and of elution, of HSA with respect to the various clays is given in Table 4, together with the calculated values of ΔG_{132}.

Table 4.
Results of the adsorption experiments of HSA from water onto the clay minerals of this study. The adsorption data for SWy-1 are expressed for two cases: (1) where the surface area is due solely to the external surfaces of the particles (40 m^2/g), and (2) where the surface area includes the interior surfaces of the smectite [19] (800 m^2/g). For comparison, the calculated values of ΔG_{132} for each experiment are listed

Mineral	mg HSA per g clay	m^2 per g clay	mg HSA per m^2 clay	ΔG_{132} (mJ/m^2)[a]
SWy-1	< 12.3[b]	40.0 (800)	< 0.31[b] (< 0.0155)	− 7.1
SWy-1:HDTMA	26.9[c]	40.0	0.67	− 26.7
Talc	3.4	2.75	1.25	− 41.4

[a] Using equation (7): 1 = HSA; 2 = clay; 3 = water.
[b] Amount still bound after four centrifugal washes with water.
[c] SWy-1:HDTMA releases no futher HSA after two washes.

The surface areas, per 300 mg of the clays used, were taken from the literature [19]. As these surface area/weight values are not exceedingly precise data, we used, for comparison, 300 mg of untreated SWy-1 and 300 mg of SWy-1 subsequently treated with HDTMA, which thus, as a first approximation, have the same surface area. HSA was used for adsorption because of its relatively large molecular size, which precludes its intercalation between the interlayer spaces of the clays.

4. RESULTS AND DISCUSSION

Any discussion of the origin of the surface properties of the minerals of this study must rest ultimately on an understanding of their crystal structures [20]. Both montmorillonite (SWy-1) and talc are layer structure materials whose layers are a composite of a central octahedral sheet bounded by two inward-pointing tetrahedral sheets. In talc, all the octahedral sites are filled by Mg, giving an ideal formula of $Mg_3Si_4O_{10}(OH)_2$. In montmorillonite, only approximately two-thirds of the octahedral sites are filled and there is cationic substitution resulting in a net negative charge on the layer which is balanced by adsorbed cations (predominantly Na in the case of SWy-1), resulting in an ideal formula of $Na_{0.33}(Al_{1.67}Mg_{0.33})Si_4O_{10}(OH)_2$. The major differences then are layer charge (0.0 for talc and 0.33 for montmorillonite, on a per Si_4O_{10} basis) and the nature of the occupancy of the octahedral sites (trioctahedral for talc, dioctahedral for montmorillonite).

In principle, the external surfaces of the individual layers (the 001 surfaces) of both minerals are the same: a hexagonal arrangement of oxygen atoms bonded to tetrahedrally coordinated silicons. The structure of the lateral edges of layer silicates is less well known and is more complex because of the existence of broken bonds and an irregular coordination about cations near the crystal boundaries. In any event, the majority of the external surface is made up of oxygen atoms from both the 001 surfaces and from the lateral edges of the crystals. The abundance of oxygen would suggest that these materials should have a substantial Lifshitz–van der Waals component to their surface properties. This is true, as is seen in Table 3 and elsewhere [15, 16]. Given the bonding of the oxygens of the 001 surfaces to the underlying silicon atoms, it is clear that there are lone-pair electrons directed outward from the individual layers. These lone-pair electrons are available for donation (as a Lewis base) and should result in a substantial value for the $\gamma^\ominus$ parameter. This is true (Table 3) for montmorillonite (28.9 mJ/m^2) but is less obviously the case for talc (2.7 mJ/m^2). The smaller value for $\gamma^\ominus$ of talc must be related to its lack of a layer charge. The $\gamma^\oplus$ values of montmorillonite and talc are not very different. The reason for the different electron-accepting properties of these two minerals is at the moment unclear.

The cation exchange of HDTMA on the surfaces of SWy-1 resulted in very dramatic changes in the surface properties. The Lifshitz–van der Waals (γ^{LW}) component changed only slightly (as might be expected), but the polar parameters ($\gamma^\oplus$ and $\gamma^\ominus$) both decreased substantially. This has the effect of rendering the surface of the SWy-1:HDTMA more hydrophobic causing an increased adsorption of organic material, as shown by the amount of HSA adsobed (Table 4). The decrease in both polar parameters was due to the shielding of the surface oxygens by alkyl groups.

In Table 4 the amounts of HSA are given, in mg per g of clay, as well as in mg per m^2 of surface area. (Since SWy-1 may completely delaminate in water, allowing each fundamental clay layer to contribute to the surface area, two values of the surface area are considered, i.e. 40 m^2/g (no delamination) and 800 m^2/g (complete delamination).) In Table 4 these results are also compared with the total interfacial free energy of attraction ΔG_{132} (where the subscript 1 stands for HSA, 2 for the clay surface, and 3 for water), found from the data given in Table 1 (for water) and Table 3 for HSA and the clay samples, using equation (7). The negative values found for ΔG_{132} indicate that there is a net interfacial *attraction* between HSA and the clay surfaces, when immersed in water. Electrostatic interactions have not been taken into account here, as they are, at close range, of a smaller order of magnitude than the interfacial interaction energies. It is only in the case of SWy-1 that some electrostatic effect may be expected (e.g. as a function of the pH and ionic strength), which could further decrease the absolute value of ΔG_{132}. On the other hand, HDTMA largely neutralizes the negative charge of SWy-1, and talc has a relatively slight (negative) surface potential ($\zeta \approx -11.2$ mV).

Table 3 shows the considerable differences in polar (acid–base) surface properties one can encounter among different clay minerals. Many swelling clay minerals, other than SWy-1, have been observed to vary widely in surface properties [16]. The great variety of surface properties seen among clay minerals makes it possible to choose a material with the most advantageous surface properties for almost any adsorption or adhesion purpose. In addition, owing to the peculiar structural configuration of many clays, such as SWy-1, one can modify a clay mineral's polar surface properties rather irreversibly, by treatment with an appropriate cationic surfactant, such as HDTMA used here. This approach further extends the possibilities of using clay minerals for purposes of adsorption of various organic solutes from water or organic solvents. While talc is clearly even more hydrophobic than SWy-1:HDTMA, so that it adsorbs the most HSA per unit surface area, the total surface area per g of talc is much smaller than that of SWy-1:HDTMA. Thus, per g of clay, SWy-1:HDTMA is a much better adsorbent for HSA, and thus probably also for most other organic solutes from water. Nevertheless, price considerations may also play a role, and an untreated clay such as talc may be less expensive on a per m² basis than a modified clay such as SWy-1:HDTMA.

5. CONCLUSIONS

Contact angle (θ) determinations can be done for swelling clay minerals (by direct contact angle measurements on clay membranes) as well as with non-swelling clay minerals (by thin-layer wicking, which yields cos θ). By θ or cos θ measurement with at least three liquids, the apolar and polar (acid–base) components and parameters could be determined on a variety of different clay minerals. From these data, the adsorption of albumin from aqueous solution onto the clay particle surfaces could be predicted. The predicted values correlate well with the observed adsorption values. There exists among clay minerals a wide variety of surface properties, varying from almost apolar to very polar. These surface properties can in addition be modulated by treatment with cationic surfactants. Thus, clay mineral surfaces can be found, or modified, to suit a wide variety of properties, applicable to a wide array of adsorption requirements.

REFERENCES

1. F. M. Fowkes, *J. Phys. Chem.* **67**, 2538 (1963).
2. F. M. Fowkes, in: *Physicochemical Aspects of Polymer Surfaces*, K. L. Mittal (Ed.), vol. 2, p. 583. Plenum Press, New York (1983).
3. E. M. Lifshitz, *Zh. Eksp. Teor. Fiz* **29**, 94 (1955).
4. M. K. Chaudhury, Ph.D. thesis, SUNY Buffalo (1984).
5. C. J. van Oss, R. J. Good and M. K. Chaudhury, *J. Colloid Interface Sci.* **111**, 378 (1986).
6. C. J. van Oss, M. K. Chaudhury and R. J. Good, *Adv. Colloid Interface Sci.* **28**, 35 (1987).
7. C. J. van Oss, M. K. Chaudhury and R. J. Good, *Chem. Rev.* **88**, 927 (1988).
8. C. J. van Oss, R. J. Good and M. K. Chaudhury, *Langmuir* **4**, 884 (1988).

9. C. J. van Oss, R. F. Giese and R. J. Good, *Langmuir* (submitted).
10. J. Kloubek, *Langmuir* **5**, 1127 (1989).
11. P. Kollman, *J. Am. Chem. Soc.* **99**, 4875 (1977).
12. P. A. Small, *J. Appl. Chem.* **3**, 71 (1953).
13. J. Th. G. Overbeek, in: *Colloid Science*, H. R. Kruyt (Ed.), vol. 1, p. 194. Elsevier, Amsterdam (1952).
14. J. Th. G. Overbeek and P. H. Wiersema, in: *Electrophoresis*, M. Bier (Ed.), vol. 2, p. 1. Academic Press, New York (1967).
15. C. J. van Oss, R. F. Giese and P. M. Costanzo, *Clays Clay Miner.* (in press).
16. R. F. Giese, P. M. Costanzo, C. J. van Oss and J. Norris, 26th Ann. Meeting Clay Minerals Soc., Sacramento, CA, Abstr. (Sept. 1989).
17. M. M. Mortland, S. Shaobai and S. A. Boyd, *Clays Clay Miner.* **5**, 581 (1986).
18. C. J. van Oss and R. J. Good, *J. Protein Chem.* **7**, 179 (1988).
19. H. van Olphen and J. J. Fripiat, *Data Handbook for Clay Materials and Other Non-Metallic Minerals.* Pergamon Press, New York (1979).
20. D. M. Moore and R. C. Reynolds, *X-Ray Diffraction and the Identification and Analysis of Clay Minerals.* Oxford University Press, New York (1989).

Acid-Base Interactions, pp. 145-169
Eds. K.L. Mittal and H.R. Anderson
©VSP 1991

Acid–base properties of carbon and graphite fiber surfaces

SHELDON P. WESSON[1,*] and RONALD E. ALLRED[2]

[1] *TRI/Princeton, P.O.B. 625, 601 Prospect Avenue, Princeton, NJ 08542, USA*
[2] *PDA Engineering, Materials Development Department, 3754 Hawkins NE, Albuquerque, NM 87109, USA*

Revised version received 9 February 1990

Abstract—Carbon and graphite fiber surfaces representing a graded series in graphitic character were analyzed by inverse gas chromatography, programmed thermal desorption, and monofilament wetting. Lewis acids and bases were used as probes to determine the effect of electrooxidation and radio frequency glow discharge plasma on surface acid–base properties. Physical adsorption isotherms computed from chromatograms were analyzed with the CAEDMON (Computed Adsorption Energy Distribution in the MONolayer) algorithm to obtain histograms of surface area fraction versus adsorptive energy. Histograms from isotherms using the Lewis base show fiber surface energetics to be bimodal, with most of the surface presenting dispersion force or weak acid–base attraction to the adsorbates.

A small surface area fraction of acid sites shows strong chemisorptive properties, and this area fraction increases with the degree of oxidation. Plasma treatment is effective at enhancing surface acidity, especially on high modulus graphite fibers that resist conventional electrooxidative processes. Analysis of chemisorption by programmed thermal desorption suggests that oxidative processes damage fiber surfaces, and that plasma treatment enhances surface acidity while causing less surface damage than excessive electrooxidation. Monofilament wetting with a basic liquid shows trends that are qualitatively in accord with those evinced by gas adsorption.

Keywords: Carbon; graphite; fiber; surface energetics; gas adsorption; wetting; plasma.

1. INTRODUCTION

Carbon fibers have found widespread use as reinforcements for high performance composites since their introduction 20 years ago. Because of that demand, there has been an extensive amount of research on processing and structure–property relationships of carbon fibers. That work has been covered in depth in numerous excellent review articles [1–8]. It is generally accepted that carbon fibers from polyacrylonitrile (PAN) precursors consist of two-dimensional turbostratic graphite crystallites arranged in ribbon or lamellar layers oriented parallel to the fiber axis at the fiber surface. Orientation and crystallite size increase with heat treatment temperature, as does the filament modulus. In contrast, pitch-based carbon fibers exhibit a preferred radial orientation of the graphite crystallites.

It is well recognized that adequate fiber–matrix adhesion is required for carbon fibers to serve as reinforcements for high performance composites. Untreated carbon fibers show little adhesion with polymeric matrix materials. There has been

---

*To whom correspondence should be addressed.

considerable research directed at developing and characterizing surface treatments for carbon fibers [4, 7, 9–29]. As a result of those studies, electrochemical oxidation is used in virtually all commercial processes for carbon fibers. The electrochemical process has proven highly successful with PAN-based carbon fibers that are treated at lower temperatures and bonded to epoxy resin systems. Higher modulus PAN-based and pitch-based carbon fibers do not demonstrate the desired level of adhesion with epoxy resins. In addition, most carbon fibers exhibit poor adhesion with high-temperature thermosetting polymers or with engineering thermoplastics. For these composite systems, a more effective carbon surface treatment is necessary.

Improving the bonding at the interface between carbon fiber and matrix polymers can be based on three fundamental adhesion mechanisms or combinations of these: (1) mechanical interlocking; (2) acid–base interactions; or (3) chemical bond formation through reactive intermediates.

Mechanical interlocking is difficult to achieve without degrading the desirable bulk filament properties extensively. Acid–base bonding requires that one bulk material be acidic and the other basic [30, 31]. To optimize bond strength, the strength of the acid and base must be matched as well. Composite matrix resins, in general, are weak bases [23, 32]. As such, weakly acidic fiber surfaces are required to optimize this bonding mechanism. Chemical bond formation is more difficult to achieve and, when possible, is based on a particular polymer chemistry.

Electrooxidative treatments form a wide range of acidic and basic oxygen-containing moieties on carbon surfaces including ethers, hydroxyls, lactones, ketones, carboxylic acids, and carbonates [19]. The electrooxidative process cannot be controlled to produce mainly one particular surface species; it cannot be tailored for forming acid–base or chemical bonds with composite matrix resins.

Radio frequency (RF) glow discharge plasmas offer a surface modification technique that can overcome the limitations of the electrooxidative process. Far-ultraviolet radiation and ion impact generated in the plasma process volatilize surface species from the solid substrate creating surface-active sites. Weak boundary layers and contaminants are also removed by these mechanisms. The gas phase contains numerous reactive species including ions, electrons, free radicals, and molecules with a variety of electronically excited states. By proper choice of gas and plasma operating parameters, surfaces may be functionalized with a narrow distribution of moieties centered around a desired chemical group.

Plasmas are used extensively for modifying polymer surfaces because the low temperatures accompanying plasma treatment do not degrade their bulk properties. There is a large volume of literature on plasma modification of solid surfaces; Refs 33–37 offer a compendium of background information on plasma technology and plasma–surface interactions. Despite the large amount of literature on plasma surface modification techniques, little has been published on plasma treatment of carbon until recently.

The first reference to plasma treatment of carbon fibers is in the patent literature by Goan [38]. Subsequent to Goan's patent, additional patents have been filed on plasma treatment of carbon fibers [39–42]. These patents entail the use of oxidizing gases (NO_2, O_2, SO_2, CO_2, NO, H_2O, and air) or ammonia. Evans and Kuwana have studied plasma amination of bulk graphite [43]. Donnet and co-workers studied plasma modification of carbon fibers [44, 45], as did Loh [46].

Allred studied plasma interactions with organic and carbon fibers [47–52] and their effects on adhesion. Recently, conferences on composite interfaces and the literature have included numerous papers on plasma treatment of carbon fibers [53–61]. These studies show that plasmas can be used to oxidize or aminate carbon surfaces effectively, with or without prior treatment. Plasma processes thus offer a means of tailoring carbon fiber surface chemistry to interact with composite matrix resins without sacrificing fiber bulk properties.

The purpose of this study was to evaluate quantitatively the effect of chemical and thermal treatments on the surface heterogeneity (the nature and extent of electron donor/acceptor character) of a series of increasingly graphitic reinforcements. The effects of a proprietary RF glow discharge plasma are compared with those of standard electrooxidative treatments used by carbon fiber manufacturers. The experimental techniques for evaluating fiber surface properties include monofilament wetting, inverse gas chromatography to quantify physical adsorption, and programmed thermal desorption to provide information about the chemisorptive properties of these substrates.

2. EXPERIMENTAL

2.1. Materials

Table 1 lists the physical properties of the three carbon and graphite reinforcement fibers that we investigated. Tensile strength, tensile modulus, and density ρ are specifications from the manufacturer. Fiber diameter d was measured by wetting with low surface energy liquids, and by optical microscopy.

Table 1.
Properties of carbon and graphite reinforcements

	Tensile modulus (GPa)	Tensile strength (MPa)	ρ (g/cm^3)	d (μm)
Hercules AS4-12K	234	3.8	1.70	7.7
Hercules HMS-4-12K	338	2.5	1.90	7.7
Tonen FT500	483	2.9	2.14	10

Samples from Hercules, Inc. were manufactured from PAN-based precursors and were obtained unsized, with a standard electrooxidative surface treatment. Pitch-based graphite specimens from Tonen Corporation were obtained unsized, with and without a standard electrooxidative treatment. Fibers were subjected to a proprietary RF glow discharge plasma (designated XA1 by PDA Engineering) that has been shown to augment the surface acid functionality of carbon fiber [51].

2.2. Inverse gas chromatography

Nickel columns 55 cm long (0.52 mm i.d.) were loaded with 4–5 g of fiber, a quantity that provided about 1 m^2 of surface for investigation. A Hewlett-Packard 5880A gas chromatograph fitted with a flame ionization detector and a D/A output board was connected to a 20 MHz 80386/80387 Micronics computer containing a 16-bit Data Translation Series 2801 A/D conversion card. Routines written in ASYST programming language collected detector voltages at frequencies between 1 and 10 Hz.

The injector and detector temperatures were maintained at 150 and 300°C, respectively. Nitrogen carrier flow rates were measured with a Gasmet flow meter and were maintained between 22 and 24 ml/min. Measured flow rates F were corrected for column temperature to obtain the effective flow rate F using

$$F = F \frac{T_{col}}{T_{amb}}. \tag{1}$$

There was no pressure drop across the columns, obviating the need for the James–Martin correction [62]. Gas hold-up times were measured with 20 μl injections of methane.

Hamilton 7101NCH syringes were used to inject 2 μl volumes of pentane (neutral probe) and t-butylamine (Lewis base); injections of t-butanol (Lewis acid) were restricted to 0.5 μl in order to obtain sharp peak fronts. Three chromatograms were obtained on each column with each probe at 30°C after conditioning the column overnight at 30°C. Chromatograms using pentane, t-butylamine and t-butanol were collected over periods of 500, 4000, and 2000 s, respectively. An additional 2 h elapsed between measurements with t-butylamine to allow the chemisorbed probe to clear from the column.

Injections performed in triplicate for a probe vapor on carbon fiber are shown in Fig. 1. The main peak is collected at high attenuation so that the area may be obtained to calibrate the detector. Attenuation is brought to zero after peak maximum to obtain the diffuse branch (the tail of the chromatogram) at maximum resolution. The isotherm is determined by computing the number of moles adsorbed per g solid n vs. adsorbate pressure p at many points along the diffuse profile. If ideal gas behavior is assumed, the adsorbate pressure at a point on the diffuse branch is given by

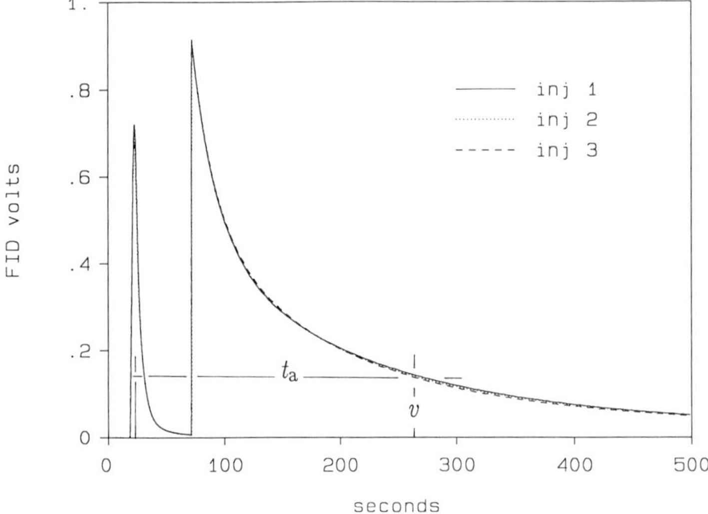

Figure 1. Chromatograms in good registry from three successive injections of t-butylamine on standard electrooxidized carbon fiber at 30°C. All fiber samples were received unsized.

$$p = \frac{n_{inj}RT}{AF}\, v, \tag{2}$$

where n_{inj} is the number of moles injected, R is the gas constant, T is the system temperature, A is the peak area, F is the corrected flow rate, and v is the detector voltage. The number of moles adsorbed per g of solid n is obtained from the relation [63] between the Henry's law constant (vapor partition coefficient) K and the adjusted retention time t_a:

$$n = Kp = \frac{n_{inj}}{g'A}\, t_a v, \tag{3}$$

where g' is the column loading in g. A value proportional to n is represented graphically by the area subtended by the lines t_a and v in Fig. 1; the corresponding adsorbate pressure is proportional to the signal height v.

The geometric specific surface $\Sigma_{geo} = 4/\rho d$ was computed for each carbon fiber sample using the values for density and diameter listed in Table 1. Adsorption volumes are displayed as n/Σ_{geo} (μmol/m^2) to normalize the isotherms for all experimental conditions except differences in adsorption energetics. Equilibrium pressures are given in kPa.

2.3. Programmed thermal desorption

Desorption polytherms were obtained by loading the column with 4–5 g of fiber and maintaining the carrier gas flow rate at 24 ml/min. Signals from the flame ionization detector and oven thermometer were collected at 2 Hz as the columns were subjected to linear temperature ramping from 30 to 300°C at 5°C per min. Parameters were entered into the chromatograph control terminal as follows: initial temperature, 30°C; initial time, 0.5 min; programme rate, 5°C per min; final temperature, 300°C; final time, 0.1 min. Detector response was monitored at zero attenuation. Thermal desorption polytherms are presented as the detector response normalized for the geometric surface area in the column (V/m^2) vs. the column temperature.

2.4. Monofilament wetting

Fiber wettability was measured by the Wilhelmy technique [64] using methylene iodide (nonpolar liquid), formamide (Lewis base), and ethylene glycol (Lewis acid). Carbon monofilaments were glued to metal hooks and suspended vertically from the working arm of a Cahn 2000 microbalance, while a precision elevator raised and lowered a liquid surface along 15 mm of fiber. A computer periodically recorded the change in apparent mass caused by wetting forces at the three-phase boundary. Advancing work of adhesion was computed for each value of apparent mass M as

$$W = \gamma_l + \frac{Mg}{\pi d}, \tag{4}$$

where γ_l is the liquid surface tension, d is the fiber diameter computed from the known adhesion tension in a low energy liquid, and g is the acceleration due to gravity. Surface tension was measured at 50.1 mN/m for methylene iodide, and was taken as 58.2 mN/m for formamide [65] and 47.7 mN/m for ethylene glycol [66]. The dispersion force component of the solid surface energy was computed from the advancing wettability in methylene iodide $W_{CH_2I_2}$ as

$$\gamma_s^d = \frac{W_{CH_2I_2}^2}{4\gamma_{CH_2I_2}}. \tag{5}$$

The dispersion force component of the advancing work of adhesion is found as

$$W^d = 2(\gamma_s^d \gamma_l^d)^{1/2}, \tag{6}$$

where the dispersion force component of liquid surface tension γ_l^d is 39.5 mN/m for formamide [65] and 30.1 mN/m for ethylene glycol [66]. The acid–base component of the work of adhesion is given by

$$W^{a-b} = W - W^d. \tag{7}$$

Values of W are reported as the average of 180 measurements per fiber using three to seven fibers for each solid/liquid combination; the maximum 95% confidence interval is 2 mJ/m^2.

3. THEORY

3.1. Analysis of heterogeneity

The monolayer analysis consists of three elements: an adsorption isotherm equation, a model for heterogeneous surfaces, and an algorithm such as CAEDMON, which uses the first two elements to extract the adsorptive energy distribution and the specific surface from isotherm data. Morrison and Ross [67] developed a virial isotherm equation for a mobile film of adsorbed gas at sub-monolayer coverage:

$$\ln p = \ln \frac{n}{\Sigma} + \frac{2nB^*}{\Sigma} + \frac{3n^2C^*}{2\Sigma^2} + \ldots + \ln \boldsymbol{K}, \tag{8}$$

where $\boldsymbol{K}$ is an integration constant, and B^* and C^* are reduced virial coefficients that describe adsorbate–adsorbate interaction in two dimensions. The specific surface Σ is given by

$$\Sigma = n_m N_a \sigma^2, \tag{9}$$

where n_m is the number of moles adsorbed per g at full monolayer coverage, N_a is Avogadro's number, and σ is the Lennard–Jones distance of closest approach for the adsorbate molecule. The integration constant in equation (8) is equivalent to Σ/K, and can be expressed in terms of U_0, the adsorptive potential of the solid surface:

$$\boldsymbol{K} = A^0 e^{-U_0/RT}, \tag{10}$$

where A^0 describes the difference in degrees of freedom between adsorbate molecules in the bulk gas and the adsorbed film. A^0 is not evaluated in the present analysis. Ross and Morrison treat the heterogeneous surface as a collection of monoenergetic patches with different adsorptive energies [68]. Patches are filled simultaneously, though not to the same density, subject to the condition that the adsorbed phase on each patch has the same chemical potential. An arbitrary range of relative adsorption energies $-RT \ln K_i$ (kJ/mol) is selected in a systematic manner [69], where K_i is the Henry's law constant for a given patch. The number of moles adsorbed per g at each pressure is obtained by summing the individual values of the number of moles per g adsorbed on each patch n_i over all the patches [70]:

$$n(p) = \sum_i n_i(p) = N_a \sigma^2 \sum_i n_{m_i} f(p/K_i), \tag{11}$$

where n_{m_i} is the number of moles adsorbed per g at full monolayer coverage on a given patch, and the function $f(p/K_i)$ is found by solving equation (11) iteratively for n_i/Σ_i. A linear operations algorithm provides a set of non-negative n_{m_i} that minimizes the relative deviation between the model isotherm and experimental data [71]. The sum of the specific surface areas of all the patches is the specific surface of the solid, and the adsorptive energy histogram is obtained as a plot of Σ_i vs. $-RT \ln K_i$.

Figure 2 shows three Gaussian distributions of varying half-widths, each constructed with 20 patches, presented as patch frequency f_i vs. $-RT \ln K_i$. Figure 3 displays model isotherms incorporating representative values of B^* and C^* and the model distributions into equation (11), plotted as fractional surface coverage θ vs. p/K. Isotherm 1, constructed with the narrowest distribution, shows adsorption occurring over a relatively narrow pressure range. Isotherms 2 and 3, computed with the wider distributions, show increased coverage in the low pressure sector due to adsorption on high energy sites. Adsorption on low energy sites diminishes coverage in the high pressure region.

Isotherms are analyzed to produce adsorption energy histograms as follows:

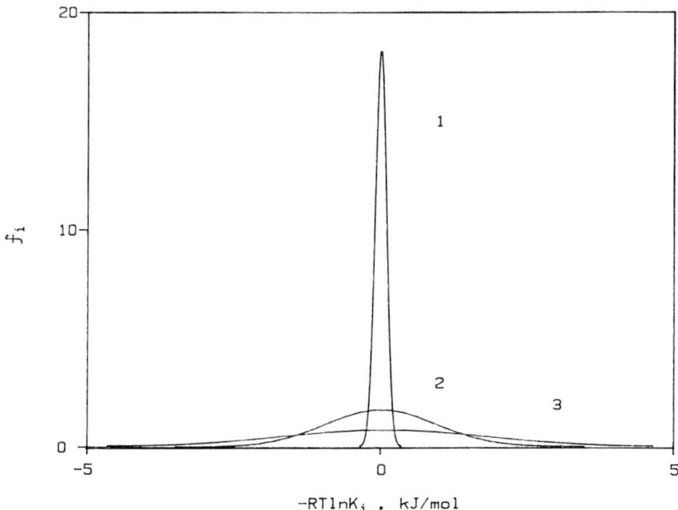

Figure 2. Model distributions of patch fequency vs. relative adsorption energy. Curves 1, 2, and 3 were generated using Gaussian distributions with increasing half-widths to represent surfaces of successively wider ranges of adsorptive energy.

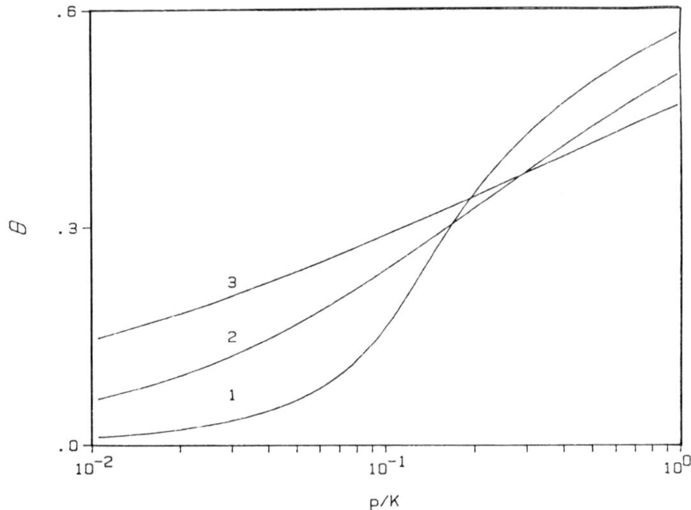

Figure 3. Monolayer adsorption on the model surfaces depicted in Fig. 2, using the virial isotherm equation with $\varepsilon/k = 345$ K and $T = 280$ K.

compute the reduced temperature $T^* = kT/\varepsilon$ using the appropriate Lennard–Jones force constant listed in Table 2; use T^* to interpolate for reduced virial coefficients B^* and C^* from the values calculated by Morrison and Ross [67]; submit the virial coefficients, a vector of adsorptive energies, and a vector of specific moles adsorbed from the isotherm to the CAEDMON algorithm; then compute Σ_i for each n_{m_i} in the vector returned by CAEDMON, using σ from Table 2 in equation (11). The range of adsorptive energies supplied to the algorithm affects the shape of the histogram and the computed value of specific surface. Σ is sensitive to the position of the lowest energy patch; the low energy bound is adjusted to bring Σ to within 10% of Σ_{geo} in the present analysis.

Table 2.
Parameters for gas adsorption analysis

	ε/k (K)	σ (nm)
Pentane	345 [72]	0.5769 [72]
t-Butylamine	310 [73]	0.5701 [74]
t-Butanol	450 (est)	0.5950 [74]

The model for adsorption on heterogeneous substrates is complex, so it is worth noting that other investigators use simpler methods to analyze chromatographic data. Schreiber and co-workers compute the specific retention volume at an arbitrary time, and describe the acid–base character of a substrate as the ratio of the retention volumes of probe acids and bases [75]. A more elaborate treatment of the one-point method used by Papirer [76] and Schultz *et al.* [77] describes nondispersive solid–vapor interactions in terms of the difference in specific retention volume between an acid or base and an alkane at equivalent relative pressure.

These methods feature the implicit assumption that one point on the chromatogram can represent adsorption on the entire surface. This may be possible for adsorption on energetically uniform surfaces, which are rarely encountered, and unlikely in the case of heterogeneous substrates, since the effect of heterogeneity on the shape of monolayer adsorption isotherms is complex. To choose any one degree of coverage as representative is simply to guess at the adsorbate pressure at which high energy sites are covered, and to assume further that only the site areas change with the surface treatment, and not the site energies.

Boudreaux and Cooper [78] transformed chromatograms to adsorptive energy distributions directly, using an analytical expression made possible by assuming that the adsorption isotherm equation is a step function. This assumption is not valid when the system temperature is above the two-dimensional critical temperature for the adsorbed film, and its practical effect is to distort the adsorption energy scale. The method does provide an index of surface heterogeneity, however, and can be used to make qualitative comparisons with less computational effort than that required to evaluate the virial adsorption isotherm equation.

4. RESULTS AND DISCUSSION

4.1. Standard modulus carbon fiber

4.1.1. Physical adsorption and wetting. Isotherms for adsorption at 30°C of pentane, a dispersion force fluid; t-butanol, an acidic probe; and t-butylamine, a Lewis base, are shown in Fig. 4. The substrate is unsized, standard electrooxidized AS4-12K, before and after treatment with the XA1 plasma. The isotherms are presented in log/log format to emphasize differences in adsorptive capacity below 10^{-4} kPa, the sector that displays the effect of adsorption on high energy sites.

Pentane engages only in dispersion force interactions with the substrate. The similarity of the isotherms before and after plasma treatment shows that the dispersion force character of the substrate was not modified, and also that the specific surface was unchanged. The adsorptive capacity of the substrate for t-butanol, the Lewis acid, is greater than that for pentane, showing that acid–base and dispersion force attraction contribute to the total solid–gas interaction. Isotherms obtained before and after plasma treatment are similar, the inference being that the surface concentration of basic sites was not altered appreciably.

The isotherm for t-butylamine on standard electrooxidized AS4 fiber shows a higher adsorptive capacity in the low pressure sector than the isotherms for the other probes, indicating that the primary effect of electrooxidative treatment is to implant high energy acid sites. The dramatic increase in sorptive capacity for the basic probe after plasma treatment shows that the surface acidity imparted by electrooxidation is augmented considerably by the plasma. We observed this effect previously in work where the C_{1s} and O_{1s} photopeaks from high resolution XPS analysis were obtained from electrooxidized fiber before and after XA1 plasma treatment [51]. The plasma increases the surface oxygen concentration, primarily at crystalline edges and defects, and converts carbonate groups imparted by electrooxidation into hydroxyl and carboxyl functionalities.

Figure 5 shows isotherms of t-butylamine on a series of AS4 specimens that were subjected to increasingly intensive electrooxidation, with 100% representing an arbitrary standard. These fibers were subsequently subjected to XPS analysis

    S. P. Wesson and R. E. Allred

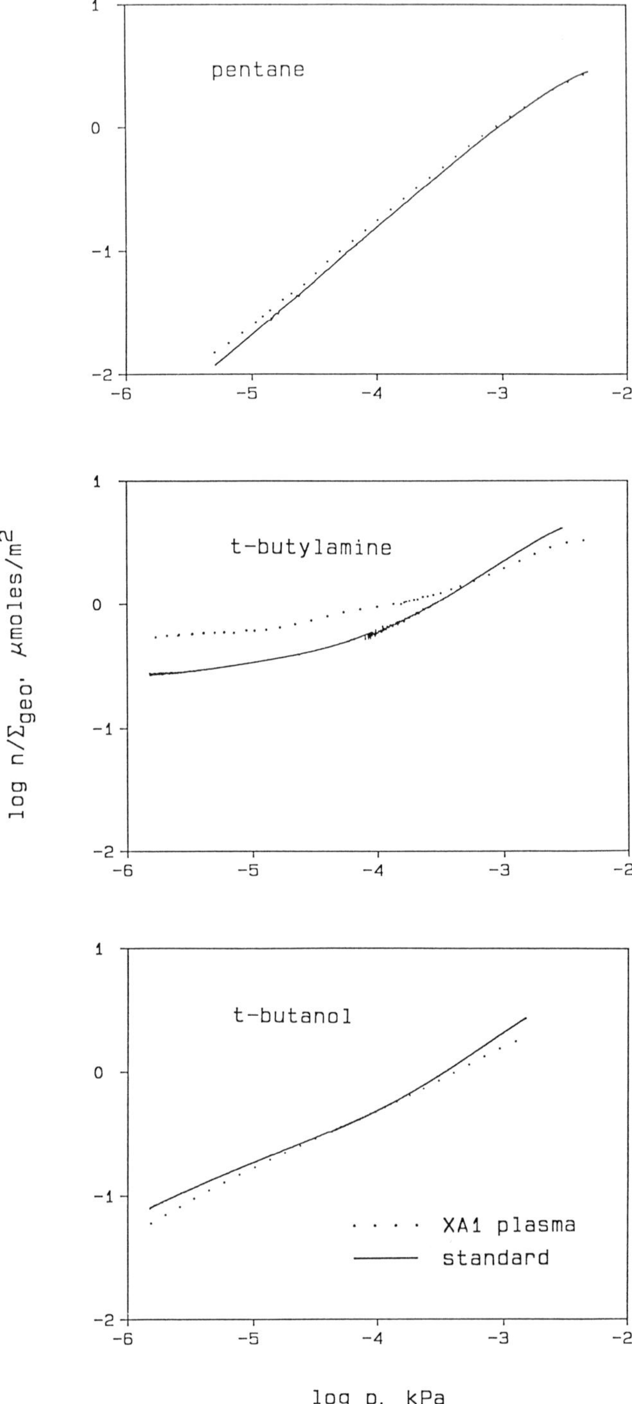

Figure 4. Monolayer adsorption isotherms on standard electrooxidized and plasma-treated AS4 carbon fiber at 30°C using pentane (dispersion force fluid), *t*-butylamine (Lewis base), and *t*-butanol (Lewis acid) as adsorbates.

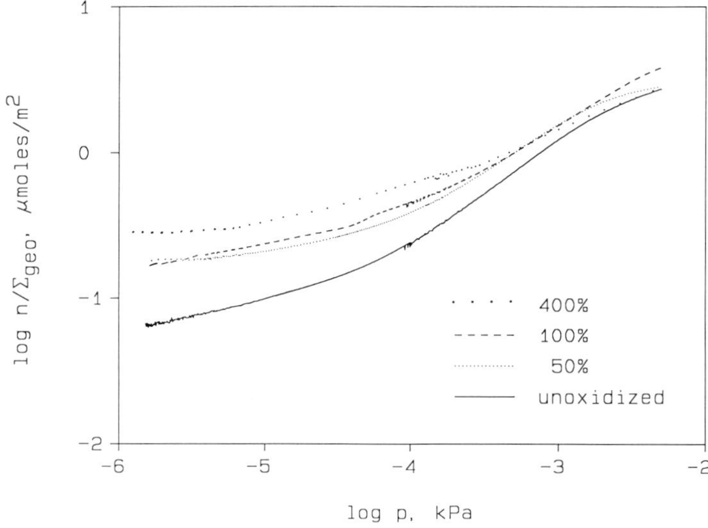

Figure 5. Adsorption of *t*-butylamine at 30°C on AS4 carbon fiber subjected to various levels of electrooxidation.

and used in adhesion studies by Maahs and Yamaki [79]. The unoxidized sample has the lowest adsorptive capacity for the basic probe, hence, the lowest surface acidity. Treatment levels of 50% and 100% are approximately equivalent in their enhancement of surface acidity. The most acidic surface results from 400% treatment, but the extra functionality is not directly proportional to the index of treatment intensity.

The results of monofilament wetting on the graded series of oxidized fibers and displayed in Table 3. The dispersion force component of the fiber surface energy computed from wetting in methylene iodide declines from 44 mJ/m² to about 33 mJ/m² with increasing oxidation, probably caused by disruption of the graphitic surface structure. The value of 31.6 mJ/m² for 100% treated fiber is significantly lower than the measurements on other samples of standard oxidized AS4 (about 34 mJ/m²). It is possible that the 100% treatment damaged the fiber surface more extensively than the treatment level would indicate.

The acid–base component of the work of adhesion in formamide, a Lewis base, is an index of surface acidity. This parameter increases from zero for the untreated specimen to 30 mJ/m² at 100% treatment, and only by an additional 10 mJ/m² at

Table 3.
Wettability of AS4 carbon fiber (mJ/m²)

Level of electrooxidation (%)	CH_2I_2		$HCONH_2$		$HOCH_2CH_2OH$	
	W	γ_s^d	W	W^{a-b}	W	W^{a-b}
0	94.5	44.5	72.2	−6.7	70.5	1.7
50	88.3	38.8	78.3	15.3	75.7	8.3
100	79.6	31.6	101.4	30.7	75.6	14.7
200	83.4	34.6	108.3	34.4	84.8	21.1
400	81.8	33.3	112.3	39.8	84.0	21.5

400% treatment. (The negative value listed for 0% treatment is an artifact, probably resulting from a high value of γ_s^d.) Surface basicity, as indicated by W^{a-b} in ethylene glycol, the Lewis acid, also shows the largest increase between 0 and 100% treatment.

The results from wetting corroborate the main inference from gas adsorption analysis, that most of the surface acid–base character is imparted during the initial stages of electrooxidation. One major difference between the two data sets is that gas adsorption indicates no significant difference in adsorptive capacity resulting from 50 and 100% treatment levels. It is interesting to note that Maahs and Yamaki reported interlaminar shear and tensile properties in unidirectionally wound carbon–carbon composites to be disproportionately high with 50% treated reinforcement fiber, when plotted against surface oxygen concentration from XPS. A rationale for these composite properties is not present in the wetting data, but chromatographic analysis suggests that the marginal adsorptive capacity imparted by increased oxidation diminishes after the 50% treatment level. We noted previously that W^{a-b} from formamide wetting is roughly proportional to the total surface oxygen concentration, while t-butylamine adsorption is sensitive mainly to the concentration of high energy acid sites [80]. The 50% treatment level may, therefore, provide a relatively high surface concentration of strongly acidic sites with minimum damage to the fiber surface structure.

One method for quantifying sorptive characteristics is to process monolayer isotherms with the CAEDMON algorithm to obtain histograms of surface area fraction vs. adsorption energy. Distributions for t-butylamine adsorption on unoxidized AS4 (see Fig. 5), standard oxidized AS4, and plasma-treated AS4 (shown in Fig. 4) are displayed in Fig. 6. The histogram for the basic probe on unoxidized AS4 is essentially unimodal, consisting mainly of low energy dispersion force interactions and a very small surface fraction (about 2%) of high energy sites. The standard electrooxidized surface shows a substantially higher surface fraction of high energy sites (9%); this high energy fraction is increased to 17% by subsequent plasma treatment. The main inference is that acid functionality implanted by electrooxidation and plasma treatment constitutes a relatively small fraction of the reinforcement surface, which is in accord with the results of other investigators [81].

Most of the errors in the analysis combine to overestimate the high energy fraction. These errors include uncertainty in the position of the lowest energy patch caused by distortion of the isotherm by column effects near the peak maximum in the chromatogram, and uncertainty in the position of the highest energy patch, which is saturated at pressures lower than the range of the detector. The assumption of mobile adsorption implicit in the CAEDMON method is reasonable for describing solid–gas interaction with the large area fraction of low energy sites, as indicated by the short retention time for most of the injected probe (see Fig. 1).

A small fraction of the basic probe takes a very long time to clear the column, however, causing skew in the chromatogram that is interpreted as a high energy site in the distribution. This may indicate localized adsorption, or chemisorption, or even condensation of the probe on the high energy surface fraction. The area of this site is computed from the number of molecules adsorbed and the effective area per molecule. Uncertainty in the amount adsorbed is a function of detector

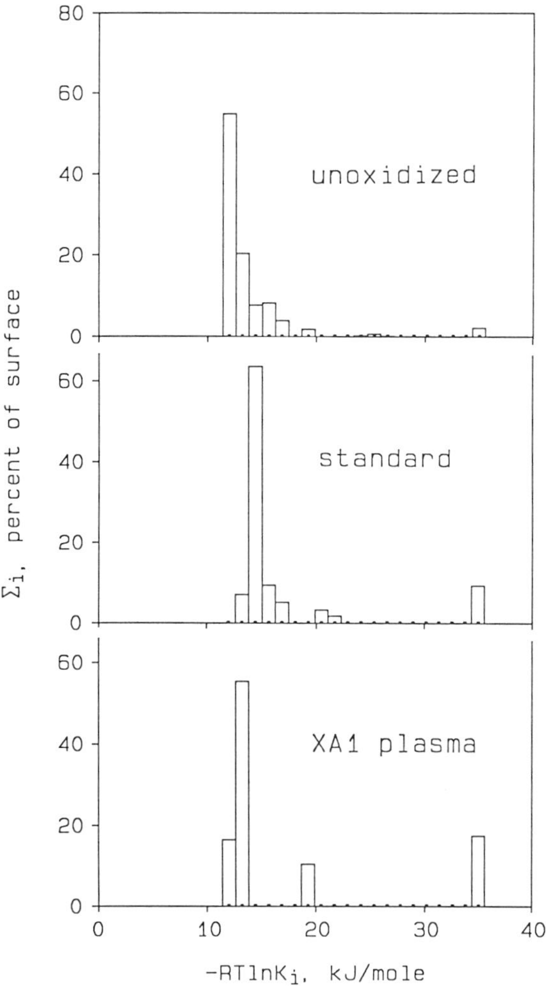

Figure 6. Histograms of surface area fraction vs. relative adsorption energy from CAEDMON analysis of *t*-butylamine adsorption at 30°C on AS4 carbon fiber with unoxidized, standard electrooxidized, and plasma-treated surfaces.

linearity, which is known to be good under these operating conditions. The effective area per molecule for the chemisorbed probe on a high energy site may be much smaller than σ^2 calculated from Table 2, however, and we estimate this uncertainty to be a factor of 2–4. The CAEDMON method could be augmented for analysis of bimodal substrates by using the virial isotherm equation for low energy adsorption and a localized adsorption isotherm equation for the high energy site. This modification is pointless, however, until the stoichiometry of acid–base complex formation is established for these systems.

4.1.2. Chemisorptive properties. We use programmed thermal desorption to obtain more detailed information about the chemisorptive capacity of carbon fiber for acidic and basic adsorbates. Figure 7 shows thermal desorption polytherms from standard electrooxidized AS4 carbon fiber. The curves drawn with widely

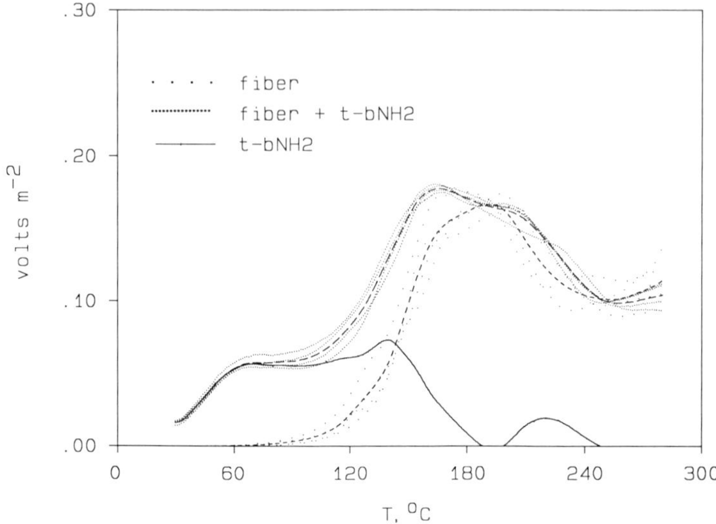

Figure 7. Thermal desorption from standard electrooxidized AS4 carbon fiber. The average polytherm for desorption of surface contaminants with no probe on the column (widely spaced dots) is subtracted from the average desorption polytherm for *t*-butylamine and contaminants (closely spaced dots) to obtain the desorption polytherm for *t*-butylamine alone (solid curve). The column was emptied and reloaded with fresh fiber after each polytherm.

spaced dots were obtained by loading the column and ramping the temperature. Three runs, each using a fresh load of fiber, were averaged to produce the lower dashed line. The curves drawn with closely spaced dots were obtained by loading the column, injecting 2 μl of *t*-butylamine at 30°C, and waiting until the physically adsorbed probe passed through the column. The temperature ramp was started when the detector response fell to the arbitrary level of 0.02% of maximum amplitude. The upper dashed curve represents the average of three runs using a fresh column load for each run. The solid line is obtained by subtracting the lower dashed curve from the upper dashed curve. The solid curve thus shows desorption of *t*-butylamine vs. temperature, with desorption of extraneous material subtracted.

The origin of this extraneous material is deduced from two observations. The desorption volume from old fiber samples is greater than that from new specimens, and the desorption volume diminishes from the outer layers to the inner layers of strand on the bobbin. The fiber evidently acts as a scavenger for organic contaminants in the atmosphere, with the level of contamination increasing with time. Removal of adsorbed contaminants may account for the increased sorptive capacity of carbon fiber substrates for krypton [15] and nitrogen [81] after heating to 300°C. All subsequent desorption polytherms in this work are presented with contaminant desorption subtracted.

The desorption polytherms of *t*-butylamine from standard electrooxidized, plasma-treated, and 400% overoxidized AS4 are displayed in Fig. 8. The desorption envelope from plasma-treated fiber is greater than that from standard electrooxidized fiber across the temperature range. The detector response for the overoxidized fiber is less than either of these below 60°C, and swells rapidly to a

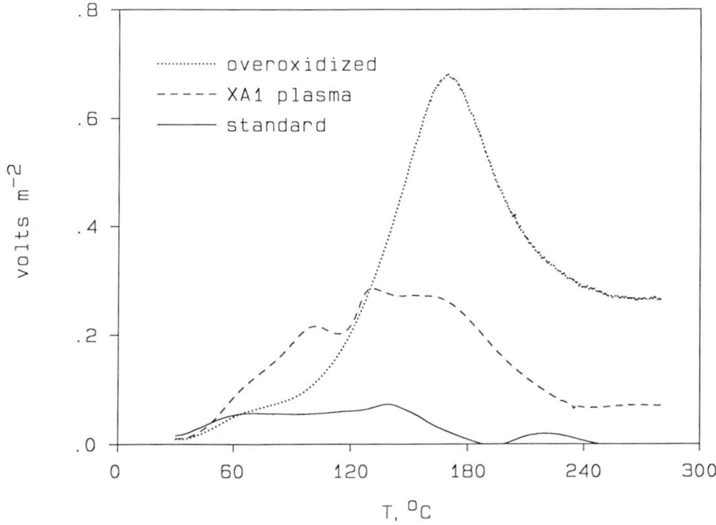

Figure 8. Thermal desorption polytherms for *t*-butylamine on AS4 carbon fiber received as standard electrooxidized, plasma-treated, and 400% overoxidized, with desorption of surface contaminants subtracted.

large maximum at 170°C. One explanation for this large peak is that overoxidation produces a large surface concentration of carboxyl groups [82] and a proportionately large chemisorptive capacity for the basic probe.

This contradicts the evidence from physical adsorption data: monolayer isotherms show that the adsorptive capacity of plasma-treated fiber is greater than that of overoxidized fiber (see Figs 4 and 5). Another possibility is that the desorption volume below 100°C is an index of the chemisorptive capacity of the substrate, while desorption above 100°C is caused by the basic probe interacting with damaged fiber in the near surface region.

The latter interpretation is corroborated in another experiment where the fiber surface was modified by thermal treatment. Standard electrooxidized AS4 was subjected to heating to 300°C under nitrogen; another sample was heated under vacuum to 600°C for 4 h after degassing under vacuum at 120°C overnight. The physical adsorption isotherms for the control and 300°C samples show a similar adsorptive capacity for the basic probe in the low pressure range, whereas the adsorptive capacity of the fiber treated at 600°C is significantly lower than the others (see Fig. 9).

Polytherms obtained from *t*-butylamine desorption are shown in Fig. 10. There is no difference in the area under the polytherms up to 90°C for the control and 300°C samples, and there is no difference in the physical adsorption characteristics at low pressure in Fig. 9. The area under the polytherm for the 600°C fiber is considerably less than that for the first two specimens, up to 90°C. We interpret the three large peaks in the polytherm above 120°C to indicate susceptibility to surface damage by the basic probe that results from thermal stress. By analogy, we interpret desorption above 100°C in Fig. 8 as a measure of the surface damage caused by chemical stress.

Figure 11 displays desorption polytherms from a specimen of overoxidized

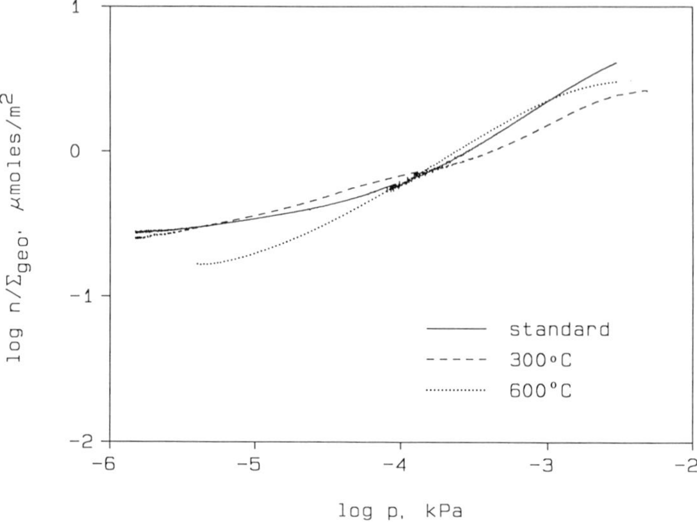

Figure 9. Isotherms for adsorption of *t*-butylamine at 30°C on standard electrooxidized AS4 carbon fiber as received, and after thermal treatment to 300°C under nitrogen, and after thermal treatment to 600°C under vacuum.

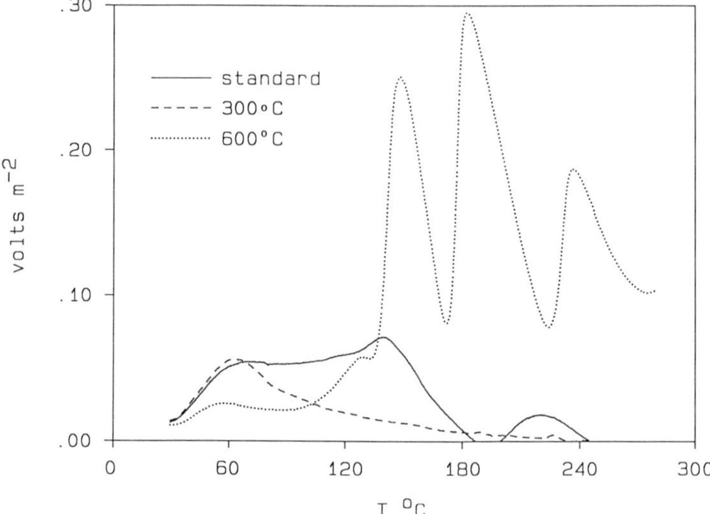

Figure 10. Thermal desorption of *t*-butylamine from the substrates in Fig. 9, with desorption of surface contaminants subtracted from the standard oxidized specimen. There was no desorption of contaminants from the thermally treated fibers.

standard modulus carbon fiber. The dashed line, obtained with no probe in the column, shows desorption of surface contaminants. The dotted curve and the solid curve denote polytherms obtained with *t*-butylamine and *t*-butanol on the column, respectively, after subtracting the curve for contaminant desorption. It is apparent that high temperature desorption occurs only when the basic probe is placed on the column; features of a *t*-butylamine polytherm above 90°C are therefore considered to index the susceptibility of the fiber surface to damage by

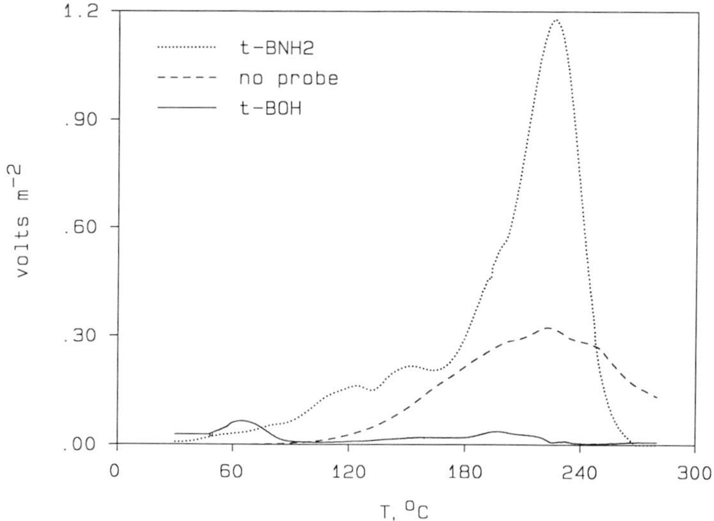

Figure 11. Thermal desorption polytherms from overoxidized carbon fiber. The dashed curve shows desorption of surface contaminants with no probe on the column. The dotted curve is the polytherm for *t*-butylamine desorption, and the solid curve shows desorption of *t*-butanol, both shown with desorption of surface contaminants subtracted.

the base. If overoxidized fiber is subjected to two consecutive ramps to 300°C, each with base on the column, the first polytherm shows high temperature desorption as in Fig. 11, but the second polytherm is similar to the dashed curve in Fig. 10. This shows that high temperature peaks are not caused by desorption from inorganic residues remaining from the oxidation process. The area under the polytherm up to 90°C, therefore, reflects the chemisorptive capacity of fiber surface acid groups, while the area above 90°C is indicative of changes in near surface fiber morphology resulting from surface treatment.

Rand and Robinson [80] reported exotherms disproportionately larger than the increase in specific surface from *n*-butylamine adsorption on carbon fiber oxidized with nitric acid. Our observations suggest that they may have observed a reaction between the adsorbate and products of surface damage, in addition to chemisorption on surface acid groups.

The main inference from our analysis, upon inspection of Fig. 8, is that plasma treatment was more efficacious at augmenting the surface acid concentration than overoxidation, while causing less physical damage to the fiber. Another important conclusion is that carbon fibers maintain a considerable chemisorptive capacity for basic adsorbates after thermal treatment to 300°C under nitrogen (see Fig. 10).

4.2. Intermediate modulus carbon fiber

4.2.1. Physical adsorption and wetting. Physical adsorption isotherms on HMS-4-12K, an intermediate modulus reinforcement, are shown in Fig. 12. Comparison of the sorptive capacity for the basic probe with the *t*-butylamine isotherm in Fig. 4 shows that standard electrooxidized HMS-4 is significantly less acidic than the equivalent AS4 surface. The surface acidity of HMS-4 is augmented considerably by plasma treatment.

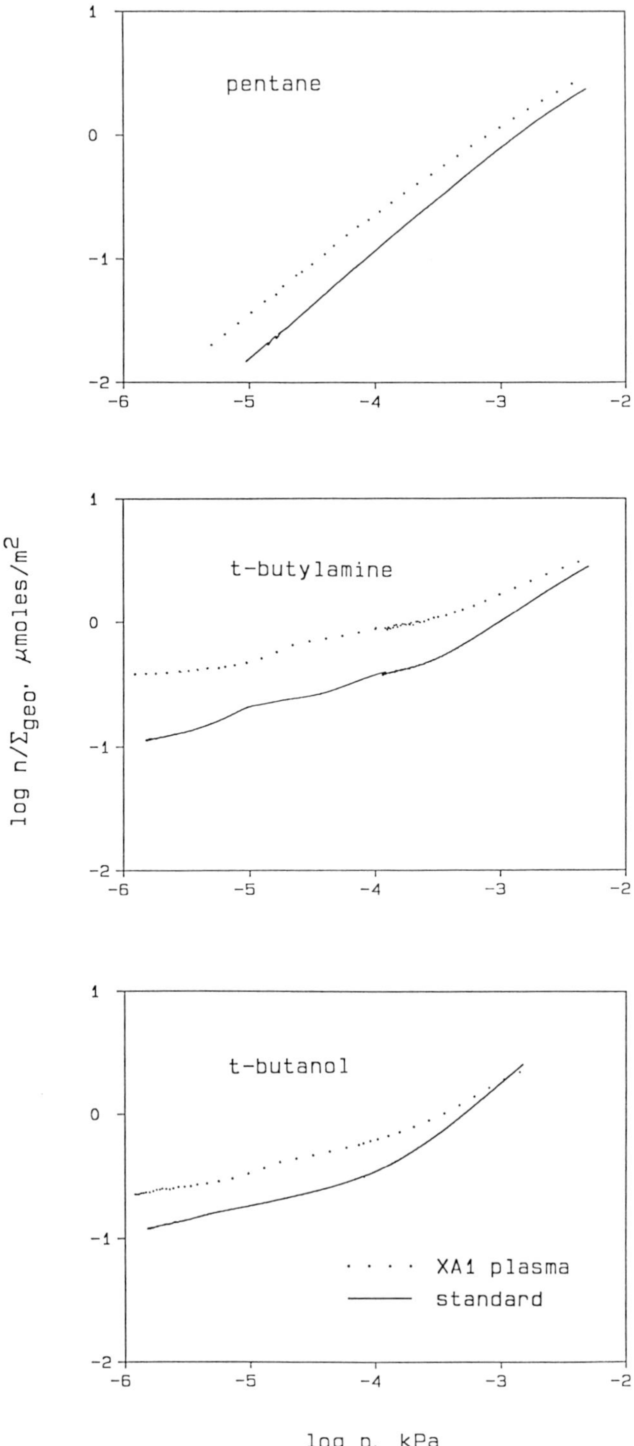

Figure 12. Isotherms for adsorption of pentane, *t*-butylamine, and *t*-butanol on standard electro-oxidized and plasma-treated HMS-4 graphite fiber at 30°C.

The adsorptive capacity of HMS-4 for *t*-butanol, the Lewis acid, is greater than that exhibited by AS4 fiber. This relatively large basic site concentration increases significantly upon plasma treatment. The contrast between AS4 and HMS-4 fiber in susceptibility to basic site formation probably reflects process variations or differences in PAN precursor chemistry.

Inspection of Table 4 shows that γ_s^d for HMS-4 fiber is 3–4 mJ/m^2 larger than that for AS4, a difference that is marginally greater than experimental uncertainty. This difference may indicate that the more highly graphitized HMS-4 (cf. bulk density values in Table 1) features a higher level of surface crystallinity than AS4 fiber. Wetting with formamide and adsorption of *t*-butylamine both show that plasma treatment rendered the surface more acidic. Wetting with the acid probe indicates the surface basicity to be somewhat greater than that of AS4. There is no increase in wettability with the acid probe upon plasma treatment, however, a result that is contradicted by *t*-butanol adsorption.

Table 4.
Wettability of HMS-4 carbon fiber (mJ/m^2)

Treatment	CH$_2$I$_2$		HCONH$_2$		HOCH$_2$CH$_2$OH	
	W	γ_s^d	W	W^{a-b}	W	W^{a-b}
Standard	86.1	36.9	95.4	19.0	85.0	19.2
XA1 plasma	85.9	36.8	109.6	33.4	85.6	20.0

4.2.2. Chemisorptive properties. Thermal desorption of *t*-butylamine from standard oxidized and plasma-treated HMS-4 is shown in Fig. 13. The chemisorptive capacity indicated by desorption below 100°C is less than that of AS4 fiber (cf with Fig. 8, noting change of scale), a result in accord with wetting and physical adsorption measurements. High temperature desorption, our index of

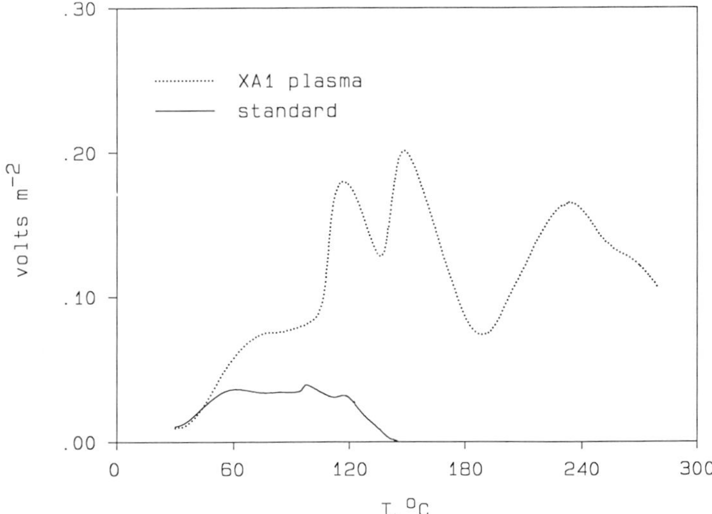

Figure 13. Thermal desorption of *t*-butylamine from standard electrooxidized and plasma-treated HMS-4 graphite fiber, with desorption of surface contaminants subtracted.

surface damage, is also significantly less than that shown by polytherms from AS4 fiber. The main inference from gas adsorption and wetting experiments is that the graphitic HMS-4 surface is more resistant to reaction with and damage by oxidative treatments.

4.3. Pitch-based graphite fiber

4.3.1. Physical adsorption and wetting. Adsorption of pentane, *t*-butylamine, and *t*-butanol on Tonen HM, a pitch-based graphite reinforcement, is displayed in Fig. 14. The specimens were received untreated, electrooxidized, and plasma-treated. (We note that this is the only series in which the plasma was applied to untreated fiber rather than a standard electrooxidized specimen.) We consider the isotherms for pentane and *t*-butanol adsorption to be equivalent within experimental error: surface treatment does not alter the dispersive and electron-donor properties of these three substrates significantly.

Electrooxidation has a minimal effect on the surface acidity of graphite fiber, as shown by the isotherms for *t*-butylamine, while plasma treatment imparts a significant level of surface acidity. This may be empirical evidence that the plasma attacks the basal plane in addition to crystallite edges. The effect of electrooxidation is limited to surface defects [83, 84], which presumably are minimal on this highly graphitized surface.

The values for γ_s^d in Table 5 are not significantly different from those reported for HMS-4 fiber, which is unexpected, considering the difference in bulk density. Changes in W^{a-b} from formamide wetting are in accord with the results of gas adsorption, but ethylene glycol is apparently sensitive to an aspect of surface basicity that *t*-butanol is not.

Table 5.
Wettability of Tonen FT500 graphite fiber (mJ/m^2)

Treatment	CH_2I_2		$HCONH_2$		$HOCH_2CH_2OH$	
	W	γ_s^d	W	W^{a-b}	W	W^{a-b}
Untreated	85.2	36.2	69.5	-6.1	68.9	3.8
Standard	84.6	35.6	82.2	7.2	76.5	11.9
XA1 plasma	88.1	38.7	92.1	13.9	83.6	16.2

4.3.2. Chemisorptive properties. The chemisorptive capacity of graphite fiber is virtually unchanged by electrooxidation, as shown by the polytherms in Fig. 15. The desorption volume of *t*-butylamine below 100°C is augmented considerably by plasma treatment. Difficulties in subtracting the curve for adsorbed contaminants preclude a precise assessment of high temperature desorption, but the general indication is that graphite fiber is less susceptible to damage by surface treatment than lower modulus reinforcements.

5. SUMMARY

The results of CAEDMON analysis of *t*-butylamine adsorption at 30°C are summarized in Table 6. $\Sigma_{chemisorption}$ is the surface area fraction assigned to the highest adsorption energy in histograms computed from monolayer isotherms.

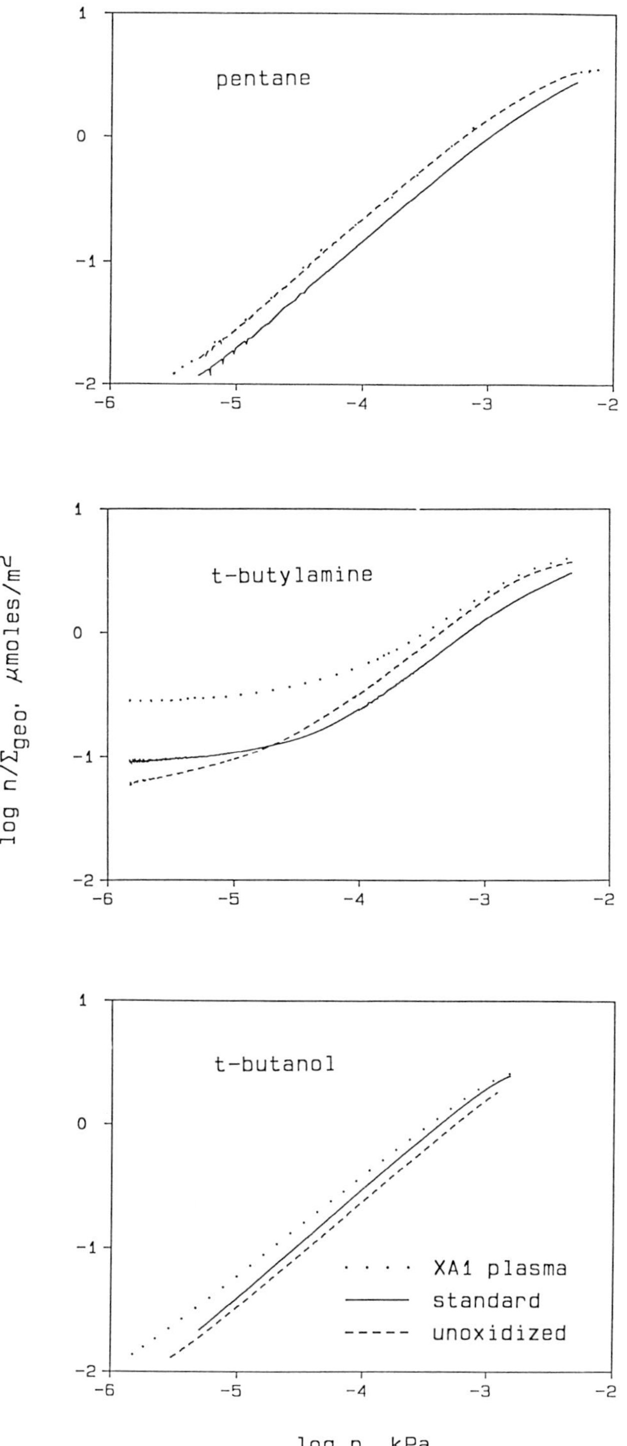

Figure 14. Isotherms for adsorption of pentane, *t*-butylamine, and *t*-butanol on unoxidized standard electrooxidized, and plasma-treated Tonen FT500 graphite fiber at 30°C.

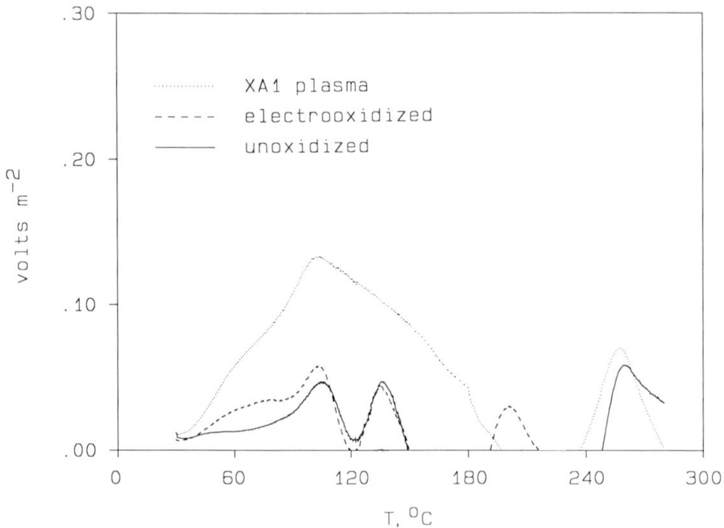

Figure 15. Thermal desorption of t-butylamine from unoxidized, standard electrooxidized, and plasma-treated Tonen FT500 graphite fiber, with desorption of surface contaminants subtracted.

Table 6.

Sorption of t-butylamine by carbon and graphite fibers

	$\Sigma_{\text{chemisorption}}$ (%)		
	Unoxidized	Standard	XA1 plasma
AS4-12K	2.2	9.4	17.4
HMS-4-12K	—	3.7	12.1
Tonen FT500	1.9	3.0	9.3
	Volume desorbed (V °C/m^2): T_{max}		
	Unoxidized	Standard	XA1 plasma
AS4-12K	0.4:50°C	3.2:70°C	5.5:75°C
HMS-4-12K	—	1.7:63°C	3.8:75°C
Tonen FT500	0.3:50°C	0.9:60°C	3.8:75°C

The results from programmed thermal desorption are presented in Table 6 as the area (V °C/m^2) of the first peak between 30 and 100°C, computed as twice the area of integration from 30°C to the peak maximum T_{max}.

The sorption data in Table 6 constitute an empirical index of fiber surface acidity. It is apparent that the effectiveness of the surface treatments diminishes with increasing graphitization, and that plasma treatment increases the level of acidity by a factor of 2–3. Plasma treatment implants acid groups on previously untreated graphite fiber, which may indicate that the XA1 plasma attacks the graphite basal plane as well as crystallite edges, or that it implants more acid sites per unit area of surface defects.

The plot of t-butylamine desorption volume below 100°C vs. $\Sigma_{\text{chemisorption}}$, the fraction of high energy surface area from physical adsorption analysis, is linear, with the intercept at the origin. This is another indication that desorption from acid sites occurs below 100°C, and that desorption at higher temperatures is an index of damage to the fiber surface resulting from surface oxidation.

Desorption volumes above 100°C in *t*-butylamine polytherms diminish with increasing graphitization, suggesting that the locus of damage is edges and defects. High temperature desorption volumes are much smaller when fibers are treated by plasma than by overoxidation, which implies that plasma augments the surface acid functionality with minimal physical damage to carbon and graphite reinforcements.

Physical adsorption and chemisorption of *t*-butylamine are not greatly affected by ramping to 300°C under nitrogen. While only one example was presented in this study, our experience with a variety of PAN-based carbon reinforcements is that ramping in an inert atmosphere never passivates oxidized or overoxidized fiber completely.

The results of formamide wetting are in qualitative accord with most of the trends observed with gas adsorption. Neither pentane adsorption nor wetting with methylene iodide showed γ_s^d to change systematically with the degree of graphitization, perhaps because of the organic contaminants strongly adsorbed on all of these substrates. The Lewis acids used for gas adsorption and wetting responded differently to surface treatment; ethylene glycol often shows increased wettability upon surface oxidation when the adsorptive capacity for *t*-butanol remains unchanged.

The main emphasis of this work was the characterization of surface acidity, primarily because these substrates show weak surface basicity, and changes in basicity with treatment are usually minimal. One interesting exception is the chemisorptive capacity for *t*-butanol that develops upon excessive electro-oxidation of PAN-based fiber (see Fig. 11). This indicates that the relative proportions of hydroxyl, carboxyl, and lactone functionalities change with the duration or intensity of electrooxidation. The detailed characterization of carbon fiber surface basicity and its effect on practical adhesion are subjects for future study.

REFERENCES

1. R. Bacon, in: *Chemistry and Physics of Carbon*, P. L. Walker, Jr. and P. A. Thrower (Eds), vol. 9, p. 1. Marcel Dekker, New York (1973).
2. W. N. Reynolds, in: *Chemistry and Physics of Carbon*, P. L. Walker, Jr. and P. A. Thrower (Eds), vol. 11, p. 1. Marcel Dekker, New York (1973).
3. R. J. Diefendorf and E. Tokarsky, *Polym. Eng. Sci.* **15**, 150 (1975).
4. D. M. Riggs, R. J. Shulford and R. W. Lewis, in: *Graphite Fibers and Composites Handbook of Composites*, G. Lubin (Ed.), p. 196. Van Nostrand Reinhold, New York (1982).
5. J. B. Donnet and R. C. Bansal, *Carbon Fibers*, p. 1. Marcel Dekker, New York (1984).
6. G. Dorey, *J. Phys. D, Appl. Phys.* **20**, 245 (1987).
7. E. Fitzer and M. Heine, in: *Fiber Reinforcements for Composite Materials*, A. R. Bunsell (Ed.), p. 73. Elsevier, New York (1988).
8. A. Oberlin and M. Guigon, in: *Fiber Reinforcements for Composite Materials*, A. R. Bunsell (Ed.), p. 149. Elsevier, New York (1988).
9. J. W. Herrick, *Proc. 12th Natl SAMPE Symp.*, Soc. Adv. Mater. Process Eng., Covina, CA, paper AC-8 (1967).
10. J. W. Herrick, *Proc. 23rd Annu. Tech. Conf.*, Soc. Plastics Ind., Reinforced Plastics/Composites Inst., paper 16-A (1968).

11. J. C. Goan and S. P. Prosen, *Interfaces in Composites*, ASTM STP 452, p. 3. Am. Soc. Testing Mater., Philadelphia, PA (1969).
12. D. W. McKee and V. J. Mimeault, in: *Chemistry and Physics of Carbon*, P. L. Walker, Jr. and P. A. Thrower (Eds), vol. 8, p. 151. Marcel Dekker, New York (1973).
13. J. C. Goan, T. W. Martin and R. Prescott, *Proc. 28th Annu. Tech. Conf.*, Soc. Plastics Ind., Reinforced Plastics/Composites Inst., paper 21-B (1973).
14. L. T. Drzal, *Carbon* **15**, 257 (1977).
15. L. T. Drzal, J. A. Mescher and D. L. Hall, *Carbon* **17**, 375 (1979).
16. P. Ehrburger and J. B. Donnet, *Philos. Trans. R. Soc. London, Ser. A* **294**, 495 (1980).
17. E. Fitzer, K.-H. Geigl, W. Huttner and R. Weiss, *Carbon* **18**, 389 (1980).
18. J. Delmonte, *Technology of Carbon and Graphite Fiber Composites*, p. 177. Van Nostrand Reinhold, New York (1981).
19. A. Ishitani, *Carbon* **19**, 269 (1981).
20. J. B. Donnet and R. C. Bansal, *Carbon Fibers*, p. 109. Marcel Dekker, New York (1984).
21. O. P. Bahl, R. B. Mathur and T. L. Dhami, *Polym. Eng. Sci.* **24**, 455 (1984).
22. A. Ishitani, in: *Molecular Characterization of Composite Interfaces*, H. Ishida and G. Kumar (Eds), p. 321. Plenum Press, New York (1985).
23. R. J. Diefendorf, in: *Tough Composite Materials—Recent Developments*, L. F. Vosteen, N. J. Johnston and L. A. Teichman (Eds), p. 191. Noyes Publications, Park Ridge, NJ (1985).
24. I. L. Kalnin and H. Jager, in: *Carbon Fibers and Their Composites*, E. Fitzer (Ed.), p. 62. Springer-Verlag, New York (1985).
25. T. A. DeVilbliss and J. P. Wightman, in: *Composite Interfaces*, H. Ishida and J. L. Koenig (Eds), p. 307. North-Holland, New York (1986).
26. E. Fitzer and R. Weiss, *Carbon* **25**, 455 (1987).
27. W. D. Bascom and L. T. Drzal, *The Surface Properties of Carbon Fibers and Their Adhesion to Organic Polymers*, NASA Contractor Report 4084, NASA Contract NAS1-17918, NASA Langley Research Center (1987).
28. G. Dorey and J. Harvey, in: *Interfaces in Polymer, Ceramic and Metal Matrix Composites*, H. Ishida (Ed.), p. 693. Elsevier, New York (1988).
29. J. B. Donnet and G. Guilpain, *Carbon* **27**, 749 (1989).
30. F. M. Fowkes, *J. Adhesion* **4**, 155 (1972).
31. F. M. Fowkes, *J. Adhesion Sci. Technol.* **1**, 7 (1987).
32. T. A. DeVilbiss and J. P. Wightman, *Surface Characterization in Composite and Titanium Bonding: Carbon Fiber Surface Treatments for Improved Adhesion to Thermoplastic Polymers*, Final Report No. 105, CAS/CHEM-20-87, Virginia Polytechnic Institute and State University, Blacksburg, VA (1987).
33. J. R. Hollahan and A. T. Bell (Eds), *Techniques and Applications of Plasma Chemistry*, p. 113 (and refs cited therein). John Wiley, New York (1974).
34. B. Chapman, *Glow Discharge Processes*. John Wiley, New York (1980).
35. M. Venugopalan (Ed.), *Reactions under Plasma Conditions*, Vols 1 and 2, John Wiley, New York (1971).
36. R. F. Baddour and R. S. Timmins (Eds), *The Application of Plasmas to Chemical Processing*. Pergamon Press, Oxford (1967).
37. H. V. Boenig, *Plasma Science and Technology*. Cornell University Press, Ithaca, NY (1982).
38. J. C. Goan, Method for improving graphite fibers for plastic reinforcement and products thereof, U.S. Patent 3,634,220 (11 Jan. 1972).
39. K. C. Hou, Surface modification of carbon fibers, U.S. Patent 3,745,104 (10 July 1973).
40. J. C. Goan, Aminated carbon fibers, U.S. Patent 3,776,829 (4 Dec. 1973).
41. H. Fujimaki *et al.*, Process for the surface treatment of carbon fibers, U.S. Patent 4,009,305 (22 Feb. 1977).
42. S. Ueno and H. Kamata, Method for imparting improved surface properties to carbon fibers and composite, U.S. Patent 4,487,880 (11 Dec. 1984).
43. J. F. Evans and T. Kuwana, *Anal. Chem.* **51**, 3 (1979).
44. J. B. Donnet, M. Brendle, T. L. Dhami and O. P. Bahl, *Carbon* **24**, 757 (1986).
45. J. B. Donnet, T. L. Dhami, S. Dong and M. Brendle, *J. Phys. D, Appl. Phys.* **20**, 269 (1987).
46. I. H. Loh, Sc. D. Thesis, Massachusetts Institute of Technology (1985).
47. R. E. Allred, H. K. Street and R. J. Martinez, *Proc. 24th Natl SAMPE Symp. Exhib.* 31, Soc. Adv. Material and Process Eng., Covina, CA (1979).

48. R. E. Allred, E. W. Merrill and D. K. Roylance, in: *Molecular Characterization of Composite Interfaces*, H. Ishida and G. Kumar (Eds), p. 333. Plenum Press, New York (1985).

49. H. H. Madden and R. E. Allred, *J. Vacuum Sci. Technol. A*, **4**, 1705 (1986).

50. H. M. Stoller and R. E. Allred, *Proc. 18th Int. SAMPE Tech. Conf.* 993, Soc. Adv. Material and Process Eng., Covina, CA (1986).

51. S. P. Wesson and R. E. Allred, in: *Inverse Gas Chromatography*, D. R. Lloyd, T. C. Ward, and H. P. Schreiber (Eds), Am. Chem. Soc. Symp. Series 391, Ch. 15, p. 204. ACS, Washington, DC (1989).

52. R. E. Allred and L. A. Harrah, *Proc. 34th Int. SAMPE Symp. Exhibition* 2559, Soc. Adv. Material Process Eng., Covina, CA (1989).

53. J. Su, X. Tao, Y. Wei, Z. Zhang and L. Liu, in: *Interfaces in Polymer, Ceramic and Metal Matrix Composites*, H. Ishida (Ed.), p. 269. Elsevier, New York (1988).

54. B. Z. Jang, H. Das, L. R. Hwang and T. C. Chang, in: *Interfaces in Polymer, Ceramic and Metal Matrix Composites*, H. Ishida, (Ed.), p. 319. Elsevier, New York (1988).

55. J. B. Donnet, S. Dong, G. Guilpain and M. Brendle, in: *Chemistry and Physics of Carbon*, P. L. Walker, Jr. and P. A. Thrower (Eds), vol. 9, p. 35. Marcel Dekker, New York (1973).

56. N. H. Sung, G. Dagli and L. Ying, *Proc. 37th Annual Conference Reinforced Plastics/Composites Institute*, 23-B. Society Plastics Industry, New York (1982).

57. T. C. Chang and B. Z. Jang, in: *Interfaces in Composite Materials*, C. G. Pantano and E. J. H. Chen (Eds), **170**, 321. Materials Research Society, Pittsburgh, PA (1990).

58. U. Gaur and T. Davidson, in: *Interfaces in Composite Materials*, C. G. Pantano and E. J. H Chen (Eds), **170**, 309. Materials Research Society, Pittsburgh, PA (1990).

59. J. A. Hinkley, W. D. Bascom and R. E. Allred, in: *Interfaces in Composite Materials*, C. G. Pantano and E. J. H. Chen (Eds), **170**, 351. Materials Research Society, Pittsburgh, PA (1990).

60. C. Jones and E. Sammann, in: *Polymer Based Molecular Composites*, D. W. Schaefer and J. E. Mark (Eds), **171**, Materials Research Society, Pittsburgh, PA (1990).

61. S. Mujin, H. Baorong, W. Yisheng, T. Ying, H. Weigiu and D. Youxian, *Composite Sci. Tech.* **34**, 353 (1989).

62. A. T. James and A. J. P. Martin, *J. Biochem.* **50**, 679 (1952).

63. M. G. Neumann, *J. Chem. Ed.* **53**, 708 (1976).

64. S. P. Wesson and J. S. Jen, *Proc. 16th Natl SAMPE Tech. Conf.* **16**, 375 (1984).

65. F. M. Fowkes, in: *Treatise on Adhesion and Adhesives*, R. L. Patrick (Ed.), vol. 1, p. 360. Marcel Dekker, New York (1967).

66. T. Wakida, H. Kawamura and J. Song, *Sen'I Gakkaishi* **43**, 384 (1984).

67. I. D. Morrison and S. Ross, *Surf. Sci.* **39**, 21 (1973).

68. S. Ross and I. D. Morrison, *Surf. Sci.* **52**, 103 (1975).

69. S. P. Wesson, J. J. Vajo and S. Ross, *J. Colloid Interface Sci.* **94**, 552 (1983).

70. R. S. Sacher and I. D. Morrison, *J. Colloid Interface Sci.* **70**, 153 (1979).

71. R. J. Hanson, *Solving Least Squares Problems*, Prentice Hall, Englewood Cliffs, NJ (1974).

72. J. O. Hirschfelder, C. F. Curtiss and R. B. Bird, *Molecular Theory of Gases and Liquids*, p. 1112. John Wiley, New York (1964).

73. *DIPPR Data Compendium of Pure Compound Properties:* NBS Standard Reference Database No. 11.

74. A. L. McClellan and H. Harnsberger, *J. Colloid Interface Sci.* **23**, 557 (1967).

75. Y. M. Boluk and H. P. Schreiber, *Polym. Composites* **7**, 295 (1986).

76. E. Papirer, in: *Composite Interfaces*, H. Ishida and J. L. Koenig (Eds), p. 203. North-Holland, New York (1986).

77. J. Schultz, L. Lavielle and C. Martin, *J. Adhesion* **23**, 45 (1987).

78. S. P. Boudreaux and W. T. Cooper, *Anal. Chem.* **59**, 353 (1987).

79. H. G. Maahs and Y. R. Yamaki, *Effect of Fiber Surface Treatment Level on Interlaminar Strengths of Carbon–Carbon Composites*, NASA Technical Memorandum 4033 (1988).

80. B. Rand and R. Robinson, *Carbon* **15**, 311 (1977).

81. B. Rand and R. Robinson, *Carbon* **15**, 257 (1977).

82. T. Takahagi and A. Ishitani, *Carbon* **22**, 43 (1984).

83. C. Kozlowski and P. M. A. Sherwood, *Carbon* **25**, 751 (1987).

84. A. A. Galuska, H. Madden and R. E. Allred, *Appl. Surf. Sci.* **32**, 253 (1988).

Acid-Base Interactions, pp. 171-190
Eds. K.L. Mittal and H. R. Anderson, Jr.
©VSP 1991

Acid–base character of carbon fiber surfaces

R. S. FARINATO,* S. S. KAMINSKI and J. L. COURTER
American Cyanamid Company, 1937 W. Main St., Stamford, CT 06904, USA

Revised version received 21 April 1990

Abstract—The acid–base character of the surfaces of commercially available carbon fibers used in advanced composites is determined using both pulsed and continuous flow microcalorimetry techniques. The carbon fibers investigated include unsized versions of AS4, IM6, IM7X, T300, and several Apollo fibers with different levels of surface treatment. All of these carbon fiber surfaces are amphoteric and energetically heterogeneous. In general, the heats of preferential adsorption in the pulsed flow mode of bases dissolved in *n*-heptane are larger than for acids. While most of the adsorbates used are reversibly and physically adsorbed onto the carbon fibers, some primary and secondary amines exhibit irreversible binding to a portion of the surface. In continuous flow experiments the adsorption heat isotherms for several bases on T300 display a sharp jump at low probe concentrations, reflecting the energetically heterogeneous nature of these surfaces.

Comparisons between the flow microcalorimetry data and other measures of the surface chemistry are made. Pulsed flow adsorption heats correlate with the amount of oxidized carbon species on the fiber surfaces as detected using ESCA, and recently reported results of inverse gas chromatography and programmed thermal desorption (S. Wesson, Textile Research Institute). Calorimetry results are also compared with fracture mechanical measures of fiber–resin adhesion in manufactured composites. Adsorption heats of both acids and bases on selected carbon fibers correlate with edge delamination and 90° flexural strengths of composites composed of these fibers in both epoxy and bismaleimide resins. This supports a causal connection between carbon fiber surface adsorption heats and a practical measure of fiber–resin adhesion.

Keywords: Carbon fiber; adhesion; acid–base; flow microcalorimetry; surface chemistry.

1. INTRODUCTION

The surface chemistry of carbons has historically been a subject of intense study [1–7]. For example, a wide variety of surface oxygen-containing species have been found on carbon, including carboxyls, lactones, phenolic OH, ketones and pyrones [8, 9]. Carbon fibers, as a subclass, have also been the subject of much effort [10–12]. This has been especially motivated by the use of carbon fibers in structural composites. A variety of thermodynamic (wetting [13–15], chromatography [16–22], calorimetry [23], absorption/desorption [24, 25], spectroscopic (infra-red [26, 27], ESCA [28–30], SIMS [31, 32]), microscopic (SEM, TEM [33], STM [34]) and chemical (titration [35], derivatization [36]) techniques have been used to investigate carbon fiber surfaces. Information about carbon fiber surface chemistry is an integral component to understanding the carbon fiber–resin interphase and its role in adhesion. Control of the surface chemistry of the

---

*To whom correspondence should be addressed.

carbon fibers and the bulk chemistry of the matrix resin constitute the main options available to the chemist for optimizing the interfacial interaction with respect to the overall performance properties of a composite. Surface treatments of carbon fibers to promote adhesion generally increase the acid–base character of the surface [10, 11, 37], for example, by oxidizing the surface. The details of the chemical and physical interactions between the fiber and the resin are, therefore, invaluable in the formulation of guidelines for proper choice of materials and chemistries.

The work reported here is on the use of flow microcalorimetry techniques to make a thermodynamic assessment of carbon fiber surface chemistry. The heats of adsorption of small probe molecules onto the carbon fiber surface are governed by the chemistry and surface density of adsorption sites on the carbon fiber. The strength and number of interactions between resin molecules and these surface adsorption sites are expected to determine the fiber–resin work of adhesion (W_a). In addition to this thermodynamic component, the strength of the fiber–resin bond would also depend on the material and rheological properties of the resin and the carbon fiber, and on the defect distribution. The flow microcalorimetric measurements were only used to assess the thermodynamic component of this interaction.

1.1. Adhesion of solids

The propagation of a crack in a solid or along an interface has been described in terms of the relation between the strain energy release rate, G, and the crack speed, v, by several authors [38–46]. Several forms of this relationship have been proposed and they all include a direct proportionality to the thermodynamic work of adhesion, W_a. While W_a is the relevant thermodynamic parameter in assessing crack propagation or joint strength [47] it is seen to be modified by dissipative (rheological) factors.

The reversible work of adhesion can be related to the surface energies of the solids in contact through the Dupré equation:

$$W_a = \gamma_1 + \gamma_2 - \gamma_{12} \tag{1}$$

where γ_1 and γ_2 are the surface energies of the solids and γ_{12} is the interfacial energy between them. Quantities related to the work of adhesion can be determined from either contact angle measurements [48, 49], calorimetry [50, 51], inverse gas phase chromatography [52–54] or infrared spectroscopy [55–57].

Most of the models for crack propagation have been based on the mechanics of elastic materials. In such cases, we would expect that an assessment of the work of adhesion would give us a predictor of the fracture strength. The carbon fiber composite resins of interest here are based on epoxy and bismaleimide chemistries. Depending on fracture conditions, these materials can display brittle fracture. The theories for predicting fracture energies for brittle materials also include a dependence on the thermodynamic (reversible) work of adhesion, but are complicated by the influence of material flaws on the joint strength [38, 39]. Still, over a limited set of fracture conditions we would expect the thermodynamic work of adhesion to be a good predictor of necessary conditions for developing fiber–resin adhesion strength.

1.2. Work of adhesion

The reversible work of adhesion, W_a, was defined in equation (1). The inter-molecular interactions responsible for W_a derive from several sources. However, the two most important contributions vis-à-vis fiber-resin adhesion are London–Lifshitz–van der Waals (dispersion) forces and Lewis acid–base (electron donor–acceptor) interactions [55, 58]. This allows the work of adhesion to be separated, for our purposes, into a dispersion and an acid–base contribution:

$$W_a = W_a^d + W_a^{ab}. \tag{2}$$

The dispersion force component of the work of adhesion is given as:

$$W_a^d = 2(\gamma_1^d \, \gamma_2^d)^{1/2} \tag{3}$$

where the γ_i^d are the dispersion force components of the surface energies of the two adhering materials. The non-dispersion component of the work of adhesion has been modeled in a variety of ways in terms of the polar and acid–base components of the surface energies [59]. The functional form of this relationship in terms of surface energy components has been under debate for some time, with the latest and most promising contributions being made by van Oss *et al.* [58] and Chang and Chen [60]. A more direct and very useful way to model the acid–base contribution has been suggested by Fowkes [61–64]. His method relates W_a^{ab} directly to the enthalpy of interaction between the two phases.

$$W_a^{ab} = k(C_A C_B + E_A E_B) \text{ (moles A-B pairs/unit area)} \tag{4}$$

where the constant k (~ 1) converts enthalpy/unit area into surface free energy and the C_i and E_i are the Drago parameters associated with the acid (A) and the base (B) involved in the interaction. Note also the need to know the surface density of sites in order to calculate W_a. The Drago representation of the enthalpy of interaction may also be supplanted by another parameterized scheme for expressing this quantity (e.g. Gutmann's acceptor and donor numbers) [65], or by a direct thermochemical measurement of this quantity.

There are four conventional methods for determining the components of the work of adhesion between two phases: wetting, calorimetry, inverse gas phase chromatography (IGC) and infrared spectroscopy.

In IGC a gas phase probe adsorbs on the substrate being investigated. For dilute probe concentrations the log of the retention volume has been related to the adsorption enthalpy [52]. A series of probes can be used to determine the components of W_a [53]. For concentrated probe IGC, in addition to the retention volume measurement, the diffuse profile of the chromatogram (after peak maximum) can be analyzed to yield an adsorption isotherm [21]. The isotherm data in combination with a model for the heterogeneous surface can be further analyzed to yield adsorption energy profiles [66, 67] which characterize the surface heterogeneity.

In contact angle measurements the wetting properties of a series of liquids of known surface energies (dispersion and acid–base components) are used to calculate the components of W_a for the substrate [59, 64]. These measurements are also possible on fibers [13, 68]. While wetting measurements are averages over all surface sites at the contact line, a limited amount of information on the surface heterogeneity can be obtained from the contact angle hysteresis (advancing vs. receding angles) if surface roughness effects can be deconvoluted from the data [69].

Infrared spectroscopy can be used to assess the energetics of reversible interactions as well as the formation of new interfacial bonds. Allara [70, 71] used reflection FTIR to study the details of the interactions between thin polymer films and aluminum oxide surfaces. Fowkes *et al.* [72] have used transmission FTIR on polymer solutions to pinpoint specific reversible interactions operating in polymer–polymer adhesion. Transmission FTIR measurements on polymer blends have also been used to gain detailed information on specific interfacial interactions which control blend morphology and interfacial adhesion [72, 73]. In the FTIR methods, as they are applied to the assessment of interfacial interactions, the shifts in IR vibrational peaks are related to specific functional group interactions. In the case of reversible interactions these IR peak shifts can often be related linearly to the enthalpy of interaction. If the number of interactions per unit surface area of contact can also be determined or estimated, then the enthalpic contribution to the interfacial free energy or the work of adhesion can be calculated.

In calorimetry the heat of preferential adsorption, Q_a, of a probe molecule from a solution onto a surface is related to W_a. Since Q_a can be measured as a function of the amount adsorbed and for different adsorbate species, some measure of the surface heterogeneity can also be obtained. To fully develop this information the adsorption isotherm must also be concurrently measured. While flow calorimeters are typically less sensitive than batch calorimeters, they offer advantages for systems which reach equilibrium in a short time. These advantages include reduced experimental time and the ability to separate heat events with significantly different equilibration rates. Calorimetry measurements tend to be more precise at low surface coverage, and concentrated probe IGC measurements are more precise at high surface coverage. This makes the two methods complementary.

1.3. Flow microcalorimetry

Two kinds of flow microcalorimetric measurements were used here. In the pulsed flow method the heats of preferential adsorption onto the carbon fibers of small slugs of dilute solutions of probe molecules in *n*-heptane were measured. In the continuous flow mode the differential heat of preferential adsorption was measured when the solution stream was suddenly incremented in adsorbate concentration. By simultaneously measuring the amount of probe adsorbed in the continuous flow experiment, the adsorption isotherm can also be determined. The amount of heat released as a function of surface excess then gives a measure of surface energy heterogeneity.

The measured quantity, the heat of preferential adsorption Q_a, was determined from the thermistor output integrated over the time of the thermal event. Q_a can be related to the excess enthalpy of the surface phase, H_s, subject to corrections due to the heats of mixing of the surface and bulk phases [74, 75]. The excess surface enthalpy is a product of the molar enthalpy of interaction of the probe with the surface adsorption site and the surface concentration of these sites. On an energetically heterogeneous surface the individual contributions must be summed.

In the pulsed flow mode the value of Q_a will also be influenced by kinetic effects [75]. In addition to the kinetics of the adsorption process, the heat diffusion kinetics must also be considered. This usually defines an upper limit for the flow rate through the calorimeter cell.

The above considerations made the pulsed flow technique inappropriate for accurate determinations of excess surface enthalpies. However, the pulsed flow technique did allow a rapid estimation of the relative adsorption site activity for a particular carbon fiber sample. This relative activity was seen to correlate with fiber–resin adhesion properties in composites as discussed below. This was roughly equivalent to equating Q_a with H_s and correlating H_s with W_a.

The wide variety of probe molecules used allowed insight to be gained about the effect of chemical functionality on the adsorption process and led to the generation of heat of adsorption 'fingerprints' for each carbon fiber. Probe species were chosen that would adsorb physically and reversibly. Because of the dilute concentrations used and the fact that heptane was already adsorbed on the surface, the dominant contribution to Q_a was expected to come from specific acid–base interactions. Indeed, the adsorption from heptane of liquids displaying only dispersion force interactions did not give measurable exotherms. This was due to the low specific surface areas of the carbon fibers measured (< 1 m^2/g) and the small differences in the surface energies of these specific probes compared to heptane. This yielded a very small exotherm which was outside the detection limits of the instrument.

The continuous flow experiments can potentially yield more quantitative results than the pulsed flow experiments if the adsorption isotherm is measured concurrently. The differential heats of adsorption in the submonolayer portion of the adsorption isotherm give information on the distribution of surface energies and on the adsorption site densities. Preliminary measurements of this type were performed.

1.4. Fiber–resin adhesion measurement

A strong motivation for measuring the surface thermodynamic properties of carbon fibers is to use those quantities as predictors and components in models of fiber–resin adhesion. There has been a good deal of effort at relating measures of the thermodynamics of the interface to measures of the interfacial bond strength [11, 76–78]. Single fiber tests have been quite popular for these kinds of correlations and have helped clarify the local microscopic model of fiber–resin adhesion. Of more direct use in composite manufacture would be a correlation of thermodynamic measurements with a bulk material property which reflected fiber–resin adhesion. One of our goals, in addition to quantifying the surface chemistry of carbon fibers, was to determine to what extent a correlation could be made between a measure of the surface chemistry and a practical measure of the fiber–resin adhesion while controlling as many other variables affecting the bulk material property as possible.

Our approach was to use well-engineered matrix resins (epoxy and bismaleimide) and fabrication techniques to control the influence of factors such as defect density, neighbor constraints, fiber volume fraction, the transverse tensile modulus of the fibers, and the tensile modulus of the matrix resin. A host of composite properties were evaluated for their usefulness in assessing fiber–resin adhesion strength [79]. Analysis of the data indicated that the best tests for evaluating fiber–resin adhesion were the 90° flexural strength, and edge delamination strength tests. Differences in the surface chemistries of several commercial carbon fibers used in some of these composites were correlated with the results of these two tests.

2. EXPERIMENTAL

Unsized carbon fibers were obtained directly from manufacturers. These included AS-4, IM6, IM7X (Hercules), T300 (Amoco), Hysol Apollo 53-750, Grafil Apollo 43-750 and 53-750 with two levels of a proprietary surface treatment (Courtaulds). The surface treatments produced different levels of surface oxidation. The geometric specific areas were ~ 0.3 m^2/g. N$_2$ BET measurements were not reproducible except for T300 which had a specific surface area of 0.46 m^2/g using this technique. The fiber surface topographies were viewed in a scanning electron microscope (Cambridge Stereoscan 100 operated at 25 kV). The AS-4, IM6 and IM7X fibers were featureless at $6000\times$. The T300 and Apollo fibers exhibited longitudinal striations and non-circular cross-sections at this magnification. The Apollo fibers with different levels of surface treatment all had qualitatively the same appearance.

Approximately 0.2–0.25 g of carbon fibers were packed in as a parallel bundle extending the length of the sample chamber (~ 16 mm). The void fraction was ~ 0.57. The fibers were exposed to vacuum at room temperature overnight, then washed with a slow flow of *n*-heptane for > 1 hour before measuring adsorption heats.

A Microscal 3V flow calorimeter (Microscal Ltd.) was used in both pulsed and continuous flow operations around 25°C. A schematic of the experimental apparatus is shown in Fig. 1. Dried *n*-heptane (Aldrich chromatography grade over 4A sieves) was used as a carrier solvent. The water content of all solvents and solutions was measured using a Karl-Fischer titrator (Photovolt) immediately

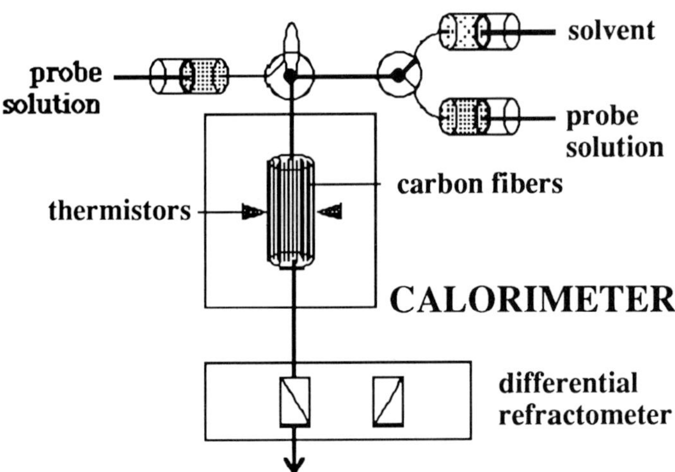

Figure 1. Schematic of the flow microcalorimeter.

before or after use. This allowed a correction to adsorption heats for differences in water content between solvent and probe solution (typically 5–30 ppm). High purity reagents (usually chromatography grade) were used for the adsorption probes. Probe solutions were 0.2 wt % in *n*-heptane for the pulsed mode experiments. Water, however, was only soluble to the extent of ~60 ppm at room temperature and saturated solutions were used as a probe. The solutions were pumped through the fiber pack using a syringe pump (Perfusor) at 0.2 ml/min in the pulsed mode. The Reynolds number at this flow rate was calculated assuming laminar flow to be ~0.0015. The flow rates were 0.1 ml/min in the continuous mode. The valve system included at 25 μl injection loop for the pulsed mode. The effluent from the microcalorimeter was passed through a Perkin-Elmer LC-25 differential refractometer to monitor solute concentration profiles.

The calorimeter bridge output was recorded and analyzed in one of two ways. Mostly, the data were recorded on an *x–t* recorder (Omega) and the signal areas were measured using a digitizing tablet (Videoplan). We also collected some data on a DEC PRO-350 microcomputer with a Micromac 4000 A/D converter and the areas were calculated using in-house software. Signal areas were calibrated in a separate experiment by applying controlled electrical pulses to a heater coil embedded in a carbon fiber sample. The precision in measured heat was ~0.04 mJ. Typical exotherms were ~4 mJ. In the pulsed mode, multiple pulses, each separated by ~5–10 min, were run to check for reversible adsorption. The average adsorption heats reported typically had a standard deviation of 10% and were the results of several different samples each with multiple pulses. These values were corrected for differences between the water content of the solvent and probe solution and are reported here as $\langle dHc \rangle$, per gram of fiber. Corrections for different water contents were made by subtracting from the measured heat a heat proportional to the difference in water contents between the probe solution and the *n*-heptane carrier solvent assuming a linear relationship between water content and adsorption heat due to water. Water contents were measured using a Karl-Fischer titrator immediately after the solutions were used. The qualitative trends cited below about the data are preserved whether this correction is applied or not.

The desorption endotherm seen in the continuous flow experiments was typically not visible in the pulsed flow experiments. This was a combined effect of the desorption rate being slower than the adsorption rate, and the small amounts of heat associated with the desorption process.

ESCA spectra of the carbon fibers [80] were taken on an HP ESCA spectrometer. Both survey and high resolution spectra were recorded. The C1*s* spectra were curve resolved to determine the relative amounts of oxidized carbon species.

Composite specimens were prepared using several of the carbon fibers and two different resins used in commercial prepregs: Cycom® 1827 (epoxy) and Cycom® 3100 (bismaleimide), both from American Cyanamid Company. Edge delamination specimens were 6 ply [$(\pm 25°)_2/90°]_s$ and were stressed in longitudinal tension. 90° flexural specimens were 16 ply [$(90°)_{16}$] and were loaded in a 4-point mode (ASTM D790 method II). The force necessary to initiate failure was measured in each instance using an Instron model 1125 with a 5000 lb capacity load cell and wedge-action grips with serrated faces.

3. RESULTS AND DISCUSSION

3.1. Pulsed flow microcalorimetry

Preferential heat of adsorption 'fingerprints' for several fibers are shown in Fig. 2(a–d). Twenty-three probes are graphed with bases on the left and acids on the right. The vertical axis is the corrected (for water) average adsorption heat per unit fiber mass of a 25 μl slug of a 0.2 wt % solution of the adsorbate in dry *n*-heptane. The concentration of the water was only 60–70 ppm due to limited solubility. Typical relative standard deviation of the averaged adsorption heats was 10%. Twenty-three probes were used: triethylamine, diethylamine, dipropylamine, *n*-butylamine (*n*-BA), *t*-butylamine (*t*-BA), *N*-methyl pyrrolidinone (NMP), 2,6-lutidine (2,6-dimethyl pyridine), pyridine (pyrid), pyridazine, 1,4-dioxane, acetonitrile, dimethyl sulfoxide (DMSO), 1,1,1,3,3,3,-hexafluoro-2-propanol (HFIP), phenol, 3-fluorophenol, 2,2,2-trifluoroethanol, α,α,α-trifluoro-*m*-cresol, pyrrole (pyrr), water, *n*-butanol (*n*-BuOH), *t*-butanol (*t*-BuOH) and iodine. In Fig. 2(a–d) the probes are ordered with the bases on the left and acids on the right. Selected values of the preferential heats of adsorption are also reported in Table 1.

Both acidic and basic probes adsorbed onto these amphoteric carbon fibers. In general, the adsorption exotherms were larger for:

(1) bases vs acids of comparable structure
    pyridine > pyrrole
    n-butylamine > *n*-butanol
    t-butylamine > *t*-butanol

(2) probes having interaction sites with less steric hindrance
    pyridine > 2,6-dimethyl pyridine
    n-butylamine > *t*-butylamine
    n-butylamine > diethylamine > triethylamine

(3) shorter alkyl substituents on an amine
    diethylamine > dipropylamine

(4) electronegative substituents on acidic probes
    3-fluorophenol > phenol
    1,1,1-trifluoroethanol > *n*-butanol

Table 1.

Carbon fiber characterization

Fiber	$\langle dHc \rangle$ (mJ/g) (25 μl of 0.2 wt % solutions)							ESCA (at. %)		
	n-BA[a]	*t*-BA	pyrid	pyrr	H_2O[b]	*n*-BuOH	*t*-BuOH	O	N	C
AS-4	2.5	—	2.9	0.8	1.7	0.4	0.0	6.4	2.4	91.2
IM-7X	3.8	1.7	2.1	0.4	2.1	0.4	0.4	8.7	3.7	87.7
T300	7.5	4.2	5.4	2.9	3.8	1.7	1.3	13.4	1.9	84.7
GA43-750	1.3	0.4	0.0	0.4	1.7	0.4	0.4			
GA53-750 (50%)	2.9	1.3	0.4	0.0	2.1	0.8	0.4			
GA53-750 (100%)	2.9	1.3	0.4	0.0	1.7	0.4	0.0			

[a] First exotherm for *n*-butylamine on virgin fiber not included.
[b] 60 ppm in heptane.

Iodine adsorption exotherms tended to be unaffected by the range of surface chemistries exhibited by our collection of samples. This many indicate predominantly an adsorption on the basal planes of the carbon fibers.

If a material was present on the carbon fiber surface which was soluble in the probe solution, a measurable exotherm due to dissolution was often seen. For example, this occurred with fibers sized with organic resins when methylene chloride was used as the probe.

In addition to these regular trends in the data, there were certain probes whose adsorption enthalpy behavior could not be easily interpreted in terms of conventional acid–base paradigms. For example, DMSO often exhibited surprisingly large adsorption exotherms (see Fig. 2). This may be a consequence of the highly polar S—O bond in this compound.

Water always adsorbed appreciably. Even though its solution concentration was over 30 times less than the other probes, its preferential adsorption exotherm was usually comparable. This prompted measuring the water content of each solution and making corrections for different amounts of water.

It might be expected that the magnitudes of the preferential adsorption exotherms should correlate with other empirical measures of the reactivity of the adsorbates. To test this we list in Table 2 the pK_a, pK_b, Drago E and C, the Gutmann AN and DN parameters for the probe molecules, when available. In Table 3 are listed, for comparison, the preferential heats of adsorption for the

Table 2.
Reactivity parameters of adsorbates

Adsorbate	$pK_a{}^a$	$E_B{}^b$	$C_B{}^b$	$E_A{}^b$	$C_A{}^b$	AN^b	DN^b
triethylamine	10.72	0.991	11.090				61.0
diethylamine	10.80	0.866	8.830				58.0
dipropylamine	10.91						
n-butylamine	10.64						
t-butylamine	10.69						57.5
NMP						13.3	27.3
2,6-lutidine	6.71						
pyridine	5.17	1.170	6.400			14.2	33.1
pyridazine	2.33						
dioxane		1.090	2.380			10.8	14.8
acetonitrile		0.886	1.340			18.9	14.1
DMSO		1.340	2.850			19.3	29.8
HFIP	9.42			5.930	0.623		
phenol	9.99			4.330	0.442		
3-fluorophenol	9.29			4.170	0.446		
trifluoroethanol				3.880	0.451		
trifluorocresol	8.95			4.480	0.530		
pyrrole	17.00			2.540	0.295		
water	7.00			2.450	0.330	54.8	18.0
n-butanol	16.07			1.120	−0.050		
t-butanol				2.040	0.300	30.0	
iodine				1.000	1.000		

^{a} pK_a values of organic materials in water at 25°C from table 5-8 in [83]; the value for *n*-butanol was calculated using Taft equation and constants in table 3-12 of [83].

^{b} Drago constants (E and C) and the donor (DN) and acceptor (AN) numbers were taken mostly from [65]. The E_A and C_A values for water were taken from [84].

R. S. *Farinato* et al.

(a)

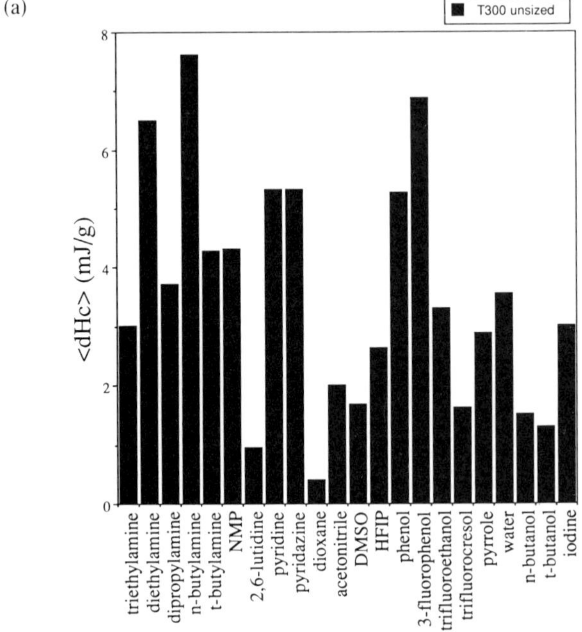

(b)

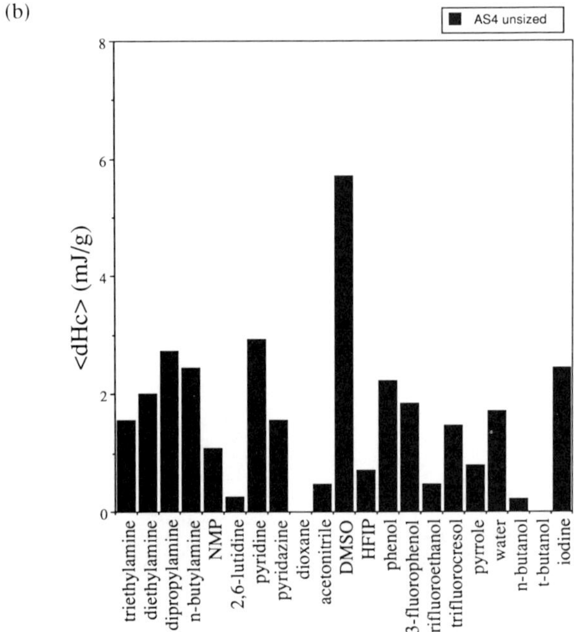

Figure 2. Pulsed flow microcalorimetry adsorption heat 'fingerprints'. Unsized carbon fiber substrates shown are: (a) T300; (b) AS4; (c) Hysol Apollo 53-750; and (d) IM7-X. Note, adsorption heats were not measured for: *n*-butylamine and *t*-butylamine on Hysol Apollo 53-750, nor *t*-butylamine on AS4.

(c)

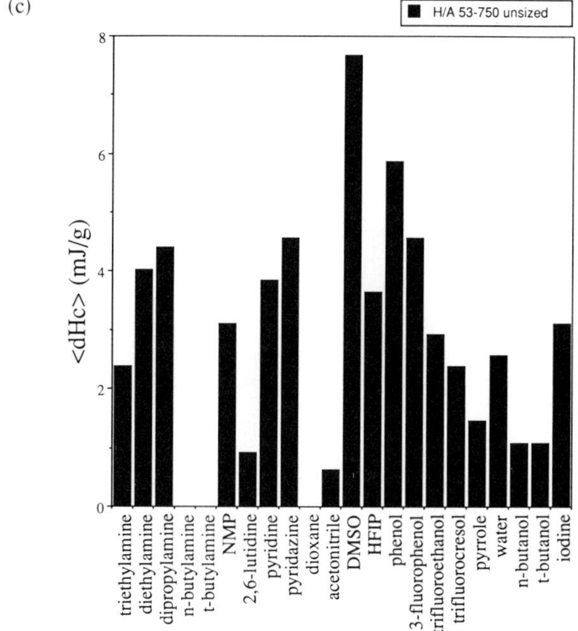

(d)

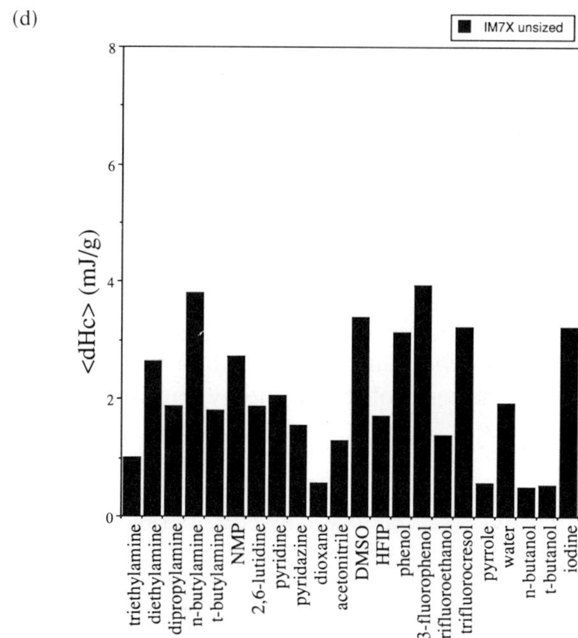

Figure 2. (*cont.*)

               R. S. Farinato et al.

Table 3.
Pulsed preferential adsorption heats

	$\langle dHc \rangle$ (mJ/g)[a]			
Adsorbate	AS4	IM7X	HA 53-750	T300
triethylamine	1.6	1.0	2.4	3.0
diethylamine	2.0	2.6	4.0	6.5
dipropylamine	2.7	1.9	4.4	3.7
n-butylamine	2.4	3.8		7.6
t-butylamine		1.8		4.3
NMP	1.1	2.7	3.1	4.3
2,6-lutidine	0.3	1.9	0.9	1.0
pyridine	2.9	2.1	3.9	5.3
pyridazine	1.6	1.6	4.6	5.3
dioxane	0.0	0.6	0.0	0.4
acetonitrile	0.5	1.3	0.6	2.0
DMSO	5.7	3.4	7.7	1.7
HFIP	0.7	1.7	3.6	2.6
phenol	2.2	3.1	5.9	5.3
3-fluorophenol	1.8	3.9	4.6	6.9
trifluoroethanol	0.5	1.4	2.9	3.3
trifluorocresol	1.5	3.2	2.4	1.6
pyrrole	0.8	0.6	1.5	2.9
water	1.7	1.9	2.6	3.6
n-butanol	0.2	0.5	1.1	1.5
t-butanol	0.0	0.5	1.1	1.3
iodine	2.4	3.2	3.1	3.0

[a] Pulses were 25 μl of 0.2 wt % adsorbate in dry *n*-heptane, except for water which was 60–70 ppm. Flow rate was 0.2 ml/min.

same probes on AS-4, IM7X, Hysol Apollo 53-750 and T300 fibers. In general, the correlations of pulsed adsorption heats with various reactivity scales were not strong. Certainly a contributing factor to this was the heterogeneous nature of the carbon fiber surfaces.

From a comparison of the pulsed adsorption heats with the limited continuous flow calorimetry results using *n*-butylamine and pyridine (Figs 3 and 4), the heats evolved with the pulses (25 μl of ~ 0.2 wt % solutions) were about 1/3–1/4 of the values at saturation adsorption which occurred at bulk solution concentrations slightly in excess of the concentrations in the pulses. Some reduction of the pulsed adsorption heats compared to the continuous flow heats can be attributed to thermal diffusion effects. A comparison of the characteristic time for thermal diffusion and the mean residence time in the carbon fiber bed illustrates this. The characteristic time, τ, for non-isothermal energy transport is given by [84]:

$$\tau = \frac{\rho C_p R^2}{\kappa} \qquad (5)$$

where ρ is the fluid density, C_p is the heat capacity at constant pressure, R is the radius of the sample chamber, and κ is the thermal conductivity of the fluid. The mean residence time, t_{res}, in the calorimeter sample chamber is:

$$t_{res} = \frac{\varepsilon \pi R^2 L}{Q} \qquad (6)$$

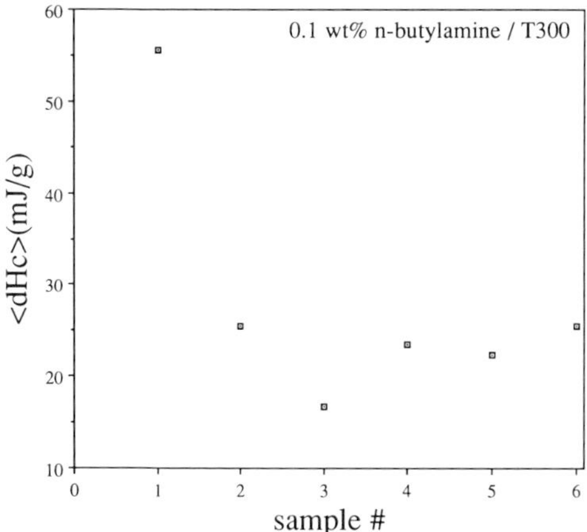

Figure 3. Continuous mode flow microcalorimetry of repeated doses of solutions of 0.1 wt % *n*-butylamine in dry *n*-heptane onto unsized T300 carbon fibers. Note that the corrected adsorption heat for the first sample on a virgin fiber is more than twice as large as the subsequent heats. The carbon fibers were allowed to desorb in pure *n*-heptane between each data point.

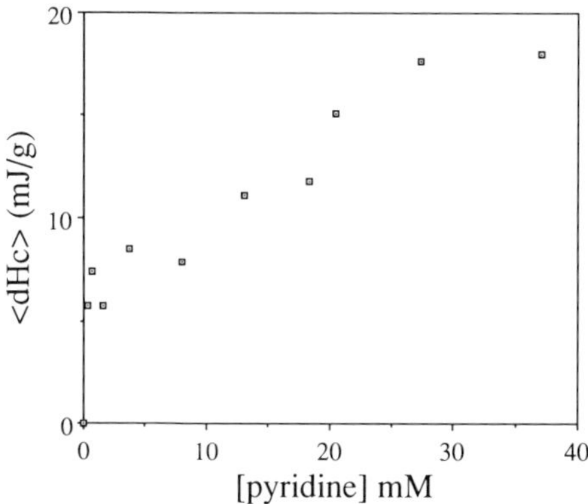

Figure 4. Continuous mode flow microcalorimetry of pyridine adsorption from dry *n*-heptane onto unsized T300 carbon fibers. The corrected (for water) adsorption heat is plotted vs the bulk solution concentration of pyridine.

where ε is the void fraction in the carbon fiber bed, L is the length of the sample chamber, and Q is the volumetric flow rate. As a first approximation we calculated the characteristic time neglecting the presence of the carbon fibers, which were expected to have a higher thermal conductivity than the *n*-heptane. For the flow rates used in the pulsed flow experiments we found $\tau = 120$ s and $t_{res} = 50$ s. While these results are only order of magnitude calculations, it can be seen that the time for the thermal event to register at the thermistors is comparable to the time for convective transport out of the system.

In each pulsed flow calorimetry experiment we were probably sampling the carbon fiber surface at some intermediate point on the adsorption isotherm, i.e. somewhere below the saturation adsorption heat value, but significantly above the low concentration regime. Since the adsorption isotherms were most likely different for the different probes, the effective sampling point on the isotherm was different for each probe. Because of the energetic heterogeneity of the surfaces, this situation would tend to blur any correlations of the pulsed adsorption heats with reactivity scales built upon strictly ideal two component thermodynamics. The potential for clarifying this situation rests with the continuous flow micro-calorimetry experiments, wherein the specific amount adsorbed is concurrently measured with the heat evolved. This would allow, for example, a calculation of the effective Drago parameters (E and C) for the adsorbents at various points along the adsorption isotherm. We plan to extend our work in this direction.

A smaller set of probes was found useful on a routine basis. These included *n*-butylamine, *t*-butylamine, pyridine, pyrrole, water, *n*-butanol and *t*-butanol. For example, the pulsed flow microcalorimetric 'fingerprints' of a set of Grafil Apollo carbon fibers with different amounts of a proprietary surface treatment, expressed as a percentage of a standard treatment, are shown in Fig. 5. The results indicate increased adsorption heats of both acidic and basic probes for '50%' surface treated fibers over untreatd fibers. This degree of surface treatment thus augments both the basic and acidic character of the surface. However, for the '100%' surface treatment, base adsorption exotherms were as large as for the '50%' surface treated fibers and the acid adsorption exotherms were measurably less. This pointed to a leveling-off of the surface chemistry changes and in fact a measurable loss of basic sites on the surface at higher levels of surface treatment.

Judging from the results shown in Fig. 2 and the selected exotherm values listed in Table 1, the ranking of fibers used in the composite study in terms of adsorption energetics would be T300 > IM-7X > AS-4.

It was obvious even from the pulsed flow experiments that the virgin carbon fiber surface consisted of a mixture of high and low energy sites. While most of the

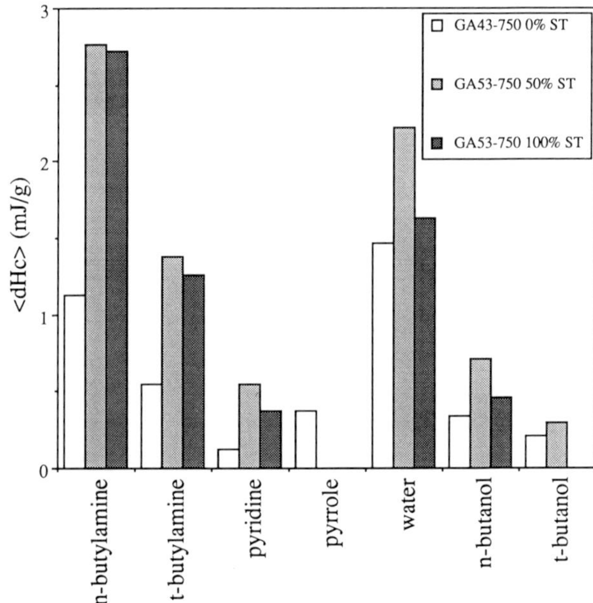

Figure 5. Pulsed flow microcalorimetry adsorption heat 'fingerprints' for selected probes. Unsized carbon fiber substrates shown are Grafil Apollo 43-750 (0% surface treatment), Grafil Apollo 53-750 (50% surface treatment) and Grafil Apollo 53-750 (100% surface treatment).

adsorbates used were reversibly and physically adsorbed on the carbon fibers, some of the primary and secondary amines displayed strong irreversible binding. This was evidenced by large first exotherms for primary and some secondary amines on virgin fibers, long desorption times (no or slow recovery to initial exotherm value after 24 h in *n*-heptane) and a solute loss from a pulse as measured with the refractometer. Figure 3 shows the large first exotherm effect on a T300 fiber with 0.1 wt % *n*-butylamine run in the continuous flow mode. In this case the sample remained overnight in *n*-heptane in the cell in between each measurement.

3.2. Continuous flow microcalorimetry

Figure 4 shows an example of continuous mode flow microcalorimetry on T300 with various concentrations of pyridine in solution. The substantial heats at the lowest concentrations measured are indicative of the energetic heterogeneity of the fiber surface. For pyridine adsorption on T300 more than 25% of the adsorption heat occurred in this low concentration limit. This number is proportional to the weighted (by the enthalpy of interaction between the pyridine and the surface sites) surface density of the high energy sites relative to the total number density of pyridine adsorption sites.

At the higher pyridine concentrations the measured heat uniformly reaches an asymptote, reminiscent of saturation coverage. An uncontrolled level of dissolved gas in our solvents led to baseline drift in the refractive index detector signals. This prevented an accurate determination of the amount of material adsorbed during the exothermic adsorption event and hence a determination of the active site fractional area. This will be the subject of future work.

3.3. ESCA

The relative atomic percentages of O, N, C for several fibers are shown in Table 1. Note that the relative amount of O on the fiber surface decreases in the order T300 > IM7X > AS4. This is the same order as that based on pulsed flow microcalorimetry adsorption heats. The pulsed adsorption heats of both acidic and basic probes generally correlate with the surface oxygen content. However, no clear correlation with the surface nitrogen content is seen.

In high resolution C1*s* spectra the peak at 286 eV (C—O) was greatly enhanced in the T300 spectrum compared to the AS4 spectrum, indicative of the higher degree of surface oxidation of the T300. A comparison of high resolution ESCA and flow microcalorimetry results for a series of carbon fibers with increasing amounts of surface oxidation will be the subject of a future publication.

3.4. Mechanical testing

Both edge delamination strength and 90° flexural strength were found to be somewhat sensitive to fiber choice. Figures 6 and 7 show comparisons of these strengths for various fiber-resin combinations. The matrix resins used in the Cycom® 1827 (epoxy) and Cycom® 3100 (bismaleimide) prepregs contain a variety of Lewis acid and base functionalities and thus they would both be expected to adsorb onto an amphoteric carbon fiber surface. The correlation between the preferential adsorption heats and mechanical test results can be seen

                                   R. S. Farinato et al.

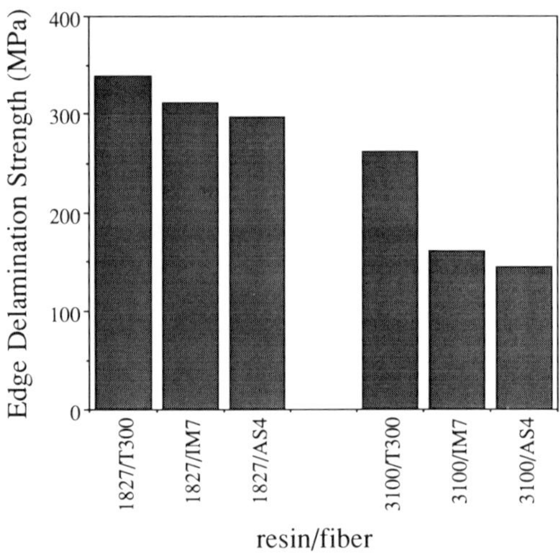

Figure 6. Edge delamination strength results for several combinations of carbon fiber (T300, IM7 and AS4) and matrix resin (Cycom® 1827 (epoxy) and Cycom® 3100 (bismaleimide) prepegs).

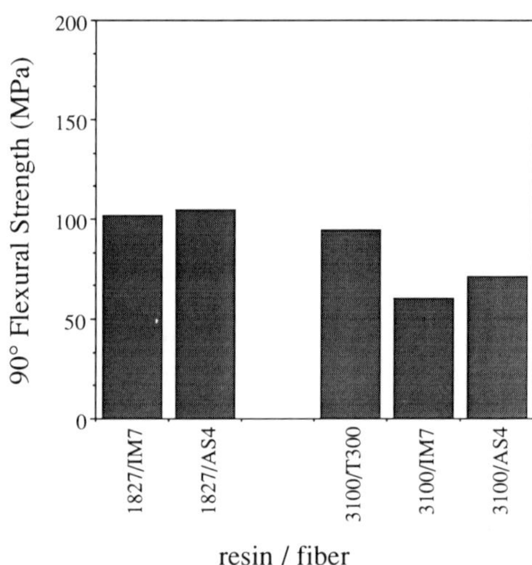

Figure 7. Ninety degree flexural strength results for several combinations of carbon fiber (T300, IM7 and AS4) and matrix resin (Cycom® 1827 (epoxy) and Cycom® 3100 (bismaleimide) prepegs).

by comparing Figs 6 and 7 with Fig. 2 and Table 1. The standard deviations in the mechanical tests were about 3% for the edge delamination strength and about 5% for the 90° flexural strength. In the bismaleimide matrix composites there was a significantly larger edge delamination strength and 90° flexural strength for the T300 fibers as compared to the AS-4 fibers. The IM7 and AS-4 fiber composites appeared equivalent. This equivalence of the IM7 and AS-4 fibers carried over to the epoxy matrix composites in both the edge delamination strengths and the 90° flexural strengths. The T300/epoxy composites showed a marginally larger (separated by about 2 S.D.) edge delamination strength compared to the AS-4/ epoxy composites.

While the effect of fiber choice was noticeable in these properties, other factors, such as matrix properties appear to have a stronger influence. For example, the bismaleimide resin was less tough and had a higher glass transition temperature than the epoxy resin. The latter can lead to higher residual thermal stresses in the edge delamination specimen. The large differences in the composite properties between the bismaleimide and epoxy composites reflected these differences.

3.5. *Comparison with IGC and PTD*

It is useful to compare different measures of the surface chemistry of the same carbon fibers. Different lots of some of the same carbon fibers reported on here have been characterized recently using inverse gase chromatography (IGC) and programmed thermal desorption (PTD) [82]. A selected summary of these results is reproduced in Table 4 along with the appropriate flow microcalorimetry results. The IGC results reported in Table 4 were derived from adsorption isotherms computed from the diffuse portions of chromatograms in concentrated probe experiments. Histograms of the surface area fraction vs. relative adsorption energy were computed from the isotherms. In Table 4 are reported the surface fraction of the high energy sites. This fraction was presumed to account for the acid–base interactions between the absorbate and the adsorbent. The PTD results were obtained by integrating the desorption polytherms from 30–90° C as an empirical index of the acid–base interactions [82].

Table 4.

Comparison of microcalorimetry with IGC and PTD

Fiber	$\langle dHc \rangle$ (mJ/g) (25 μl of 0.2 wt %)				IGC[a]		PTD[a]	
	n-BA	t-BA	n-BuOH	t-BuOH	t-BA	t-BuOH	t-BA	t-BuOH
AS-4	2.5	—	0.4	0.0	5.2	4.0	5.4	0.0
IM7-X	3.8	1.7	0.4	0.4	9.2	4.0	5.6	0.3
T300	7.5	4.2	1.7	1.3	26.0	—	17.0	11.0

[a] See text for description of units.

While it is difficult to make a quantitative comparison between the three kinds of experiments, the general ranking of the carbon fibers with respect to surface energetics is the same for each method.

4. SUMMARY

Preferential heat of adsorption 'fingerprints' of acidic and basic probes onto carbon fibers measured using pulsed flow microcalorimetry are valuable in characterizing their surface chemistry. Adsorption exotherms are, for the most part, influenced by probe chemistry in expected ways. Virgin carbon fiber surfaces are amphoteric and energetically heterogeneous as judged from both pulsed and continuous flow microcalorimetric experiments.

The preferential heats of adsorption of selected probes are used to rank carbon fibers with respect to the potential work of adhesion in a fiber–resin interaction. In the case of the most energetically different carbon fibers this ranking correlates with edge delamination strengths of carbon fiber composites made with both epoxy and bismaleimide resins and with the 90° flexural strengths of composites made with a bismaleimide matrix resin.

The ranking of carbon fiber surface energies using flow microcalorimetry is the same as the ranking using IGC and PTD. The adsorption heats of acids and bases also correlates with the level of surface oxidation as detected using ESCA.

Acknowledgements

We gratefully acknowledge the ESCA measurements performed by J. Brinen and S. Greenhouse, the computer programs written by G. Walters and the BET measurements performed by R. Guiles, all of American Cyanamid Company.

REFERENCES

1. E. M. Dannenberg, Carbon blacks, in: *Encyclopedia of Chemical Technology*, M. Grayson (Ed.), vol. 4. John Wiley and Sons, New York (1978).
2. J.-B. Donnet and A. Voet, *Carbon Black: Physics, Chemistry and Elastomer Reinforcement*, Marcel Dekker, New York (1976).
3. G. D. Parfitt and K. S. W. Sing (Eds.), *Characterization of Powder Surfaces*, Academic Press, New York (1976).
4. D. Rivin, *Rubber Chem. Technol.* **44**, 307–343 (1971).
5. A. J. Groszek, *Proc. R. Soc. Lond.* **A314**, 473–498 (1970).
6. A. J. Groszek, *Carbon* **25**, 717–722 (1987).
7. A. J. Groszek, *Carbon* **27**, 33–39 (1989).
8. H. P. Boehm, *Ang. Chem., Int. Ed.* **5**, 533–544 (1966).
9. E. Papirer, S. Li and J.-B. Donnet, *Carbon* **25**, 243–247 (1987).
10. J.-B. Donnet and R. C. Bansal, *Carbon Fibers*, Marcel Dekker, New York (1984).
11. W. D. Bascom and L. T. Drzal, *NASA Contractor Report 4084*, Contract NAS1-17918, 1–91 (1987).
12. L. T. Drzal, in: *Treatise on Adhesion and Adhesives*, R. L. Patrick (Ed.), vol. 5, pp. 21–43. Marcel Dekker, New York (1981).
13. B. Miller, *Text. Sci. Technol.* **7**, (Absorbency), 121–147 (1985).
14. J. Schultz, K. Tsutsumi and J.-B. Donnet, *J. Colloid Interface Sci.* **59**, 272 and 277 (1977).
15. W. J. Lee, J. C. Seferis and J. C. Berg, *Polym. Composites* **9**, 36–41 (1988).
16. J. Schultz, L. Lavielle and C. Martin, *J. Chim Phys.* **84**, 231–237 (1987).
17. J. Schultz, L. Lavielle and C. Martin, *J. Adhesion* **23**, 45–60 (1987).
18. A. J. Vukov and D. G. Gray, *Langmuir* **4**, 743–748 (1988).
19. Various articles in: *ACS Symp. Ser. 391: Inverse Gas Chromatography Characterization of Polymers and Other Materials*, D. R. Lloyd, T. C. Ward, H. P. Schreiber and C. C. Pizana (Eds). American Chemical Society, Washington D.C. (1989).
20. A. E. Bolvari and T. C. Ward, *ACS Polym. Mater. Sci. Eng.* **58**, 655–659 (1988).
21. S. P. Wesson and R. E. Allred, *ACS Polym. Mater. Sci. Eng.* **58**, 650–654 (1988).
22. M.-F. Grenier-Loustalot, Y. Borthomieu and P. Grenier, *Surface Interface Anal.* **14**, 187–193 (1989).
23. B. Rand and R. Robinson, *Carbon* **15**, 311–315 (1977).
24. L. T. Drzal, *Carbon* **15**, 129 (1977).
25. I. M. Ismail, *Carbon* **25**, 653–662 (1987).
26. A. Garton and J. H. Daly, *Polym. Composites* **6**, 195–200 (1985).
27. C. Sellitti, J. L. Koenig and H. Ishida, in: *Interfaces in Polymer, Ceramic, and Metal Matrix Composites*, H. Ishida (Ed.), pp. 163–178. Elsevier, New York (1988).
28. J. Harvey, C. Kozlowski, P. M. A. Sherwood, *J. Mater. Sci.* **22**, 1585–1596 (1987).
29. T. Takahagi and A. Ishitani, *Carbon* **26**, 389–396 (1988).
30. P. Denison, F. R. Jones and J. F. Watts, *J. Mater. Sci.* **20**, 4647 (1985).
31. J. E. Castle and J. F. Watts, in: *Interfaces in Polymer, Ceramic, and Metal Matrix Composites*, H. Ishida (Ed.), pp. 57–71. Elsevier, New York (1988).

32. D. Briggs, in: *Interfacial Phenomena in Composite Materials '89*, F. R. Jones (Ed.), pp. 13–14. Butterworths, London (1989).
33. D. J. Johnson, *Chem. Ind.* 692–698 (Sept. 18, 1982) and references therein.
34. W. P. Hoffman, V. B. Elings and J. A. Gurley, *Carbon* **26**, 754–757 (1988).
35. P. Ehrburger and J.-B. Donnet, *Phil. Trans. R. Soc. Lond.* **A294**, 495–505 (1980).
36. P. Denison and F. R. Jones, in: *Interfaces in Polymer, Ceramic, and Metal Matrix Composites*, H. Ishida (Ed.), pp. 77–86. Elsevier, New York (1988).
37. E. Fitzer and R. Weiss, *Carbon* **25**, 455–467 (1987).
38. D. Maugis, *J. Mater. Sci.* **20**, 3041–73 (1985).
39. D. Maugis, in: *Microscopic Aspects of Adhesion and Lubrication*, J. M. Georges (Ed.), pp. 221–252. Elsevier, Amsterdam (1982).
40. J. Cognard, in: *Adhesion '87, Plast. Rub. Inst. 3rd Int. Conf., York, England*, paper no. 31, pp. 1–20 (1987).
41. A. N. Gent and J. Schultz, *J. Adhesion* **3**, 281 (1973).
42. A. N. Gent and J. Schultz, *J. Chim. Phys.* **70**, 708–716 (1973).
43. J. Cognard, *Int. J. Adhesion Adhesives* **8**, 93–99 (1988).
44. P. T. Reynolds and D. E. Packham, *Mater. Sci. Tech.* **3**, 1047–50 (1987).
45. A. Carré and J. Schultz, *J. Adhesion* **18**, 135–156 (1984).
46. A. Carré and J. Schultz, *J. Adhesion* **18**, 207–216 (1984).
47. K. L. Mittal, in: *Adhesion Science and Technology*, L.-H. Lee (Ed.), Part A, pp. 129–168. Plenum Press, New York (1975).
48. W. A. Zisman, in: *Adhesion and Cohesion*, P. Weiss (Ed.), p. 176. Elsevier, Amsterdam (1962).
49. F. M. Fowkes, in: *Treatise on Adhesion and Adhesives*, R. L. Patrick (Ed.), Vol. 1. Marcel Dekker, New York (1967).
50. S. T. Joslin and F. M. Fowkes, *Ind. Eng. Chem. Prod. Res. Dev.* **24**, 369–375 (1985).
51. F. M. Fowkes, K. L. Jones, G. Li and T. B. Lloyd, *Energy Fuels* **3**, 97–105 (1989).
52. J. Schultz, L. Lavielle and C. Martin, *Adhesion International 1987: Proceedings of the 10th Annual Meeting of the Adhesion Society*, L. H. Sharpe (Ed.), pp. 513–528. Gordon and Breach, New York (1987).
53. F. Chen, *Macromolecules* **21**, 1640–1643 (1988).
54. S. Wesson and R. E. Allred, in: *ACS Symposium Series 391: Inverse Gas Chromatography, Characterization of Polymers and Other Materials*, D. R. Lloyd, T. C. Ward, H. P. Schreiber and C. C. Pizana (Eds.), Ch. 15, pp. 203–216. American Chemical Society, Washington, D.C. (1989).
55. F. M. Fowkes, in: *Adhesion and Adsorption of Polymers*, L.-H. Lee (Ed.), Part A, pp. 43–52. Plenum Press, New York (1980).
56. F. M. Fowkes, D. O. Tischler, J. A. Wolfe, L. A. Lannigan, C. M. Ademu-John and M. J. Halliwell, *J. Polym. Sci.: Polym. Chem.* **22**, 547–566 (1984).
57. F. M. Fowkes, in: *ACS Symposium Series 199: Industrial Applications of Surface Analysis*, L. A. Casper and C. J. Powell (Eds.), Ch. 5, pp. 69–87. American Chemical Society, Washington D.C. (1982).
58. C. J. van Oss, M. J. Chaudhury and R. J. Good, *Chem. Rev.* **88**, 927–941 (1988).
59. S. Wu, *Polymer Interface and Adhesion*. Marcel Dekker, New York (1982).
60. W. V. Chang and F. Chen, *ACS Polym. Mater. Sci. Eng.* **61**, 607–613 (1989).
61. F. M. Fowkes, *J. Phys. Chem.* **66**, 382 (1962).
62. F. M. Fowkes, *Ind. Eng. Chem.* **56**, 40 (1964).
63. F. M. Fowkes and M. A. Mostafa, *Ind. Eng. Chem. Prod. Res. Dev.* **17**, 3–7 (1978).
64. F. M. Fowkes, *J. Adhesion Sci. Technol.* **1**, 7–27 (1987).
65. W. B. Jensen, *The Lewis Acid-Base Concepts*. Wiley-Interscience, New York (1980).
66. S. Ross and I. D. Morrison, *Surface Sci.* **52**, 103–119 (1975).
67. R. S. Sacher and I. D. Morrison, *J. Colloid Interface Sci.* **70**, 153–166. (1979).
68. J. C. Berg, in: *Composite Systems from Natural and Synthetic Polymers*, L. Salmen, A. de Ruvo, J. C. Seferis and E. B. Stark (Eds.), pp. 23–46. Elsevier Science, Amsterdam (1986).
69. R. E. Johnson, Jr. and R. H. Dettre, in: *Surface and Colloid Science*, E. Matijevic and F. R. Eirich (Eds.), Vol. 2, pp. 85–153. Wiley-Interscience, New York (1989).
70. D. L. Allara, in: *ACS Symposium Series 199: Industrial Applications of Surface Analysis*, L. A. Caspar and C. J. Powell (Eds), Ch. 3, pp. 31–47. American Chemical Society, Washington D.C. (1982).

71. D. L. Allara, in: *Adhesion and Adsorption of Polymers*, L.-H. Lee (Ed.), Part B, pp. 751–756. Plenum Press, New York (1980).
72. J. L. Koenig, *Adv. Polym. Sci.* **54**, 87–147 (1983).
73. M. M. Coleman and P. C. Painter, *Appl. Spectroscopy Rev.* **20**, 255–346 (1984).
74. G. W. Woodbury Jr. and L. A. Noll, *Colloids Surfaces* **8**, 1–15 (1983).
75. G. W. Woodbury Jr. and L. A. Noll, *Colloids Surfaces* **28**, 233–245 (1987).
76. T. Gendron, M. Waterbury and L. T. Drzal, *ONR Report AD-A185 966 (Contract # N00014-86-K-0393)* (1987).
77. M. Narkis, E. H. Chen and R. B. Pipes, *Polym. Composites* **9**, 245–250 (1988).
78. M. R. Piggott, *Composites Sci. Technol.* **30**, 295–306 (1987).
79. S. Kaminski and J. Courter (American Cyanamid Company), unpublished results (1989).
80. ESCA spectra were taken by J. Brinen and S. Greenhouse (American Cyanamid Company).
81. R. B. Bird, W. E. Stewart and E. N. Lightfoot, *Transport Phenomena*, p. 315. John Wiley & Sons, New York (1960).
82. S. Wesson, paper presented at the 13th Ann. Adhesion Soc. Meeting, Savannah GA (Feb. 1990).
83. J. A. Dean, *Lange's Handbook of Chemistry*, 13th ed. McGraw-Hill, New York (1985).
84. R. S. Drago, L. B. Parr and C. S. Chamberlain, *J. Am. Chem. Soc.*, **99**, 3202–3209 (1977).

Acid-Base Interactions, pp. 191-205
Eds. K.L. Mittal and H.R. Anderson, Jr.
©VSP 1991

Influence of surface chemistry and surface morphology on the acid–base interaction capacities of glass fibers and silicas

EUGÈNE PAPIRER* and HENRI BALARD
*Centre de Recherches sur la Physicochimie des Surfaces Solides (CNRS),
24 Avenue du Président Kennedy, 68200 Mulhouse, France*

Revised version received 8 March 1990

Abstract—Inverse gas chromatography is utilized for the study of the surface properties of glass fibers treated with either a silane or a titanate coupling agent, and silicas before and after treatment with alcohols, diols, and polyethyleneglycols. The dispersive component of the surface energy is easily assessed from the retention data of *n*-alkanes, and the acid–base interaction capacity is estimated by injecting probes of known acid–base properties. Besides the decisive role of acid–base interactions in the formation of the interface, it is shown that the surface morphology should not be ignored in order to achieve a satisfactory description of the interaction capacity of a given solid surface.

Keywords: Glass fibers; silicas; modification; surface properties; acid–base interaction; surface morphology; inverse gas chromatography.

1. INTRODUCTION

Since the pioneering work of Fowkes [1] on the determination and significance of the surface energy components, i.e. dispersive (γ_S^D) and specific (γ_S^{SP}) components, much progress has been made in understanding the behavior of solid surfaces when in contact with either gaseous, liquid, or solid environments. In particular, the adhesion phenomena could be examined on a more fundamental level. By making an acceptable hypothesis, γ_S^D has become accessible to measurement through readily performable methods such as contact angle measurements of liquid drops deposited on flat solid surfaces. On powdered solids, however, the contact angle method is not valid, or at least it needs considerable care and precaution in order to obtain meaningful results [2]. Therefore, different routes were explored, such as the determination and interpretation of gas adsorption isotherms [3]. But this procedure is very time-consuming. More recently, inverse gas chromatography (IGC) [4, 5] was applied with success to the characterization of the surface properties of divided solids. It is partly the aim of this paper to demonstrate the potential of this method in following very small changes in γ_S^D induced by controlled chemical surface modifications. IGC also provides information on the acid–base characteristics of the solid supports, provided that an appropriate procedure has been worked out for the evaluation

---

*To whom correspondence should be addressed.

means that depending on its partner, a molecule such as acetone, for instance, will be able to exchange acidic (due to the electron deficiency of the carbonyl carbon), or base-like (due to the high electron density of the oxygen atom) interactions, or both.

When injecting polar probes into a GC column containing polar fillings, one obviously records only one peak, or V_N, per probe. Yet V_N accounts for both London and specific interactions, and hence a procedure has to be established to separate out the two types of contribution. Such a procedure is illustrated schematically in Fig. 3.

In this graph, ΔG_a° of various probes is plotted against the vapor pressure (P_0) of the probes. The choice of P_0 is based on the fact that P_0 is a relevant quantity for GC experiments and also because it is related to the surface energy of the solutes [18]. All points representative of alkanes fall on a line, whereas those corresponding to polar probes are located above that line.

By definition, their departure from the alkane line will be I_{SP}, the specific interaction parameter. To go one step further, the acidity (acceptor number: AN) and basicity (donor number: DN) of the solid may be evaluated by applying an expression which is an extension of Gutmann's equation:

$$\Delta H^{SP} \approx \frac{(AN)_1\,(DN)_2}{100} + \frac{(DN)_1\,(AN)_2}{100}, \qquad (6)$$

where ΔH^{SP} is the specific interaction enthalpy (calculated from the variation of I_{SP} with temperature) and subscripts 1 and 2 refer to the solid surface and the probe, respectively.

The results concerning glass fibers have already been published [9]. It was found that the silane treatment definitely increases the basicity of the glass fiber, an increase which of course favors the acid–base interactions with the acidic phenolic resin; whereas the titanate treatment causes the opposite effect, which is detrimental to adhesion.

This earlier study demonstrates that acid–base interactions indeed play a

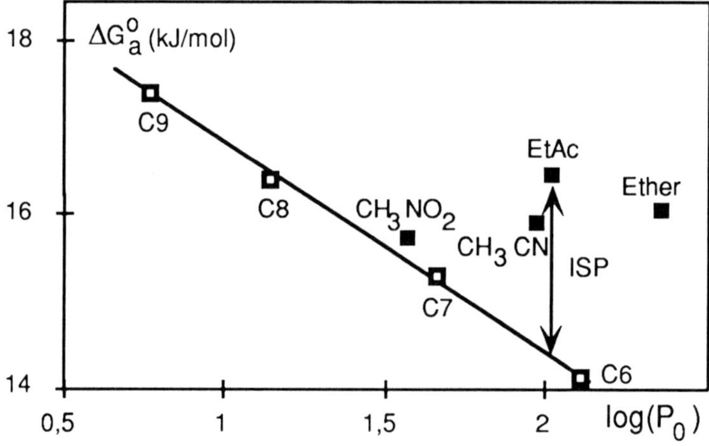

Figure 3. Example of the determination of the specific interaction parameter (I_{SP}) by IGC, by plotting ΔG_a° versus the vapour pressure of the solutes injected in the column of glass fibers.

Hence taking into account Fowkes's relation for the calculation of the work of adhesion through the dispersion forces only and making the required change in units [transforming ΔG units (kJ/mol) into surface energy units (mJ/m^2)], one obtains the expression

$$\gamma_{S}^{D} = \frac{1}{4\gamma_{CH_2}}\left[\frac{\Delta G_{a}^{CH_2}}{Na_{CH_2}}\right]^2. \tag{5}$$

In this expression, everything is known (γ_{CH_2}: the surface energy of molecules made up of CH_2 groups, i.e. polyethylene; N: Avogadro's number; a: the area of an adsorbed $-CH_2-$ group) or measurable ($\Delta G_{a}^{CH_2}$) as illustrated below, except the quantity of interest, i.e. γ_{S}^{D}.

Typical plots are shown in Fig. 1, which relates (ΔG_{a}^{o}) to the number n_C of carbon atoms of the probes. From the slopes of these lines, $\Delta G_{a}^{CH_2}$ is determined and γ_{S}^{D} values of initial, silane- and titanate-treated fibers are calculated. The γ_{S}^{D} values are, respectively, 30.1 mJ/m^2 for initial fibers, 42.9 mJ/m^2 for silane-treated fibers and 26.3 mJ/m^2 for titanate-treated fibers. These results indicate that γ_{S}^{D} does not change appreciably after silane or titanate treatment. Yet, the behaviors of these fibers in a phenolic resin are quite different, as seen in the electron micrographs of fractured surfaces of composite materials (Fig. 2) and from the results of mechanical tests (Table 1).

Whereas silane treatment leads to a clear increase in the adhesion properties of the fiber, the titanate treatment causes the opposite effect. This difference possibly stems from the different acid–base interaction capacities of the treated fibers. To determine the acid–base characteristics of the fibers by IGC, one may think of injecting probes of given acid or base properties.

Acidity and/or basicity of organic probes may be defined by the semi-empirical scales of Drago *et al.* [16] or Gutmann [27]. We choose Gutmann's scale essentially for two reasons. First, because a sufficient number of data for volatile probes suited for IGC experiments are available, and second, because Gutmann's approach considers the amphoteric nature of most polar probes. This

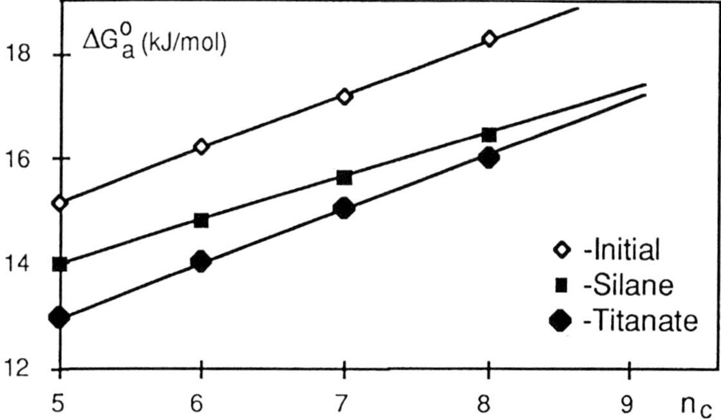

Figure 1. Variation of the alkane retention volume with the number of carbon atoms in the alkanes injected in the column containing glass fibers.

of the chromatographic data. Such a procedure was developed and will be illustrated by taking some earlier results [6] obtained on glass fibers.

Whereas acid–base interactions explain a series of important phenomena, such as polymer adsorption [7], filler dispersion in a polymer matrix, and the formation of an interface in composite materials [8], several recent investigations point to the necessity of considering the morphological characteristics of the entities in contact (the solid surface and polymer, for example) in order to obtain a molecular description of an interface.

This paper is divided into three sections. First we will underline the correct predictions of Fowkes's approach, i.e. the importance of the acid–base interaction for glass fiber–resin adhesion and hence the mechanical performance of this composite. Then we will consider the progressive decrease of the specific interactions capacity of silicas grafted with alkyl chains or chains possessing a hydroxyl end group. An example will also be given where surface acid–base properties are controlled by the grafting of polyethylene glycols. Finally, it will be shown that besides energetic considerations, one indeed has to take into account the molecular architecture in order to describe more precisely the formation of an interface.

2. METHODS

Most of the experimental procedures have been described in detail in the literature [5] and only general trends will be given here.

2.1. *Modification of solid surfaces*

Glass fibers (Rockwool) from Dannemark were modified inside the GC column by injecting the appropriate reagents (γ-aminopropyltriethoxysilane, and isopropyltridecylbenzene sulfonyl titanate). Silane (A1100) was from Union Carbide and titanate (TTBS-95) was from Kenrich Petrochemicals. For the silane treatment, an aqueous solution (2%) was injected in excess (equivalent to about 15 monolayers) in the column maintained at 40°C, in order to avoid water evaporation. Then the column was heated at 120°C, for 24 h, under a helium flow to eliminate the non-reacted silane. The treatment with titanate was done in a very similar manner, but with a 2% isopropanol solution. Larger amounts of treated glass fibers, for the preparation of composite materials, were prepared the same way but using ordinary laboratory glass equipment. Composite materials were obtained by first dispersing 2% fibers in water and then submitting the dispersion to centrifugation (4000 rpm). After drying, at 100°C, the glass bed was impregnated with a phenolic resin, and the mix was cured at 150°C, under 8 MPa. Elongation measurements were performed on composite samples using an Instron machine. From Instron recording, the modulus and tensile strength were calculated. Glass–resin adhesion was also qualitatively estimated from the fracture surface of the composite, using scanning electron microscopy [9]. Silica was grafted with alkyl chains by reacting in an autoclave, at 180°C, the surface acidic silanol groups with alcohols [10] or diols [11] of increasing chain length. Silica samples were also treated with polyethylene glycols (PEG) simply by contacting the powder with the necessary amount of melted PEG, at 200°C, in the absence of air [12].

2.2. Inverse gas chromatography

Inverse gas chromatography experiments were performed with a commercial model GC (Intersmat 121FL) fitted with a flame ionization detector [13].

2.3. NMR

The measurements were made in the laboratory of Professor Legrand (ESPCI, Paris), using a Bruker spectrometer and the cross-polarization magic angle spinning technique [14].

3. RESULTS

3.1. Acid–base interactions between glass fibers and phenolic resin

The surface properties of Rockwool (Dannemark) glass fibers are most difficult to evaluate, except by IGC, as will be seen. In the first step, γ_S^D was measured, according to the method of Dorris and Gray [4], by injecting a homologous series of n-alkanes in the column containing the glass fibers. The retention times and retention volumes V_N (volume of carrier gas necessary to push the probe through the column) were determined. Intuitively already, it is understandable that a larger V_N will correspond to a higher affinity of the fibers for the alkanes. Thermodynamically, affinity means standard variation of the free energy of adsorption ΔG_a°. ΔG_a° and V_N are related by

$$\Delta G_a^\circ = -RT \ln V_N + C, \tag{1}$$

where C is a constant depending on the choice of the reference state of the adsorbed alkane. ΔG_a, expressed in kJ/mol, and V_N (in cm^3) are related by

$$\Delta G_a^\circ = -RT \ln 8.186 \times 10^4 \frac{V_N}{TmS\rho}. \tag{2}$$

In this expression, T is the temperature of measurement, m is the amount (in g) of fibers in the column, S is their specific surface area (0.3 m^2/g), and ρ is the density of the fibers. The chosen reference state is that of Kemball and Rideal [15]. IGC measurements are easily performed at various temperatures and, consequently, from the variation of ΔG_a° with temperature, the enthalpy of adsorption is computed:

$$\Delta H_a = -T^2 \frac{\partial}{\partial T}(\Delta G_a^\circ/T). \tag{3}$$

Further, since ΔG_a° (or $RT \ln V_N$) varies linearly with the number of carbon atoms of the n-alkanes, it becomes possible to define an incremental value $\Delta G_a^{CH_2}$ which no longer depends on the arbitrary choice of the standard reference state of the adsorbed alkane:

$$\Delta G_a^{CH_2} = -RT \ln \frac{V_{N+1}}{V_N}, \tag{4}$$

where V_{N+1} and V_N are the net retention volumes of n-alkanes having respectively $N+1$ and N atoms of carbon. Furthermore, it is a common observation that ΔG_a° and the enthalpy (ΔH) are directly proportional for alkane adsorption [4].

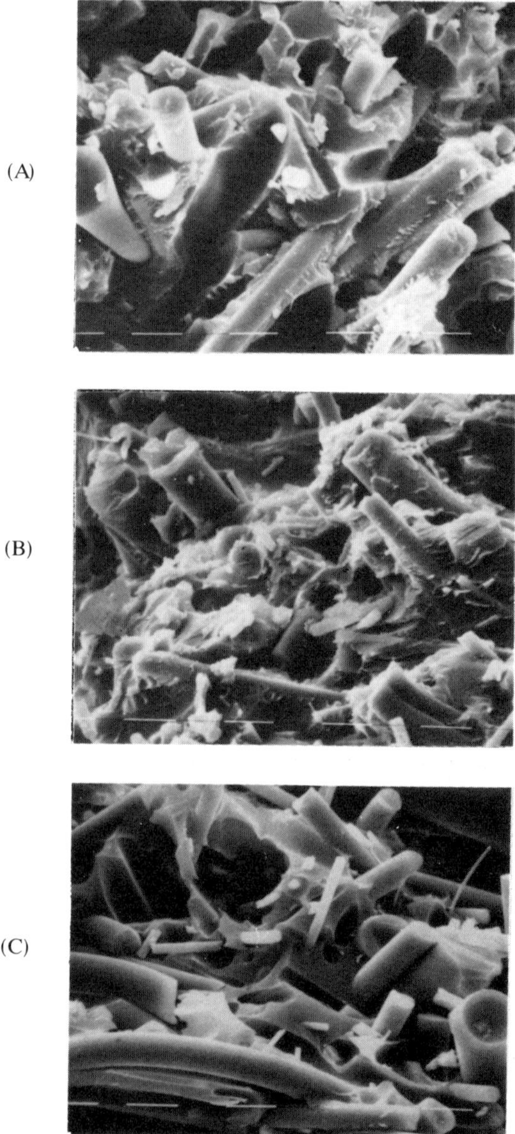

Figure 2. Comparison of the fracture surfaces of glass fiber–phenolic resin composites. (A) Untreated; (B) silane-treated; (C) titanate-treated.

Table 1.
Tensile properties of glass fiber–phenolic resin composites

Treatment	Fiber content (volume fraction, %)	Modulus (MPa)	Tensile strength (MPa)
No	50	0.5	26
Titanate	46	0.3	16
Silane	39	1.3	63

decisive role in adhesion and underlines the high sensitivity of IGC, capable of revealing surface properties of solids which cannot be easily analyzed by another technique. In the following sections, it will be seen that IGC is useful for demonstrating phenomena occurring at a molecular level on a solid surface.

3.2. Modification of the acid–base interaction potential of silica by grafting PEG

Like glass fibers, the surfaces of silicas carry hydroxyl and siloxane groups conferring on them acid–base properties [19, 20] which can be modulated using appropriate modification procedures, one of these being the grafting of poly-ethylene glycols (PEGs) [12]. Since PEG contains ether linkages, it is a potential molecule first to increase or preserve, when grafted on silica, the base-like character of the oxide, and secondly to decrease drastically its acidic character. This is shown in Table 2 by the enthalpies of adsorption of either an acidic probe ($CHCl_3$) or a basic probe (diethyl ether).

The sample reference indicates the nature of the silica (Aerosil 300, from Degussa), the molecular weight of the graft, and the grafting ratio expressed as a weight percentage: for instance, A300/4000/7 means that silica Aerosil 300 has been grafted with PEG having a molecular weight of 4000, the grafting ratio being equal to 7%. n_{UM} represents the number of monomer units per unit surface area. As stated above, the non-modified silica interacts with $CHCl_3$, but much more strongly with ether, the interaction being so efficient that it takes too long a time to perform the IGC measurement! This means that the surface is rather acidic. Upon treatment with PEG, the enthalpy of adsorption of ether drops significantly even for relatively low surface coverages (n_{UM}). As n_{UM} increases, one approaches a ΔH value close to that of the vaporization of ether, demon-strating that ether no longer exchanges specific interactions with the silica surface. On the other hand, the base-like character is not significantly altered, the enthalpy of interaction of $CHCl_3$ becoming very close to that measured on pure PEG.

One decisive factor for the change in surface properties is obviously the chemical nature of the graft, but the surface coverage is also important. The influence of this last parameter will be shown in the following section.

Table 2.

Enthalpies of adsorption of acidic or basic probes on PEG-modified silica

Sample	n_{UM}	ΔH of adsorption (kJ/mol)	
		$CHCl_3$	Diethyl ether
A300/—/0	0.0	43	—
A300/2000/5	2.6	40	55
A300/4000/7	3.7	36	51
A300/2000/17	8.9	37	39
A300/2000/56	29.5	35	21
A300/10 000/57	30.0	37	23
A300/4000/72	37.9	36	28
PEG	—	37	—
ΔH vaporization	—	29	29.5

3.3. Suppression of the acid–base interaction potential of silica by the grafting of alkyl chains

To demonstrate the influence of the surface morphology on the formation of an interface, three types of silica were selected: a pyrogenic silica (Aerosil 130 from Degussa) called silica A, a precipitated silica (Z175 from Rhône Poulenc) denoted silica P, and a lamellar crystalline silica (silica L) obtained after acid attack on a synthetic Kenyaite (Laboratoire des Matériaux Minéraux, Ecole Nationale Supérieure de Chimie, Mulhouse). These samples have comparable surface areas and have been studied extensively in this laboratory [21].

The main differences between these three samples may be summarized as follows: silica A exhibits a rather smooth surface on a molecular level and is covered by silanol groups distributed randomly all over the surface; silica P, on the contrary, has a rugose surface with silanols concentrated in islands, and remaining flexible silicic acid surface chains. Silica L is made by the superposition of silicic tetrahedral layers separated by a distance of about 15 Å. This sample is microporous and it is expected that the accessibility of the inner hydroxyl groups to gaseous probes will depend not only on the size, but also on the flexibility of the probes.

To investigate first the influence of the surface morphology, the behaviors of silica A and silica P will be compared.

Both samples were submitted to grafting with diols of increasing chain length. Diols are capable of binding to the silica surface by both ends, i.e. a diesterification reaction cannot be excluded provided that the configuration of diols and surface hydroxyl groups on silica can be matched. In other words, the study of the grafting of diols may be considered as a way to explore the surface morphology of silicas on a molecular level. Moreover, if both ends of the diol are linked to the surface, then silica should lose most of its acid–base interaction potential. This potential may be estimated by the enthalpy of adsorption of solutes having H-bonding properties. These probes were selected because their enthalpy of interaction with residual hydroxyl groups (silanol and alcohol groups) is important and may thus be more easily evidenced.

Figure 4 displays the enthalpies of adsorption (calculated from the variation of ΔG_a° with temperature) of a series of alcohols on modified silica A. It is apparent that depending on the oddness of the number of carbons on the grafted diol, ΔH takes alternatively higher and lower values—higher values for grafted diols having an odd number of carbon atoms. For shorter (C_5) grafts, it is probable that the adsorbing alcohol probe also interacts directly with the silica surface and not just with the graft, and hence the ΔH value is complex. In other words, short grafts are unable to shield the silica surface entirely. The main result suggests, however, that diols with an even number of carbon atoms preferentially form diesters.

The number of mono- and diester bonds was determined by ^{13}C CPMAS NMR spectroscopy. Figure 5 shows the variation of the ratio (R) of the number of diester bonds to the number of monester bonds formed on silica A and silica P. It is seen that indeed R follows, for silica A, the same trend as ΔH adsorption of alcohol probes on the grafted silica. The point to be stressed is that silica P behaves differently and does not discriminate between odd and even diols, both

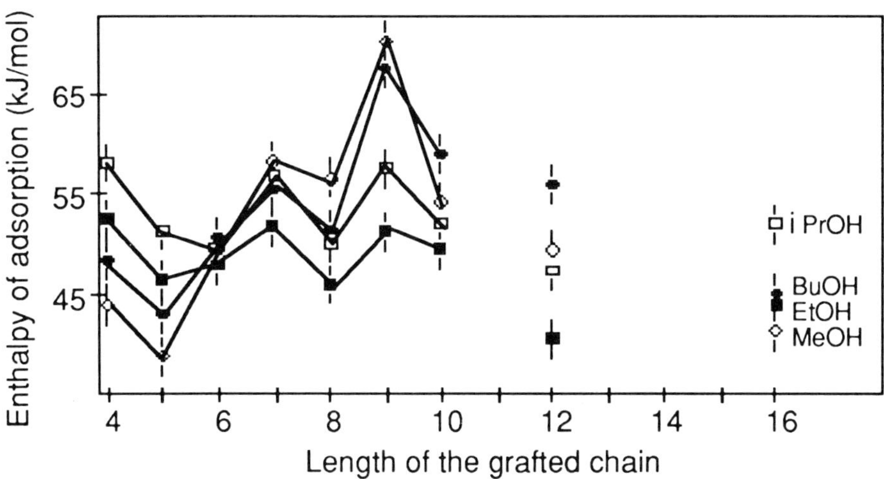

Figure 4. Variation of the enthalpy of adsorption of alcohol probes on silica Aerosil 130 (silica A) modified by esterification with diols.

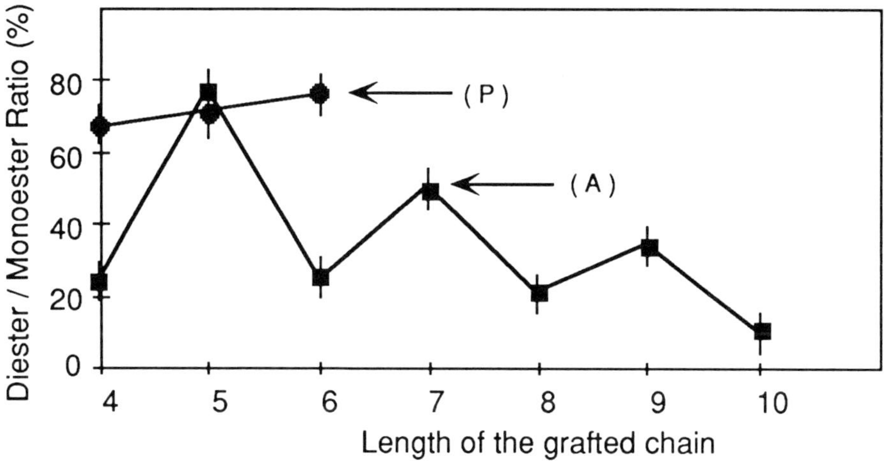

Figure 5. Variation of the ratio R (diesterification/monoesterification) for diols grafted on silica Aerosil 130 (silica A) and precipitated silica Z175 (silica P).

of which essentially form diesters. This behavior is accounted for by the larger number of silanol groups (about 8 for P instead of 3 Si—OH/nm^2 for A) and by the existence of silicic acid chains, which are less rigid than the silanols belonging to a silica structure and permit a more random orientation of the silicon–oxygen bonds in the space, i.e. a better accessibility. However, there is another reason based on the conformation of the grafted chains on the various types of silica.

From energetic considerations, it is highly probable that the grafted alkane chains will enter a *trans–trans* conformation, which allows optimal interactions with the surface (the chains lie flat on the surface) and between the grafts (leading to an optimal surface ordering). Such a possibility is only offered by the smooth surface of silica A. Figure 6 illustrates this situation: it is seen further that

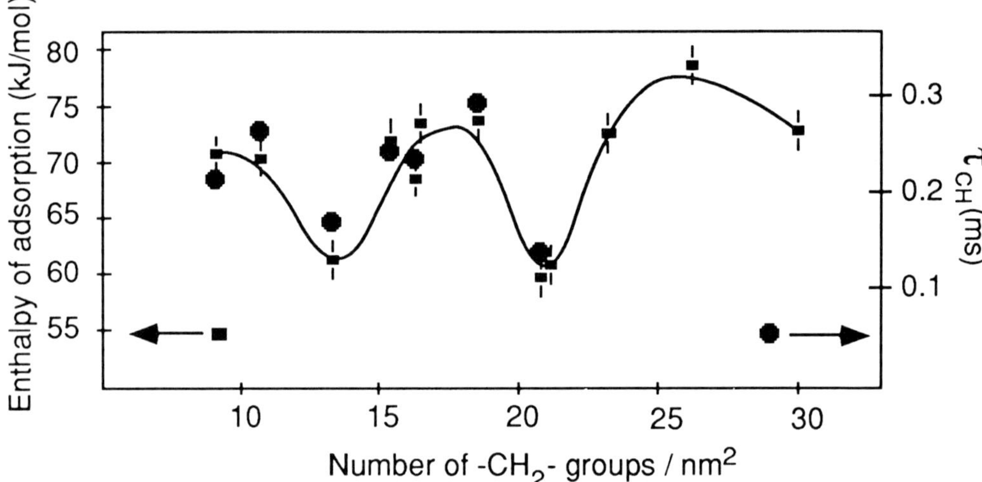

Figure 9. Variation of the enthalpy of adsorption of butanol and of the time constant of NMR cross-polarization (τ_{CH}) of the terminal $-CH_3$ with the number of methylene groups per unit surface area (nm²).

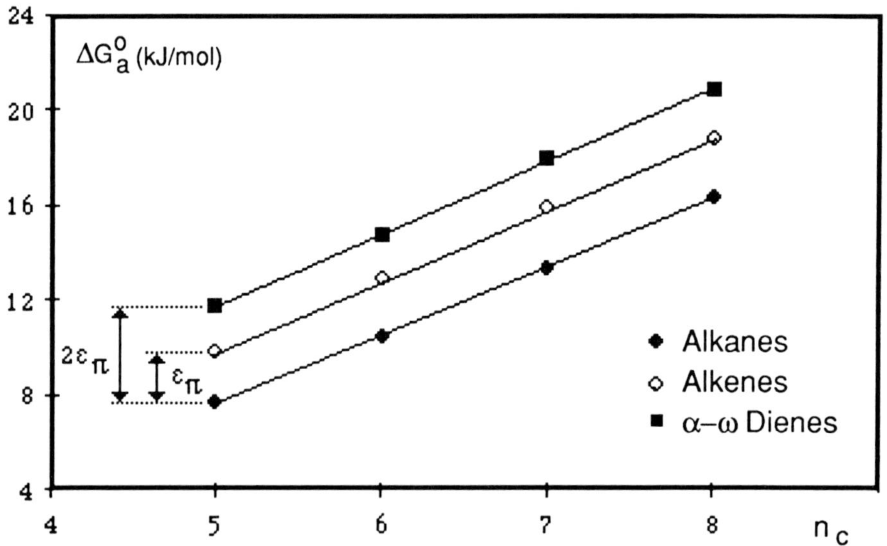

Figure 10. Variation of the free energy of adsorption (ΔG_a°) of alkanes, alkenes, and $\alpha-\omega$ dienes on the surface of Aerosil 130 (silica A).

comparison, shows the variation in free enthalpy of adsorption of alkanes, alkenes, and $\alpha-\omega$ dienes with increasing numbers of C atoms, on the better-defined planar surface of silica A.

As verified earlier with glass fibers, the plot of ΔG_a° of alkanes versus their number of C atoms is linear. Similar observations were made with alkenes and $\alpha-\omega$ dienes. Obviously the higher ΔG_a° values of alkenes are accounted for by the specific interaction capacity of the acidic silica surface with double bonds, coverages equal to 15 and 21 $-CH_2-$ groups/nm^2, respectively. Taking 0.06 nm^2 as a mean value of the cross-section of an adsorbed $-CH_2-$ group, it is probable that these two minima correspond to one or two monolayers of $-CH_2-$, respectively. Such a situation is represented schematically in Fig. 8, which, moreover, shows the silica surface covered with short chains which are unable to cover it entirely.

If this scheme is correct, one would predict that movements of the terminal $-CH_3$ group of the grafted alkyl chain will be more restricted when the surface coverage corresponds exactly to the organized one or two monolayer situation. This hypothesis may be verified, calling on high resolution solid-state NMR, by measuring the cross-polarization time constant (τ_{CH}) which reflects the mobility of the methyl end group.

Figure 9 displays, on the one hand, the NMR results and, on the other hand, the enthalpy of adsorption (ΔH) of n-butanol. Both curves show minima values for the same surface coverages: the lower values of ΔH are accounted for by the lack of access of the highly packed monolayers of $-CH_2-$ groups.

These observations made on a molecularly flat silica A surface do not hold for the tortuous surface of silica P, for the reasons given earlier, in particular the accessibility and conformation of the surface interacting groups.

The last section of this paper will analyze a situation where steric hindrance is foreseeable: the case of a lamellar crystalline silica, having a lamellar distance of 15 Å. Small linear hydrocarbon probes will be able to insert between lamellae, whereas bulky branched alkanes will not be capable of doing so, as shown elsewhere [22].

In addition, the interaction of this silica with alkenes and $\alpha-\omega$ dienes was investigated in order to check their acid–base characteristics, or at least their acidic characteristics since unsaturations act as bases. Figure 10, for the sake of

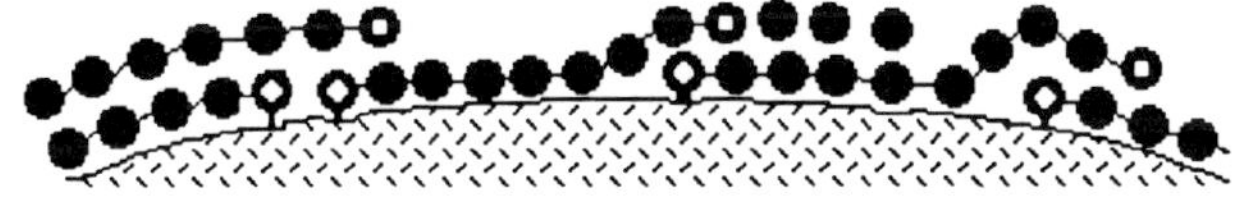

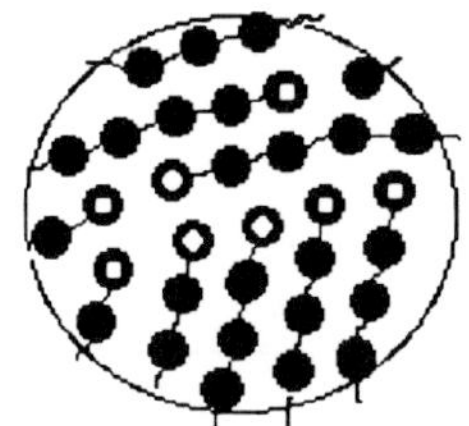

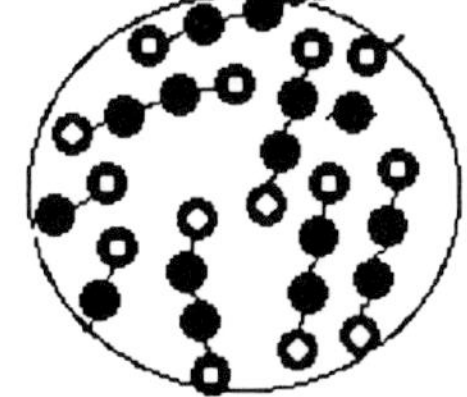

Figure 8. Schematic representation of the surface coverage of silica Aerosil 130 by $-CH_2-$ groups.

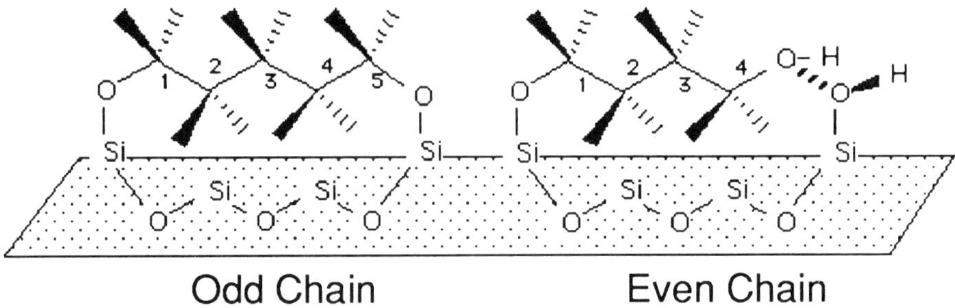

Figure 6. Schematic representation of the conformation of diols grafted on the quasi-planar surface of Aerosil, according to the oddness of the chains.

for obvious reasons, an odd number of carbon atoms are favorable for diesterification.

This part of the study brings in a new concept which underlines the influence of morphology of both interacting entities. Whereas acid–base interaction analysis is certainly an important contribution to the understanding of the interface formation of organic molecules in contact with a mineral oxide, morphological aspects should not be ignored.

A further example will be given to illustrate the influence of the solid surface on the conformation of grafted alkyl chains. The grafting of silica with alkyl chains is easily performed by an esterification reaction of the silica surface by alcohols. Between 1.8 and 2.2 chains per nm^2 are thus fixed on the oxide surface. It is first established that silica A covered with alkyl chains loses largely its acid–base interaction capacity. Second, when measuring γ_S^D as a function of the surface coverage ratio of methylene groups, an unexpected behavior is observed: γ_S^D does not decrease smoothly so as to reach the value of a solid entirely covered by $-CH_2-$ groups, i.e. polyethylene, but oscillates between about 32 and 38 mJ/m^2 (Fig. 7). Two minimum values of γ_S^D are recorded for surface

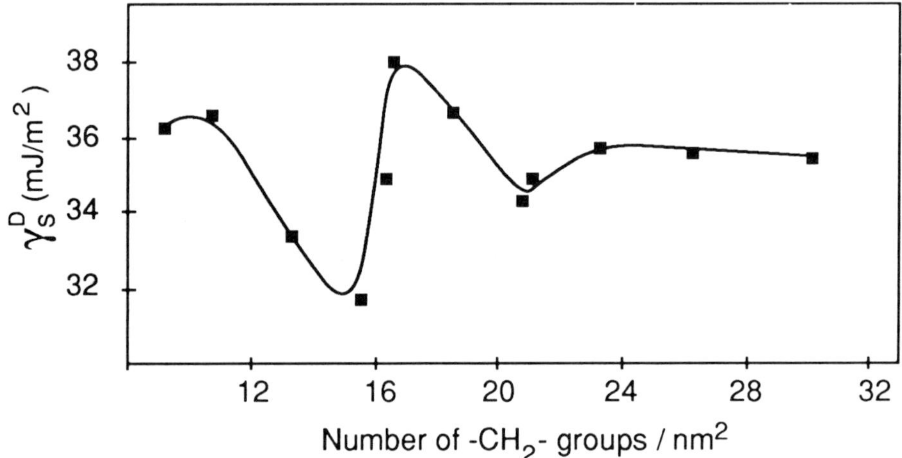

Figure 7. Variation of γ_S^D of silica Aerosil 130 (silica A) with the degree of surface coverage expressed as methylene groups per unit surface area (nm^2).

but an interesting observation is that the ΔG_a of α–ω dienes is twice that of alkenes. This result is understandable, since dienes have two double bonds capable of undergoing specific interactions with the surface acidic silanol groups of silica.

For crystalline silica (Fig. 11), the situation is much more complex [23]. As before, each family of probes: alkanes, alkenes, and α–ω dienes, is characterized by a continuous line when ΔG_a° is plotted against n_C. But the three lines are no longer parallel. Moreover, dienes apparently interact less than alkenes, even less than alkanes having n_C values higher than 6.

The explanation for these different behaviors is once more given by steric considerations: alkanes are able to insert between layers of crystalline silica. This will lead to optimum interaction. Alkenes will do almost the same and the existence of the double bond, even though it is more rigid, will cause the expected increase in ΔG_a. For α–ω dienes, the interaction becomes more difficult owing to the rather rigid end groups of these molecules—end groups which cannot interact with the adsorbing sites located between the lamellae. Figure 12 shows schematically the difference between the interaction capacity of alkane and alkene end groups of hydrocarbon probes used to test the surface properties of the crystalline silica. The radius of gyration of the end group is significantly greater for an unsaturated group than for a methyl group and, therefore, hinders the diffusion of such an unsaturated group between the lamellae of the crystalline silica.

4. CONCLUSION

This study highlights the significance of acid–base interactions which appreciably contribute to the formation of efficient composite materials (glass fiber–phenolic

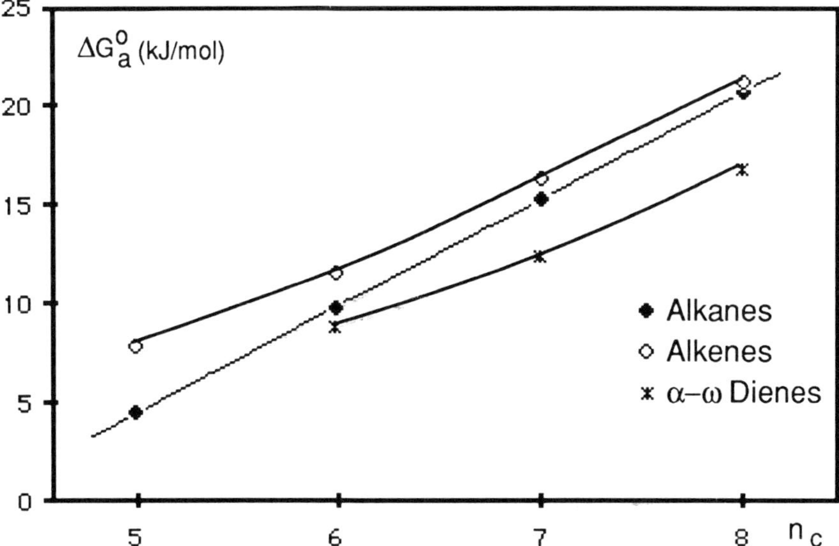

Figure 11. Variation of the free energy of adsorption (ΔG_a°) of alkanes, alkenes, and α–ω dienes on lamellar silica (H-Kenyaite).

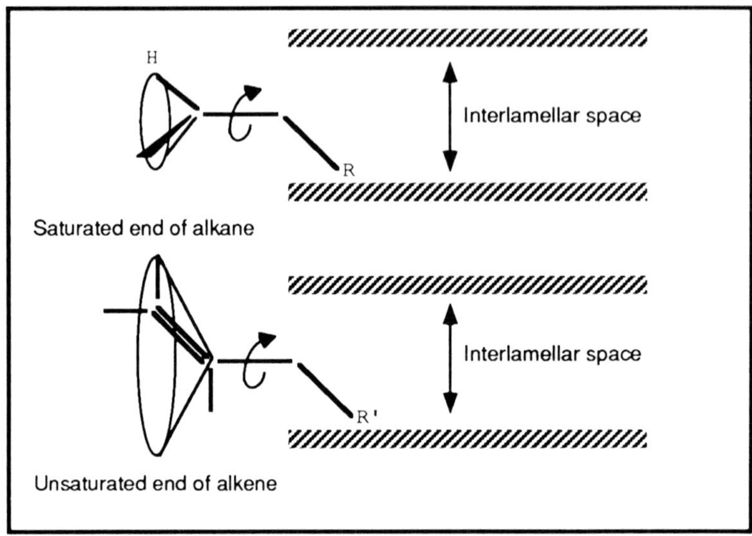

Figure 12. Schematic comparison of the adsorption of alkanes and alkenes on the lamellar silica.

resin). It also shows that the intensity of acid–base interactions may be varied by controlled chemical modifications of fibers or powders such as silicas. Finally, it appears that a satisfactory description of the formation and properties of an interface requires knowledge of the morphology of the entities involved in inter-action.

Acknowledgements

The authors thank their collaborators C. Saint Flour, M. Sidqi, G. Ligner, and J. Jagiello, who contributed to the present study. They also thank Professor A. P. Legrand for NMR measurements.

REFERENCES

1. F. M. Fowkes, *Ind. Eng. Chem.* **56**, 40 (1964).
2. Z. Kessaïssia, E. Papirer and J. B. Donnet, *J. Colloid Interface Sci.* **79**, 257 (1981).
3. B. Bilinski and E. Chibowski, *Powder Technol.* **35**, 39 (1983).
4. G. M. Dorris and D. G. Gray, *J. Colloid Interface Sci.* **71**, 93 (1979).
5. E. Papirer, in: *Composite Interfaces*, H. Ishida and J. L. Koenig (Eds), p. 203. North Holland, New York (1986).
6. C. Saint Flour and E. Papirer, *Ind. Eng. Chem.* **21**, 666 (1982).
7. F. M. Fowkes and M. A. Mostafa, *Ind. Eng. Chem.* **17**, 3 (1978).
8. F. M. Fowkes, *J. Adhesion Sci. Technol.* **1**, 7 (1987).
9. C. Saint Flour and E. Papirer, *J. Colloid Interface Sci.* **91**, 69 (1983).
10. M. Sidqi, H. Balard, E. Papirer, A. Tuel, H. Hommel and A. P. Legrand, *Chromatographia* **27**, 311 (1989).
11. H. Balard, M. Sidqi, E. Papirer, J. B. Donnet, A. Tuel, H. Hommel and A. P. Legrand, *Chromatographia* **25**, 712 (1988).
12. E. Papirer, H. Balard, Y. Rahmani, A. P. Legrand, L. Facchini and H. Hommel, *Chromatographia* **23**, 639 (1987).
13. E. Papirer, A. Vidal and H. Balard, in: *Inverse Gas Chromatography*, D. Lloyd, T. Ward, H. Schreiber and C. Pizana (Eds), ACS Symposium Series No. 391, p. 248. American Chemical Society, Washington (1989).

14. H. Ben Ouada, H. Hommel, A. P. Legrand, H. Balard and E. Papirer, *J. Colloid Interface Sci.* **122**, 441 (1988).
15. C. Kemball and E. K. Rideal, *Proc. R. Soc. London, Ser. A* **187**, 53 (1946).
16. R. S. Drago, G. C. Vogel and T. E. Needham, *J. Am. Chem. Soc.* **93**, 6014 (1971).
17. V. Gutmann, *The Donor–Acceptor Approach to Molecular Interactions.* Plenum Press, New York (1983).
18. M. Nardin and E. Papirer, *J. Colloid Interface Sci.* (in press).
19. G. Ligner, A. Vidal, H. Balard and E. Papirer, *J. Colloid Interface Sci.* **133**, 200 (1989).
20. E. Papirer, H. Balard and A. Vidal, *Eur. Polym. J.* **24**, 783 (1988).
21. G. Ligner, Ph.D. Thesis, Université de Haute Alsace, Mulhouse, France (1989).
22. M. Sidqi, G. Ligner, J. Jagiello, H. Balard and E. Papirer, *Chromatographia* **28**, 588 (1989).
23. G. Ligner, M. Sidqi, J. Jagiello, H. Balard and E. Papirer, *Chromatographia* **39**, 35 (1990).

Acid-Base Interactions, pp. 207-225
Eds. K.L. Mittal and H.R. Anderson, Jr.
©VSP 1991

A method of determining the surface acidity of polymeric and metallic materials and its application to lap shear adhesion

ELISABETH J. BERGER

Polymers Department, General Motors Research Laboratories, 30500 Mound Road, Warren, MI 48090, USA

Revised version received 5 March 1990

Abstract—A method has been developed to measure the surface acidity of solids using the contact angles of seven probe liquids. The geometric mean model was used to calculate the surface free energy. Then the value of the solid polarity, from the geometric mean, was compared with the polarity values calculated using the geometric mean dispersive component and the contact angles of two Lewis acids (liquefied phenol and glycerol) and two Lewis bases (formamide and aniline.) The deviation between these values was used to determine a value for the acidity, referred to as *D*. *D* was measured for a series of polymeric and metallic materials. Lap shear joints were fabricated and tested using these substrates and two adhesives, a urethane and an epoxy. The acidity of the substrate surfaces was found to affect the lap shear joint strength.

Keywords: Acid–base interactions; surface acidity; surface energetics; lap shear adhesion; polymer surfaces.

1. INTRODUCTION

The joining of materials is at the heart of many manufacturing technologies. This is especially true in the automotive industry. Whether the joints are mechanical or adhesive, if the components of an automobile or any other structure do not stay attached to each other, the structure will not function. Traditionally, mechanical fastening or welding has been used for most load-bearing joints. However, the development of stronger, more durable, and more easily-processed structural adhesives is changing that tradition. Adhesive bonds distribute stresses more evenly within a joint, can be less labor-intensive, and can be used in places where mechanical fastening is difficult.

The use of structural adhesives is growing rapidly in the automotive industry. To take full advantage of this technology, the requirements for a strong and durable bond must be better understood from the perspectives of the adhesive, the substrates, the design of the joints, and the automotive service environment. Adding to the challenge of structural adhesive bonding is the increased interest in and use of new substrate materials as alternatives to conventional steel. These include polymeric materials, such as thermoplastic polyolefins or polymer blends, as well as new metallic materials, such as electroplated galvanized steel. Both the bulk mechanical properties of the new materials and their surface chemistry must be investigated to determine which variables are important in an adhesive joint. The stiffness values of the substrates are one parameter affecting the bond. The work reported herein looks at another aspect of the adhesive bond—the surface energetics and acid–base properties of substrate and adhesive—and investigates whether or not these surface properties affect our ability to form a strong and durable adhesive bond.

Surface energetics have long been thought to play a role in adhesive bonds, and there are many theories of the relationship between surface energetics and adhesion [1]. The critical surface tension [2], the geometric mean [3], the harmonic mean [4], and combinations of these have all been investigated, and have all been found to be useful for certain systems. In previous work in this laboratory [5], the geometric mean approximation has been found to be the most helpful model for describing the total surface free energy and the division between polar and dispersive components of the surface free energy. This approximation assumes that the surface free energy is divisible into polar and dispersive components:

$$\gamma^{t} = \gamma^{p} + \gamma^{d}, \tag{1}$$

and that the surface free energy of a solid can be calculated from contact angle data. Here γ is the surface free energy, and t, p, and d refer to total, polar, and dispersive, respectively. 'Polar' is used here for forces arising from all specific interactions, including hydrogen bonding, induced dipole, and acid–base interactions, and does not refer to dipole–dipole interactions alone. To determine the surface free energy experimentally via the geometric mean, the contact angles of several liquids for which γ_{lv}^{p} and γ_{lv}^{d} are known are used to calculate the works of adhesion, W_{a}, of the liquids on the solid substrate of interest using the equation

$$W_{a} = \gamma_{lv}(1 + \cos\theta), \tag{2}$$

where lv refers to the liquid–vapor interface, and θ is the contact angle. The geometric mean equation

$$W_{a} = 2\{(\gamma_{lv}^{d}\gamma_{s}^{d})^{1/2} + (\gamma_{lv}^{p}\gamma_{s}^{p})^{1/2}\} \tag{3}$$

(the subscript s refers to the solid surface) can be rearranged to

$$\frac{W_{a}}{2(\gamma_{lv}^{d})^{1/2}} = (\gamma_{s}^{d})^{1/2} + (\gamma_{s}^{p})^{1/2}\left[\frac{(\gamma_{lv}^{p})}{(\gamma_{lv}^{d})}\right]^{1/2} \tag{4}$$

and a straight line plotted which has slope $(\gamma_{s}^{p})^{1/2}$ and intercept $(\gamma_{s}^{d})^{1/2}$. A match between the polar and dispersive components of a substrate and a wetting liquid has been found to be advantageous in many systems, including fiber–matrix adhesion in composites [6] and adhesion of tapes to various substrates [7].

However, there are also cases in which the theory described above does not predict the observed trends in adhesion strength. The major theoretical objection to the geometric mean is that it treats all specific interactions the same, without differentiating between dipole–dipole, hydrogen bonding, or acid–base interactions [8]. Since the values of the surface energies calculated from the geometric mean depend on the liquids used as probes, a change of probe liquids may give different values. This is not accounted for in the geometric mean theory. This gives us the options of completely doing away with the geometric mean approximation, or of using it as a model, recognizing that it is not a fundamental theory, and utilizing it for the information it can give, while interpreting its conclusions with care.

Besides the geometric mean, another approach of great interest in the literature is acid–base theory, which states that interactions between an acidic substrate and a basic adhesive or other wetting liquid (or vice versa) are very beneficial to adhesion [9–11]. For example, when the surface energetics of two reinforcement fibers are similar, if one is acidic and the other basic, the acidic one will be wetted better by a basic epoxy resin [5]. When the Lewis definition of acids and bases (electron-pair acceptors and donors, respectively) is used, the concept of acid–base interactions also extends to hydrogen bonding and most other specific interactions, encompassing the forces referred to as 'polar' by the geometric mean.

However, the acidity of a surface, particularly a solid surface, is difficult to quantify. There are several semi-empirical methods in the literature, including infrared spectroscopy and calorimetric determinations of the Drago coefficients [12, 13], which are themselves semi-empirical parameters. These techniques are fairly complicated experimentally and computationally, and usually measure bulk properties rather than surface properties, although some work has been done on using attenuated total reflectance IR as a surface probe [8]. Because of these deficiencies, a new method is needed which is directed towards surface acidities and is experimentally straightforward.

A surface acidity determination based on contact angles meets these requirements. Contact angles are restricted to probing surfaces and surface interactions with wetting liquids. Contact angles can be measured either using a contact angle goniometer for flat solid surfaces or a wetting balance for fibers or thin films. While fairly rigorous standards of cleanliness and care in sample preparation are necessary, the actual measurements are not difficult. However, there is little available information [9] on the conversion of contact angle measurements into surface acidity values.

In this work, contact angles of seven probe liquids on a series of solid surfaces have been used to determine the geometric mean approximation to the surface free energy. Then values for four of the liquids, two Lewis acids (glycerol and liquefied phenol) and two Lewis bases (formamide and aniline), were used to develop a means of calculating an empirical value for the total acidity of a surface. While the method of calculation required several iterations to develop into a useful form, it is not a difficult calculation. This acidity value, referred to as D, was measured for a number of surfaces of interest to the automotive industry, including thermoplastic body panel materials such as nylon blends or thermoplastic polyolefins, and other polymeric materials and metals. Lap shear strengths were measured for these materials, and the importance of surface acidity to the adhesive bonds is discussed.

2. EXPERIMENTAL

2.1. Materials

A list of substrate materials is given in Table 1. The polymeric substrates were cleaned by solvent wiping with either methylene chloride or acetone, or by being agitated in a detergent solution for 30 min, followed by distilled water rinsing and air-drying. An abbreviated postcure cycle of 30 min at 121°C, cool to room

Table 1.
A listing of the substrate materials used

Designation	Material	Surface preparation	Comments
CRS	Cold-rolled steel	Vapor-blasted, acetone-wiped	—
GALV	Electrogalvanized steel	Acetone-wiped	—
NPB	Thermoplastic blend of nylon 6,6 and polyphenylene oxide	—	—
NPB-dry	—	CH_2Cl_2 wipe	Dried immediately after cutting with a diamond saw
NPB-damp	—	CH_2Cl_2 wipe, acetone wipe, or detergent wash	Stored damp after cutting with a diamond saw
NPB-amb	—	CH_2Cl_2 wipe, acetone wipe, or detergent wash	Stored in ambient conditions after cutting
NPB–PC	—	CH_2Cl_2 wipe or acetone wipe	NPB-dry, solvent-wiped, then post-cured
PMMA	Poly(methyl methacrylate)	CH_2Cl_2 wipe, acetone wipe, or detergent wash	—
NUN	Nylon 6,6	CH_2Cl_2 wipe	Stored in ambient conditions
N43	Nylon 6,6	CH_2Cl_2 wipe	43% glass reinforced, stored in ambient conditions
TPO-A	Thermoplastic polyolefin	CH_2Cl_2 wipe, acetone wipe, or detergent wash	'Paintable' grade TPO, manufacturer A
TPO-B	Thermoplastic polyolefin	CH_2Cl_2 wipe, acetone wipe, or detergent wash	'Paintable' grade TPO, manufacturer B

temperature, and 60 min at 177°C was used for NPB–PC (nylon–polyphenylene oxide blend, postcured) samples to simulate the changes in NPB after going through a paint oven.

The formulations and cure cycles of Epoxy A and Epoxy B, the two resins which were used to investigate the reproducibility of the D parameter, are listed in Table 2. Flat sheets of the resins were cured with the top surfaces molded against air. Ten sheets each were molded of each of the two formulations. These sheets were molded from five different batches of each resin mix. The cured sheets were stored covered, and care was taken to avoid contaminating the surfaces with fingerprints. The sheets were cut or broken into small pieces, to fit on the goniometer stage.

The adhesives used to bond the lap shear samples and their cure schedules are also given in Table 2. The wetting liquids are listed in Table 3 and were used as received.

2.2. Contact angles

Contact angles of each of the wetting liquids on the substrates were measured using a contact angle goniometer (Rame-Hart, Inc.). Contact angle measurements

Table 2.
Epoxy formulations and adhesives used

Designation	Description	Cure schedule
UA	Commercial, two-part urethane adhesive	20 min, 90°C
EA	Commercial, two-part acid-cured, epoxy adhesive	20 min, 90°C
Epoxy A	100 parts DGEBA,[a] 23.5 parts isophorone diamine (IPD)	2 h, 80°C 3 h, 125°C
Epoxy B	100 parts DGEBA, 4 parts 2-ethyl-4-methyl-imidazole (EMI-24)	4 h, 60°C 2 h, 150°C

[a] Diglycidyl ether of bisphenol A.

Table 3.
Wetting liquids used and their surface energies (in ergs/cm^2)

Liquid	γ_{lv}^d	γ_{lv}^p	γ_{lv}^t
Aniline[a]	41.2	2.0	43.2
Bromonaphthalene[b]	44.8	0.0	44.8
Formamide[b]	31.8	25.7	57.5
Glycerol[b]	33.9	29.8	63.7
Methylene iodide[b]	49.8	2.4	52.2
Phenol, liquefied[c]	37.8	2.6	40.4
Water[b]	21.7	50.2	71.9

[a] γ_{lv}^d and γ_{lv}^p values determined from contact angle on TPO-A.

[b] From ref. [5].

[c] 12% water. γ_{lv}^d and γ_{lv}^p values determined from contact angle on TPO-A.

were made on cut or broken pieces of the substrates, with measurements being made on at least two samples of substrate for each wetting liquid, a total of at least 14 samples of each substrate per surface chemistry determination. About 12 contact angles per liquid were measured for each substrate. For the two formulations used for the reproducibility studies, two researchers measured the contact angles on half the sheets each, to check for operator variability.

2.3. Lap shear surface preparation and testing

Lap shear specimens were made from coupons 2.5 cm by 10.0 cm, bonded in fixtures with a 0.030 cm spacing wire and 1.2 cm bond length overlap. Ten specimens were made and measured for each substrate–adhesive combination. The polymeric surfaces to be bonded with the urethane adhesive were abraded with Scotchbrite (3M), then either wiped with the appropriate solvent (methylene chloride or acetone) or washed in a 5% solution of detergent for 30 min, rinsed several times with distilled water, and allowed to dry. The cold-rolled steel was vapor-degreased with 1,1,1-trichloroethane and grit-blasted with Novecite 200 in water, then wiped with acetone, and the galvanized steel was wiped with acetone. The NPB–PC surfaces to be bonded with the epoxy adhesive (EA) were not abraded. After curing, all samples were left in the fixtures overnight to allow for temperature equilibration. Thermoplastic polyolefin (TPO-A) bonded with a urethane adhesive (UA) was postcured for 4 h at 100°C. All other samples bonded with UA went through a simulated paint cycle:

> 30 min at 121°C, cool to room temperature;
> 30 min at 149°C, cool;
> 60 min at 177°C, cool;
> 15 min at 149°C, cool;
> 35 min at 163°C, cool;
> 30 min at 154°C, cool.

Half the samples were soaked in 60°C water for 1 week. Samples were tested on an Instron Testing Machine (model TTC) at a crosshead speed of 0.127 cm/min at room temperature.

3. RESULTS AND DISCUSSION

3.1. *The effect of substrate stiffness on lap shear strengths*

The lap shear results given in Table 4 were obtained by others in this laboratory [14]. These are polymeric or metallic substrates bonded with UA and then post-cured. Also in Table 4 are values for the stiffness of the substrates, from the same source. Failure modes were either adhesional or mixed adhesional and cohesional, with the unfilled nylon 6,6 (NUN) samples failing by stock breaks. (Failures were said to be adhesional if visual inspection showed the failure to be between the substrate and the adhesive, cohesional if the failure was within the adhesive, and by stock break if the substrate failed.) If the dry strength values are plotted against the natural logarithm of the stiffness, an excellent correlation is found, as is seen in Fig. 1. The lap shear test has long been recognized as being sensitive to materials properties such as stiffness, since the stiffness determines the type and extent of bending and thus the stress mode [15]. For a given load, a lap shear bond of a more flexible substrate will experience more peel stress and less shear or tensile. The 0.99 r^2 correlation coefficient shows that, for this set of data, stiffness is the key factor influencing dry lap shear strengths.

However, for the lap shear strengths after a 1-week water soak at 60°C, stiffness does not appear to be as definitive (Fig. 2). Failure modes were adhesional or mixed adhesional and cohesional, with some of the NUN samples failing by stock

Table 4.

Lap shear strength and stiffness for automotive body panel materials bonded with UA, from ref. [14]

Sample[a]	Dry strength (kPa)	Wet strength (kPa)	Stiffness (kN/mm)
CRS	20 100 ± 2000	9800 ± 900	1.86
N43	14 700 ± 800	11 500 ± 2000	0.50
NUN	9700 ± 1100	5600 ± 500	0.14
GALV	7300 ± 800	2100 ± 700	0.12
NPB	5600 ± 400	3500 ± 800	0.075

[a] Sample notation is defined in Table 1.

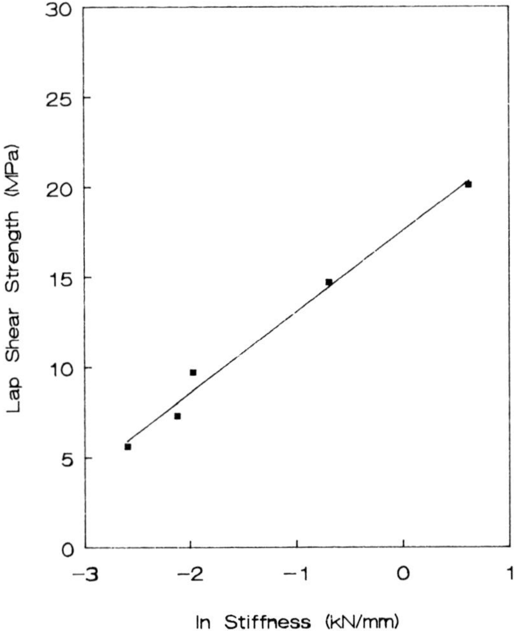

Figure 1. Effect of substrate stiffness on dry lap shear strength for samples bonded with a urethane adhesive. Data from ref. [14].

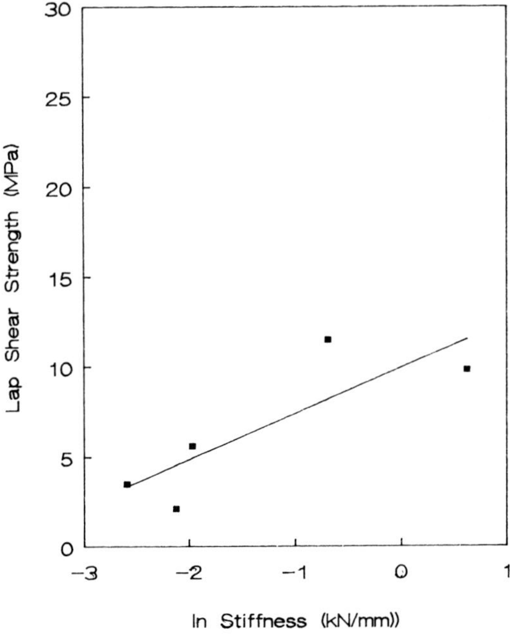

Figure 2. Effect of substrate stiffness on the lap shear strength after a 1-week, 60°C water soak for samples bonded with a urethane adhesive. Data from ref. [14].

break. There is still a correlation, but the r^2 correlation coefficient is only 0.69, indicating that other factors must be important. The stiffness values used in this correlation were measured on dry substrates, not on water-soaked substrates. If the stiffness values for NPB and the nylons decrease after the water soak, as might be expected, the correlation is even worse. Although there is still a correlation between the bond strength and stiffness after water soak, the decrease of the strong (0.99 r^2) correlation for the dry strengths to the weaker correlation after water soak (0.69 r^2) suggests that surface chemistry and interactions between the adhesive and substrate are now playing a more important role.

3.2. Surface energetics

3.2.1. Determination of γ_{lv}^{d} and γ_{lv}^{p} for aniline and liquefied phenol. In order to measure γ_{lv}^{p} and γ_{lv}^{d} for aniline and phenol, a non-polar surface is needed. The contact angles of five liquids (water, bromonaphthalene, methylene iodide, formamide, and glycerol) were used to calculate the geometric mean approximation to the surface free energy for TPO-A wiped with methylene chloride. The polar and dispersive components of the surface energy for these liquids are in the literature [5, 16], but the polar and dispersive components for aniline and liquefied phenol are not. The total surface free energy for TPO-A was calculated to be 30.2 ergs/cm^2, with a polar component of less than 0.01 ergs/cm^2. Therefore, it is an excellent surface to use in determining the polar and dispersive components [16]. If γ_{s}^{p} is zero, equation (3) reduces to

$$W_{a} = 2(\gamma_{lv}^{d} \gamma_{s}^{d})^{1/2}. \tag{5}$$

W_{a} can be calculated from the contact angle of the liquid of interest on the non-polar substrate (equation (2)) and γ_{s}^{d} equals the total surface free energy if there is no polarity. Aniline was found to have a polar component of 2.0 ergs/cm^2 out of a total surface energy of 43.2 ergs/cm^2, and phenol has a polar component of 2.6 out of a total of 40.4 ergs/cm^2.

3.2.2. Surface energetics of substrates. The same five liquids, plus aniline and phenol, were then used to calculate the surface energetics of the other substrate materials, via the geometric mean. The works of adhesion for the seven liquids against the substrates were calculated, and the $W_{a}/2(\gamma_{lv}^{d})^{1/2}$ values were plotted against the square root of the ratio of the liquid polar component to the dispersive component, as described in equation (4). Figures 3 and 4 show these plots for two of the substrates. Figure 3 is NPB-amb wiped with methylene chloride, which has a dispersive component (the square of the intercept) of 40.0 ergs/cm^2 and a polar component (the square of the slope) of 0.0 ergs/cm^2. The slope and intercept are determined by linear regression. The same type of plot is shown for NUN in Fig. 4, giving a dispersive component of 33.1 ergs/cm^2 and a polar component of 8.5 ergs/cm^2. The polar and dispersive components of the surface free energies for all of the substrates are given in Table 5. The error values are from propagation of the standard error calculated by the linear regression.

Figures 3 and 4 also demonstrate one of the problems with the geometric mean approximation. The slope and intercept are calculated from a linear regression, but there is considerable deviation about the line. For the NPB-amb plot, it is clear

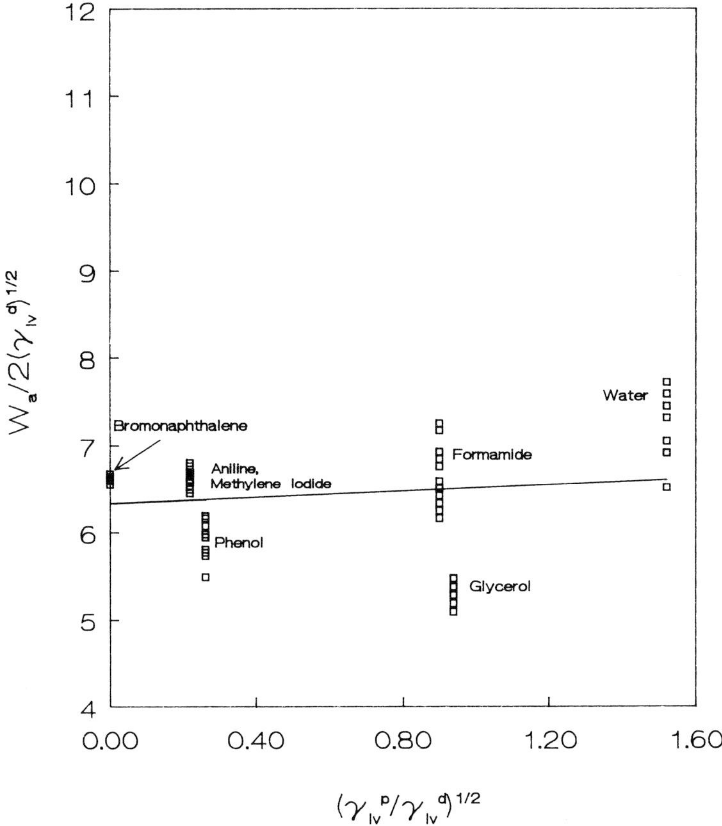

Figure 3. Determination of the geometric mean approximation to the surface free energy of NPB-amb wiped with methylene chloride. The slope is the square root of the polar component, and the intercept is the square root of the dispersive component.

that a regression must give a very small slope. It is also clear that glycerol, for example, does not interact with this sample as the geometric mean would predict. On the NUN plot, the bromonaphthalene values are above the line, and the liquefied phenol values are significantly below the line. Thus, the geometric mean is seen to depend on the particular liquids used as probes, as discussed above. It is possible for a solid surface, which has a polar component measured by the geometric mean to be zero, to still have significant interactions with polar liquids. This suggests that a means of describing the nature of these interactions in terms of Lewis acids and bases would be useful.

3.2.3. Effect of surface energetics on the adhesive bonds. Lap shear strengths for several other polymeric substrates were measured. The results, both before and after a 1-week, 60°C water soak, are in Table 6. As mentioned above, it has been found in some systems that the better the match between the polar and dispersive components of the surface free energy for a substrate and an adhesive, the stronger and more durable will be the bond between them [6, 7]. This is visualized by plotting the square roots of the polar and dispersive components of the surface

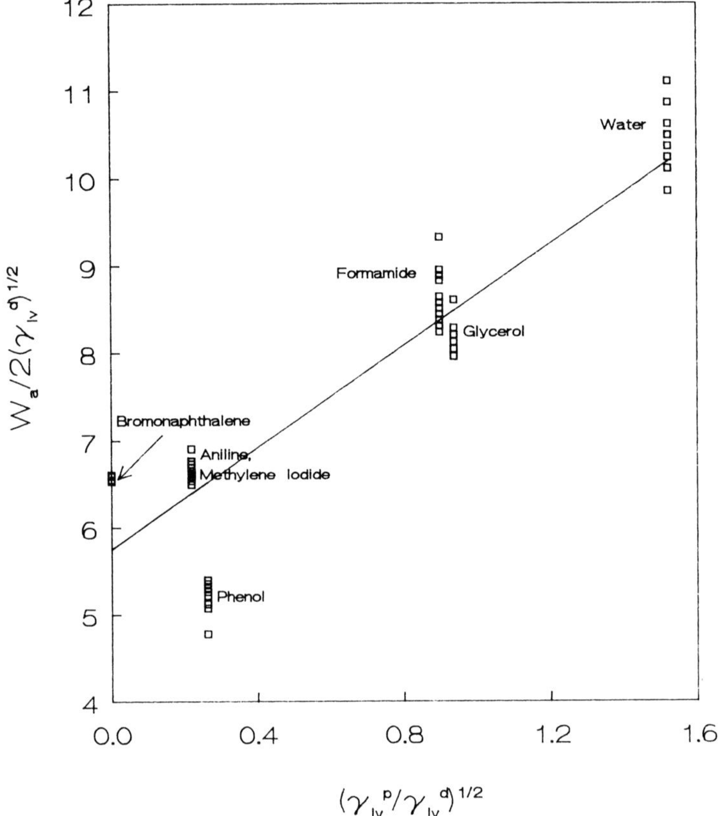

Figure 4. Determination of the geometric mean approximation to the surface free energy of unfilled nylon 6,6, wiped with methylene chloride. The slope is the square root of the polar component, and the intercept is the square root of the dispersive component.

free energy against each other, for the adhesive and the substrates. A circle is drawn for each adhesive and substrate system, using the coordinates of the adhesive and the substrate to define the diameter of the circle. Figure 5 is an example of this type of plot for the systems listed in Table 6. The labels in Fig. 5 are the lap shear strengths after water soak. This theory predicts that smaller circles will have greater bond strengths.

This analysis goes on to consider the effect of immersing the bond in an environment used to test for durability, such as water. The components of the surface energy for that environment are also plotted. The difference of the square of the radius of one of the circles, R_0, subtracted from the square of the distance between the center of a circle and the water point, R, has been equated with the Griffith critical fracture energy, γ_G. In some systems [6, 7], γ_G has been found to have a positive correlation with the adhesive strength after durability testing. Figure 6 is a plot of the lap shear strength after a 1-week, 60°C water soak against the Griffith critical fracture energy $(R^2 - R_0^2)$. Instead of the positive correlation expected, the best line through the data has a negative slope, indicating that the more favorable the surface energetics, the *worse* the bonding. This shows that this

Table 5.
Surface energies and acidities of automotive materials

Substrate	Surface preparation	γ_s^d	γ_s^p (ergs/cm^2)	γ_s^t	D (ergs/cm^2)$^{1/2}$
CRS	Acetone	35.4 ± 1.9	15.0 ± 0.3	50.4 ± 1.9	-1.4 ± 1.6
GALV	Acetone	33.9 ± 1.4	0.2 ± 0.1	34.1 ± 1.4	-0.8 ± 1.2
NPB-damp	CH$_2$Cl$_2$	43.0 ± 2.5	0.6 ± 0.2	43.6 ± 2.5	10.9 ± 2.0
NPB-dry	CH$_2$Cl$_2$	37.7 ± 2.8	0.7 ± 0.2	38.4 ± 2.8	7.6 ± 2.3
NPB–PC	Acetone	34.9 ± 2.7	1.1 ± 0.2	36.0 ± 2.7	12.7 ± 2.3
NPB-PC	CH$_2$Cl$_2$	38.6 ± 2.9	0.0 ± 0.1	38.6 ± 2.9	6.2 ± 2.3
NPB-amb	Acetone	31.0 ± 2.2	11.3 ± 0.4	42.3 ± 2.2	8.2 ± 1.9
NPB-amb	CH$_2$Cl$_2$	40.0 ± 2.1	0.0 ± 0.1	40.0 ± 2.1	9.2 ± 1.7
NPB-amb	Detergent	34.6 ± 1.5	9.6 ± 0.2	44.2 ± 1.5	7.2 ± 1.2
PMMA	Acetone	37.2 ± 1.6	6.7 ± 0.2	43.9 ± 1.6	3.5 ± 1.3
PMMA	CH$_2$Cl$_2$	37.6 ± 1.8	4.1 ± 0.2	41.7 ± 1.8	4.3 ± 1.4
PMMA	Detergent	38.6 ± 1.3	4.3 ± 0.2	42.9 ± 1.3	4.8 ± 1.1
N43	CH$_2$Cl$_2$	31.7 ± 2.4	7.6 ± 0.4	39.3 ± 2.4	13.0 ± 2.0
NUN	CH$_2$Cl$_2$	33.1 ± 2.2	8.5 ± 0.4	41.6 ± 2.2	13.1 ± 1.9
TPO-A	Acetone	31.3 ± 1.2	0.5 ± 0.1	31.8 ± 1.2	2.1 ± 1.0
TPO-A	CH$_2$Cl$_2$	30.2 ± 2.0	0.0 ± 0.02	30.2 ± 2.0	0.4 ± 1.7
TPO-A	Detergent	29.0 ± 1.3	0.3 ± 0.1	29.3 ± 1.3	0.8 ± 1.1
TPO-B	Acetone	26.5 ± 1.8	0.8 ± 0.1	27.3 ± 1.8	0.5 ± 1.6
TPO-B	CH$_2$Cl$_2$	29.3 ± 2.2	0.0 ± 0.04	29.3 ± 2.2	3.3 ± 1.9
TPO-B	Detergent	24.9 ± 1.5	0.9 ± 0.1	25.8 ± 1.5	-0.4 ± 1.4
UA		34.3 ± 2.5	4.4 ± 0.3	38.7 ± 2.5	3.8 ± 2.1
EA		42.6 ± 2.6	0.0 ± 0.05	42.6 ± 2.6	2.6 ± 2.1

Table 6.
Lap shear strengths for coupons bonded with UA, before and after a 1-week, 60°C water soak

Substrate	Surface preparation	Dry strength (kPa)	Failure mode, dry	Wet strength (kPa)	Failure mode, wet
CRS[a]	Acetone	$20\,100 \pm 2000$	Mixed[b]	9800 ± 900	Adh[c]
GALV[a]	Acetone	7300 ± 800	Adh	2100 ± 700	Adh
N43[a]	CH$_2$Cl$_2$	$14\,700 \pm 800$	Adh	$11\,500 \pm 2000$	Mixed
NPB-damp[a]	CH$_2$Cl$_2$	5600 ± 400	Mixed	3500 ± 800	Mixed
NPB-dry	CH$_2$Cl$_2$	8600 ± 550	Adh	3300 ± 500	Adh
TPO-A	CH$_2$Cl$_2$	1500 ± 90	Adh	1200 ± 100	Adh
NUN[a]	CH$_2$Cl$_2$	9700 ± 1100	Stock[d]	5600 ± 500	Stock, adh

[a] Lap shear data from ref. [14].
[b] Mixed adhesional and cohesional failure.
[c] Adhesional failure.
[d] Stock break.

theory is inadequate for describing these systems, and suggests that another means of analyzing their interactions must be found. Acid–base theory offers an alternative methodology, if a way of quantifying the acidity of the surfaces is available.

3.3. Calculation of surface acidity

In the initial work in this laboratory on characterizing the acidity of surfaces with contact angles [5], the contact angles of aniline and phenol on a given surface were

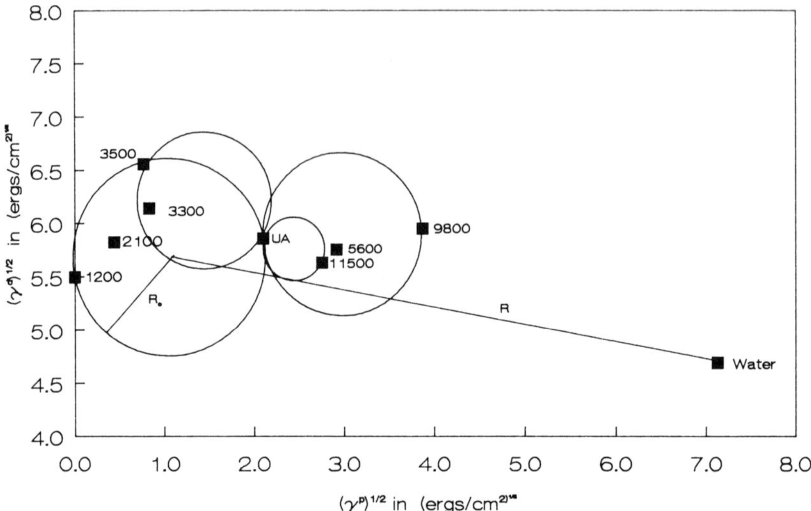

Figure 5. Surface energies of substrates and adhesive, with wet strength values in kPa, for samples bonded with a urethane adhesive. The circles describe the closeness between the polar and dispersive components of the substrates and adhesives. Not all the circles were drawn, for clarity.

compared with each other, as were those of formamide and glycerol. The reasoning was that aniline and phenol have similar surface energies, as do formamide and glycerol. With no specific interactions (i.e. no interactions which are specific to the two materials involved), aniline's contact angle should be similar to phenol's, and formamide's should be similar to glycerol's. There are small differences in the surface energies, with the base aniline having a slightly higher surface energy than the acid phenol, and the base formamide having a slightly lower surface energy than the acid glycerol. Since lower surface energy liquids normally have lower contact angles on a given substrate (again neglecting specific interactions), the base being lower in one case should cancel the effect of the acid being lower in the other. Thus, if both bases wet better, the substrate is said to be acidic, and if both acids wet better, the substrate is said to be basic.

The first attempt to quantify this was by the ratio of the works of adhesion of the bases to that of the acids. However, the calculated values were not sufficiently different from each other to give a good basis of comparison. Another objection to this approach is that it did not use all the information available in the data. The dispersive and polar breakdown of the probe liquid surface energies should give information about the degree of their possible acid–base interactions, information which was neglected in the first approach. If the work of adhesion is assumed to be

$$W_a = W_a^d + W_a^{ab}, \tag{6}$$

with W_a^d the dispersive component described by

$$W_a^d = 2(\gamma_{lv}^d \gamma_s^d)^{1/2} \tag{7}$$

and W_a^{ab} the acid–base component, then the part of the work of adhesion which is due to dispersive interactions alone can be subtracted from the total W_a to give W_a^{ab} [12]:

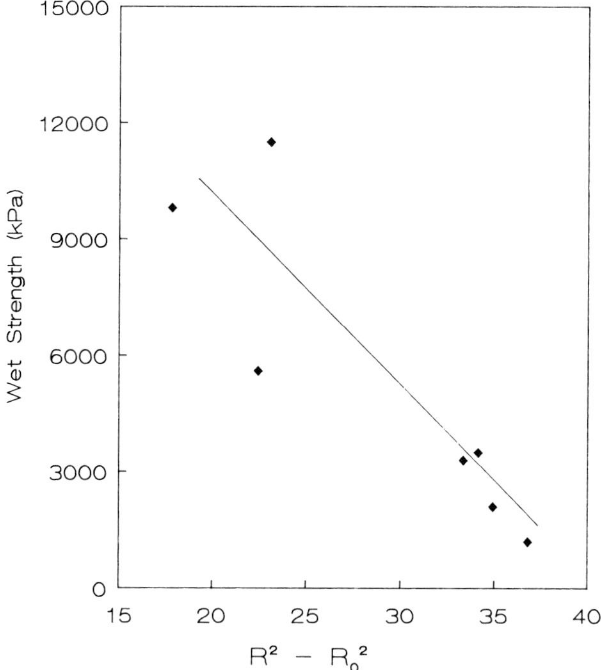

Figure 6. Lap shear strengths after a 1-week, 60°C water soak plotted against the difference of squared radius of circle describing the match of polar and dispersive components for substrates, and squared distance to adhesive (Fig. 5).

$$W_a^{ab} = W_a - W_a^d = W_a - [2(\gamma_{lv}^d)^{1/2}(\gamma_s^d)^{1/2}]. \tag{8}$$

A more successful attempt at quantifying the surface acidity utilized this separation of the work of adhesion into two components. The γ_s^d value used was that calculated from the geometric mean approach described above. The W_a^{ab} for each of the four liquids was determined. In order to avoid skewing the values by the differences in the polarities of the probe liquids, the W_a^{ab} values were divided by the square root of the polar component of each wetting liquid. Then the difference between the bases and the acids was calculated.

The procedure is given below, as it is calculated for NUN and methylene chloride-wiped NPB (NPB–MC).

(1) Measure the contact angles of the seven liquids (aniline, formamide, liquefied phenol, glycerol, water, methylene iodide, and bromonaphthalene) on the substrate of interest. Calculate W_a for each liquid (equation (2)).

	W_a (ergs/cm^2)	
	NUN	*NPB–MC*
Aniline	84.8	85.6
Formamide	97.2	75.5
Liquefied phenol	64.3	72.8
Glycerol	94.7	62.1
Water	97.0	67.2
Methylene iodide	88.0	93.3
Bromonaphthalene	94.5	88.6

(2) Calculate the geometric mean approximation to the surface free energy of the substrate (equation (3)).

	Surface free energy (ergs/cm^2)	
Component	*NUN*	*NPB–MC*
Dispersive	33.1	40.0
Polar	8.5	0.0
Total	41.6	40.0

(3) Using γ_s^d from step (2), calculate and subtract W_a^d from the total W_a for each of the four acid–base probes (aniline, formamide, liquefied phenol, and glycerol), giving a value of W_a^{ab} for each of the four liquids (equation (8)).

	W_a^{ab} (ergs/cm^2)	
	NUN	*NPB–MC*
Aniline	10.9	4.4
Formamide	32.3	4.2
Liquefied phenol	− 6.4	− 5.0
Glycerol	27.7	− 11.5

The negative value for the phenol on the nylon is a reflection of how little the phenol interacts with the solid surface compared to what would be predicted by the geometric mean. This is seen by how far below the line the phenol values are in Fig. 4. The same phenomenon is seen for the phenol and the glycerol in Fig. 3.

(4) Divide each W_a^{ab} by the square root of γ_{lv}^p for that liquid. Recognizing from equations (3) and (8) that

$$\frac{W_a^{ab}}{[\gamma_{lv}^p]^{1/2}} = 2[\gamma_s^p]^{1/2}, \tag{9}$$

this gives twice the value of the square root of the polar component of the substrate surface free energy *as it would be calculated from its interactions with each of the four liquids individually.* If there were no acid–base interactions and the geometric mean measured the surface energetics perfectly, all of these values would be the same. The fact that they are not, as can be seen below, gives a measure of the acidity of the system.

	$2(\gamma_s^p)^{1/2}$ (ergs/cm^2)$^{1/2}$	
	NUN	*NPB–MC*
Aniline	7.8	3.1
Formamide	6.4	0.8
Liquefied phenol	− 4.0	− 3.1
Glycerol	5.1	− 2.1

(5) Subtract $2[\gamma_s^p(\text{probe})^{1/2}]$ for the two acidic probes from $2[\gamma_s^p(\text{probe})^{1/2}]$ for the two basic probes:

$$D \equiv 2[\gamma_s^p(\text{aniline})^{1/2} + \gamma_s^p(\text{formamide})^{1/2}] \\ - 2[\gamma_s^p(\text{phenol})^{1/2} + \gamma_s^p(\text{glycerol})^{1/2}]. \tag{10}$$

The result is the characteristic quantity D, a measure of the surface acidity. For NUN, $D = 13.1$ (ergs/cm^2)$^{1/2}$, and for NPB–MC, $D = 9.2$ (ergs/cm^2)$^{1/2}$. The more acidic the material, the higher its value of D. Initially, it appears contradictory that a surface such as NPB–MC, which has a 0.0 ergs/cm^2 γ_s^p as measured by the geometric mean, can have a high value of D. Inspection of Fig. 3 helps to reconcile this. The regression line is flat, but the phenol and glycerol values are significantly below it. The formamide values straddle the line, and the aniline values are just above it. Therefore, the acids interact much less than is predicted by the geometric mean, and the bases interact more than is predicted. This indicates that the surface has a repulsion for acids and an affinity for bases, and is, therefore, acidic. Calculated D values for the surface acidity of the automotive materials characterized are listed in Table 5.

3.3.1. Reproducibility of the D *parameter.* There are many places 'errors' can enter into the D calculation. The first is from variations in the surfaces themselves. This is not an error as such, but a reflection of the physical nature of the system. If all deviations could be attributed to this, the 'error' values assigned to the calculation would be a further description of the system, rather than a comment on the technique. However, aside from this source, other sources of error exist. Contact angles of perfectly flat, homogeneous systems have reproducibilities of about $\pm 1°$. The surfaces of the polymeric and metallic materials measured in this work, and of most samples, are not perfectly flat on either a microscopic or a macroscopic scale. Combining these physical surface variations with chemical heterogeneity can lead to contact angle variations of several degrees. The surface tensions of the liquids have reproducibilities of about 0.2 ergs/cm^2. And of course, the theory is built upon approximations, rather than fundamental truths. All of these contribute to the variation in the value of the surface acidity. The error values in Table 5 come from propagation of the standard error of the linear regression used to calculate the surface energies, which depends both on the error in the contact angle measurements and the error intrinsic in the geometric mean approximation.

In order to evaluate the variability due to all factors and to confirm the reproducibility of the D parameter, ten samples were made of each of two epoxy resin formulations (Table 2). This involved making five batches of each resin at five different times, and curing two samples at a time. The samples thereby took into account any variability in mix procedure and cure, as well as random deviations in ambient conditions. The seven liquids listed above were used to measure the surface energetics and acidities of the two formulations. The contact angles were measured for these 20 samples by two different researchers, to take operator variability into account.

The results of the reproducibility measurements on epoxies A and B are in Table 7. Formulation A is a DGEBA epoxy cured with the aliphatic amine IPD, and formulation B is a DGEBA epoxy cured with the catalyst EMI-24. The set of D values measured for both formulations have 95% confidence intervals of less than ± 1 (ergs/cm^2)$^{1/2}$, about half the value calculated from the propagated error. Since the D parameter is an intrinsic and not an extrinsic variable, the variability should be compared with the range of likely values and not with an individual value. The range of measured values in Table 5 is the -1.4 to 13.1 (ergs/cm^2)$^{1/2}$, so a variability of less than 1 (ergs/cm^2)$^{1/2}$ allows considerable differentiation between values. The values of surface free energy and surface acidity were also

Table 7.
Reproducibility of surface energies and acidities

Epoxy	Sample	Batch	Operator	γ_s^d	γ_s^p (ergs/cm^2)	γ_s^t	D (ergs/cm^2)$^{1/2}$
A	1	1	EJB	36.6	6.2	42.8	4.2
	2	1	MAB	36.5	7.4	43.9	4.6
	3	2	EJB	37.5	6.9	44.4	4.8
	4	2	MAB	38.1	7.5	45.6	3.3
	5	3	EJB	38.1	5.6	43.7	3.2
	6	3	MAB	38.7	5.4	44.1	2.8
	7	4	EJB	35.6	3.9	39.5	2.1
	8	4	MAB	34.8	6.6	41.4	1.2
	9	5	MAB	38.0	6.4	44.4	3.1
	10	5	EJB	38.4	4.5	42.9	3.0
Mean				37.2	6.0	43.3	3.2
95% confidence interval				0.9	0.8	1.2	0.7
B	1	1	EJB	37.6	3.1	40.7	2.3
	2	1	MAB	38.1	3.2	41.3	2.0
	3	2	MAB	38.3	3.6	41.9	1.6
	4	2	EJB	38.1	1.4	39.5	1.1
	5	3	EJB	37.6	1.8	39.4	1.6
	6	3	MAB	38.6	1.6	40.2	0.5
	7	4	EJB	35.1	4.0	39.1	1.9
	8	4	MAB	36.5	4.3	40.8	1.3
	9	5	EJB	37.6	3.3	40.9	2.3
	10	5	MAB	38.6	3.7	42.3	0.2
Mean				37.6	3.0	40.6	1.5
95% confidence interval				0.7	0.7	0.7	0.5

checked for reproducibility as a function of the operator. Kolmogorov–Smirnov statistics [17] show no effect of operator on either the D values or the surface energies.

3.4. Acidity of automotive materials

3.4.1. Effect of substrate materials. It can be seen in Table 5 that nylon is the most acidic of the substrates, with D values of about 13 (ergs/cm^2)$^{1/2}$ for both glass-reinforced nylon 6,6 (N43) and NUN wiped with methylene chloride. PMMA was included not as an automotive material, but as a polymer previously determined to be basic [8]. Its D values of about 4 (ergs/cm^2)$^{1/2}$ are much lower than the nylon or most of the NPB values, as would be expected. TPO-A and TPO-B have the lowest values of D of the polymeric materials, which is rather surprising. This is possibly because of basic contributions from the π-electrons [18] or from additives which make these materials more 'paintable'. More probably, the low values of D for the TPOs are an artifact of the calculation procedure, because of the low values of W_a^{ab} of the TPOs. Since TPO was used to measure γ_{lv}^d for the aniline and phenol, those two liquids will necessarily have zero deviation from the geometric mean, and may be skewing the calculation for the TPOs. The metallic substrates, cold-rolled steel and electrogalvanized steel, have the lowest values measured, making them the most basic. The metal oxides on their surfaces are expected to be basic, and there may also be a contribution from free metallic electrons.

NPB has D values ranging from 6.2 to 12.7 $(ergs/cm^2)^{1/2}$ depending on the handling and surface treatment. One variable affecting D for NPB was the storage condition. The sample designated 'NPB-damp' was stored damp after cutting with the diamond saw, while the 'NPB-dry' sample was dried immediately after cutting. For a methylene chloride wipe, damp storage increases the acidity from $D = 7.6$ to $D = 10.9$ $(ergs/cm^2)^{1/2}$. Postcuring the NPB also has an effect on its acidity, probably because of oxidation of the surface. The methylene chloride-wiped, then postcured samples had D values of 6.2 $(ergs/cm^2)^{1/2}$, and the D values of the acetone-wiped, postcured samples were 12.7 $(ergs/cm^2)^{1/2}$. The effect of postcure *within* the bond area may well be different from the effect on the surface of the unbonded coupon, since the unbonded coupon is exposed to the atmosphere, and the bond area is not.

3.5. Correlation of lap shear results with acidity

3.5.1. Lap shear specimens bonded with the urethane adhesive (UA). The dry strength of this series of lap shear measurements is shown in Fig. 1 to depend primarily on substrate stiffness, but in Fig. 2 the strength after a 1-week, 60°C water soak is seen to depend on other variables as well as the stiffness. The effect of acidity on the lap shear strength after water soak for these lap shear data, plus other lap shear data taken for this investigation, is shown in Fig. 7 for substrates bonded with UA (Table 6). The wet strength is plotted against the absolute

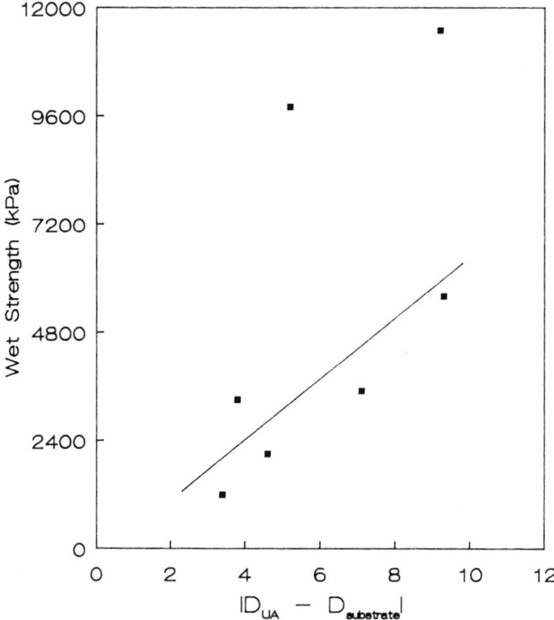

Figure 7. Effect of acidity on the lap shear strengths after a 1-week, 60°C water soak for samples bonded with a urethane adhesive. The x-axis is the absolute difference between the D of the adhesive and the D of the substrate.

difference between the D value of the substrate and that of the adhesive (3.8 (ergs/ cm^2)$^{1/2}$). (The acidity of the adhesive was determined on a surface molded against air. While the surfaces may change when molded against the substrate, the difference will probably be small. This is, however, a possible source of error for which it is not possible to account.) All bond breaks were either adhesional or mixed adhesional and cohesional. Two groups of data are seen, with the two stiffest materials—cold-rolled steel and reinforced nylon—having strengths well above those of the other materials. For the group of less stiff materials (TPO-A, galvanized steel, NPB (-dry and -damp), and unfilled nylon), there is a correlation between the wet strength and the difference of the substrate and adhesive acidities, even taking into consideration the questionable values of the TPO-A. Since this is the absolute difference, it appears that UA can behave as either an acid or a base. As a two-part urethane adhesive, UA will have a variety of functional groups available for bonding, including the initial isocyanate and hydroxyl functionalities, and amine and carbonyl groups of the polyurethane. While UA is predominately basic, having a D value of 3.8 (ergs/cm^2)$^{1/2}$, this variety of molecular groups allows for attractions with either acidic or basic substrates. As the substrates become increasingly acidic or basic relative to UA, the intermolecular interactions increase and the bonding is more durable.

3.5.2. Bonds with the epoxy adhesive (EA). One difficulty with interpreting the data in Fig. 7 is the range of stiffness values represented. Few substrates can be found which have large differences in surface chemistry, but the same stiffness. However, NPB wiped with acetone, then postcured has a D value of 12.7 (ergs/ cm^2)$^{1/2}$, while NPB wiped with methylene chloride, then postcured has a D value of 6.2 (ergs/cm^2)$^{1/2}$. The stiffness of the two samples should be the same. EA is more basic than either of the NPB–PC samples or the urethane adhesive, with $D = 2.6$ (ergs/cm^2)$^{1/2}$), which gives differences between the substrates and adhesive of 10.1 and 3.6 (ergs/cm^2)$^{1/2}$. Bonding the two NPB–PC samples with the EA, then, allows the separation of the effect of surface acidity from the effect of stiffness.

The lap shear specimens were tested, and all failures were adhesional. The lap shear strength of the acetone-wiped NPB–PC/EA system with the difference in D values of 10.1 (ergs/cm^2)$^{1/2}$ is 3700 ± 250 kPa, while for the methylene chloride-wiped NPB–PC/EA with the D difference 3.6 (ergs/cm^2)$^{1/2}$, the lap shear strength is 1800 ± 200 kPa. The acetone-wiped NPB–PC is more acidic, and, therefore, better able to interact with the basic adhesive. Thus, when stiffness is not a factor, acidity can be seen to be an important influence on lap shear strengths.

4. CONCLUDING REMARKS

There are many important variables in structural adhesive bonding. These include joint design as well as bulk material properties such as stiffness and relative moduli of the adhesive and the substrate, together with surface properties such as morphology and surface chemistry. The results presented herein indicate that acid–base interactions are an important part of surface chemistry as it relates to adhesive bonding. The ability to measure the acidity of solid surfaces using an experimentally straightforward technique such as contact angles and a simple calculation scheme such as the one described above allows acid–base interactions to be utilized in the design of adhesive systems. Knowing the acidity of a substrate makes it possible to select an adhesive which may have advantageous interactions with that substrate. An acidic substrate will interact favorably with a basic adhesive and vice versa. The ability to change the surface chemistry and acid–base properties of a substrate by materials handling or surface preparation allows acid–base interactions to be optimized. Conversely, a poorly stored or prepared substrate can cause unexpected changes in surface chemistry and thus lead to poor bonding. Understanding the acid–base interactions between adhesives and substrates will allow us to control an important part of adhesive bonding technology.

Acknowledgements

I would like to thank Dr. J. A. Schroeder for advice and assistance with sample preparation and many helpful discussions, Dr. R. T. Foister and Dr. T. J. Dearlove for helpful discussions, and Mr. K. S. Snavely and Mr. M. A. Buffa for assistance with sample preparation and testing.

REFERENCES

1. K. L. Mittal, in: *Adhesion Science and Technology*, L. H. Lee (Ed.), Part A, pp. 129–168. Plenum Press, New York (1975).
2. W. A. Zisman, *Adv. Chem. Ser.* **43**, 1 (1964).
3. D. H. Kaelble, *J. Adhesion* **2**, 66 (1970).
4. S. Wu, *Polymer Interface and Adhesion*, pp. 178–181. Marcel Dekker, New York (1982).
5. E. J. Berger, in: *Advanced Composites III: Expanding the Technology*, Proceedings of the ASM/ESD Advanced Composites Conference and Exposition, pp. 309–322. ASM International, Metals Park, OH (1987).
6. E. J. Berger and Y. Eckstein, in: *Adhesive Joints: Formation, Characteristics, and Testing*, K. L. Mittal (Ed.), pp. 51–65. Plenum Press, New York (1984).
7. T. Smith, *J. Adhesion* **11**, 243 (1980).
8. F. M. Fowkes, D. O. Tischler, J. A. Wolfe, L. A. Lannigan, C. M. Ademu-John and M. J. Halliwell, *J. Polym. Sci., Polym. Chem. Ed.* **22**, 547 (1984).
9. F. M. Fowkes and S. Maruchi, *Org. Coatings Appl. Polym. Sci. Proc.* **37**, 605 (1977).
10. J. C. Bolger, in: *Adhesion Aspects of Polymeric Coatings*, K. L. Mittal (Ed.), pp. 3–18. Plenum Press, New York (1983).
11. F. M. Fowkes, *J. Adhesion Sci. Technol.* **1**, 7 (1987).
12. R. S. Drago, G. C. Vogel and T. E. Needham, *J. Am. Chem. Soc.* **93**, 6014 (1971).
13. R. S. Drago, L. B. Parr and C. S. Chamberlain, *J. Am. Chem. Soc.* **99**, 3203 (1977).
14. J. A. Schroeder, General Motors Research Laboratories, private communication (1989).
15. A. H. Landrock, *Adhesives Technology Handbook*, pp. 35–37. Noyes Publications, Park Ridge, NJ (1985).
16. F. M. Fowkes, *Ind. Eng. Chem.* **56**, 40 (1964).
17. W. J. Conover, *Practical Nonparametric Statistics*, 2nd edn, pp. 368–371. John Wiley, New York (1971).
18. W. B. Jensen, *Chem. Rev.* **78**, 1 (1978).

Part 3: Applications of Acid-Base Interactions

Acid-Base Interactions, pp. 229-241
Eds. K.L. Mittal and H.R. Anderson, Jr.
©VSP 1991

Acid–base interactions in wetting

GEORGE M. WHITESIDES,* HANS A. BIEBUYCK, JOHN P. FOLKERS
and KEVIN L. PRIME
Department of Chemistry, Harvard University, Cambridge, MA 02138, USA

Revised version received 27 August 1990

Abstract—The study of the ionization of carboxylic acid groups at the interface between organic solids and water demonstrates broad similarities to the ionizations of these groups in homogeneous aqueous solution, but with important systematic differences. Creation of a charged group from a neutral one by protonation or deprotonation (whether $-NH_3^+$ from $-NH_2$ or $-CO_2^-$ from $-CO_2H$) at the interface between surface-functionalized polyethylene and water is more difficult than that in homogeneous aqueous solution. This difference is probably related to the low effective dielectric constant of the interface ($\varepsilon \approx 9$) relative to water ($\varepsilon \approx 80$). It is not known to what extent this difference in ε (and in other properties of the interphase, considered as a thin solvent phase) is reflected in the stability of the organic ions relative to their neutral forms in the interphase and in solution, and to what extent in differences in the concentration of H^+ and OH^- in the interphase and in solution. Self-assembled monolayers (SAMs)—especially of terminally functionalized alkanethiols ($HS(CH_2)_nX$) adsorbed on gold—provide model systems with relatively well-ordered structures that are useful in establishing the fundamentals of ionization of protic acids and bases at the interface between organic solids and water. These systems, coupled with new analytical methods such as photoacoustic calorimetry (PAC) and contact angle titration, may make it possible to disentangle some of the complex puzzles presented by proton-transfer reactions in the environment of the organic solid–water interphase.

Keywords: Contact angle titration; photoacoustic calorimetry; polyethylene carboxylic acid; contact angle; polymer surfaces; wetting; self-assembled monolayers.

1. INTRODUCTION

Interaction of two condensed phases across an interface or interphase (the latter is usually a more appropriate word for organic solids) depends on the details of chemical interactions at/within that interphase. In exploring the details of the influence of the chemistry of the interphase on its properties, it is particularly useful to examine functional groups capable of undergoing acid–base reactions. Proton transfer between acids and bases in solution is the best understood class of organic reactions [1]. These reactions are usually fast: Systems consisting of acids and bases are often at thermodynamic equilibrium, and their energies can thus be analyzed in believable detail. Manipulation of the properties of the solid–water interphase through proton transfer to and from suitable functional groups in the interphase thus offers the opportunity to use simple, well-understood chemistry to influence, cleanly and selectively, one aspect of the molecular composition of the interphase, while leaving others—morphology, many interfacial mechanical properties, the character of non-acidic and non-basic species— largely or entirely unchanged.

---

*To whom correspondence should be addressed.

Understanding the acidity of organic functional groups at interfaces is important in surface science in at least two contexts. First, it can help to rationalize and control the properties of materials: wetting, adhesion, surface electrical conductivity, tribology, and others of more purely technological interest. Second, it can help to define the fundamental physical properties and 'character' of the interphase—as distinct from any bulk phases present—by using equilibrium proton-transfer reactions as probes. The most relevant properties of an interphase from the vantage of controlling the ionization of groups contained in it would be the capacity of the interphase to support charge (either on an ionizable functional group or as a diffusible species, especially H^+ or OH^-), the accessibility of acids and bases in the interphase to potential proton donors and acceptors, and the stability of the interphase to changes in its state of protonation.

In this paper, we will show how the titration of acidic functional groups in the interphase yields useful information about the properties of the interphase: we will focus on the use of contact angle titration and photoacoustic calorimetry (PAC) to study polyethylene carboxylic acid and its derivatives. We will also illustrate the effects of titratable groups in the interphase upon interfacial reconstruction.

2. THE PHYSICAL-ORGANIC CHEMISTRY OF INTERFACIAL ACIDITY

Measurement of the acidity of functional groups at the interphase between an organic solid and water is substantially more complicated than the corresponding measurement in homogeneous aqueous solution. We list below, without greatly detailed elaboration, some of the factors that contribute to this complexity (Fig. 1). For specificity, we will illustrate points using 'polyethylene carboxylic acid' ($PE-CO_2H$), a material about which we have accumulated a substantial body of background information [2–10]. $PE-CO_2H$ is prepared by oxidation of low-density polyethylene film with chromic acid solution under specified conditions.

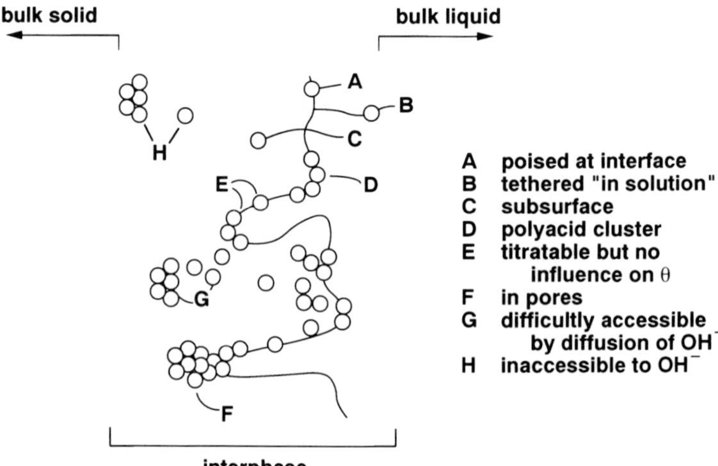

Figure 1. Schematic examples of different types of environments for an ionizable acidic functional group (O) in the interphase between an organic solid and water.

It has carboxylic acid and ketone or aldehyde groups as the only functionality in the oxidatively transformed interphase [2, 4].

(1) *The interphase is a finite, heterogeneous region.* The concept of a homogeneous, well-defined 'surface' (in the sense of the gold (111) crystal face) is not a good description of the functionalized region between the bulk polymer and homogeneous solution. This interfacial region is rough and heterogeneous on many scales [4].

(2) *Local dielectric constant.* The dielectric constant at any point in the interphase will be intermediate between that of the bulk polyethylene (~2) and bulk water (80). The local dielectric constant in the interphase may, of course, vary substantially from point to point, depending on the local structure in the interphase.*

(3) *Local polyacidic interactions.* If the local volumetric density (the equivalent of the concentration in homogeneous solution) of carboxylic acid groups in the interphase is high, hydrogen bonding between them may serve to stabilize the protonated form of the system, and thus to increase the concentration of base required at equilibrium to remove protons [12, 13]. Similarly, if this density is high, charge–charge interactions between carboxylate anions (especially when poorly screened in a medium of low dielectric constant) will be unfavorable and may substantially hinder the introduction of further negative charges into the interphase, once ionization has started [14].

(4) *Local dipole arrays.* Ordered regions of the polymer could align dipoles originating in individual functional groups in ways that would be energetically favorable or unfavorable. The ability of the functional groups to relax to conformations having a lower energy would depend on the local viscosity. The mobility of functional groups within the interphase is another parameter that is not clearly understood in these systems, although movement of groups between the interphase and the bulk is well documented [7, 15].

(5) *Interfacial water.* Water close to interfaces appears, in some circumstances, to have properties sufficiently different from water in the bulk homogeneous liquid phase that 'interfacial water' can be considered to be a distinct liquid [16]. The extent to which water in, for example, a pore in polyethylene lined with carboxylic acid groups would have the same dielectric constant as water in the bulk solution is hard to judge.

(6) *Local concentrations of hydronium and hydroxide ions.* The pK_a of an acid in solution is defined in terms of concentrations [equation (1)].

---

*The low interfacial dielectric constant of the interphase arises from the chemical heterogeneity in the interphase, which can be seen as a mixed 'solution' of functionalized and unfunctionalized polyethylene and water. Thus, this effect will be observed irrespective of the actual dielectric constant of the interfacial water. For experimental evidence pertaining to the dielectric constant of the interphase between $PE-CO_2H$ and water, see ref. [4]. For a discussion of the properties of interfacial liquids, see ref. [11].

$$K_a = \frac{[\text{H}^+][\text{A}^-]}{[\text{HA}]}. \tag{1}$$

The concentration of hydronium and hydroxide ions in homogeneous aqueous solution is a reasonably well-defined quantity over a wide range of concentrations. The corresponding 'concentration' in the vicinity of a functional group in an interphase is presently impossible to measure. Any measurement of $[\text{H}^+]$ or $[\text{OH}^-]$ in this type of environment can probably only be an estimate based on a relative scale. One can, in principle, assume the behavior of a reference acid in the interphase and measure the proton donor ability of another acid of interest relative to that reference. Assuming that the group being examined and the reference group experience the same local environment—an assumption that is probably incorrect in most cases—it is at least possible using this procedure to *order* acidities in the interphase.

Note that this procedure for comparing acidities is substantially less certain than that used in constructing acidity functions [1, 17] or in estimating acidities in non-aqueous or mixed aqueous–organic solvents [1, 18]. In both of these problems in measuring acidities in non-aqueous but still homogeneous solution, it is possible to extend the relative scales of acidity so that they are referenced directly to measurements of aqueous solutions in which 'pK_a' and 'pH' have their normal meanings. The strategies of using series of overlapping reference indicators (when constructing acidity functions) and of using the glass membrane electrode as a reproducible, if thermodynamically ill-defined, reference (as one technique for measuring acidity in non-aqueous solutions) have no direct counterpart in surface chemistry. It is thus not possible to measure the concentration or thermodynamic activity of hydronium ion or hydroxide ion in an interphase. Since both the medium in which the interfacial proton-transfer reactions occur—the interphase—and the concentrations of ions in the interphase are undefined, it is not possible to define a value of 'pK_a' for a functional group in an interphase using currently available techniques (Fig. 2). It may, however, be possible to define the value of the *solution* pH at which the functional group in the interphase is half-protonated [4, 6, 8, 9]. This value—which we call the $pK_{1/2}$—is a reproducible and useful number, but is less interpretable than the pK_a.

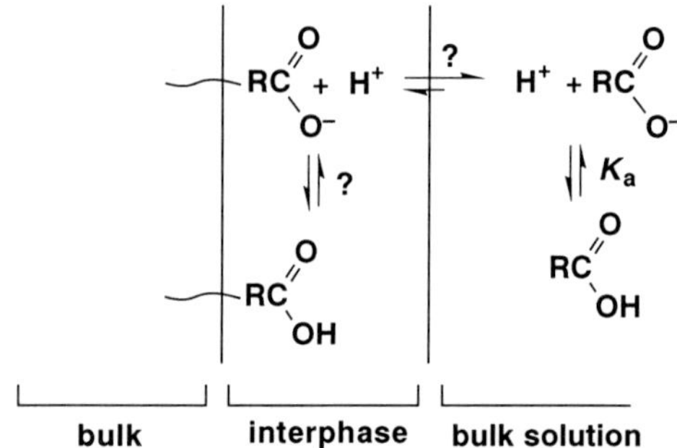

Figure 2. The estimation of a value of acidity in an interphase requires knowing both the concentrations of H^+ and OH^- in the interphase and the concentrations of unionized and ionized functional groups. In the absence of complete information concerning all the species involved in ionization, it is only possible to talk about the *solution* pH at which *interfacial* ionization occurs.

(7) *Surface reconstruction: changes in interfacial structure on ionization.* The mobility of functional groups in polymers varies with (*inter alia*) the temperature, the molecular weight and molecular weight distribution of the polymer chains, and the extent to which the system is plasticized by low-molecular-weight components including the solvent [15]. Parameters such as local interfacial viscosity for PE–CO_2H and its derivatives would be expected to be different in protonated and unprotonated forms [19]. Although, in principle, these parameters can be measured experimentally, only limited measurements have been made [3]. The extent to which the positions of functional groups in the interphase change, or to which the gross morphology of the interphase changes, and the influence of these changes on the ionization and swelling of functional groups during acid–base reactions remain to be established.

3. MEASUREMENT OF SURFACE ACIDITY

A wide range of experimental techniques have been applied with varying degrees of success to the measurement of the extent of ionization of groups at surfaces [4–6, 8–10, 20–23]. These measurements are often technically difficult, as well as difficult to interpret. Any method requiring the equivalent of a 'pH electrode' is probably not practical for measuring acidity in the microscopic volume of the interphase. Other electrochemical measurements are similarly limited, although techniques such as the measurement of electrophoretic mobility can provide some information about the density of fixed charge on the surface as a function of the pH [24]. As a result, most measurements of surface acidity have relied on spectroscopic examination of chromophoric groups [4–6, 23]. Infrared (IR) spectroscopy is, in general, difficult to apply in aqueous solutions, and Raman spectroscopy (aside from surface-enhanced Raman) does not have the sensitivity required for most studies of interfaces. Ultraviolet, visible, and fluorescence measurements are complicated by scattering from the interphase, and often by absorption by the bulk phases. The difficulty in identifying and introducing an appropriate chromophore for each spectroscopic technique often limits the usefulness of that technique.

We, and others, have experimented with protocols to avoid some of these problems. The experiments were sometimes practical, but the results were often ambiguous. For example, in an effort to apply IR spectroscopy in studying the ionization of interfacial acids, we equilibrated PE–CO_2H film against aqueous solutions having known values of pH, removed the film from the water, blotted it until the majority of the superficial water had been removed, and measured the IR spectrum of the CO_2H and CO_2^- groups [4]. How closely the ratio of CO_2H to CO_2^- measured spectroscopically using this technique at any value of pH (and especially at high and low values of the pH) corresponds to that obtained when the film is in contact with the solution is difficult to establish absolutely, although the results were in surprisingly good agreement with those obtained using other methods, when comparisons were possible.

We have recently explored two new methods to try to provide procedures for measuring acidity that complement the existing methods. *Contact angle titration* measures the contact angle of drops of buffered aqueous solution on surfaces containing an ionizable functionality as a function of the pH of these drops [4–6,

8–10, 20–22]. *Photoacoustic calorimetry* (PAC) measures the photoacoustic signal from a surface as a function of the pH [23, 25]. Neither technique is a panacea, but each provides valuable information. Contact angle titration is amenable to a wide variety of surfaces, and is independent of the optical properties of these surfaces. PAC circumvents many of the optical problems associated with light scattering at rough surfaces.

The origin of contact angle titration was the elementary hypothesis that when functional groups ionize at a solid–water interface, their hydrophilicity (and the hydrophilicity of the interphase) increases. This change would be reflected in a change in the solid–liquid interfacial free energy, γ_{SL}, and thus, via the familiar Young's equation, in a change in cos θ [equation (2); γ_{SV} and γ_{LV} are, respectively, the interfacial free energies of the solid–vapor and solid–liquid interfaces].

$$\cos \theta = \frac{\gamma_{SV} - \gamma_{SL}}{\gamma_{LV}}. \tag{2}$$

A plot of θ (or of cos θ) vs. pH does, in fact, show an inflection resembling that expected for the titration of a functional group in solution (Fig. 3). This value of the solution pH at which the inflection is centered has, in several instances, been correlated with ionization of the interfacial functional group [4–6, 8, 20]. Thus, contact angle titration is, at least, a useful *qualitative* technique for examining certain types of ionizations.

The most important generalization to emerge from application of contact angle titration to derivatives of PE–CO$_2$H is that the apparent acidities of functional groups in these interphases differ markedly from those in solution [4–6, 8]. This difference is uniformly in the sense expected if transformation of a neutral to a charged group were more difficult in the interphase than in solution. For most of the systems examined, the absolute value of the difference between $pK_{1/2}$ and pK_a is 2–4 units, although in related studies of self-assembled monolayers, it

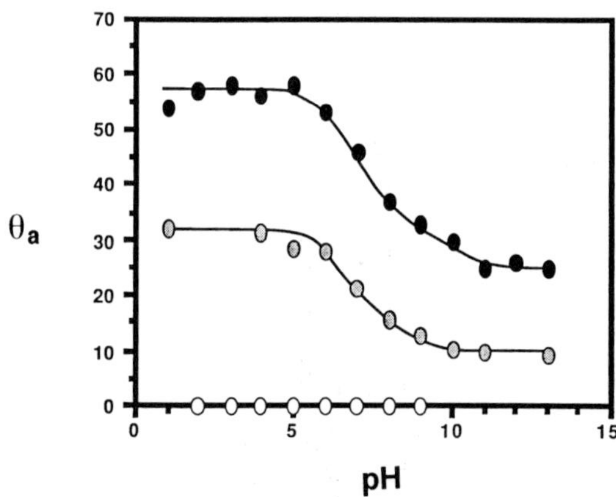

Figure 3. Contact angle titrations of three organic solids having interfacial CO$_2$H groups: (●) = PE–CO$_2$H; (◉) = Si/O$_2$Si(CH$_2$)$_{15}$CO$_2$H; (○) = AuS(CH$_2$)$_{10}$CO$_2$H. The 'pH' in this plot is the pH of the buffered aqueous solution in contact with the sample, and the 'θ_a' is the advancing contact angle. Although it is not possible to titrate a SAM of HS(CH$_2$)$_{10}$CO$_2$H on gold under air, it is possible when the sample is immersed in a non-polar solvent, e.g. cyclo-octane.

appears that carboxylic acids present in relatively low 'concentration' in a non-polar surface may not ionize until the pH of the aqueous solution is greater than 10 [21, 22].

As our exploration of the phenomena contributing to contact angle titration has continued, the subject has become more complex. It appears, for example, that the plateau in the titration curve observed for polyethylene carboxylic acid may be due to a limit in the hydrophilicity of the material (Fig. 4) [8]. This limit, in turn, is probably dependent on the procedure used for derivatization. We still have not resolved the relative contributions of thermodynamic and kinetic factors to the observed contact angle. For example, is the value of the advancing contact angle set, at least in part, by reactive spreading? That is, does the edge of a drop of alkaline water expanding on $PE-CO_2H$ (or a carboxylic acid-terminated SAM) continue beyond some hypothetical 'equilibrium' value, driven by reaction of hydroxide ion with the carboxylic groups? The advancing and receding contact angles of water on $PE-CO_2H$ and its derivatives and on carboxylic acid-terminated SAMs on gold have different values [8, 9, 21]: What is the origin of this hysteresis, and what information can be derived from it? How important is the presence of a condensed water film on that part of the surface not covered by the drop [8]?

PAC is a spectroscopic technique that has characteristics of potentially great value in surface chemistry. In this technique, the sample is irradiated with a short pulse of light, and the dissipation of the energy absorbed by a chromophore into the solution as heat is detected in the form of an acoustic wave in the liquid film using a pressure transducer (Fig. 5) [25]. PAC is very sensitive, especially for systems showing weak absorbances due to sample against a very weakly

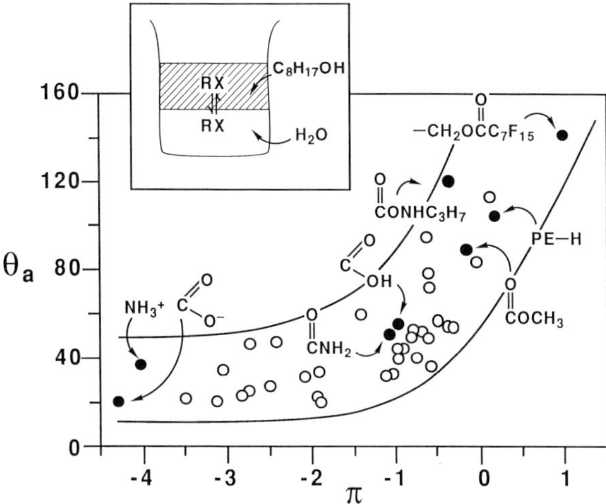

Figure 4. A plot of the advancing contact angle, θ_a, of water on various functional derivatives of $PE-CO_2H$ against the hydrophilicity of the functional groups shows a limit at high values of hydrophilicity. The hydrophilicity is given by the Hansch parameter, π. This parameter is defined as $\pi_X = \log P_X - \log P_H$, where P_H is the partition coefficient of the parent compound, R–H, between two solvents, and P_X is the partition coefficient of the substituted compound, R–X. The values shown here were derived using benzene as the parent compound, and water and 1-octanol as the solvents (see ref. [27]).

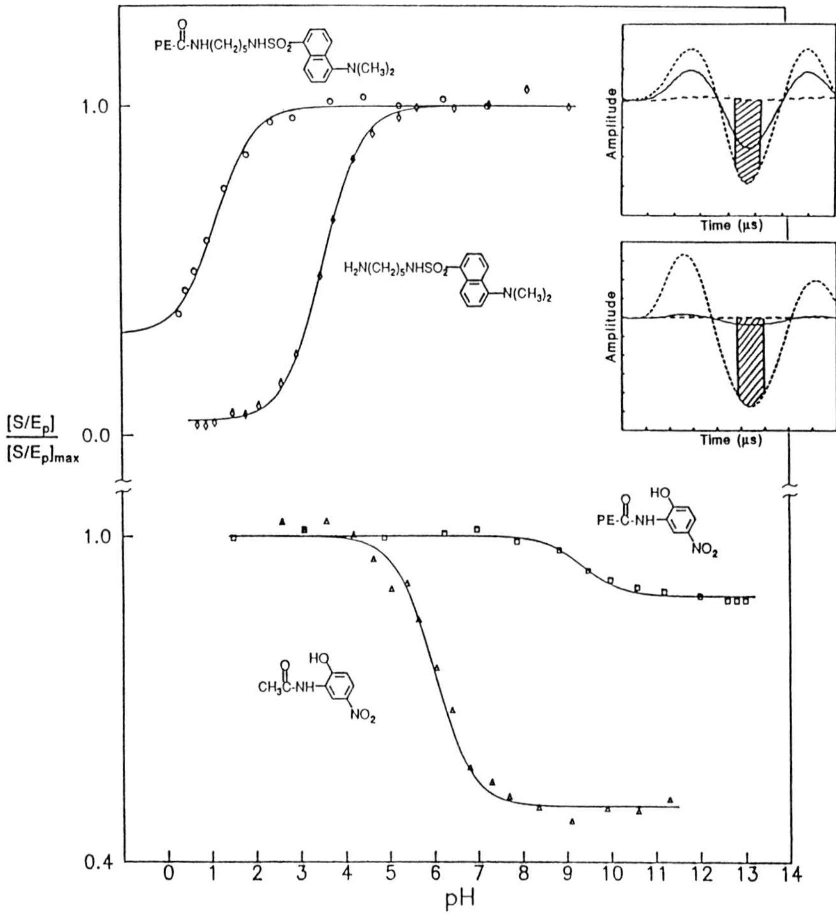

Figure 5. Representative titration curves for chromophoric acids and bases at a polyethylene–water interphase, and comparison with curves for the same groups in water. The vertical axis is the normalized intensity of the photoacoustic signal. The inset shows representative photoacoustic waves arising from dansyl groups at low (———) and high (- - - - - -) values of pH, and from the background (– – – –). The upper inset shows signals arising from dansyl groups attached to polyethylene; the lower inset shows signals arising from dansyl groups in aqueous solution. The value of pH is the pH of the buffered aqueous solution in which the derivatized polyethylene is immersed or in which the dansyl derivative is dissolved. S is the integrated intensity of the observed pressure wave (shown as the cross-hatched region in the inset); E_p is the power of the laser pulse.

absorbing backgroud. PAC is thus ideal for examining a strongly absorbing functional group located in the interphase between optically transparent polyethylene and water, because the technique has an intrinsically low background in this system and is also insensitive to scattering of light at the solid–liquid interface or by crystalline heterogeneity in the polyethylene.

We believe that PAC, when applicable, provides the best spectroscopic technique now available for analyzing the acidity of interfacial functional groups. Application of PAC to acidic and basic derivatives of $PE\text{-}CO_2H$ has given values of $pK_{1/2}$ in agreement with those from contact angle titration [23].

PAC has, so far, been applied to a limited range of problems in surface acidity.

Nonetheless, in several cases, the normalized dependence of the PAC signal on the pH for similar functional groups in the interphase between polyethylene and water and in homogeneous aqueous solution has also suggested two additional, important facts. First, the titratable groups seem to comprise a single population: they fit a single titration curve with a single value of $pK_{1/2}$ [23]. Second, a significant, non-zero PAC signal is observed for values of pH where no signal is expected, based on our previous observations [5]. The origin of this background is not presently understood.

4. THE INFLUENCE OF FUNCTIONAL GROUP ACIDITY ON SURFACE RECONSTRUCTION

One of the interesting characteristics of $PE-CO_2H$ and its derivatives is that the surface of these materials reconstructs on heating (and probably also reconstructs on mechanical deformation or on swelling with organic solvents) [7]. When the temperature of $PE-CO_2H$ is raised to a value close to the melting point of the polymer ($T \simeq 100°C$), functional groups migrate from the region of the interphase that determines the contact angle into deeper regions of the polymer [7, 15]. In these regions, the functional groups may still be observed by penetrating forms of spectroscopy (IR [4, 6], fluorescence [5], and to some extent XPS [4]), but they no longer influence wetting by water.

This reconstrution of the surface involves migration of a functional group from a near-surface region to the interior of the polymer. Particularly when the functional group is polar and in contact with water, this movement might be anticipated to be energetically unfavorable, mainly owing to the loss of solvation energy when the functional group moves from water to hydrocarbon. As expected, using this argument, the rate of thermal reconstruction of $PE-CO_2H$ is much slower when the sample is in contact with aqueous alkaline solution and the carboxylic acid groups are present in ionized form than when the carboxylic acids are protonated and the sample is dry [7, 15].

A remarkable example of the coupling of wetting to the ionization of functional groups is presented by the amide formed by reaction of PE-COCl with anthranilic acid (PE-anthranilamide) (Fig. 6) [20]. In contact with acid, this material is very hydrophobic; in contact with base, it becomes very hydrophilic. The difference between these two states of the material appears to be a small conformational change in the anthranilamide moiety (Fig. 7). At high pH, the surface arranges the anthranilamide groups to expose CO_2^- moieties; at low pH, only C–H bonds are exposed.

5. ACIDITY OF INTERFACIAL GROUPS. RESULTS FROM STUDIES OF $PE-CO_2H$ AND DERIVATIVES, AND COMPARISONS WITH SELF-ASSEMBLED MONOLAYERS

Protonation or deprotonation of functional groups at interfaces changes the atomic/molecular properties of these groups, and consequently changes the macroscopic properties of the surfaces. To what extent is it now possible to predict the acidity of functional groups in interphases from their properties in solution? Can one predict the influence of protonation or deprotonation on a macroscopic property such as wetting?

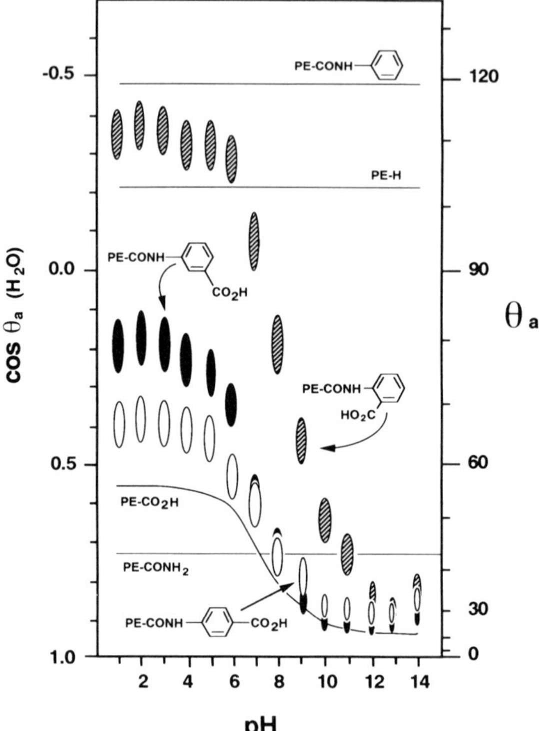

Figure 6. Representative pH titration curves for a number of aminobenzoic acid derivatives of PE–CO_2H. The vertical axis is proportional to the cosine of the advancing contact angle of water, cos θ_a(H_2O), on these surfaces. This measure, and not simply the contact angle θ_a, is used because, from Young's equation, cos θ is directly proportional to the net free energy of interaction of the contacting liquid and the polymer interphase. The size of the symbols represents the variability in the contact angle measurements. The horizontal lines represent the contact angles of several analogous but non-titratable derivatives of polyethylene. The solid curve is the contact-angle titration curve obtained from PE–CO_2H.

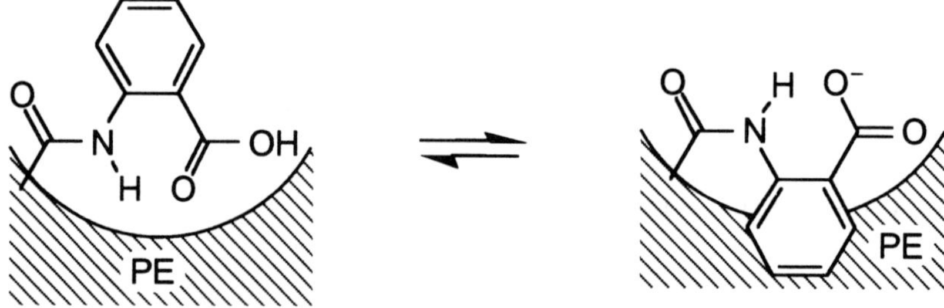

Figure 7. Hypothesized change in conformation of 'polyethylene–anthranilamide' on going from low to high pH.

In broad terms, it is possible to predict trends and approximate magnitudes. It is not possible to predict accurate values for p$K_{1/2}$ or contact angle based on a knowledge of pK_a, or even to rationalize satisfactorily many of the trends. Deprotonation of a carboxylic acid by aqueous hydroxide ion and protonation of an amine by hydronium ion both require higher concentrations of the base or acid by two to four orders of magnitude when the functional group is present in the interphase between PE–CO$_2$H or its derivatives and water than when that functional group is simply dissolved in water [4–6, 8, 20]. This effect could be due to a relatively low local dielectric constant in the interphase, or to a low concentration of hydroxide ion (hydronium ion) in the interphase (both relative to bulk solution), or to a combination of the two. Many of the properties of interfacial groups on PE–CO$_2$H suggest an interfacial dielectric constant of approximately 10 [4]. This value would tend to discriminate energetically against any charged species—whether covalently incorporated into the interphase or entering it by diffusion—relative to the bulk solution.

Polyacidity in the interphase could take two forms. First, during ionization of carboxylic acids by base there could be an accumulation of concentrations of weakly screened negative charges sufficiently high that unfavorable coulombic interactions would require increased concentrations of hydroxide to achieve complete ionization. Second, the carboxylic acid groups might be sufficiently close together that hydrogen bonds would stabilize them in protonated form and thus, also, would increase the difficulty of removing protons. Although the majority of carboxylic acid groups in PE–CO$_2$H are clearly hydrogen bonded to one another (by infrared analysis) [4], polyacidity probably does not contribute to the apparent low acidity of the carboxylic acid groups in this material. Increasing the ionic strength of the aqueous solution in contact with PE–CO$_2$H does not seem to change the curves obtained by contact angle titration [4]. This observation supports the assertion that polyacidity is not a major contributor to the acidity of carboxylic acid groups on PE–CO$_2$H. This traditional test is, however, one that has been developed for use in homogeneous solution with soluble ions, and its applicability to the interphase is open to some question for two reasons. First, the solutions used in contact angle titrations are buffered and are already at a relatively high ionic strength (especially at low and high values of pH). Second, added metal halide salts might themselves be excluded from the interface, and might even be excluded preferentially relative to hydronium or hydroxide ion.

Comparison of the highly disordered solid–water interphase presented by PE–CO$_2$H and the more ordered interfaces presented by self-assembled monolayers suggests that the two are governed by similar rules. The carboxylic acid groups in SAMs also require contact with a strongly basic solution to accomplish their ionization [21]. (Corresponding studies with amine-terminated SAMs are complicated by the high reactivity of the amino group, which leads to ready contamination of the surfaces [26].) It is not possible to define a value of p$K_{1/2}$ for CO$_2$H groups incorporated into SAMs, since no plateau in the curve of cos θ vs. pH is observed at high values of pH [21]. The magnitude of the shift in *onset* of ionization can, however, be defined, and can be larger for SAMs than for PE–CO$_2$H. Since this shift toward high values of pH becomes larger as the carboxylic acid groups become more dilute at the interface [21], the most probable explanation for the low apparent acidity of these groups is a low local dielectric constant at the monolayer–water interface. To what extent the low local dielectric constant influences the energetics of converting CO$_2$H to CO$_2^-$ groups, and to what extent it influences the partitioning of hydroxide ion or hydronium ion between the interface and the bulk solution, remains to be established.

The only macroscopic property that correlates well with ionization of interfacial groups is wettability by water. As expected, converting neutral to charged groups at the interface increases the wettability of the interface by water. In very general terms, the contact angle of water on derivatives of $PE-CO_2H$ correlates with the hydrophilicity of the functional groups (as reflected in their Hansch π values—group additivity parameters related to the free energy of transfer of the group between water and octanol) [27, 28]. The correlation is, however, odd, in the sense that the influence of functional group polarity on wetting seems to 'saturate' at high values of polarity [8]. Although increasing the polarity of the group X in a series of materials having the composition PE–X decreases the contact angle for groups of low and intermediate polarity, a number of highly polar groups give rise to similar, non-zero values of the contact angle [8]. We have offered a hypothesis to explain these data based on the idea that, for highly polar interfacial functional groups and at high relative humidity, the surface will consist mostly of adsorbed water, independent of the polarity of the groups at the interface. We have not, so far, devised a direct test of this hypothesis.

6. SUMMARY

The behavior of acidic and basic functional groups at interfaces is an apparently simple subject with many complex features. As with many simple reactions, it is difficult to disguise ignorance. We know that these functional groups follow patterns of reactivity similar to those observed in solution, but that there are important systematic differences between behavior in the interphase and in solution. There is no sound theoretical understanding of these differences, and, at the moment, no experimental method applicable to the measurement of those properties—interfacial thermodynamic activities of hydronium and hydroxide ion, microscopic dielectric constant and solvating capacity of the interphase, structure, properties, and thermodynamic properties of the three-phase solid–liquid–vapor region at the edge of a reactive liquid drop on an acidic or basic surface—that would be most helpful in resolving these issues.

New experimental techniques are, however, now appearing that promise to renew the study of interfacial acid–base reactions. SAMs are structurally well-defined systems that will serve as tractable models for these studies [21, 22]. PAC and contact angle titration offer new approaches to determining the extent of ionization, and to correlations of ionization and wetting. Microfabrication may make possible the construction of microelectrodes, and the study of local concentrations [29]. X-Ray techniques are offering increasing resolution in the study of the structure of interfaces between condensed phases [30, 31].

The simplest subjects may teach the most profound lessons in science. We hope that study of simple acid–base reactions at solid–water interfaces will provide fundamental information concerning these ubiquitous systems.

Acknowledgements

This work was supported in part by the Office of Naval Research and the Defense Advanced Research Projects Agency. The work summarized here is the result of research carried out by colleagues whose names are given in the references.

REFERENCES

1. R. P. Bell, *The Proton in Chemistry*. Cornell University Press, Ithaca, New York (1959); C. H. Bamford and C. F. H. Tipper (Eds), *Comprehensive Chemical Kinetics*, vol. 8. Elsevier, Amsterdam (1977); H. L. Finston and A. C. Rychtman, *A New View of Current Acid–Base Theories*. John Wiley, New York (1982); M. Eigen, *Angew. Chem. Int. Ed.* **3**, 1 (1964); W. P. Jencks, *Acc. Chem. Res.* **9**, 425 (1976).
2. J. R. Rasmussen, E. R. Stedronsky and G. M. Whitesides, *J. Am. Chem. Soc.* **99**, 4736 (1977).
3. J. R. Rasmussen, D. E. Bergbreiter and G. M. Whitesides, *J. Am. Chem. Soc.* **99**, 4746 (1977).
4. S. R. Holmes-Farley, R. H. Reamey, T. J. McCarthy, J. Deutch and G. M. Whitesides, *Langmuir* **1**, 725 (1985).
5. S. R. Holmes-Farley and G. M. Whitesides, *Langmuir* **2**, 266 (1986).
6. S. R. Holmes-Farley and G. M. Whitesides, *Langmuir* **3**, 62 (1987).
7. S. R. Holmes-Farley, R. G. Nuzzo, T. J. McCarthy and G. M. Whitesides, *Langmuir* **3**, 799 (1987).
8. S. R. Holmes-Farley, C. D. Bain and G. M. Whitesides, *Langmuir* **4**, 921 (1988).
9. G. S. Ferguson and G. M. Whitesides, *Chemtracts* **1**, 171 (1988).
10. G. M. Whitesides and P. E. Laibinis, *Langmuir* **6**, 87 (1990).
11. J. N. Israelachvili, *Intermolecular and Surface Forces*. Academic Press, London (1985).
12. J. Rebek, Jr., R. J. Duff and W. E. Gordon, *J. Am. Chem. Soc.* **108**, 6068 (1986).
13. C. L. Perrin, T. J. Dwyer, J. Rebek, Jr. and R. J. Duff, *J. Am. Chem. Soc.* **112**, 3122 (1990).
14. H. P. Gregor and M. Frederick, *J. Polym. Sci.* **23**, 454 (1957).
15. G. S. Ferguson and G. M. Whitesides, in: *Modern Approaches to Wettability: Theory and Applications*, G. Loeb and M. Schrader (Eds). Plenum Press, New York (in press).
16. B. V. Derjaguin and N. V. Churaev, in: *Fluid Interfacial Phenomena*, C. A. Croxton (Ed.), pp. 663–738. John Wiley, New York (1986).
17. C. H. Rochester, *Acidity Functions*; A. T. Blomquist, Ser. Ed; Organic Chemistry, Vol. 17; Academic Press: New York (1970); C. H. Rochester, *Q. Rev. Chem. Soc.* **20**, 511 (1966); R. A. Cox and K. Yates, *Can. J. Chem.* **61**, 2225 (1983).
18. K. M. Dyumaev and B. A. Korolev, *Russ. Chem. Rev.* **49**, 1021 (1980); W. S. Matthews, J. E. Bares, J. E. Bartmess, F. G. Bordwell, F. J. Cornforth, G. E. Drucker, Z. Margolin, R. J. McCallum, G. J. McCollum and N. R. Vanier, *J. Am. Chem. Soc.* **97**, 7006 (1975).
19. C. L. Rice and R. Whitehead, *J. Phys. Chem.* **69**, 4017 (1965).
20. M. D. Wilson, and G. M. Whitesides, *J. Am. Chem. Soc.* **110**, 8718 (1988).
21. C. D. Bain and G. M. Whitesides, *Langmuir* **5**, 1370 (1989).
22. S. R. Wasserman, Y.-T. Tao and G. M. Whitesides, *Langmuir* **5**, 1074 (1989).
23. L. Zhang, M. A. Shulman, G. M. Whitesides and J. J. Grabowski, *J. Am. Chem. Soc.* **112**, 7069 (1990).
24. S. S. Dukhin and B. V. Derjaguin, in: *Surface and Colloid Science*, vol. 7, pp. 1–336. Interscience, New York (1974); A. M. James, in: *Surface and Colloid Science*, R. J. Good and R. R. Stromberg (Eds), vol. 11, pp. 121–185. Plenum Press, New York (1979).
25. L. J. Rothberg, J. D. Simon, M. Bernstein and K. S. Peters, *J. Am. Chem. Soc.* **105**, 3454 (1983); S. E. Braslavsky, R. M. Ellul, R. G. Weiss, H. Al-Ekabi and K. Schaffrier, *Tetrahedron* **39**, 1909 (1983); T. J. Burkey, M. Majewski and D. Griller, *J. Am. Chem. Soc.* **108**, 2218 (1986).
26. K. L. Prime and G. M. Whitesides, unpublished results (1990).
27. C. Hansch, A. Leo, A. S. M. Ungar, K. H. Kim, D. Nikaitani and E. J. Lien, *J. Med. Chem.* **16**, 1207 (1973).
28. M. D. Wilson, G. S. Ferguson and G. M. Whitesides, *J. Am. Chem. Soc.* **112**, 1244 (1990).
29. P. E. Laibinis, J. J. Hickman, M. S. Wrighton and G. M. Whitesides, *Science (Washington, DC)* **245**, 845 (1989); J. J. Hickman, C. Zou, D. Ofer, P. D. Harvey, M. S. Wrighton, P. E. Laibinis, C. D. Bain and G. M. Whitesides, *J. Am. Chem. Soc.* **111**, 7271 (1989).
30. S. R. Wasserman, G. M. Whitesides, I. M. Tidswell, B. M. Ocko, P. S. Pershan and J. D. Axe, *J. Am. Chem. Soc.* **111**, 5852 (1989); I. M. Tidswell, B. M. Ocko, P. S. Pershan, S. R. Wasserman, G. M. Whitesides and J. D. Axe, *Phys. Rev. B* **41**, 1111 (1990); M. V. Baker, I. M. Tidswell, T. A. Rabedeau, P. S. Pershan and G. M. Whitesides, *Langmuir* (submitted).
31. M. J. Bedzyk, M. G. Bommarito, M. Caffrey and T. L. Penner, *Science (Washington, DC)* **248**, 52 (1990).

Acid-Base Interactions, pp. 243-256
Eds. K.L. Mittal and H.R. Anderson, Jr.
©VSP 1991

Acid–base interfaces in fiber-reinforced polymer composites

DAVID W. DWIGHT,[1,*] FREDERICK M. FOWKES,[1] DAVID A. COLE,[1] MARY JO KULP,[1,†] PHILIPPE J. SABAT,[2,‡] LAWRENCE SALVATI, Jr.[3] and T. CLARE HUANG[4,§]

[1] *Chemistry Department #6, Lehigh University, Bethlehem, PA 18015, USA*

[2] *Engineering Science and Mechanics Department, Virginia Tech, Blacksburg, VA 24061, USA*

[3] *Perkin-Elmer Corp., 5 Progress St., Edison, NJ 08820, USA*

[4] *PPG Industries Fiberglass Research Center, Pittsburgh, PA 15230, USA*

Revised version received 23 April 1990

Abstract—The role of Lewis acid–base interactions at the fiber-matrix interface in composites is studied with both glass and Teflon fibers. In the glass fiber case, surface chemistry is modified with amino-, methacryloxy- and glycidoxy-silane coupling agents (A-1100, A-174 and A-187, respectively). Silane adsorption mechanisms as well as the properties of filament-wound, unidirectional epoxy and polyester composites are explained by a combination of X-ray photoelectron spectroscopy (XPS), scanning electron microscopy (SEM), and flow microcalorimetry. The heats of adsorption of pyridine and phenol prove that the coupling agents add acidic sites to the glass fiber surface as well as stronger basic sites. The subsequent adhesion of the matrix polymers and the short beam shear strengths of composites are explained on this basis. The Teflon fibers are first etched with sodium naphthalene solutions, and then sequentially hydroborated and acetylated, producing approximately mono-functional hydroxyl (acidic) and ester (basic) groups on the surfaces, as determined by XPS, FTIR, and electrophoretic mobility analyses. Composites prepared with the acetylated fibers and a chlorinated polyvinyl chloride (acidic) matrix are superior in tensile properties, and SEM fractography shows PTFE fibrillation, indicative of good fiber–matrix adhesion and stress transfer, in this case only.

Keywords: Acid–base interactions; adhesion; surface modification; glass- and Teflon-fiber composites; XPS; mechanical properties.

1. INTRODUCTION

Interfacial forces arise primarily from London–Lifshitz and Lewis acid/Lewis base interactions [1]. London–Lifshitz interactions include a polarizability component as well as the ubiquitous dispersion forces [2]. Acid–base interactions, exemplified by the classical hydrogen bond, are specific interactions between an electron donor (Lewis base) and an electron acceptor (Lewis acid). Previously Fowkes *et al.* [3] have demonstrated that calorimetric and spectroscopic measurements can be used to characterize a material's acidic and/or basic character.

Adhesive bond strength is directly proportional to the thermodynamic work of adhesion [4]. Also, the work of adhesion is very important in polymeric composites for optimization of molecular interactions between fibers and the surrounding

---

*To whom correspondence should be addressed.

†Present address: Lubrizol Corporation, 29400 Lakeland Boulevard, Wickliffe, OH 44092-2298, USA.

‡Present address: 1 Rue de la reine Cecile Pruniers, Bouchemaine, France.

§Present address: Pennzoil Technology Center, P.O. Box 7569, The Woodlands, TX 77387, USA.

polymer matrix. Although adhesion between polymer matrices and the reinforcing fibers can arise from primary chemical bonds, from London–Lifshitz interactions across the interface or from mechanical interlocking, acid–base interactions are the most generally useful chemical forces that a technologist has available to *modify* interfacial bonding. That principle has been demonstrated in several studies relating interfacial acid–base interactions to mechanical properties of various *polymers filled with particles* including silica [5], zinc and wollastonite [6], calcium carbonate [7], carbon black [8], and alumina and glass [9] powders.

This paper demonstrates the effect of Lewis acid–base interactions upon stress transfer at interfaces in *fiber-reinforced composites*, with both continuous (glass) and chopped (Teflon) fibers. The fiber mechanical properties are significantly different, as are the types of surface modification strategies and the morphologies and volume fractions in the composites. In the glass fiber case, surface chemistry was modified with silane coupling agents. Numerous studies of mechanical properties of composites or laminates have implied that silanes form primary chemical bonds with both glass substrates and polymeric resins [10]. We are not aware of any results that also include direct thermodynamic and spectroscopic evidence of primary chemical bond formation. The magnitude of acid–base interactions at interfaces is sufficient to explain the enhancements of mechanical properties produced by silanes. Furthermore, most studies [11] have assumed that the silanes form perfect mono- or multi-layers of uniform thickness, but direct experimental verification is lacking.

The Teflon fibers were first etched, then a reaction sequence was used to produce either 'monofunctional' acidic or basic groups on the surfaces. One of the first publications on XPS applied to polymers [12] elucidated the surface chemistry of sodium-etched polytetrafluorethylene (PTFE), as well as subsequent reactions of the etched surface and correlations with wettability. Later it was shown how the sodium-etched PTFE surface could be reacted to produce various monofunctional surfaces [13]. However, neither the acid–base nature of the surfaces nor their adhesive or composite properties have been reported.

Surface chemistry and acid–base character were determined by X-ray photoelectron spectroscopy (XPS or ESCA), infrared spectroscopy (FTIR), flow microcalorimetry, and electrophoretic mobility. In the glass fiber case, high volume fraction, continuous fiber, unidirectional composites were fabricated by filament winding using polyester or epoxy matrix polymers, and mechanical properties were evaluated with three-point bending (short beam shear) tests. Low volume fraction, filled chlorinated polyvinylchloride (CPVC) composites were made with the chopped PTFE fibers. Correlations were obtained between fiber surface chemistry and the properties of the composites.

2. EXPERIMENTAL SECTION

The E-glass fibers were surface treated with silane coupling agents and with a mixture of silane coupling agent and polyvinylacetate (PVAc) film former immediately after the glass fibers were drawn from a forming bushing in PPG Industries Fiberglass Research Center equipment. Aminopropyl-, methacryloxy-, and glycidoxy-silanes (Union Carbide A-1100, A-174 and A-187, respectively) were used in separate experiments. Aqueous solutions containing ≈ 0.5 wt% of the coupling agent at a pH adjusted to ≈ 4.5 with acetic acid were applied to the fibers by a kiss-coating technique, wrapped onto a paper core and dried at 120°C for 6–12 h. This method yielded a loading of ≈ 0.2 wt%.

Using essentially the same procedures, E-glass fibers were produced and sized with A-1100 and A-174 by Vetrotex Saint-Gobain (France). Unidirectional composites were produced on commercial filament winding equipment. Short beam shear specimens of $19 \times 6 \times 3$ mm dimensions were cut with a diamond saw from plates $(310 \times 210 \times 3$ mm) obtained from filament winding. Two matrix resins, epoxy (Ciba-Geigy Araldite F) and polyester (Rhone Poulenc Stratyl A) were used with standard cure cycles. Short beam shear testing was done according to ASTM 2344-76 using a crosshead speed of 0.13 cm/min on an Instron Model 1125. Ten specimens of each type were tested to ensure reproducibility. For each matrix, four different interphases were investigated: (1) bare fiber; (2 and 3) silane coupling agent *on the fiber* or *in the resin*; and (4) silicone (polydimethyl siloxane) lubricant on the fiber. It should be noted that no film formers were used on the E-glass fibers.

As-received, the chopped Teflon fibers were brown due to the presence of degraded cellulosic material—a by-product of the fiber spinning process. This extraneous material was oxidatively removed from the fibers using refluxing H_2SO_4 to which HNO_3 was added [14]. XPS spectra of the bleached fibers showed essentially pure Teflon. 'Etching' reactions were carried out with sodium naphthalene dispersions in diglyme, producing jet black fibers with a variety of reactive functional groups on the surface. Hydroxylation was carried out on the etched fibers using borane in tetrahydrofuran which was followed by alkaline hydrogen peroxide [15]. Esterification was effected by treating some of the hydroxylated fibers with acetic anhydride [16] in the presence of 4-pyrrolidino-pyridine and pyridine [17]. Each of the three types of fiber was suspended in a 1,2-dichloroethane solution of an acidic polymer, chlorinated polyvinyl chloride (CPVC). After drawing down onto a Ferrotype plate with a Gardner 'doctor' blade and evaporating the solvent, 20 vol.% fiber composite films ≈ 0.5 mm thick were obtained. Strips (285 mm $\times$ 95 mm) were tested in tension on the Instron at a strain rate of 0.25 cm/min. Three to five samples of each type were tested, and all the data are displayed as 'scatter bands' on the stress-strain diagrams.

The angle-dependent X-ray photoelectron spectroscopy (ADXPS) experiments [18] were carried out either on a Perkin-Elmer model 5000LS (Teflon and PPG glass fibers) or a Kratos XSAM-800 (Vetrotex St.-Gobain fibers) X-ray photoelectron spectrometers using polychromatic magnesium anodes operating at 400 W. The glass fibers were mounted in a multilayer, parallel array on glass microscope slide coverslips. The Teflon fibers were analyzed as packed powders on double-sided adhesive tape.

Flow microcalorimetry (a Microscal with sensitivity in the microcalorie range) coupled with a downstream Perkin-Elmer LC-75 UV detector was used to determine the strength and concentration of acidic and basic surface sites by measuring the heats of adsorption of dilute cyclohexane solutions of pyridine and of phenol on the untreated and treated E-glass fibers. The response of the Microscal to changes in the bed temperature was calibrated by heating a ground glass filled bed with a resistive heating coil. The number of moles adsorbed during a given time interval was determined by computing the integrated UV signal during adsorption and subtracting the corresponding value for a blank experiment where the solution bypassed the calorimeter. The concentration of sites was calculated as the number of moles of solute adsorbed per gram of sample which assumes the formation of 1:1 complexes. The molar heat as a function of coverage was obtained by dividing the heat per given time interval by the corresponding moles of solute adsorbed. Phenol and pyridine solutions of ≈ 20 mmol concentration were each prepared in cyclohexane having a moisture content of about 8 ppm. The glass fibers were dried for 5 h at 50°C in a vacuum oven. The removal of water is very important in these studies as it is amphoteric and can interact with both the glass fibers and the probes. Approximately 0.15 g of the 14 μm diameter glass fibers 45 cm in length were placed in the calorimeter cell as a random coil.

The Teflon fibers were too low in surface area for calorimetry, so acid–base character was evaluated by electrophoresis in HPLC grade tetrahydrofuran using a Coulter Electronics, Inc. Delsa 440 instrument. No special care was taken to exclude moisture; it was not expected to influence the sign of the surface potential and the magnitudes were not subjected to interpretation. FTIR spectra were collected on a Mattson Sirius 100 Fourier transform infrared spectrometer using the diffuse reflectance accessories. Scanning electron microscopy was carried out on the Autoscan model from ETEC Corporation.

3. RESULTS AND DISCUSSION

3.1. *Glass fiber surfaces and coupling agent adsorption*

The essential results of our analysis of the high-resolution XPS spectra and angular-dependent depth profiles of E-glass plates and fibers with and without silane coupling agent coatings are elaborated in reference [18]. A comparison of the XPS analyses of some of the PPG and St.-Gobain glass fiber specimens is shown in Table 1. There is a close similarity between them both before and after silane adsorption. The difference in carbon contamination level on the two bare fiber types obscures quantitative comparisons, but the results lead us to believe that both glass fibers are basic due to the presence of aluminum and calcium oxide network stoppers. Perhaps the St.-Gobain fibers are more basic because of the greater aluminum content and absence of boron. The most important similarities emerge upon comparing the changes in surface chemistry produced by A-1100 aminosilane adsorption. In both cases: (1) an identical, broad nitrogen peak appears (previously interpreted as a mixture of quaternary and un-ionized amine [18]); (2) carbon concentration increases due to the propyl groups of the silane; and (3) aluminum, calcium and oxygen concentrations decrease and the oxygen peak changes in shape and position—indicating the presence of a silane overlayer on some, but not all of the glass atoms, and that the silane oxygens are different in oxidation state from the glass oxygens. Thus we assume that both glass fiber types treated with aminosilane will have qualitatively the same acid–base character. Furthermore, the silane overlayers must be no more than a monolayer in many patches because the glass substrate elements still appear in the XPS spectra. (Note that when a polydimethyl siloxane (PDMS) lubricant alone was applied to E-glass fibers, the glass elements were suppressed completely from the XPS spectrum, and the results were characteristic of the PDMS treatment alone.) Also, the angular

Table 1.

Quantitative XPS parameters of PPG and St.-Gobain E-glass fiber specimens

Specimen		Element and orbital						
		$O(1s)$	$N(1s)$	$Ca(2p)$	$C(1s)$	$Si(2p)$	$B(1s)$	$Al(2p)$
PPG								
E-glass	BE	532.5	400.2	348.4	(285.0)	102.4	192.4	74.2
bare	W	3.5	2.0	2.5	2.3	2.4	2.1	2.4
	%	43.2	0.8	3.5	33.8	12.1	1.7	2.5
E-glass	BE	533.0	399.0	348.4	(")	101.8	191.8	74.8
+ A-1100	W	2.6	3.3	2.1	2.4	2.4	—	1.9
	%	23.5	4.3	1.5	56.5	8.0	1.2	1.0
St.-Gobain								
E-glass	BE	531.8	—	347.9	(")	102.2	—	74.0
bare	W	3.2	—	3.0	3.1	2.8	—	2.7
	%	41.8	0	4.7	20.0	15.3	0	18.1
E-glass	BE	532.1	400.2	347.9	(")	102.6	—	73.8
+ A-1100	W	2.9	4.0	3.2	3.1	2.8	—	2.5
	%	35.3	4.1	1.3	39.7	15.7	0	3.9
E-glass	BE	532.3	—	—	(")	102.0	—	—
+ PDMS	W	1.9	—	—	2.0	2.1	—	—
	%	12.1	0	0	77.2	10.7	0	0

Carbon was assigned to BE = 285.0 eV for charge calibration.
BE is binding energy (eV).
W is peak width at half height (eV).
% is atomic percent.

dependent XPS depth profiles previously reported [18] showed little difference between the silane overlayers and the adventitious hydrocarbon contamination overlayers present on the bare glass fibers.

Evidence for gross heterogeneity of the silane overlayers is shown in the photomicrographs in Fig. 1. Clearly there are 'islands' of silane on both the PPG and St.-Gobain E-glass fibers. We were unable to prove that there was no silane adsorbed on the areas of glass fiber in between the islands, but we conclude that there are some bare patches of glass based on a combination of the SEM, ADXPS and calorimetric results.

The molar heats of adsorption of pyridine, a measure of the strength of the acidic surface sites, and the concentration of acidic surface sites in micromoles per gram of glass fibers are shown in Table 2. The bare (water-sized) glass is seen to have very weakly acidic surface sites, for the heat of adsorption of pyridine is only about 1 kJ/mol. Treatment with the A-1100 aminosilane raised the surface acidity about four-fold. Similar though smaller heat of adsorption of pyridine is observed on glass fibers treated with A-174 methacryloxysilane. The heat of adsorption of pyridine on the silane coating are much too low to be silanol groups, for SiOH is about as acidic as phenol (≈ 30 kJ/mol interaction with pyridine). The observed acidity may be due to siloxane ($-Si-O-Si-O-$) groups [19]. The treatment of glass fiber with A-1100 aminosilane plus polyvinylacetate (PVAc) reduced the acid strength of the glass fibers to very low values, indicating that the film former now dominates the surface chemistry.

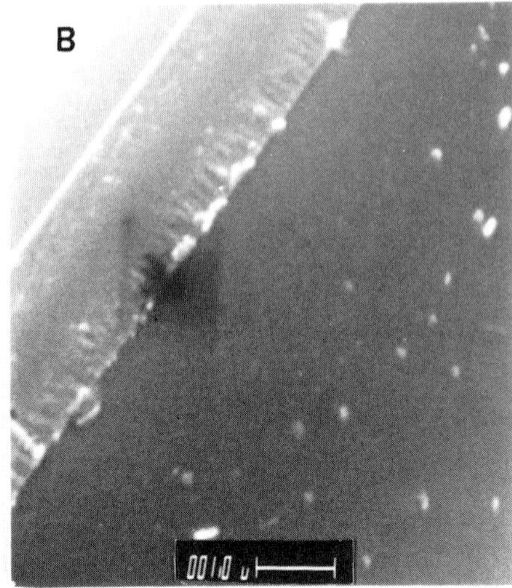

Figure 1. SEM photomicrographs showing side views of E-glass fibers upon which A-1100 amino-silane was applied. A. On the St.-Gobain glass fibers, large silane island is apparent in the middle of the left hand fiber, an another small silane patch can be seen in the lower right hand corner of the photomicrograph. B. At higher magnification, part of a long silane strip on a PPG fiber is shown. These features were found in every photomicrograph.

Table 2.

PPG E-glass fiber acidic sites probed by pyridine adsorption in flow microcalorimetry

Fiber surface	Concentration of sites (μmol/g)	Heat of adsorption (kJ/mol)
Bare	4.6	1.3
A-1100 treated	4.4	5.7
A-174 treated	4.8	3.2
A-1100 + PVAc treated	5.9	0.8

Table 3 summarizes the titration with dilute phenol in flow micro-calorimetry, probing the basic surface sites of various PPG E-glass fiber specimens. The bare (water-sized) glass is seen to have a surface concentration of basic sites about twice that of acidic sites, and the strength of the basic sites is about three times the strength of acidic sites. The silane coupling agents produce about a 50% increase in the basic strength of surface sites. The observed basic strength of the A-1100 aminosilane is no greater than the other silanes. The amine groups would have heats of acid-base interaction several times those observed, and this evidence suggests that in the period of some weeks after treatment the amine must have reacted with carbon dioxide to form the carbonate, a weaker base. Unfortunately, FTIR analyses were not sensitive enough to detect any coupling agent. The PVAc film-forming polymer dramatically enhanced the surface basicity, providing a two-to three-fold increase in surface concentration of basic surface sites, as well as a five- to seven-fold increase in strength. The PVAc forms multilayers, and the phenol may be detecting sites that are not at the top surface.

Table 3.
PPG E-glass fiber basic sites probed by phenol adsorption in flow microcalorimetry

Fiber surface	Concentration of sites (μmol/g)	Heat of adsorption (kJ/mol)
Bare	9.5	4.2
A-1100 treated	5.3	6.7
A-174 treated	7.2	5.0
A-187 treated	5.3	6.8
A-1100 + PVAc treated	15.4	36
A-174 + PVAc treated	15.5	37
A-187 + PVAc treated	15.6	34

From the correlation of the XPS and microcalorimetric results, we conclude that the PPG and St.-Gobain E-glass fibers both have predominantly weak basic surface sites. Silane coupling agents add both stronger basic sites, as well as somewhat weaker acidic sites. This helps to shed light on the reason why silanes have been so popular as glass coupling agents, for reactivity with both acidic and basic resins should be enhanced by their presence.

3.2 Unidirectional fiberglass composite properties

Figures 2 and 3 summarize the results of the three-point bending tests on St.-Gobain E-glass composites with polyester and epoxy matrix polymers. First note that there were very different failure mechanisms in the composite beams. Four modes of failure were observed: (1) shear; (2) fracture perpendicular to the fiber direction, called bending failure; (3) microbuckling under the point where the load was applied, and (4) fiber–matrix interfacial slipping. Failure of the short beams occurred by one or a combination of these mechanisms depending on the matrix and fiber properties and on the quality of the fiber–matrix adhesion.

For the samples made of polyester resin, Fig. 2 indicates that only the beams with the fibers coated with silane fail in pure shear. The samples made with polyester resin containing silane fail mainly in shear with some bending. In that case, there is evidence obtained by imaging SIMS [20] that the silane coupling agent segregates to the glass–polyester interphase. However, in the case without coupling agent, the samples do not fail in bending because internal slipping was the predominant failure mechanism. Without silane on the fiber or in the matrix, the fiber–polyester matrix adhesion seems to be very poor—not much better than the case with fibers coated with silicone lubricant. The difference in the latter two stress-strain curves may be due simply to a small increment over the silicone viscous slipping in stress transfer capability provided by the London–Lifshitz interaction at the bare glass–polyester interface. However, the acidic sites provided by the silane coupling agent are available to form strong bonds with the basic ester groups in the matrix resin. This appears to be responsible for the five-fold improvement in composite shear strength.

Because the epoxy matrix is more ductile than the polyester, it was impossible to obtain a pure shear failure, as can be seen in Fig. 3. The most striking observation is that the fiber–matrix adhesion was much better for bare glass–epoxy than

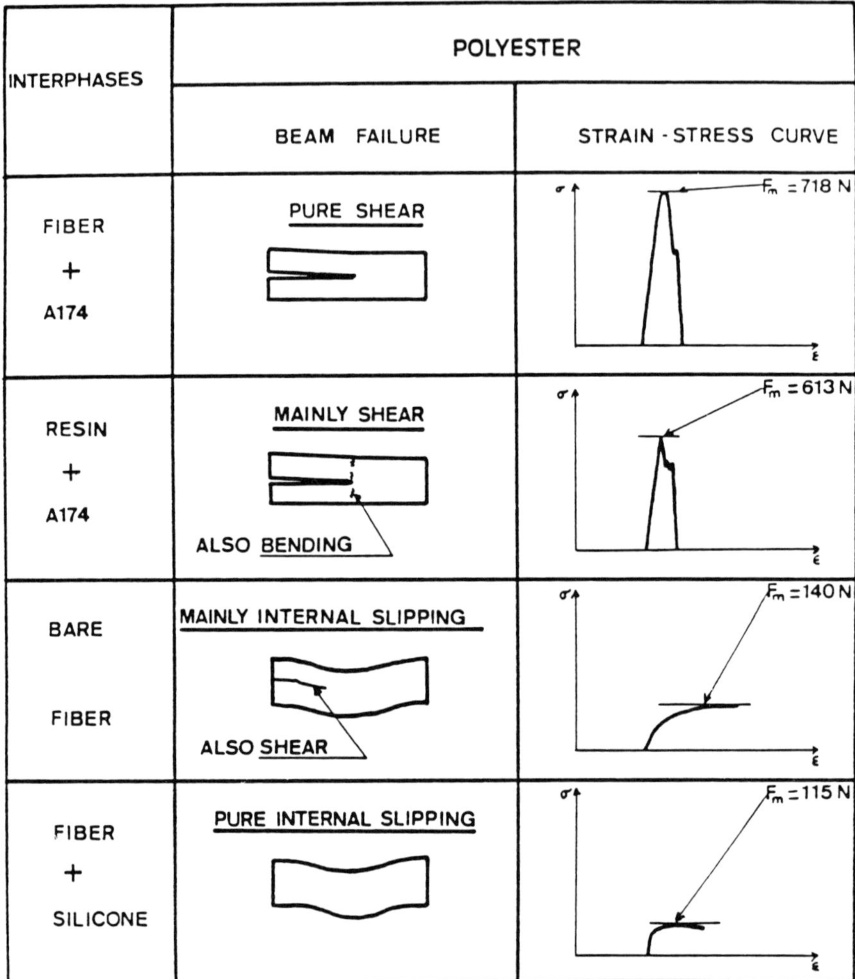

Figure 2. Schematic diagram summarizing the results of the short beam shear tests on unidirectional, E-glass fiber composites with polyester resin matrix. The second column contains sketches of cross-sections of the short beam shear specimens, illustrating the principal mode of failure. Corresponding compressive stress strain diagrams are sketched in the third column with the failure loads indicated.

for bare glass–polyester, as no internal slipping occurred with the epoxy-bare fiber sample. Again this can be understood in terms of Lewis acid–base interactions between the acidic epoxy matrix and basic bare glass surfaces. Obviously these interactions are enhanced by the presence of silane coupling agent on the glass surface, as predicted by the microcalorimetry data.

SEM fractography illustrated in Fig. 4 appears to indicate some areas where the resin and glass fibers are still in contact while other areas debonded during failure of the specimen. It is tempting to conclude that the areas that retain fiber–matrix adhesion are those where there exist silane patches such as shown in Fig. 1.

3.3. Teflon fiber surface chemistry and composite properties

In this example, three surfaces of differing acid–base character were synthesized

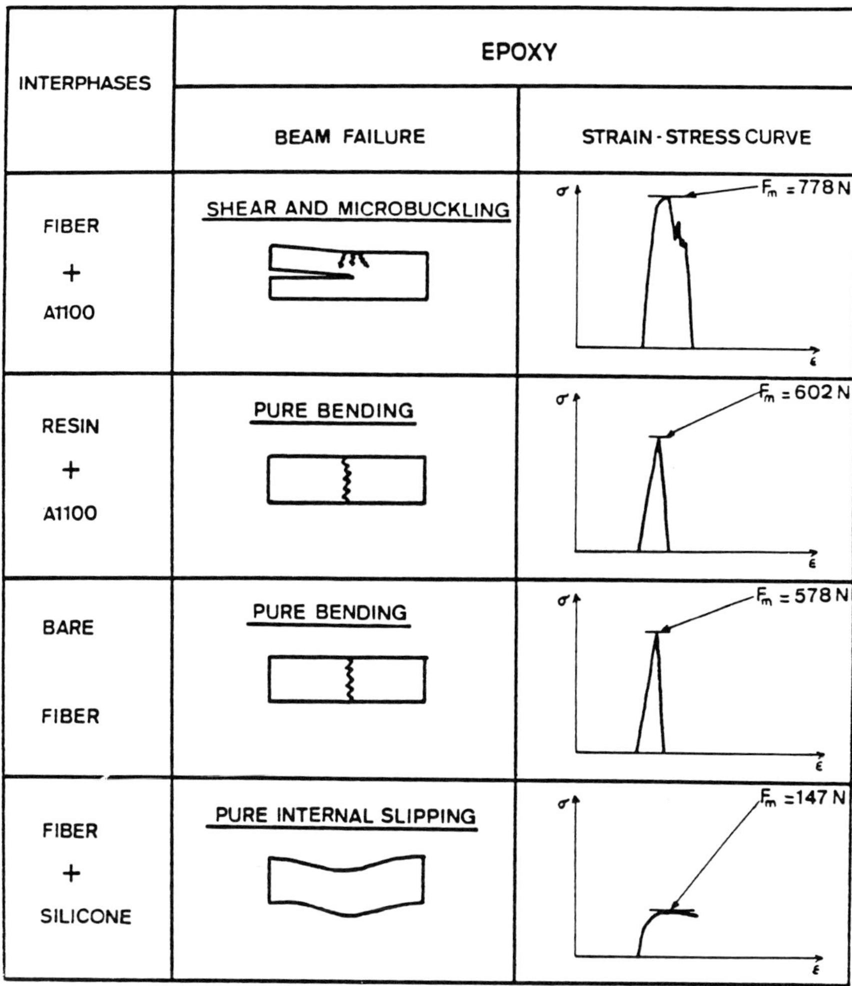

Figure 3. Schematic diagram summarizing the results of the short beam shear tests on unidirectional, E-glass fiber composites with epoxy resin matrix. The second column contains sketches of cross-sections of the short beam shear specimens, illustrating the principal mode of failure. Corresponding compressive stress strain diagrams are sketched in the third column with the failure loads indicated.

through a sequence of organic chemical reactions. Infrared and electron spectroscopies were used to determine the reaction product chemistries. The FTIR spectra juxtaposed in Fig. 5 shows removal of C=C absorption in the 1600–1670 cm^{-1} region and a sharpening of the peak in the O—H stretching region at 3200–3600 cm^{-1} upon hydroxylation. A marked decrease in absorption in the O—H stretching region along with an increase in absorption in the carbonyl stretching region around 1700–1760 cm^{-1} occurs after acetylation.

Figure 6 compares the XPS carbon and oxygen spectra of the hydroxylated and acetylated fibers. The broadening of the carbon peak both to the left ($\underline{C}$=O and right ($\underline{C}$—H) in the latter is consistent with replacement of the hydroxyls with esters. This is corroborated by the broadening of the oxygen peak on the right hand side, indicating the presence of new $\underline{O}$=C bonding.

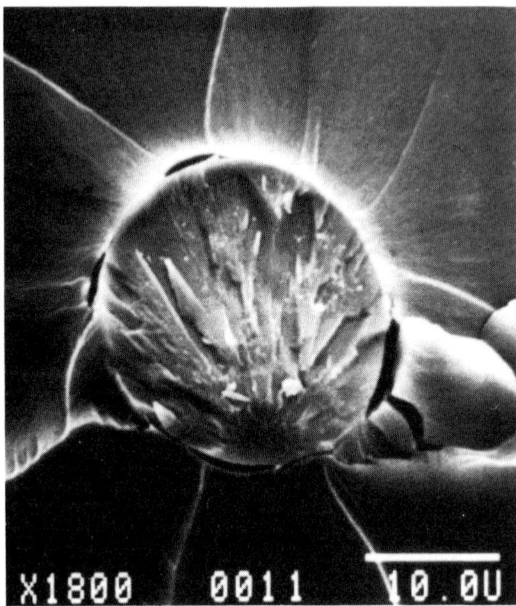

Figure 4. SEM photomicrograph of a fracture surface of a polyester composite made with A-174 methacryloxysilane treatment on the E-glass fiber. Note that there are regions where fiber-matrix delamination has occurred, but there are also locations where strong adhesion is retained.

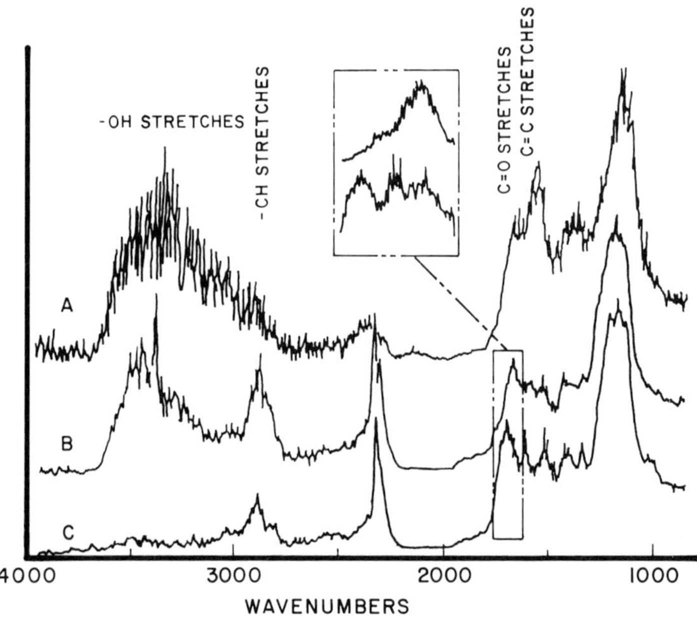

Figure 5. Diffuse reflectance FTIR spectra of chopped and bleached Teflon fibers after: (A) etching with sodium naphthalene in digylme, (B) hydroxylation with borane, and (C) esterification with acetic anhydride. The *y*-axis is relative absorbance in arbitrary units.

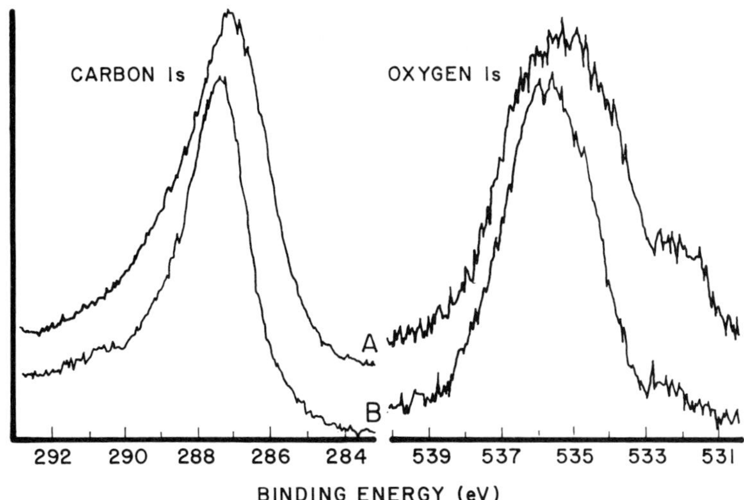

Figure 6. XPS high resolution spectra of chopped and bleached Teflon fibers after etching and: (A) esterification, and (B) hydroxylation. The *y*-axis is relative intensity in arbitrary units.

Thus both analytical techniques indicate qualitatively that we were successful in our attempts to introduce selectively hydroxyl or ester functional groups onto the chopped Teflon fiber surfaces. However, the analyses are only qualitative, and it is impossible to tell whether there are small amounts of other functional groups mixed with the primary constituents.

Electrophoretic mobilities of the hydroxylated and acetylated chopped Teflon fiber surfaces in tetrahydrofuran (THF) averaged -45 and $+31$ m/s/V/m $\times 10^{-6}$ respectively. Interpreting these results qualitatively in terms of electron donor–acceptor interactions between the particle surfaces and the solvent [21], the hydroxylated Teflon is more acidic than THF and the acetylated Teflon is more basic than THF. Unfortunately, attempts to determine the electrophoretic mobilities in methylene chloride, an acidic solvent, have been anomalous and irreproducible. Still we expect that there will be greater interfacial interaction between the acidic polymer matrix, chlorinated polyvinyl chloride (CPVC) and the acetylated (basic) Teflon fiber surfaces. That this was indeed the case is clear in the properties of the respective composite materials.

Stress–strain curves for the CPVC composites (Fig. 7) show that the properties are essentially identical whether the surfaces were perfluorinated or hydroxylated, proving that merely introducing polarity at the interface is not helpful when it creates an acid–base interaction. However, a marked increase in tensile strength and toughness (area under the stress–strain curve) is apparent when basic ester groups on the Teflon fiber surface interact with the acidic CPVC matrix.

SEM fractography confirmed that conclusion: Teflon fibrillation, indicative of good adhesion and stress transfer, was observed only for the acid–base paired interface. Clean interfacial failure occurred in the other two cases with neutral and acidic Teflon fiber surfaces. As illustrated in Fig. 8A, there is an obvious inter-facial separation, leaving an intact Teflon fiber protruding out of a 'socket' in the CPVC matrix. However, in the fractography of composites with acetylated

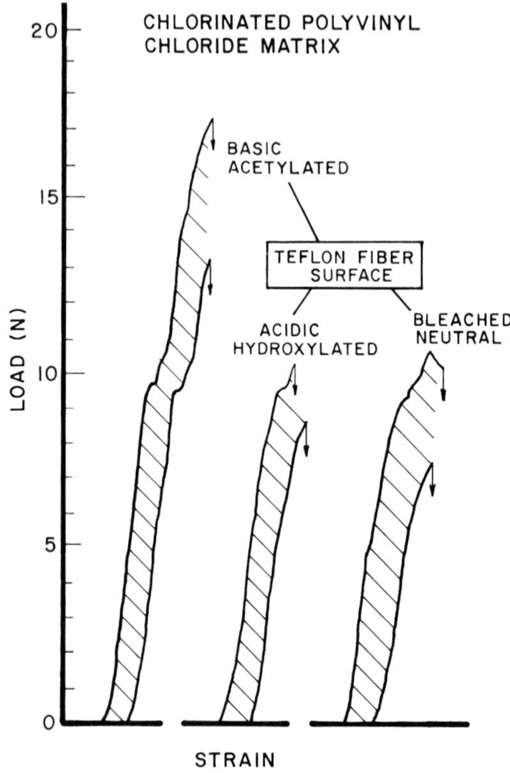

Figure 7. A side-by-side comparison of the three tensile stress-strain curves from composites made from the three types of Teflon fiber with a chlorinated polyvinylchloride (CPVC) matrix. The 'scatter bands' are bounded by the highest and lowest curves obtained from sets of five tensile specimens.

surfaces, no intact Teflon fibers could be found. Instead, as Fig. 8B illustrates, failure surfaces showed the interior of the Teflon fibers, where extensive sub-micron fibrillation (characteristic of PTFE deformation and failure), accompanied crack propagation through the composite. Clearly there was effective stress transfer from the matrix into the fibers in this case.

Based upon the fractography, the differences in the stress-strain curves can be explained in terms of micromechanics: Below a load of about 10 N, the tensile properties are dominated by the CPVC matrix, plus a small contribution from the Teflon fibers. London–Lifshitz forces do transfer some stress from the CPVC even into the control and the hydroxylated Teflon fibers, but not enough to initiate microfibrillation. The additional stress transferred by the acid–base interaction between the acetylated fibers and CPVC may be detectable in the higher and more uniform slope (modulus) of the stress–strain curve below the initial yield point. When deformation and cracking in the matrix occur in the latter composite, the load is taken up by the more ductile Teflon fibers, and significant additional force is necessary to generate the microfibrillation. In the former two composites, however, the first yielding in the matrix initiates fiber–matrix interfacial failure, and the entire specimen fractures.

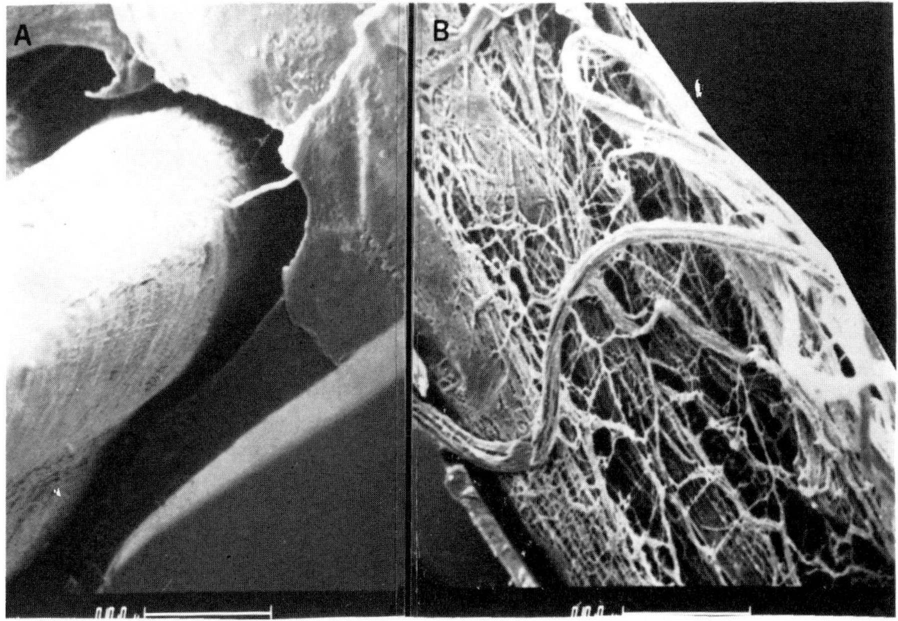

Figure 8. SEM photomicrographs of the fracture surfaces of two of the CPVC/Teflon fiber composite tensile specimens. In (A) with bleached (pure Teflon) surfaces, there is clear interfacial separation between fiber and matrix. However, in (B) with ester groups on the surface to participate in Lewis acid–base interactions with the CPVC matrix, stress was transferred into the Teflon fibers causing the extensive internal microfibrillation observed.

4. CONCLUSIONS

These two fibrous composite studies demonstrate the importance of acid–base interfacial interactions in the understanding and manipulation of stress transfer through fundamental interfacial forces. Among the details uncovered during these studies are:

(1) Glass fiber surfaces—primarily basic (oxides of aluminum and calcium).

(2) Silane coupling agents—adsorbed in patches: silanol interactions with the basic surface sites. Added both acidic (siloxane) sites and new, stronger basic sites (groups that terminate the organic 'tails' on the coupling agents).

(3) Unidirectional filament-wound composites—bare glass fibers (basic) bond to (acidic) epoxy matrix but not polyester). Coupling agents provide acidic sites for interaction with the basic polyester functional groups, as well as stronger basic sites to enhance interaction with the acidic epoxy sites.

(4) Chopped Teflon fiber composites—monofunctionalized with hydroxyl surfaces (acidic) or esterified surfaces (basic). Electrophoretic mobility in tetrahydrofuran corroborated acid/base nature. Tensile properties of (acidic) CPVC-matrix composites are *double* with basic fiber surfaces.

Acknowledgement

We appreciate the financial support of part of this work by PPG Industries, Inc.

REFERENCES

1. F. M. Fowkes, *J. Adhesion Sci. Technol.* **1**, 7 (1987).
2. C. J. van Oss, R. J. Good and M. K. Chaudhury, *J. Colloid Interface Sci.* **111**, 378 (1986).
3. F. M. Fowkes, D. O. Tischler, J. A. Wolfe, L. A. Lannigan, C. M. Ademu-John and M. J. Halliwell, *J. Polymer Sci.: Polymer Chem. Ed.* **22**, 547 (1984).
4. A. N. Gent and J. Schultz, *J. Adhesion* **3**, 281 (1972); D. Maugis, *J. Mater. Sci.* **5**, 679 (1986); K. L. Mittal, in: *Adhesion Science and Technology*, L. H. Lee (Ed.), Part A, pp. 129–168. Plenum Press, New York (1975).
5. J. A. Manson and J. T. Williams, *ACS Org. Coat. Plast. Chem. Preprints* **42**, 175 (1980).
6. J. A. Manson, J.-S. Lin and A. Tiburcio, *ACS Org. Coat. Plast. Chem. Preprints* **46**, 121 (1982).
7. M. J. Marmo, M. A. Mostafa, H. Jinnai, F. M. Fowkes and J. A. Manson, *Ind. Eng. Chem., Prod. Res. Dev.* **15**, 206 (1976).
8. F. M. Fowkes, D. C. McCarthy and D. O. Tischler, in: *Molecular Characterization of Composite Interfaces*, H. Ishida and G. Kumar (Eds), p. 401. Plenum Press, New York (1985).
9. F. M. Fowkes, D. W. Dwight, J. A. Manson, T. B. Lloyd, D. O. Tischler and B. A. Shah, *Mat. Res. Soc. Symp. Proc.* **119**, 223 (1988).
10. E. P. Plueddemann, *Silane Coupling Agents*. Plenum Press, New York (1982).
11. Y. Eckstein, *J. Adhesion Sci. Technol.* **3**, 337 (1989).
12. D. W. Dwight and W. M. Riggs, *J. Colloid Interface Sci.* **47**, 650 (1974).
13. C. A. Costello and T. J. McCarthy, *Macromolecules* **20**, 2819 (1987).
14. DuPont Bulletin TF-2, May 1978.
15. H. C. Brown and B. C. S. Rao, *J. Am. Chem. Soc.* **82**, 681 (1960); H. C. Brown and W. Korytnyk, *J. Am Chem. Soc.* **82**, 3866 (1960); H. C. Brown and A. W. Moerikofer, *J. Am. Chem. Soc.* **84**, 1478 (1962).
16. F. A. Carey and R. J. Sundberg, *Advanced Organic Chemistry, 2nd Ed., Part B: Reactions and Synthesis*. Plenum Press, New York (1983).
17. G. Hofle, W. Steglich and H. Vorbruggen, *Angew. Chem. Int. Engl. Ed.* **17**, 569 (1978).
18. F. M. Fowkes, D. W. Dwight, D. A. Cole and T. C. Huang, *J. Non-Cryst. Solids* **120**, 47 (1990).
19. S. Ross and N. Nguyen, *Langmuir* **4**, 1188 (1988).
20. D. W. Dwight, A. Brown and J. C. Vickerman, unpublished results: using a Vacuum Generators SIMS with gallium small spot ion gun, mass peak 73 (trimethyl silyl ion) was used to map the position of the coupling agent with respect to the glass fibers, which were mapped with the mass 23 (sodium ion) peak.
21. F. M. Fowkes, H. Jinnai, M. A. Mostafa, F. W. Anderson and R. J. Moore, in: *Colloids and Surfaces in Reprographic Technology*, M. Hair and M. D. Croucher (Eds), ACS Symposium Series #200, pp. 307–324. American Chemical Society, Washington, DC (1982).

Acid-Base Interactions, pp. 257-272
Eds. K.L. Mittal and H.R. Anderson, Jr.
©VSP 1991

Acid–base interactions in the interpretation of aramid composite and fabric mechanics

B. J. BRISCOE* and D. R. WILLIAMS

Department of Chemical Engineering and Chemical Technology, Imperial College, London, SW7 2BY, UK

Revised version received 7 August 1990

Abstract—This paper considers two facets of the role of surface properties in the performance and manufacture of aramid fabric–thermoset composites. The main factors considered are the intrinsic adhesion between the fibre and the matrix in the composite, and the fibre–fibre frictional interactions during its manufacture. The former is elucidated by using essentially the Fowkes acid–base approach and the latter by examining the mutual interfacial shear characteristics of the aramid fibres.

A number of aramid fibre surfaces have been prepared with a range of interfacial shear strengths and acid–base properties. The interfacial shear strength behaviour of these fibres was qualitatively determined by measuring the compressive properties of aramid fabrics, whilst the acid–base properties of the fibres were determined from wettability experiments. These fabrics were then incorporated into an epoxy resin to form laminates.

Contrary to the predictions based on the interfacial energetics, the toughest aramid composites, as determined by mode 1 fracture mechanics, were the laminates whose fibres had the lowest potential acid–base interactions, and were thus the least likely to have the greatest interfacial adhesion. This increased toughness has been attributed to enhanced fibre bridging across the crack zone which resulted in additional volumetric energy dissipation during fracture. This phenomenon is apparently promoted by the low interfacial shear strength between the fibres in the fabric during manufacture. Thus, for these aramid fibre laminates, any improvements in the toughness due to increased interfacial interactions were effectively masked by the more significant decreases in the volumetric energy dissipated. These changes are suggested to be a result of changes in fibre/fabric structure in the laminate which originate during manufacture.

Keywords: Acid–base; aramid; composites; interfacial shear strength; surfaces; friction; fibres.

1. INTRODUCTION

It is a great pleasure to provide a contribution to this Symposium† which honours Professor Fowkes's many contributions to surface science and, in particular, his seminal work on the topic of acid–base interactions in interfacial phenomena. The present paper adopts the Fowkes's models to evaluate from first principles the quality of the adhesion which is likely to be developed between a fabric and a matrix within a composite. The fibres studied were a range of specifically modified and characterized aramid materials which were ultimately incorporated within an epoxy resin matrix to form laminates.

The experimental data to be presented and discussed indicate that it is apparently not the fibre–matrix adhesion *per se* which influences the following facet of the mechanical properties of the resulting composite: interlaminar

---

*To whom correspondence should be addressed.

†Symposium held to honour Professor Fowkes at the 64th A.C.S. Colloid and Surface Science Meeting held at Lehigh University, June 18–20, 1990.

fracture. Rather, it seems that the surface modification of the fibre influences the migration characteristics of the filaments within the weave during fabrication. This process is controlled, in part, by the fibre–fibre friction. As a result, fibre bridges are created in the wake of the interlaminar crack which toughen the composite by effectively increasing the volumetric viscoelastic work associated with the fracture process. The natural question emerges as to what correlation might exist between the potential adhesion characteristics of the fibres in a resin matrix as determined by the acid–base methods, the fibre friction, the volumetric work, and the total fracture energy. This question is the topic of the present paper.

1.1. Composite interfaces

The primary role of the interface in composite materials is to transfer the load from the 'stiff' anisotropic high modulus fibres to the 'softer' isotropic matrix materials. The adhesion between fibre and matrix defines the efficiency of this stress transfer process. However, the fundamental nature of the processes which provide adhesion between the phases is not yet fully understood.

The pioneering work of Fowkes and his co-workers [1–5] has demonstrated that a critical component of the interfacial adhesion between two phases is due to specific acid–base interactions. Whilst the success of this approach has been reported for filled polymer systems and for flat interfaces, the application of these theories to fibre-reinforced composites has been less frequently reported. It is this established acid–base approach which provides in this study, a basis for making a firm prediction regarding the potential fibre–matrix adhesion in the composite.

Fowkes's studies have shown that specific chemical interactions which occur at interfaces can provide an important contribution to the interfacial adhesion. These specific interactions are well modelled by acid–base theories which consider the existence of weak chemical bonds with typical energies of tens of kJ/mol. These interactions involve both charge redistribution and charge sharing between the acidic and the basic species. For example, Fowkes has shown that silica powder is highly acidic owing to the presence of the hydroxyl groups, whilst polymethylmethacrylate was shown to be basic owing to the basic properties of the carbonyl oxygen atom.

This paper describes a series of experiments involving chemically modified 'Kevlar' fibres and weaves which are ultimately incorporated into an epoxy matrix. The acid–base characteristics of these fibres have been evaulated using established routes [1, 6] and this part of the study is not dealt with in detail. The subsidiary experiments described and reviewed in more detail include the friction characteristics of these fibres, the mechanical properties of the weaves, and the fracture properties of the manufactured laminates.

The paper will conclude by suggesting that it is the volumetric viscoelastic work associated with the fracture process, rather than the intrinsic adhesion of the fibre–matrix interface, which is the more significant contributor to the enhanced fracture toughness of these aramid composites. Specifically, it will be argued that fibre–fibre friction leads to a composite structure which exhibits a larger decrease in the volumetric viscoelastic energy than the corresponding increase due to acid–base interactions *per se* during interlaminar fracture. To pursue this argument, it is first necessary to consider briefly the empirical models of the fracture energy for multiphase polymeric materials.

1.2. Research strategy

The research strategy adopted in this study is based upon well-established models of the contact adhesion of viscoelastic solids. These are typified by the work of Gent and Petrich [7], Maugis [8], and Andrews and Kinloch [9]. These workers have empirically concluded that, to a good approximation, the total adhesive fracture energy, G, for a material, is the product of the thermodynamic interfacial work, W, and a viscoelastic volume work, ϕ, which is dissipated in the crack zone:

$$G = W\phi. \tag{1}$$

The study to be described tacitly assumes that a major fraction of the rupture process will correspond to interfacial fracture between the fibre and the matrix. Clearly simple geometric constraints will not allow a crack to propagate entirely at the interface. Observations of the fracture surfaces indicate that a significant amount of apparent interfacial failure occurs in these systems.

In this study the interfacial work is ascribed as being due primarily to acid–base interactions in line with the findings of Fowkes and his co-workers. This characteristic is defined in terms of the acid–base component of the work of adhesion, W^{a-b}, and is to be experimentally deduced using fibre wetting experiments for a range of aramid fibres with differing surface properties. These measurements then allow the interfacial fracture energies of epoxy composite materials containing these fibres to be qualitatively ranked on the basis of the measurements of W^{a-b} for an assumed constant viscoelastic volumetric work.

The other component of the research plan is to measure the fracture energy of the epoxy–aramid materials directly using classical experimental fracture mechanics. These results would allow a direct comparison between the experimental study and results based on the empirical model. Finally, the validity of these two differing routes could then be evaluated. Figure 1 outlines the basic elements of the study to be reported.

1.3. Interfacial shear strength and friction of polymers

The interfacial shear properties of polymeric surfaces or fibres are of fundamental importance in any system of contacting bodies which involves the movement of one surface over another. The primary quantity which characterizes such movements is the interfacial shear strength, τ_i. The interfacial shear strength may be described as the interfacial force per unit area of contact, or, alternatively, as the energy dissipated per unit area of contact per unit sliding length. In the following sections the nature of friction and its relationship to the interfacial shear strength will be discussed.

The frictional interactions between solid bodies are generally the result of two main mechanisms for energy dissipation. The first mechanism involves the dissipation of energy in the interfacial layers between the two phases, and results from interfacial shearing of the bodies at the points of contact. The second

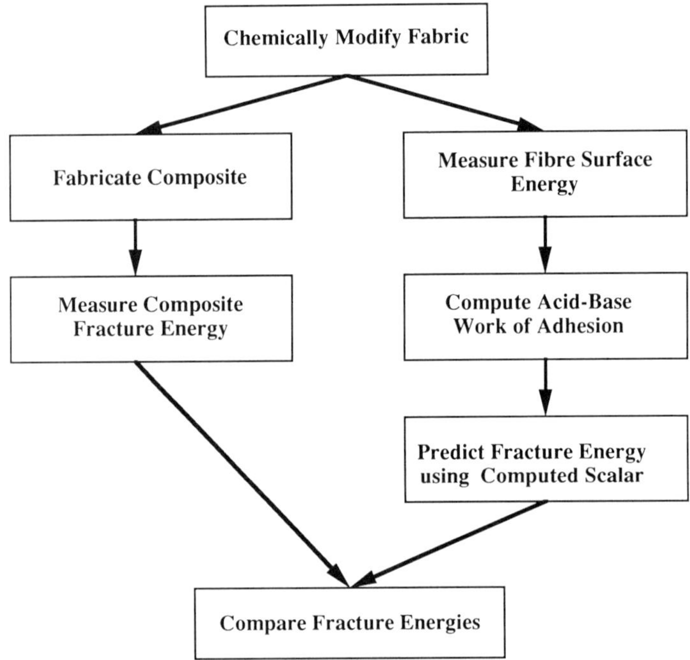

Figure 1. Schematic diagram of the Research Strategy.

mechanism involves the energy dissipation in relatively large sub-surface volumes within the interacting bodies. Energy dissipation in this case is the result of the surface asperities of the harder solid deforming the surface of the softer solid phase. Localized energy losses occur due to viscous, viscoelastic, or plastic processes. This model of interfacial friction has been reviewed in detail by Briscoe [10] and by Briscoe and Tabor [11].

These two mechanisms result in the adhesion and the bulk deformation components of the total frictional work. Both processes are generally treated as independent and as additive energy dissipation mechanisms [11, 12] and may also be distinguished by the differing sizes of the energy dissipation region. In the case of adhesional friction, the zone for energy dissipation will correspond to the distances through which surface stresses propagate and will thus be generally less than 10 nm in thickness, whilst for ploughing type deformations, a greater depth is expected and depends on the size of the bodies and the contact area. Figure 2 shows the two primary zones for energy dissipation for a rigid asperity on soft and hard polymer substrates.

The relative importance of these two mechanisms is dependent on the chemical and physical properties of the materials, the contact geometry, as well as the surface roughness and the type of relative motion between the bodies. For relatively smooth surfaces, such as high modulus fibres, in contact with a glassy matrix phase, the adhesional component of the friction would be expected to be the dominant term. Similarly, this component will dominate the friction processes for fibre–fibre contacts in weave. It is the latter which is regarded as important during the fabrication of the composite. It is also the same parameter which prescribes the interfacial shear strength of the fibre–matrix interface.

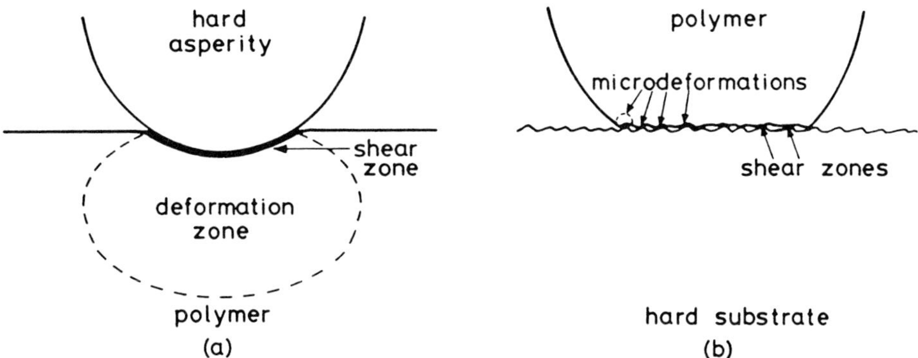

Figure 2. Schematic diagrams of a rigid asperity sliding against a soft polymer (a) and of a soft asperity sliding against a hard polymer (b). Energy is dissipated in the interfacial shear zone as well as in the bulk deformation zone.

1.4. *The interfacial component of friction*

The contribution to the interfacial friction due to adhesion is given by

$$F = A\tau_i, \tag{2}$$

where F is the frictional force per unit area and τ_i is the interfacial or adhesional shear stress.

The quantity τ_i is sensitive to the chemical composition of the fibre surface and also to the surface morphology for an inter-filament contact. The contact area, A, is defined by the surface topography but to a lesser extent by the gross contact geometry, the normal load, W, and the bulk mechanical properties of the fibres. For the present systems this property will be a relatively low strain characteristic such as a time-dependent transverse elastic modulus. Other factors such as the auto-adhesion of the contacts may also be important. Within the epoxy matrix the actual contact area is equal to the geometric contact area.

The main point, however, is that *surface* chemical modifications will influence the friction mainly as a result of the changes introduced in τ_i. Topographical modifications will also influence the magnitude of A. Later in this study we have attributed the changes in the fibre–fibre friction solely to the influence of τ_i.

1.5. *Factors affecting* τ_i

The most important variable affecting τ_i is the mean contact pressure, P, which is defined as the normal load W per unit area A. A number of studies of τ_i as a function of P for polymers have been reported [11, 13–15]. Briscoe *et al.* [13] have demonstrated that this dependence is well modelled by the expression

$$\tau_i = \tau_0 + \alpha P \tag{3}$$

where τ_0 is the interfacial shear strength at zero contact pressure and α is the pressure coefficient of the shear strength.

Values for τ_0 and α have been reviewed in detail by Briscoe [10]. Alternatively, equation (3) may be rewritten to explicitly define the coefficient of friction [10]:

$$\mu = \tau_0/P + \alpha. \tag{4}$$

For large contact pressures, this equation reduces to

$$\mu \approx \alpha. \tag{5}$$

Only a few studies of the relationship between the morphology and chemistry of the substrate and the values of τ_i have been undertaken. Amuzu *et al.* [14, 15] have reported on the variations in τ_i for a homologous series of poly *n*-alkylmethacrylates which were rationalized in terms of the molecular engagements of protruding segments of the different size pendant alkyl groups. Pooley and Tabor [16] have interpreted the interfacial shear properties in terms of the molecular profile or 'roughness' of the polymer molecules. PTFE has a smooth helical structure which can be perturbed by the introduction of perfluoromethyl groups in place of the fluorine atoms. The frictional forces of PTFE were observed to increase significantly with the introduction of these large substituents. Smith [17] has reported on the effects of polymer conformation and morphology on the interfacial shear behaviour of a wide range of polymeric films.

1.6. *Fibre friction studies*

Two experimental techniques have been developed which allow the friction of a single monofilament to be investigated. These techniques involve a monofilament in a point contact geometry or in an extended line contact geometry [18–20]. It is established that fibre–fibre friction plays a crucial role in the mechanics of fabrics [16, 21].

The original seminal work on the modelling of fabric properties as a function of the geometry of fabric weave was reported by Pierce [22]. The basic nature of this model remains unchanged, though it has since been modified by a number of other workers. It forms the basis of most analyses of specific fabric mechanical properties. The mechanical properties of fabrics are mathematically described in terms of the yarn spacing, the weave angle, the yarn diameters, the crimp heights, and the thread spacings. Research on the tensile and bending properties of fabrics has been carried out by a relatively small number of workers: Hearle, Grosberg, Leaf, and Abbot. The tensile characteristics of typical fabrics have been analysed by Hearle *et al.* [23], as well as by Grosberg and Swani [24] and Leaf [25]. The bending properties of fabrics have been studied by Grosberg and co-workers [24–26] as well as by Gibson and Postle [27]. In the cases of both tensile and bending deformations, these researchers have concluded that inter-fibre and inter-yarn friction is a very significant factor in the mechanical response of the fabrics. The ability of fibres and yarns to migrate, and thus accommodate the imposed stresses, depends directly on the frictional forces associated with the fibre interactions. However, none of the models describes directly the effects of fibre friction, but rather include this behaviour within a general hysteresis parameter.

Fibre migration has been considered by a number of workers in the context of twisted yarn systems. These models have been reviewed by Motamedi [28], who points out that none of these models accounts for the inter-fibre frictional forces. He has demonstrated the importance of frictional effects in determining the mechanical behaviour of 'Kevlar' fabrics under indentation loading. The effects of friction were successfully modelled on the basis of changes in the effective tensile modulus of the fibre assembly. We have also shown that the frictional properties of similar fabrics with or without lubricants markedly affect the static modulus and also the ballistic energy dissipation capacity. Low friction at the filament junctions reduces the tensile modulus of the yarns and also the effective modulus of the fabric [29].

2. THE EXPERIMENTAL STUDY

2.1. Modification of the fibre surface properties

Three surface-modified aramid fibres were prepared with a range of surface properties. These modifications were designed to yield a range of acid–base characteristics and interfacial shear strengths.

Fibres with the lowest interfacial shear strength and acid–base nature were created by coating the fibres with poly(dimethyl siloxane) (PDMS), a well-known lubricant. Fibres with an increased interfacial shear strength relative to the 'as received' 'Kevlar' fibres were prepared by modifying the functional chemistry of the fibre surface. Two fibre systems of this type are reported in this study. The first aramid fibre sample involved the grafting of diol groups onto the fibre surface via a solution reaction graft route [30], whilst the second technique involved the chemical etching of amorphous materials from the fibre surface, resulting in a 'clean' poly(phenylene terephthalamide) surface. Both of these modifications were confirmed by X-ray photoelectron spectroscopy. The etched/cleaned samples exhibited an enhanced surface nitrogen content due to the removal of amorphous oxidized material, whilst the diol-treated material clearly showed the presence of diol oxygen groups on the fibre surface [30].

The results shown in Table 1 are based on the advancing contact angle results obtained with water using the Wilhelmy balance technique [6]. These measurements utilized single aramid monofilaments attached to a Cahn Model 322 wetting balance. Wetting results with hexadecane, methylene iodide, and water were used to estimate the fibre surface energetics. The works of adhesion with the various liquids were simply calculated using Young's equation with the advancing contact angle data. The non-dispersive component of the work of adhesion was ascribed to acid–base interactions in line with the Fowkes approach and is given by W^{a-b} shown in Table 1.

The results in Table 1 yield a clear ordering of the potential acid–base interactions for the various aramid fibres produced. It is presumed that the epoxy prepolymer is chemically amphoteric owing to the presence of a significant

Table 1.

Surface energetics of modified 'Kevlar 49' fibres

'Kevlar' fibres	W^{H_2O} (mJ/m^2)	W^{a-b} (mJ/m^2)
PDMS coated[a]	48	0
'As received'	95.0	39.6
Diol graft	112.0	53.5
Etched/cleaned	125.0	60.2

[a] Literature values for PDMS [31].

number of hydroxyl groups. It is thus predicted that this will result in the same ordering of acid–base activity for the fibres with an epoxy, as exhibited by the amphoteric water used in the fibre wetting experiments.

2.2. Composite fabrication

Each layer of aramid fabric (4 × 1 satin weave) was melt-pressed with four layers of an epoxy film (Ciba-Geigy 913). The aramid fabrics were then fabricated into axisymmetric 8 ply [0, 90] laminates in an autoclave using a standard 120°C cure cycle. The fabricated laminates had a nominal fibre volume of 55% and a thickness of 2 mm. Ultrasonic C scanning results revealed no evidence of flaws in the tested materials. Mode I fracture was initiated in the inter-laminar plane of the laminate normal to the [0, 90] axes of the prepregs via a foil precrack. The low strain flexural modulus and the loss tangent of these prepared materials were indistinguishable within experimental error.

2.3. Experimental fracture mechanics

The mechanical toughness of these materials was then determined via mode I fracture using double cantilever beam specimens. It was observed during fracture that significant toughening of the material with respect to fracture occurred owing to migration of fibres from the fabric layers resulting in fibre bridging in the crack zone [32]. Figures 3, 4, and 5 illustrate the fracture behaviour of the laminates manufactured from the fibres with the highest and lowest interfacial shear strengths. Four samples were tested for each laminate system and the respective data are shown below. The important feature to note is the maximum values for the fracture energy, G_{IC}, recorded. The low shear strength, low acid–base content fibres have a fracture energy over two times *greater* than the high fibre shear strength, high acid–base character fibres. The data obviously contradict the simple notion that a high strength interface directly results in a tough composite.

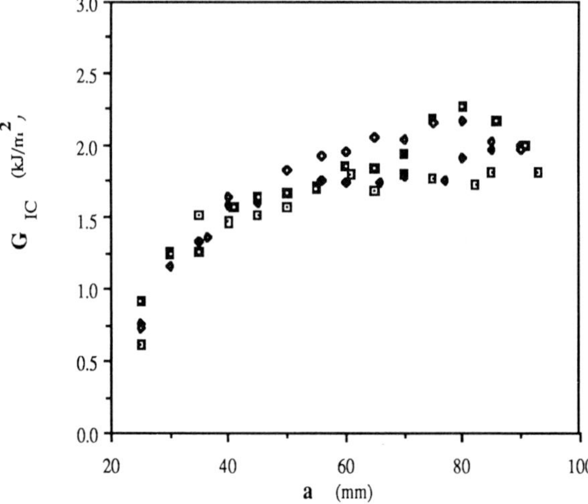

Figure 3. G_{IC} vs. crack length, *a*, for a laminate made from 'as received' 'Kevlar' fibres. Results are shown for multiple samples.

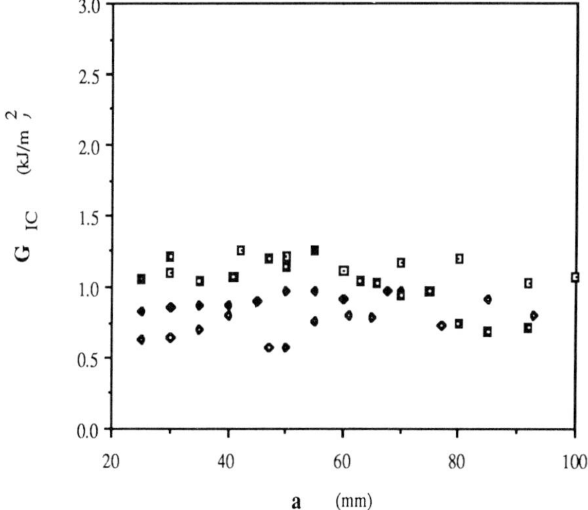

Figure 4. G_{IC} vs. crack length, *a*, for a laminate made from etched/cleaned 'Kevlar' fibres. Results are shown for multiple samples.

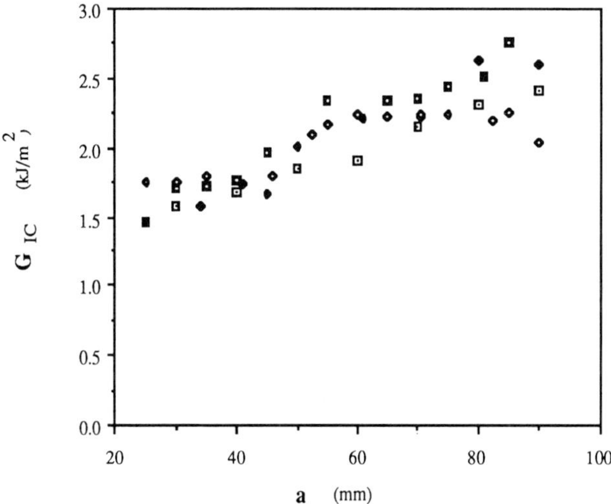

Figure 5. G_{IC} vs. crack length, *a*, for a laminate made from PDMS coated 'Kevlar' fibres. Results are shown for multiple samples.

These figures also indicate an apparent increase in the fracture energy with increasing crack length. This behaviour is ascribed to fibre bridging which occurs in the wake of the crack as it propagates through the inter-laminar regions. A detailed analysis of the fracture data has allowed the size of the fibre bridging zone, a_b, to be estimated. This estimation was accomplished by plotting a reduced bending force as a function of the crack length followed by a subsequent extrapolation to zero load, and is detailed elsewhere [32]. Such plots are shown in Figs 6 and 7. The results obtained using this analysis are summarized in Table 2 and clearly indicate that a_b varies significantly amongst the various samples. The experimental uncertainty in the estimation of a_b is between 1 and 2 mm.

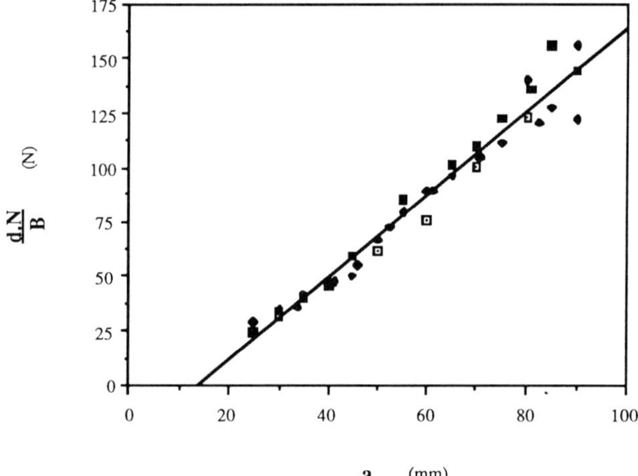

Figure 6. Reduced bending force as a function of crack length, a, for a laminate fabricated from 'as received' Kevlar. Results are shown for multiple samples.

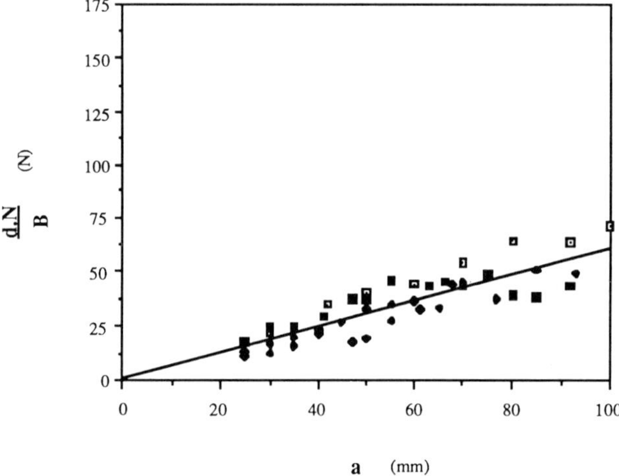

Figure 7. Reduced bending force as a function of crack length, a, for a laminate fabricated from etched/cleaned Kevlar. Results are shown for multiple samples.

Table 2.
The effective crack zone length, a_b, for various 'Kevlar'–epoxy laminates. R denotes the coefficient of fit for the linear regression data analysis [32]

'Kevlar' fibres	a_b (mm)	R
'As received'	13	0.97
Poly(dimethyl siloxane) coated	14	0.98
Diol grafted	10	0.99
Etched/cleaned	-2	0.87

2.4. Fabric compression studies

The compressional properties of a number of the surface-modified 'Kevlar' fabrics were measured using an Instron testing machine. Eight layers of the four different 'Kevlar' fabrics, used in the fabrications of laminates, were horizontally stacked together. The fabric stack was then compressed against a flat platen using a small circular indentor of 25 mm in diameter, with a flat face. The rate of compressive straining was defined by the movement of the indentor attached to the instrument crosshead (0.5 mm/min). The initial separation between the platen and the indentor was 3.5 mm and the fabric stack was strained until a load of 60 kg was recorded. This level of loading corresponds to a mean pressure of about 1.2 MPa. The impregnated fabrics used to fabricate the composite panels are subjected to a mean pressure of 0.6 MPa during autoclaving. The fabric samples were then unloaded at the same crosshead speed so that both the loading and the unloading compressive properties of the fabric were measured. The results obtained are shown in Fig. 8 as a function of the compressive strain and stress. The fabrics were loaded until a strain of 45% was recorded and then the samples were unloaded at the same strain rate.

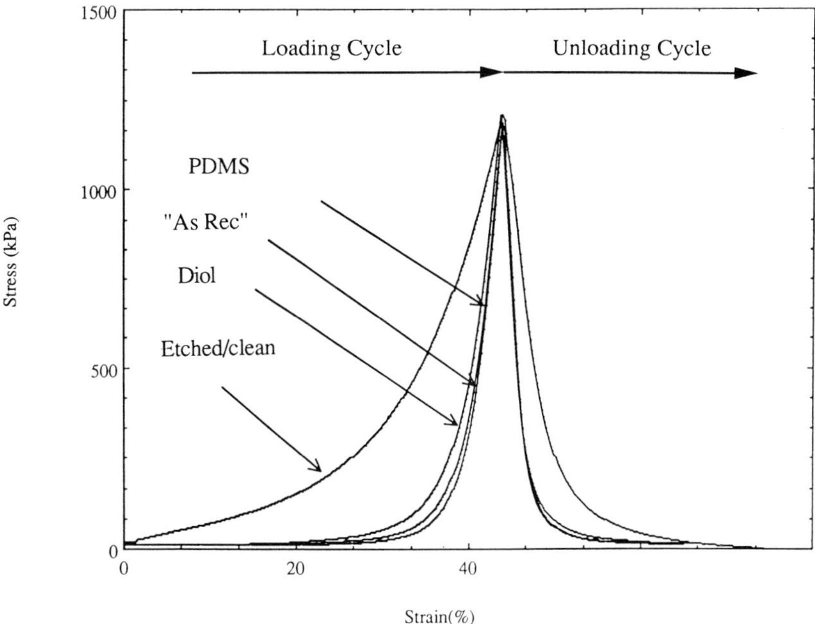

Figure 8. Compressive loading and unloading cycle for a range of surface modified 'Kevlar' fabrics.

3. DISCUSSION

This study has adopted the acid–base models to provide a firm basis for predicting the quality of the adhesion created between a range of surface-modified fibres and an epoxy resin matrix. The Fowkes's models have provided this vehicle and allowed a ranking of the expected strength of the fibre–matrix adhesion to be generated with confidence, and these are given in Table 1. The changes introduced in the fibre surface not only influence their acid–base character, but also the fibre–fibre friction and fibre topography. A further effect of the surface modification of the fibres may be to influence the quality of the wetting by the epoxy, and this could affect the defect concentration in the laminates. In addition, the fibre–matrix shear strength is also undoubtably changed.

The fracture studies produced at first sight the unexpected result that the potentially *stronger* fibre–matrix interfaces produced the *weakest* composites. We have interpreted this result as arising from the influence of the changes in the fibre–fibre friction during processing on the *structure* of the resulting composite.

It is this resultant change in structure which is argued to lead to changes in the net volumetric energy associated with the fracture. The results obtained for the size of the fibre bridging zone directly indicate a decrease in the size of this zone as the potential for adhesion between the matrix and fibres increases. The PDMS-treated material exhibited the largest damage zone and the etched/cleaned material exhibited the smallest zone, which is in accordance with the expected trends in the volumetric work. The surfaces with the lowest acid–base character have, it turns out, the lowest fibre–fibre friction in the weave during manufacture. This low friction appears to promote fibre migration during manufacture and this condition facilitates fibre bridging effects. It is these fibre bridges which we conjecture lead to the formation of a stronger composite. Whilst it has been argued that the changes in the fibre surfaces have resulted in changes in the fibre/fabric structure, the mechanism for such changes is not immediately obvious. In the following section, the significance of the interfacial shear strength of the fibres as a potential factor in controlling the fibre/fabric structure is addressed.

3.1. Measurement of the fibre interfacial shear strength

As was reviewed earlier in this paper, a number of methods are available for measuring the fibre friction, and thus the fibre interfacial shear strength. Although the single and multiple-contact single fibre friction experiments are fundamentally the most sound method for estimating τ_i, both the experiments and the data analysis are difficult. Therefore a simpler, albeit less quantitative, method was used in this study.

It was previously established that the mechanical properties of fabrics are directly related to the inter-fibre friction, although an analytical relationship between these properties has not been reported. In this section, the compressive properties of surface-modified aramid fabrics are reported as a qualitative measure of the fibre interfacial shear strength. The compressive deformation of aramid fabric is a particularly appropriate method to sense fibre–fibre friction, since the fabrics undergo a similar deformation during the fabrication of laminate specimens in the autoclave.

Figure 8 shows that the compressive properties of the various 'Kevlar' fabrics are distinctly different. The shape of the loading curve indicates that the etched/cleaned fabric is a significantly more compliant system than the other fabrics. The other three fabrics show a steeper gradient in the load vs. strain curve, indicating that these materials are effectively stiffer systems mechanically. This type of behaviour is normally associated with fibre assemblies in which the lubricated nature of the inter-fibre contact improves the effective stress transmission within the system. Using the compliance of the fibre assembly as a qualitative measure of the inter-fibre friction, a ranking order of the frictional coefficients, and thus interfacial shear strengths, can be estimated for the fibres:

$$\text{Etched/cleaned} \gg \text{diol} > \text{'as received'} > \text{PDMS}.$$

Therefore, it is clear that the surface modifications have affected the interfacial rheology and the friction properties, as is reflected in the compression properties of the fabrics. Furthermore, the order of fabric compliance in compression is consistent with the τ_i rankings anticipated in Table 3.

Table 3.
List of predicted values of the interfacial shear strength, τ_i, increasing in magnitude from the top of the list, based on ref. [17]

Poly(dimethyl siloxane) coated	10^3–10^4 Pa	↓ increasing τ_i
'As received' fibres	10^8 Pa	
Diol grafted	10^8 Pa	
Etched/cleaned	10^8–10^9 Pa	↑ decreasing τ_i

Thus, the qualitative predictions shown in Table 3 are in general agreement with the experimental interfacial shear strength results based on the fabric compression properties. At this stage it is appropriate to consider how the interfacial shear strength of the fibres might relate to the occurrence of fibre bridging during composite fracture.

By considering the dependence of τ_i on factors such as polymer morphology, bulk yield stress, and structure as established by Smith [17], it is possible to postulate a qualitative ranking of τ_i for the various aramid fibre surfaces prepared. This list is shown in Table 3. For most of the materials in this list their ranking will primarily reflect the bulk shear properties of the material. 'Kevlar' is known to have a skin/core structure, with highly oriented crystalline material in the core regions and amorphous polymer in the outer skin regions of the fibres. In the case of the grafted 'Kevlar' fibres, the removal of amorphous material from the surface of the fibres during the reaction [33] would be expected to expose crystalline regions of fibre surface with increased values of τ_i.

Two explanations for the variation in the extent of fibre bridging are immediately available. A simple pragmatic argument may be advanced on the basis of the fabric compression experiments. Fabrics used in the formation of laminates are typically exposed to uniaxial and hydrostatic pressures of between 0.5 and 1.0 MPa during autoclaved fabrication. The fabric compression results shown in Fig. 8 indicate that for this pressure range the normal compressive strain of the various surface-treated fabrics varied significantly. The lubricated fabrics underwent a more significant compressive strain deformation than the etched/cleaned materials for a specific compressive loading. A direct consequence of this observation is that the lubricated material will have accommodated significant fibre migration, across the fabric interply zones, during compression. It is argued that this same process occurs during manufacture and thus accounts for the variations in the extent of fibre bridging. An alternative explanation is that the low interfacial shear stress systems will allow more fibre pull-out and migration *during* fracture. It is also possible that the potential void size and concentration are higher in the least compressible fabric system. The higher void concentration would thus provide more points of stress concentration in the system.

3.2. Acid–base interactions and τ_i: the empirical correlation

A number of attempts have been reported for relating the interfacial mechanical behaviour of polymer films with the molecular structure of the polymer. However, these models are based on molecular topography rather than on the chemical activity of the polymer. No attempts have been reported to introduce the chemical interactions and/or acid–base concepts into this area.

Table 4 shows the pressure coefficient of the shear strength, α, and various surface chemical parameters for a range of polymer surfaces. The contact angles of water and methylene iodide were used to estimate γ^d, γ^{a-b}, and W^{a-b} using the same approach as that reported by Fowkes and Maruchi [1] with literature data [31].

Table 4.
Pressure coefficient of the shear strength, α [10], the interfacial shear strength τ_i [10], the dispersive, γ^d [31], and the acid–base, γ^{a-b} [1], components of the surface energy and the acid–base component of the work of adhesion with water, W^{a-b} [1]

Polymer film	τ_i (10^6 Pa)	α	γ^d (mJ/m²)	γ^{a-b} (mJ/m²)	W^{a-b} (mJ/m²)
Poly(dimethyl siloxane)	0.1	0.01	20	0	0
Low density polyethylene	6	0.14	33	1	2
High density polyethylene	2.5	0.10	33	1	2
Polyethylenetetrafluoride	1	0.08	19	2	5
Polypropylene	5	0.17	30	1	2
Polymethylmethacrylate	10	0.36	39	10	26
Polyvinyl chloride	11	0.57	41.5	9	16
Polystyrene	4	0.45	42	4.2	10
Nylon 6	12	0.51	39	15.5	38

Whilst the results in Table 4 do not show a clear quantitative relationship between α and τ_i and the surface chemical parameters tabulated, a number of qualitative conculsions are, however, evident. First the pressure coefficient of the shear strength, α, is not strongly dependent on the dispersive component of the surface energy, γ^d. More significantly, the polymers which exhibit the largest values of α also exhibit the largest values for acid–base interactions, as quantified by both γ^{a-b} and W^{a-b}. It is thus tentatively suggested that the magnitude of the pressure dependence of the interfacial shear strength, and hence the friction coefficient, gives a better correlation with the likely presence of acid–base interactions with the polymer surface.

This paper has examined the correlations between acid–base character and fibre–fibre friction and the data are consistent with this view. Full confirmation of the value of this argument will require extensive and detailed fractography as well as a detailed examination of the structures of the composites. This study is currently in progress.

4. CONCLUSION

This study highlights both the value of the acid–base approach and its potential limitations. Acid–base theory provides a basis for predicting the quality of solid–solid interfaces and how this quality may be changed through surface modification. The application does, however, generally presuppose that the quality of this interfacial adhesion will be the governing variable in a particular process and that other mechanisms for energy dissipation remain invariant. In this particular study it has been identified that both the quality of the interfacial adhesion *and* the viscoelastic volumetric work associated with fracture varied with the introduction of fibre surface treatment. More significantly, variations in the energies associated with the volumetric dissipation processes during inter-laminar fracture were more significant than predicted variations in the quality of this interfacial adhesion. We have envisaged that the magnitude of the volumetric work will be a function of the structure or the size of the dissipation zone associated with the crack tip or due to a change in the dissipation character of the crack zone *per se.* In this study we have thus deduced that the fracture process is not directly influenced by the interface quality, but rather by the structure of the system, and hence the size and the dissipative character of the deformation zone associated with the crack. Certain important facets of this structure, such as the creation of fibre bridging, we have tentatively deduced as being controlled by the fibre–fibre friction during manufacture. It is possible that the fibre–matrix friction may influence the dissipative character of the matrix adjacent to and deformed by the crack tip. We also recognize that the fibre–fibre frictional character may also control the defect concentration in the composite by influencing the access of the resin to the interior of the fabric during fabrication. Amongst the various alternatives suggested we favour, on balance, the structure or size of the deformation region as the most credible effect which has an established precedent. The C scan results and the flexural results are not consistent with a high flaw or void content; the flaws may, of course, be beyond detection. Similarly, as there is no significant change in the bulk viscoelastic character of the composites we may tentatively conclude that it is unlikely that the dissipative character of the matrix per unit volume has been altered. The viscoelastic properties measured were, of course, for low imposed strains and it is possible that additional volumetric work may be dissipated at high strains via fibre–matrix migration processes.

All the processes described above are in principle affected by the frictional characteristics of the fibre surfaces. As a matter of fact, the interface friction is apparently inversely correlated with the extent of acid–base interactions at the matrix–fibre interfaces. This correlation, which we suggest is a general one, seems to be the basis of the observed inverse interrelationship between the predicted extent of the acid–base interactions and the fracture toughness of aramid composites. In some manner, of which various options have been proposed, it is the frictional characteristics of the fibres, rather than their surface energy, which appear to have the major influence on the fracture toughness of the present systems.

Acknowledgements

The generous financial support provided by Materials Research Laboratory (Australia) for DRW is gratefully acknowledged, as well as the financial assistance provided by Du Pont (USA) for this research programme. The provision of materials by Ciba-Geigy (UK) and Du Pont (USA) is also acknowledged.

REFERENCES

1. F. M. Fowkes and S. Maruchi, *ACS Org. Coat. Plast. Chem. Prepr.* **37**, 605, (1977).
2. F. M. Fowkes and M. A. Mostafa, *Ind. Eng. Chem. Prod. Res. Dev.* **17**, 3 (1978).
3. F. M. Fowkes, in: *Adhesion and Adsorption of Polymers*, L. H. Lee (Ed.), Part A, p. 43. Plenum Press, New York (1980).
4. F. M. Fowkes, in: *Physicochemical Aspects of Polymer Surfaces*, K. L. Mittal (Ed.), vol. 2, p. 583. Plenum Press, New York (1983).
5. F. M. Fowkes, D. O. Tischler, J. A. Wolfe, L. A. Lannigan, C. M. Ademu-John and M. J. Halliwell, *J. Polym. Sci., Chem. Ed.* **22**, 547 (1984).
6. B. Miller, P. Muri and S. Hedvat, *Colloids Surf.* **6**, 49–61 (1983).
7. A. N. Gent and R. P. Petrich, *Proc. R. Soc. London, Ser. A* **310**, 433 (1969).
8. D. Maugis, in: *Microscopic Aspects of Adhesion and Lubrication*, J. M. Georges (Ed.). Elsevier, Amsterdam (1982).
9. E. H. Andrews and A. J. Kinloch, *Proc. R. Soc. London, Ser. A* **332**, 385 (1972).
10. B. J. Briscoe, in: *Friction and Wear of Composite Materials*, K. Freidrich (Ed.), p. 25. Elsevier, Amsterdam (1985).
11. B. J. Briscoe and D. Tabor, in: *Polymer Surfaces*, D. T. Clark and W. J. Feast (Eds), p. 1. John Wiley, Chichester (1978).
12. F. P. Bowden and D. Tabor, *The Friction and Lubrication of Solids*, Part 1. Clarendon Press, Oxford (1950).
13. B. J. Briscoe, B. Scruton and F. R. Willis, *Proc. R. Soc. London, Ser. A* **333**, 99 (1973).
14. J. K. A. Amuzu, B. J. Briscoe and D. Tabor, *Trans. ASLE* **20**, 152 (1977).
15. J. K. A. Amuzu, B. J. Briscoe and D. Tabor, *Trans. ASLE* **20**, 354 (1977).
16. C. M. Pooley and D. Tabor, *Proc. R. Soc. London, Ser. A* **329**, 251 (1972).
17. A. C. Smith, Ph.D. Dissertation, Imperial College, London (1979).
18. I. C. Roselman and D. Tabor, *J. Phys. D, Appl. Phys.* **9**, 2517 (1976).
19. M. W. Pascoe and D. Tabor, *Research* **8**, S15 (1955).
20. S. L. Kremnitzer, Ph.D. Dissertation, University of Cambridge, Cambridge (1978).
21. M. J. Schick (Ed.), in: *Surface Characteristics of Fibres and Textiles*, Fibre Science Series, vol. 1, Chap. 1. Marcel Dekker, New York (1975).
22. F. T. Pierce, *J. Textile Inst.* **28**, T45 (1937).
23. J. W. S. Hearle, P. Grosberg and S. Backer, *Structural Mechanics of Fibres, Yarns and Fabrics*, vol. 1. Wiley Interscience, New York (1969).
24. P. Grosberg and N. M. Swani, *Textile Res. J.* **36**, 332 (1966).
25. G. A. V. Leaf, in: *Mechanics of Flexible Fibre Assemblies*, W. J. S. Hearle, P. Thwaites and C. L. Amirbayat (Eds), NATO Advanced Study Institute Series. Sijthoff and Noordhoff, Alphen aan den Rijn (1980).
26. C. M. Abbott, P. Grosberg and G. A. V. Leaf, *Textile. Res. J.* **41**, 345 (1971).
27. V. L. Gibson and R. Postle, *Textile Res. J.* **48**, 14 (1978).
28. F. M. Motamedi, Ph.D. Dissertation, Imperial College, London (1989).
29. B. J. Briscoe, F. M. Motamedi and D. R. Williams, *Proceedings of Technology and Tomorrow Fibres*, Textile Institute, Coventry (1988).
30. B. J. Briscoe and D. R. Williams, in: *Proceedings of the Third International Conference on Composite Interfaces*, H. Ishida (Ed.), p. 67. Plenum Press, New York (1990).
31. S. Wu, *Polymer Interface and Adhesion*. Marcel Dekker, New York (1982).
32. B. J. Briscoe and D. R. Williams. Submitted to *Composites Science and Technology*.
33. M. Takayanagi, S. Ueta, W. Y. Lei and K. Koga, *Polym. J.* **19**, 467–474 (1987).

Acid-Base Interactions, pp. 273-285
Eds. K.L. Mittal and H.R. Anderson, Jr.
©VSP 1991

Specific interactions and their effect on the properties of filled polymers

H. P. SCHREIBER* and F. ST. GERMAIN

*Department of Chemical Engineering, Ecole Polytechnique, PO Box 6079, Stn. A.,
Montreal, Quebec H3C 3A7, Canada*

Revised version received 16 January 1990

Abstract—The surface characteristics of olefinic polymers and of $CaCO_3$, TiO_2, and carbon black particles have been determined by inverse gas chromatography. Cold, microwave plasma techniques were used to modify the surface properties of the particulates. The mechanical properties of blends using the various fillers were determined. It was found that adhesion at the polymer–filler interface exerts a considerable influence on the mechanical responses. In a non-polar polymer host (LLDPE), adhesion is promoted by minimizing the acid or base properties of the fillers. In an acidic matrix (EVA), interfacial bonding is strongly promoted by the existence of acid–base interactions. Thus, acid–base coupling is identified as an important control factor in targeting the selection of fillers to be used with specified matrix polymers.

Keywords: Filled polymers; acid–base interactions; surfaces; microwave plasmas.

1. INTRODUCTION

Multicomponent polymer systems are in special vogue. The combined benefits of performance and cost have extended the use of these materials into virtually all aspects of industrial and private activity. Polyblends, polyalloys, composites, and filled and pigmented polymers account for increasing percentages of the overall market for polymers. In all of these compositions of matter, the interphase at contacts between two or more constituents of the system is a particularly important region, control over which is acknowledged to be indispensable to the incisive formulation and use of the systems. Surface and interfacial phenomena have been shown to affect the dispersion of minor phases in polymer matrices [1, 2], and to influence the processability of polymer blends and composites [3, 4], as well as their mechanical properties [5, 6]. By inference the durability of these substances is also affected, because interfacial effects are intimately linked with questions of thermodynamic miscibility, and hence with the stability of systems under use conditions.

Few have done more than Fred Fowkes to raise the consciousness of academic as well as industrial scientists and engineers to the need for understanding and controlling surface and interfacial phenomena in polymer systems. Particularly significant has been his recent insistence on characterizing non-dispersion force interactions by the concepts of Lewis acids and bases. He has favored the Drago approach to acid–base theory, and has devised calorimetric methods [7, 8] to

---

*To whom correspondence should be addressed.

evaluate the parameters of acidity and basicity (E_A, C_A, E_B, and C_B) for organic molecules, dyes, and powders frequently involved in polymer systems. This route to the characterization of interactions has been able to explain the adsorption behavior of polymers onto powders from a variety of solvents [9]; other aspects of polymer behavior, for example polymer solubility, have also been examined purposefully by the acid–base route [10]. In short, Fowkes's introduction of acid–base theory into the surface and interface phenomenology of polymer systems appears to have major potentials for clarifying the behavior of these complex compositions.

We have, for some time, followed closely the work reported from Fowkes's laboratory. His approach has been adapted and to some extent modified to fit the terms of our experimental methodologies. Our objectives often call for the determination of acid–base interaction values for polymers, and non-volatile additives to polymer systems, preferably over wide ranges of temperatures. The technique of inverse gas chromatography (IGC) has proven to be particularly attractive for this purpose [1, 2, 11]. It is operated conveniently over requisite temperature ranges, and is readily adapted to provide acid–base interaction data relative to the selection of some reference set of acid, base, or amphipathic vapor probes. A logical extension of our concern for the inherent acid–base interaction potential of polymers and other components of complex systems is an interest in controllably adjusting these potentials to suit specific purposes. Electrical discharges and more particularly cold, large-volume microwave plasmas (LMPs) [12] are the routes chosen to effect such changes. Recent work [13] has demonstrated the power of LMP processes in modifying polymer surface phenomena such as adhesion and vapor transmission.

The lines of activity, referred to above, form the basis of the present paper. Here, LMP methodology has been extended to cope with the vexing problem of modifying the surfaces of powders used as fillers or pigments in polymer complexes. The IGC route has been used to characterize the surface properties of host polymers as well as of variously surface-modified additives. The resulting information on acid–base interaction potentials has been used to interpret some of the mechanical properties of filled polymer compounds. The value of fundamental advances by Fowkes and his co-workers has thereby been confirmed anew.

2. BACKGROUND INFORMATION

2.1. IGC and surface characteristics of non-volatiles

The IGC method has enjoyed rapid advances, both in techniques and in its range of application [11]. One of its original attractions was the inherent link between the basic datum of IGC—the specific retention volume, V_n—and the activity coefficient of a highly diluted vapor front in contact with some unknown, solid stationary phase, e.g. a polymer. Formal thermodynamic information on the interaction between stationary and mobile phases could be won provided a state of equilibrium existed in the chromatographic column. This condition generally precludes the use of materials containing active strong polar forces. Some time ago, we applied the Drago theory of acids and bases [14] to the IGC experiment, showing [1, 2] that a relative index of acidity and basicity could be obtained for polymeric or other solid stationary phases, provided the vapor used to probe the surface had been characterized as a Lewis (Drago) acid or base. The relative index of acid or base interaction, Ω, was defined according to which of the reference vapors gave the higher retention volume. Thus, for an acidic solid, where $(V_n)_b > (V_n)_a$,

$$\Omega = 1 - (V_n)_b / (V_n)_a < 0, \tag{1}$$

while for a base, where the retention volume for the acidic probe exceeds that for the basic vapor,

$$\Omega = 1(V_n)_a / (V_n)_b - 1 > 0. \tag{2}$$

Though lacking the formality of thermodynamic functions, comparative values of Ω nevertheless have proven to be useful in a number of practical cases, such as in the selection of plasticizers for PVC and other polymers.

Very recently, an alternative approach to the characterization of polymer and other solid surfaces has employed Gutmann's theory of acids and bases [15]. Though inherently no more exact than the Drago theory, its use by the research group at Mulhouse, France [16, 17] has led to an attractive link between V_n and the surface energy of solids. When only dispersion-force vapors are used in the IGC experiment, and provided the dimensions of the vapor phase molecules are known, it can be shown that

$$RT \ln V_n = 2N(\gamma_l^d)^{1/2} a (\gamma_s^d)^{1/2} + C. \tag{3}$$

Here N is the Avogadro number, a is the cross-sectional area of the vapor molecule, and the γs are the dispersion-force contributions to the surface energies of the vapor and solid stationary phases. A plot of the left-hand side of equation (3) against the product $a(\gamma_l^d)^{1/2}$ must be linear with a slope proportional to the solid surface energy. Gutmann, however, has made available specifications for a variety of organic vapors which are rated according to their acceptor and donor numbers (AN and DN). When these vapors are used with a solid containing non-dispersive forces, then the V_n datum should fall off the linear plot generated by the non-polar probes. In our simplified use of this concept, we define the perpendicular distance from the linear reference line of a given V_n datum as the solid's AN or DN. Thus, using chloroform as the reference acid and THF as the reference base vapors, the AN and DN of an adsorbent surface would be given, respectively, by the V_n for THF and chloroform. The procedure used here has been introduced in an earlier communication [18], to which the current paper serves as an extension. As in ref. [18], we also introduce a single numerical index of acidity/basicity, K_i, defined as the difference $DN - AN$. K_i thus parallels the Ω value, being positive for basic solids, negative for acids, and near zero for neutral or amphipathic materials.

2.2. LMP procedures for surface modification

Cold plasma procedures have become important factors in advanced technologies, for example, in microelectronics. The subject of frequent reviews [19, 20], plasmas operated at reduced pressures and high frequencies are able to effect surface modification by various routes. These include surface cleaning/ablation (when noble gases are used to generate the glow discharge), implantation of chemical groups into the surface and particle encapsulation by the deposition of plasma polymer moieties, when the appropriate inorganic and organic vapors are used to generate the plasma. The reactors used in plasma operations most often are appropriate for the treatment of planar surfaces, and by that token are less successful for use with particulates. A modification to the reactor vessel, making it more consistent with the use of powders, was described in ref. [18]. This unit was employed in the present work to surface-modify $CaCO_3$, TiO_2, and carbon black particles for subsequent use with polymer matrices which included linear low-density polyethylene (LLDPE) and an ethylene-vinyl acetate copolymer (EVA).

3. EXPERIMENTAL

3.1. Materials

The LLDPE used as host polymer was an Exxon hexene copolymer with a melt index of 2.4 g/10 min. Its molecular weight parameters, from size exclusion chromatography with trichlorobenzene (TCB) as solvent at 135°C, were $M_n = 0.51 \times 10^5$ and $M_w = 1.68 \times 10^5$. The EVA was a CIL, Inc., commercially available polymer with VA = 28 mol%. All of the fillers were commercially marketed products. The $CaCO_3$ was obtained from ICI Ltd, the rutile from Tioxide Canada Inc., while the furnace black was a product of Columbia Carbon. Surface areas for the solids were obtained from BET isotherm determinations, using nitrogen at its boiling point as the adsorbate. These results, along with IGC determinations of acid–base interaction potentials, are reported in Table 1.

The required IGC measurements followed routines fully described in ref. [18]. Linear alkane vapors were used to establish the applicability of equation (3) to the materials of concern, while chloroform and THF were the reference acid and base, respectively. The AN, DN and cross-sectional areas for these materials are well known [17, 18]. The applicability of equation (3) is confirmed in Fig. 1, which shows the appropriate representation for the $CaCO_3$ filler (curve A) and for the LLDPE host (curve B). The difference in slopes immediately shows the filler to have a higher γ_s^d than the LLDPE, as confirmed in Table 1. Figure 1 shows the polymer to be essentially neutral, although the placement of the THF point suggests very slight acidity (again see Table 1). In contrast, the filler is significantly basic, its retention volume for the reference acid being much greater than that for THF. An inspection of Table 1 shows the EVA to be an acidic polymer matrix. The

Table 1.

Primary materials used in this study [Matrix polymers: LLDPE, EVA (28% VA); particulates: rutile TiO_2, $CaCO_3$, Black Pearls carbon black]

	A (m^2/g)	γ_s^d (mJ/m^2)	AN	DN	K_i
LLPDE	—	28.8	0.3	0.0	−0.3
EVA	—	29.5	6.0	4.4	−1.6
TiO_2	8.4	53.0	9.2	6.5	−2.7
$CaCO_3$	3.3	44.6	1.3	8.7	7.4
Carbon black	960	42.8	11.6	8.0	−3.6

A is the BET surface area; γ_s^d is the dispersion force surface energy; AN, DN, K_i are IGC parameters.

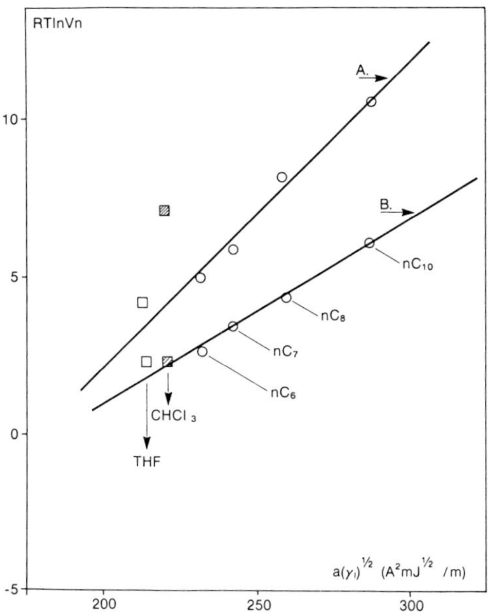

Figure 1. Retention volume as function of the product of vapor probe size, *a*, and its surface energy. (A) $CaCO_3$ filler; (B) LLDPE polymer. All data at 30°C. Coordinates as defined by equation (3).

carbon black and the rutile, with a surface coating as supplied by the manufacturer, are also acids, while, as already noted, the $CaCO_3$ is a strong base.

3.2. Plasma treatments and their effects

Apparatus and procedures described in ref. [18] were used to surface-modify the three particulates according to the steps outlined in Table 2. All plasma treatments were done at microwave frequencies of 2.45 GHz. Each treatment involved some 3–5 g of powder, which was initially evacuated to a residual pressure of less than 10^{-3} Torr. Five treatments were applied, as detailed in the table. The Ar was intended to clean and activate surfaces. CH_4 plasmas were intended to reduce surface energies and to modify acid–base behavior by the deposition of a plasma-olefin polymer. Treatment 3 was a combination of the first two, the polymerization step being carried out on a freshly activated filler surface. An acidic fluorocarbon,

Table 2.
Plasma treatment (all treatments used frequency = 2.45 GHz)

No.	Gas	P (W)	p (Torr)	t (min)	F (ml/min)
1	Ar	200	1.0	2.0	11.0
2	CH_4	200	0.8	2.0	11.0
3	1 followed by 2				
4	C_2F_4	200	0.7	3.0	11.0
5	NH_3	200	1.0	2.0	11.0

Operating parameters for plasma treatment are: P = power feed to reactor; p = dynamic gas pressure; t = treatment time; F = gas flow rate.

C_2F_4, was the monomer in treatment 4, the intention being to deposit an acidic, low energy plasma fluoropolymer on the particulates. Finally, treatment 5 aimed at generating basic N—H linkages on the particle surfaces by plasma-treating these with ammonia as monomer gas. The power supplied, the dynamic pressures during treatment, the monomer flow rates, and the treatment times are given in Table 2. Following treatment, the reactor was let down to atmospheric pressure with the monomer gas used during treatment. The powders were withdrawn after about 30 min residence time, then placed in desiccators for further use.

3.3. Sample preparation and analysis

Filled LLDPE and EVA compounds were prepared with a Brabender mixer. Either untreated or plasma-modified fillers were added up to concentrations of 20 wt%, mixing being continued until the attainment of steady-state torque readings (typically 6–10 min). The mixing temperature was 200°C, the mixing speed 50 rpm, and in all cases thermal stability was promoted by the addition of 0.3 wt% hindered phenol stabilizer. Earlier work [18] has shown that these mixing procedures lead to steady states of dispersion, but not necessarily to optimum dispersions, that state being dependent on the exact matrix–additive combination. Filled and unfilled polymers were compression-molded at 200°C, immediately quenched in cold running water, and then stored under ambient conditions for at least 72 h prior to further use. Specifically noted was stress–strain evaluation, using an Instron tester, and a separation speed of 25 mm/min. All reported data are averages of five separate determinations.

4. RESULTS AND DISCUSSION

4.1. Effects of plasma modification

The results of plasma treatment are reported individually for the three particulates in Tables 3, 4, and 5. Table 3 refers to $CaCO_3$. The effect of Ar treatment is noted principally in the γ_s^d value. This increases by some 3 mJ/m^2, suggesting the loss of adsorbed gases due to the plasma exposure. The slight rise in surface area is consistent with this. The filler remains strongly basic, although both AN and DN indices have dropped slightly as a result of the surface treatment. Treatment 2 produces very radical results, and these are intensified by the combination of treatments 1 and 2. The polymerization of methane appears to have been successful in sharply modifying the acid–base character of this solid. In both cases,

Table 3.

Effect of plasma treatment on particulate $CaCO_3$ properties

Plasma	A (m^2/g)	γ_s^d (mJ/m^2)	AN	DN	K_i
—	3.3	44.6	1.3	8.7	7.4
1. Ar	3.7	47.5	0.7	7.4	6.7
2. CH$_4$	3.0	30.3	0.4	0.9	0.5
3. (1+2)	2.7	27.2	0.3	0.4	0.1
4. C$_2$F$_4$	2.7	26.8	2.7	1.9	−0.8
5. NH$_3$	3.5	45.5	0.8	9.1	8.3

but particularly following treatment 3, the $CaCO_3$ is nearly non-polar (neutral), its original surface apparently shielded by a layer of plasma–polyolefin. The low surface energies of these samples are again consistent with the above suggestions. Slight downward shifts in the surface areas also result from these treatments. Plasma treatment with fluorocarbon vapor, intended to produce a plasma–fluoropolymer overcoat, shifts the surface interaction potential to the acidic side, with a further decrease in surface energy. Finally, the ammonia treatment increases the surface basicity beyond that of the control sample, without significantly altering either the surface energy or the surface area. The enhanced basicity is assumed to arise from an implantation of N—H linkages in the surface structure of the filler. Although the polymerization procedures (Nos 2, 3, and 4) lead to anticipated results, it is not possible to deduce, from the available data alone, how complete the surface modification has been. Even with the redesign of the plasma reactor, it seems unlikely that particle agglomeration can be avoided fully during plasma treatment, so that only a fraction of the available surface may be affected by the selected plasma procedures. The reported results, therefore, are unlikely to represent the full range of surface changes that can be realized by the plasma route.

The response of TiO_2 to plasma treatments, summarized in Table 4, is analogous to that of the $CaCO_3$. Surface cleaning by treatment 1 somewhat increases the inherent acidity of this surface, also slightly raising the dispersion-force surface energy. Plasmapolymerization of methane leads to sharp reductions in the surface energy and to milder acid interaction potentials. However, the treatments are less successful in producing neutral surfaces than was the case with the calcium carbonate. It was noted that all plasma treatments produced a slight darkening or graying in the rutile. Some crystallographic changes in the material may be responsible, and these, in turn, may influence the surface analyses. Tetrafluoroethylene plasma treatment has virtually no effect on the K_i value: the pigment remains moderately acidic. However, the strong reduction in surface energy and the pronounced loss of surface activity, reported by the lower AN and DN indices, confirm the production of a low-energy fluoropolymer on the rutile surface. The ammonia treatment here succeeds in shifting the overall interaction potential to the basic side without, however, greatly reducing the AN value of the pigment. None of the treatments affect the surface area, indicating the solid to be non-porous and not to have been sintered by the discharge.

The range of interaction values (K_i) seen in Table 5 is more restricted than in the preceding cases. Apparently, while the carbon black responds to plasma

Table 4.

Effect of plasma treatment on particulate (TiO_2) properties

Plasma	A (m²/g)	γ_s^d (mJ/m²)	AN	DN	K_i
—	8.4	53.0	9.2	6.5	− 2.7
1. Ar	8.1	55.0	9.6	6.5	− 3.1
2. CH_4	8.0	37.9	3.5	2.0	− 1.5
3. (1 + 2)	8.1	31.5	2.2	1.4	− 1.2
4. C_2F_4	7.8	26.1	5.6	3.0	− 2.6
5. NH_3	8.3	50.2	7.5	9.9	2.4

Table 5.

Effect of plasma treatment on particulate (carbon black) properties

Plasma	A (m²/g)	γ_s^d (mJ/m²)	AN	DN	K_i
—	960	42.8	11.6	8.0	−3.6
1. Ar	975	44.0	10.2	7.0	−3.2
2. CH_4	890	40.3	5.7	4.6	−1.1
3. (1 + 2)	905	37.6	3.9	3.0	−0.9
4. C_2F_4	880	38.5	7.7	4.6	−3.1
5. NH_3	955	41.9	8.0	9.1	1.1

exposure in much the same way as rutile, the effectiveness of plasma treatment is generally reduced. For example, ammonia treatment produces amphipathic or mildly basic surfaces, with the K_i shifting to the positive side but without a strong reduction in the pigment's AN. One possible reason for this is the very high surface area of the solid; this may lead to stronger interparticle association and, therefore, to less complete coverage by plasma-induced moieties.

A further, comprehensive view of the surface changes brought about by the selected plasma treatments may be obtained by comparing the final and initial surface energies, and the final and initial acid–base interaction potentials of the solids. This is done by means of bar graphs in Figs 2 and 3. The first of these considers the surface energy ratio. Evidently, plasma cleaning (Ar) and the implantation of polar (amino) linkages have little effect on the dispersion-force surface energy of particulates, regardless of their chemistry or their specific surface area. Of course, this does not necessarily mean that the total surface energy remains unaffected; in the case of ammonia treatments, increases in the polar energy contribution to γ_s would be expected. Dynamic contact angle measurements needed to resolve the issue are to be generated in forthcoming research. The capability of plasma treatments to lower the surface energy by the deposition of selected plasma–polymers is well illustrated by the methane, the combined Ar/methane, and the C_2F_4 treatments. The effectiveness of these procedures appears to depend on the substrate, and may be an inverse function of

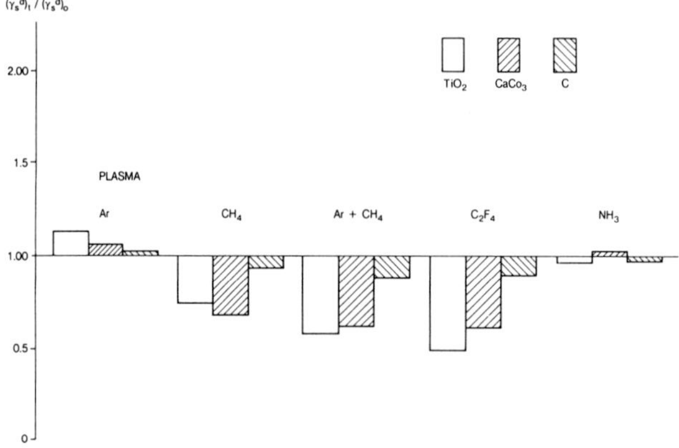

Figure 2. Effect of plasma treatments on the dispersion-force surface energy of particulates. Surface energy ratios are for treated vs. control solids.

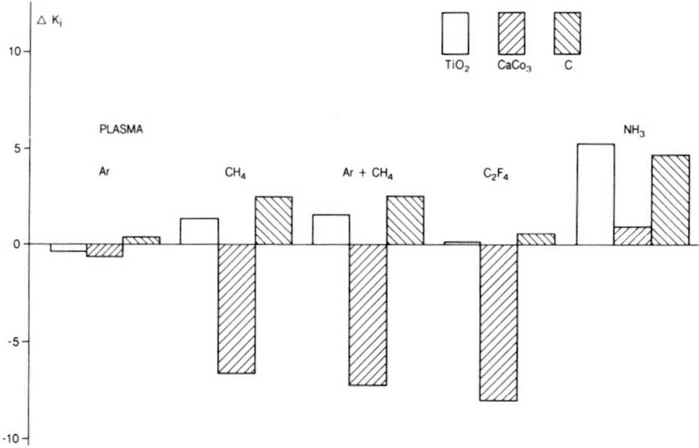

Figure 3. Effect of plasma treatment on the acid–base interaction potential of particulates. ΔK_i is difference between treated and control solids.

the specific surface area. The latter is suggested by the relatively low response level of the carbon black.

Figure 3 shows the difference between the K_i values of treated and untreated solids. Perhaps most important is the observation that, in principle, plasma procedures seem to offer the possibility of altering the interaction potential of a given powder in predetermined directions. The inherent acidity of (the present) rutile may be enhanced or shifted to the basic range; the strong base, $CaCO_3$, may be further strengthened (NH_3 plasma), or shifted toward and into the acid range; in the carbon black, the direction of change is always toward basicity, but that is unlikely to be a fundamental restriction. Rather, it is connected, in all likelihood, to the selection of the weak acid vapor (C_2F_4) to represent the acid category. Given the apparent potential of the plasma route, it should then be possible to control the surface properties of particulates so as to optimize the mechanical properties of compounds containing them. This matter is discussed further in the next section.

4.2. Properties of filled polymers

The effects of modifying particulate surfaces have been illustrated in terms of the mechanical properties of the filled LLDPE and EVA compounds. Young's modulus has been used to represent the mechanical properties in the low deformation range, while the elongation at rupture or ductility of compounds has been selected to represent mechanicals at large deformations. The response of LLDPE to the presence of variously plasma-treated fillers is illustrated in Fig. 4. The filler is $CaCO_3$. The pattern of responses here was followed closely when the other particulates were present. For convenience, data for methane-treated filler (treatment 2) are omitted from the figure, since these superimpose almost exactly on the results for samples containing fillers modified by the combined Ar, CH_4 plasmas.

Addition of unmodified filler raises the modulus, as may be expected. Modification of the acid–base interaction characteristics, and of the surface energies, accounts for major changes in the filler's effect on modulus. Treatments 1, 2, and 3 are most effective in increasing the modulus of the compounds, while treatments 4 and 5 lead to opposing results. Indeed, fillers coated by plasma–fluoropolymer

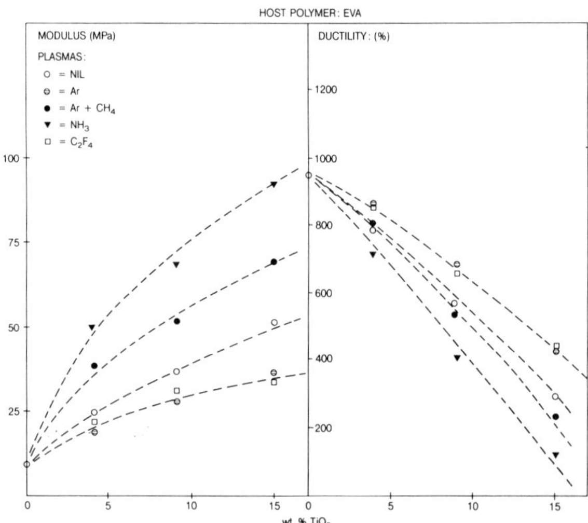

Figure 4. Modulus and ductility of filled LLDPE compounds: effects of plasma-treating the $CaCO_3$ filler.

produce very little increment in the modulus even at relatively high concentrations. An inspection of the ductility data will show trends which are opposite to those shown by the moduli. Again as may be expected, the presence of fillers reduces the elastic deformability of compounds, but the reduction was greatest when the fillers involved were exposed to Ar and CH_4 plasmas, and was less severe when the particulate surfaces were exposed to fluorocarbon or ammonia plasmas. The results strongly suggest that interfacial adhesion is a dominant reason for the observed variations. Modulus increments call for strong bonding at the polymer–filler interface. Ar cleaning, designed to eliminate weakly bonded adsorbates, and the deposition of a plasma–polymer resembling the host are reasonable strategies for enhancing that bond. With LLDPE as host, acid–base coupling of course is impossible. Thus, an increase in the filler surface polarity, due to the ammonia treatment, is a countervailing step, as is the presence of the acidic plasma–fluoropolymer. In addition, this plasma–polymer also has a sufficiently low surface energy to impede wetting of the filler by the host polymer. High ductility, on the other hand, is promoted by slippage at the polymer–filler interface [1, 2, 21]. It is not surprising then, that the surface modifications favoring one effect work to the detriment of the other.

In the case of filled LLDPE, a non-polar matrix, strong responses in the mechanical properties arise from a reduction in the non-dispersive forces of the filler surfaces. Quite different observations might be expected when EVA is the polymer host. As shown in Table 1, this polymer is a mild acid but it retains a significant base interaction potential, as indicated by its DN of 4.4. The properties of filled EVA compounds are illustrated in Fig. 5, with rutile as the pigmenting solid. Again the modulus is increased by the particulate, but in this case optimum results were obtained when the rutile was modified by ammonia plasmas. The rutile in question, of course, is basic with a $K_i = 2.4$. Acid–base coupling is strongly promoted in this case, with consequences as observed.

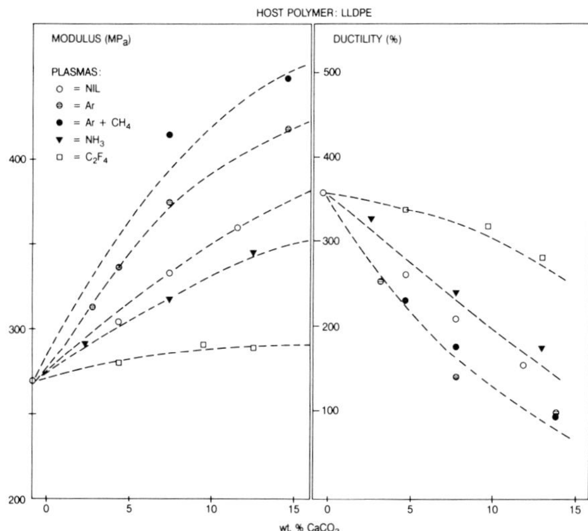

Figure 5. Modulus and ductility of filled EVA compounds: effects of plasma-treating the TiO_2 filler.

The poor ductility of this sample is consistent with the preceding rationalizations. An additional persuasive argument for the important role of acid–base interactions in this polymer host is the performance of compounds containing acidic rutiles. In particular, fluorocarbon-modified pigment leads to very low modulus increments and correspondingly lower losses in ductility. Acid–acid repulsion and, in the case of fluorocarbon-modified TiO_2, limited wettability by the host polymer may be cited as leading contributors to the observed results. Our results clearly suggest the existence of correlations between the mechanical characteristics of filled polymers and the dominant acid–base indices of the system's constituents. We cannot yet evaluate the more subtle contributions to these apparent relationships arising from the subordinate interaction potentials. It would certainly appear reasonable to suggest that even more pronounced modifications in the modulus and ductility could have been realized in the present instance if pigment treatments had been able to confer pure acidity or basicity to the rutiles (i.e. specimens with AN or DN = 0). Work aimed at clarifying these potentially important questions is in progress.

An attempt made to generalize the dependence of the mechanical property parameters on the dominant balance of acid–base interactions in EVA compounds leads to the representation of Fig. 6. Here the modulus and ductility of filled and unfilled (control) polymers are shown as a function of the difference between the K_i values for the host polymer and filler. Each of the fillers is shown in this scheme, in their original and in their plasma-modified surface states. For simplicity, the plot is restricted to compounds at 5 wt% filler content. The difference, ΔK_i, is given a positive sign to indicate acid–base coupling, and a negative index in those cases where acid–acid repulsion is in effect. The functional relationships are quite evident. Acid–base coupling consistently enhances the relative gain in compound modulus and equally consistently reduces ductility. Conversely, acid–acid repulsion generates favorable elongational properties, while reducing the gains in elastic moduli.

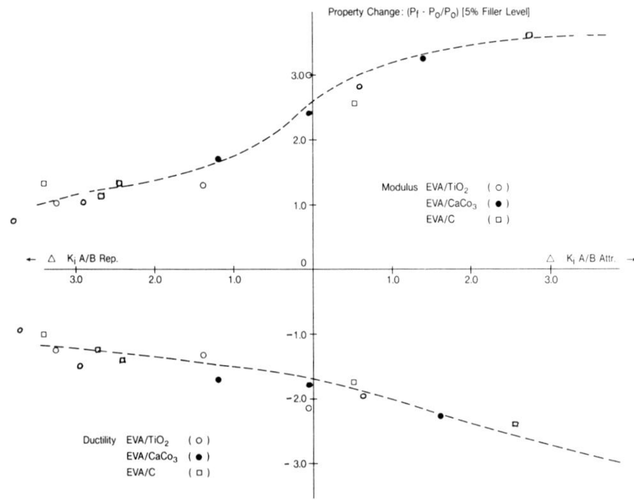

Figure 6. Effect of acid–base interactions on the moduli and ductilities of filled EVA compounds. Filler level was 5 wt%. Property change expressed as $(P_f - P_0/P_0)$.

Although at present, the generalization was limited to an acidic polymer matrix, it is reasonable to assume that relationships similar to those given in Fig. 6 will be generated for other filled polymer systems with active acid–base functionalities. Fowkes's concern for the role of acid–base interactions in polymer systems appears to be well founded. They are clearly one of the important factors to take into account when choosing materials for the formulation of multicomponent polymer systems, and when designing surface conditioning procedures for additives to be used in systems with targeted mechanical properties.

5. CONCLUSIONS

(1) Acid–base interaction characteristics of matrix polymers and a series of fillers have been determined by IGC.
(2) Cold, microwave plasma treatments have been applied to modify the surface interaction states of the fillers.
(3) The mechanical properties of filled LLDPE and EVA compounds were found to be sensitive to interfacial interactions, notably to their dependence on adhesion at the polymer–filler interface. In non-polar LLDPE compounds, surface energy balances appear particularly important. In an acidic EVA matrix, acid–base interactions are dominant in determining mechanical property responses.
(4) Strong bonding at the polymer–filler interface favors increments in elastic moduli of the filled compounds, but at the cost of reduced elongational properties at rupture.

Acknowledgement

This work was supported by grants received from the Natural Sciences and Engineering Research Council, Canada.

REFERENCES

1. M. Y. Boluk and H. P. Schreiber, *Polym. Composites* **7**, 295 (1986).
2. M. Y. Boluk and H. P. Schreiber, *Polym. Composites* **10**, 215 (1989).
3. T. M. Malik, P. J. Carreau, M. Grmela and A. Dufresne, *Polym. Composites* **9**, 412 (1988).
4. E. Trujillo, C. Richard and H. P. Schreiber, *Polym. Composites* **9**, 426 (1988).
5. M. Ratzsch, in: *Interfaces in Polymer, Ceramic and Metal Matrix Composites*, H. Ishida (Ed.), p. 425. Elsevier, New York (1988).
6. H. Ishida, *Polym. Composites* **5**, 101 (1984).
7. F. M. Fowkes, D. C. McCarthy and D. O. Tischler, in: *Molecular Characterization of Composite Interfaces*, H. Ishida and G. Kumar (Eds), p. 401. Plenum Press, New York (1985).
8. D. C. McCarthy and F. M. Fowkes, 40th Annu. Conf. Reinforced Plastics/Composites Institute, paper 17-D, Soc. Plast. Ind., (1985).
9. F. M. Fowkes and M. A. Mostafa, *Ind. Eng. Chem. Prod. Res. Dev.* **17**, 3 (1978).
10. F. M. Fowkes, *J. Adhesion Sci. Technol.* **1**, 7 (1987).
11. D. R. Lloyd, T. C. Ward and H. P. Schreiber (Eds), *Inverse Gas Chromatography*, ACS Symposium Series No. 391. American Chemical Society, Washington, DC (1989).
12. M. R. Wertheimer, J. E. Klemberg-Sapieha and H. P. Schreiber, *Thin Solid Films* **115**, 109 (1984).
13. Y. Klemberg-Sapieha, A. Migdal, M. R. Wertheimer and H. P. Schreiber, in: *Interfaces in Polymer, Ceramic and Metal Matrix Composites*, H. Ishida (Ed.), p. 583. Elsevier, New York (1988).
14. R. S. Drago, G. C. Vogel and T. E. Needham, *J. Am. Chem. Soc.* **93**, 6014 (1971).
15. V. Gutmann, *The Donor–Acceptor Approach to Molecular Interactions*. Plenum Press, New York (1983).
16. C. Saint-Flour and E. Papirer, *Ind. Eng. Chem. Prod. Res. Dev.* **21**, 666 (1982).
17. J. Schultz and L. Lavielle, in: *Inverse Gas Chromatography*, D. R. Lloyd, T. C. Ward and H. P. Schreiber (Eds), ACS Symposium Series No. 391, p. 185. American Chemical Society, Washington, DC (1989).
18. H. P. Schreiber, J.-M. Viau, A. Fetoui and Zhuo Deng, *Polym. Eng. Sci.* **30**, 263 (1990).
19. H. V. Boenig (Ed.), *Advances in Low-Temperature Plasma Chemstry, Technology and Applications*, p. 153. Technomic, Lancaster, PA (1984).
20. M. Shen and A. T. Bell (Eds), *Plasma Polymerization*, ACS Symposium Series No. 108. American Chemical Society, Washington, DC (1979).
21. L. Nicolais and M. Narkis, *Polym. Eng. Sci.* **11**, 194 (1971).

Acid-Base Interactions, pp. 287-301
Eds. K.L. Mittal and H.R. Anderson, Jr.
©VSP 1991

Adhesion of polyimides to ceramic substrates: role of acid–base interactions

T. S. OH,[1] L. P. BUCHWALTER[2,*] and J. KIM[1]
[1] *IBM, T. J. Watson Research Center, Yorktown Heights, NY 10598, USA*
[2] *IBM General Technology Division, Hopewell Jct., NY 12533, USA*

Revised version received 18 January 1990

Abstract—Adhesion of polyimides to ceramic substrates such as SiO_2, Al_2O_3, and MgO, and interfacial interactions were studied using XPS, SEM, and the peel test. The peel strength of polyimides on SiO_2 and Al_2O_3 is almost identical and higher than that on MgO at the same polyimide thickness. Contrary to the failure within the polyimides on SiO_2 and Al_2O_3, Mg was found on the peeled PMDA–ODA acid-derived polyimide surface, implying weakening of MgO by interfacial reactions with polyamic acid. With the neutral polyamic ethyl ester, the locus of failure on MgO was changed to the apparent weak boundary layer of the ester-derived polyimide. On SiO_2 and Al_2O_3 the peel crack propagated with a discontinuous stick–slip process. The constant interspacing between transverse stick–slip striations on the peeled polyimide surfaces has confirmed that plastic bending is the major energy dissipation process with a minimal contribution from tensile loading.

Keywords: Polyimides; ceramic substrates; interfacial interactions.

1. INTRODUCTION

Adhesion of polyimides to ceramic substrates is of great interest, as the importance of polyimides has been increasingly emphasized in the microelectronics industry because of their several advantageous properties such as low dielectric constant and easy processibility [1–3]. Polyimide films are usually produced by spin-coating polyimide precursors onto a substrate and curing at elevated temperatures. Although it has been reported that the peel strength of polyimide to SiO_2 was increased [4] and additional reactions with metals [5] were observed during thermal curing of polyimide at 400°C, initial interfacial reactions may also play a significant role in the bonding characteristics [6–8]. The importance of acid–base interactions between polymers and ceramic substrates to adhesion strength has been revealed by Fowkes *et al.* [8]. Adhesion of the acidic polymer chlorinated polyvinyl chloride (CPVC) to the basic surface of sodium glass was stronger than that of the basic polymer polymethyl methacrylate (PMMA), and such adhesion behavior could be reversed by acid treatment of the glass surface [8, 9]. Thus, to understand the adhesion behavior of polyimides on different ceramic substrates, the initial acid-base interactions between a polyimide precursor and a substrate should be considered. In the present study, we have examined the adhesion strength and peeling behavior of polyimides on various ceramic substrates, and have correlated the adhesion with the acid–base nature of the substrate surfaces and polyimide precursors.

---

*To whom correspondence should be addressed.

2. EXPERIMENTAL PROCEDURE

Ceramic substrates used in this study were (0001) sapphire (Al_2O_3), (001) single-crystal magnesia (MgO), and amorphous fused silica (SiO_2). All the substrates were obtained with a surface polish to 0.025 μm finish, cleaned with isopropyl alcohol (IPA), and characterized using X-ray photoelectron spectroscopy (XPS) prior to spin coating of the polyimide precursor. To study the interfacial reaction mechanisms between polyimides and various metal oxides, two different *N*-methyl-pyrrolidone (NMP) solutions of polyimide precursors, pyromellitic dianhydride (PMDA)–oxydianiline (ODA) polyamic acid and polyamic ethyl ester were used. Polyamic ethyl ester cures to polyimide, as does the polyamic acid. The only difference observed in the films exposed to 400°C bake is the lack of low binding energy (BE) nitrogen $1s$ species in the XPS data of the ester-derived PMDA–ODA imide [10]. Figure 1 shows the respective structures of the two PMDA–ODA polyimide precursors used in this study. It is the carboxylic acid group of the PMDA–ODA acid which was found to interact with the basic copper oxide surface, even to the degree of migrating copper oxide particles into the PMDA–ODA film with the aid of the solvent (NMP). This strong reaction was not observed with PMDA–ODA ester [2, 3, 11].

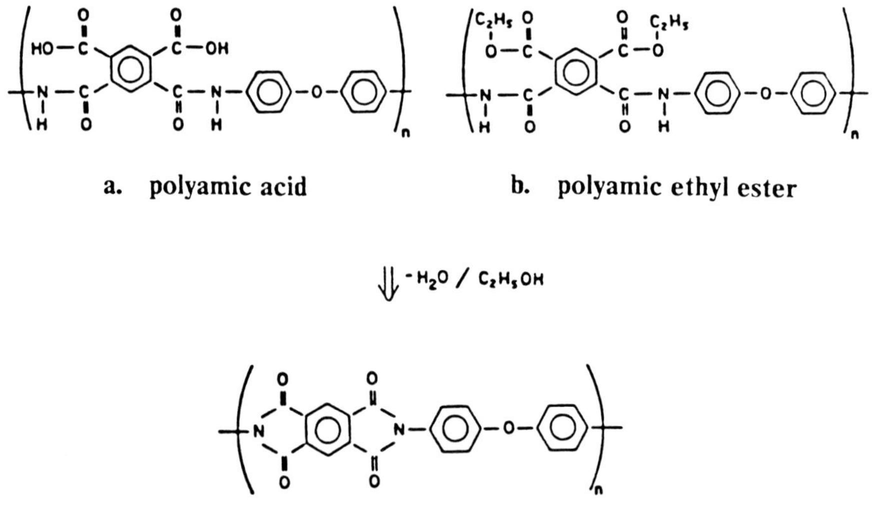

a.   polyamic acid   **b.   polyamic ethyl ester**

c.   polyimide

Figure 1. Molecular structures of (a) PMDA–ODA acid, (b) PMDA–ODA ester and (c) polyimide upon imidization. Note that these two precursors yield the same polyimide upon curing.

Polyimide precursors were spin-coated on ceramic substrates at 3000 rpm and cured at 85°C for 30 min to remove solvent. Multiple coatings were applied to obtain the desired thickness of polyimide films. Final curing was conducted at 150°C for 30 min, 200°C for 30 min, 300°C for 30 min, and 400°C for 40 min in a nitrogen atmosphere. The peel strengths of the polyimide films were measured with 5 mm-wide peel strips using a 90° peeling angle at a rate of 4 mm/min. The locus of failure was characterized using XPS and scanning electron microscopy

(SEM) on both peeled polyimide and ceramic failure surfaces. The XPS analysis was performed with a Surface Science Laboratories SSX-100-05 small spot unit. The details of the analysis method are described elsewhere [12].

3. RESULTS AND DISCUSSION

3.1. Surface analysis of ceramic substrates

Table 1 shows the XPS analysis results of the ceramic surfaces before and after IPA cleaning. It is clear from Table 1 that IPA cleaning is effective in removing organic carbon on SiO_2, but not much change was observed on Al_2O_3 or MgO surfaces after the IPA exposure. Al_2O_3 occasionally has fluorine contamination before IPA cleaning, as observed in sample 2. However, the peel strength of 60 g/mm was obtained with 35 μm-thick PMDA–ODA polyimide, compared with 60 g/mm with 30 μm-thick polyimide on non-contaminated substrates. Thus, there is no apparent difference in the measured peel strength with fluorine contamination.

Table 1.
XPS analysis of ceramic surfaces before and after IPA[a] cleaning

Ceramic substrate[b]	Composition (%) before						Composition (%) after					
	C	O	Mg	Al	Si	F	C	O	Mg	Al	Si	F
Al_2O_3 1	28	44	—	28	—	—						
2	25	45	—	29	—	1	21	54	—	23	—	2
3	30	42	—	28	—	—						
MgO 1	47	25	28	—	—	—						
2	54	25	21	—	—	—	54	30	17	—	—	—
3	51	26	23	—	—	—						
SiO_2 1	42	35	—	—	23	—						
2	42	34	—	—	24	—	16	62	—	—	22	—
3	45	35	—	—	20	—						

[a] Isopropyl alcohol.
[b] 1, 2 and 3 are different substrates.

3.2. Adhesion strength of polyimides on ceramic substrates

Figures 2 and 3 show the peel strengths of the polyimides, as a function of the polyimide film thickness, on various ceramic substrates. As revealed in the figures, the peel strengths of the polyimide films on Al_2O_3 and SiO_2 substrates are always higher than those on MgO, independent of the polyimide precursor used. The peel test also shows that the peel strengths of the ester-derived polyimide are consistently higher than those of the acid-derived polyimide. As discussed elsewhere [10], both of these precursors give the PMDA–ODA polyimide upon curing, with only a minor difference in the chemistry (no low BE nitrogen $1s$ peak in the case of PMDA–ODA ester). The molecular weight of the PMDA–ODA ester is about twice that of the PMDA–ODA acid which is about 28 000 [13, 14]. The Young's modulus and the yield stress, which affect the measured peel strength [4, 15], are not expected to change significantly in this molecular weight range or with the slight difference in the PMDA–ODA ester chemistry as compared with the acid-derived polyimide [14]. Therefore comparison of the peel strengths (Figs 2 and 3)

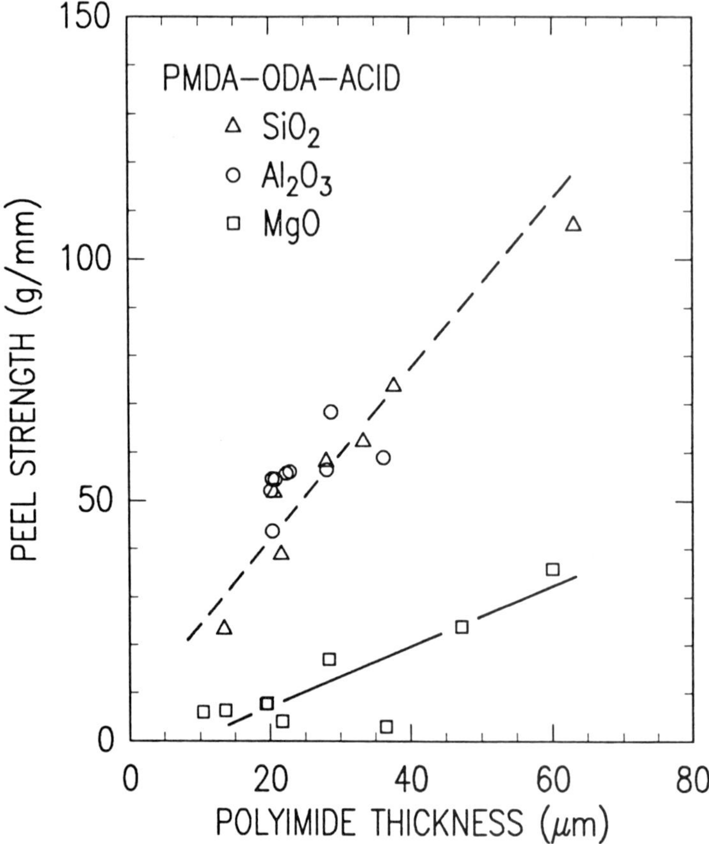

Figure 2. Peel strength of PMDA–ODA acid-derived polyimide, as a function of the film thickness, on SiO_2, Al_2O_3, and MgO substrates.

suggests that better energy dissipation with the more ductile ester-derived poly-imide films (80–125% fracture strain compared with 40–60% in acid-derived polyimide [16]) contributes to the small improvement in the peel strength on SiO_2 and Al_2O_3. However, the large enhancement of peel strength on MgO with the PMDA–ODA ester cannot be solely due to the difference in the mechanical properties of the polyimides. With the different locus of failure for each polyimide on MgO, such an increase in the peel strength can be attributed mainly to the different interfacial reaction mechanisms, as will be discussed in more detail in the next section.

As illustrated in Figs 2 and 3, the peel strengths of polyimides on ceramic sub-strates were enhanced with an increase of the film thickness up to 50 μm. With bending of the peel strip as the major energy dissipation process, the measured peel strength generally increases with the film thickness due to the increased deforming capacity of the film, reaches a maximum, and then decreases with a further increase in thickness [17–19]. However, the bending behavior, i.e. the thickness dependence of the peel strength, is strongly dependent on practical adhesion, which is governed by fundamental adhesion and residual stresses due to thermal expansion mismatch and solvent drying [20, 21], as well as to the

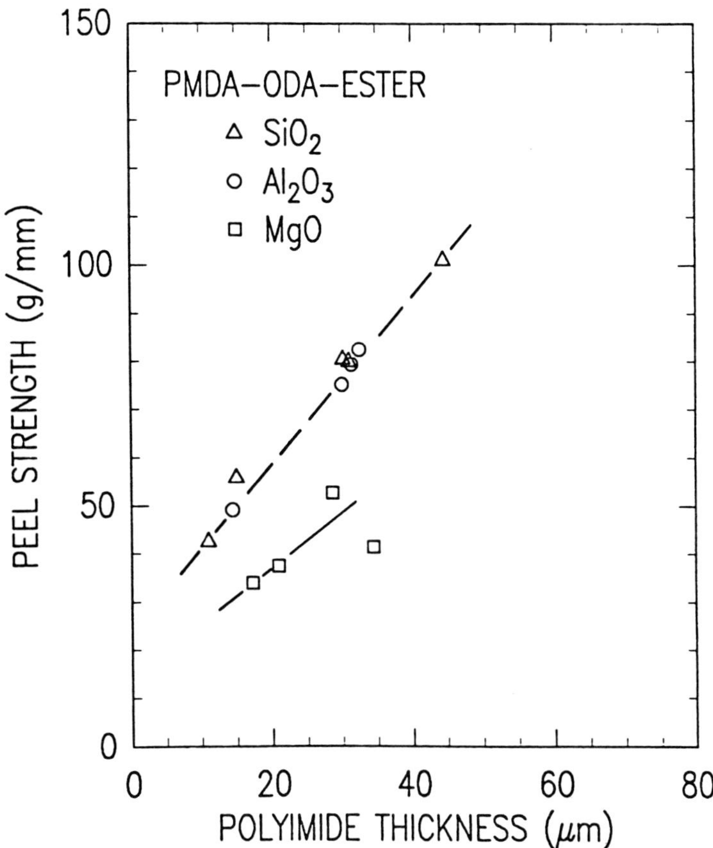

Figure 3. Peel strength of PMDA–ODA ester-derived polyimide, as a function of the film thickness on SiO$_2$, Al$_2$O$_3$ and MgO substrates.

mechanical properties of the polyimide film and the substrate. The scanning electron micrograph of a silica substrate (Fig. 4), on which 62 μm-thick acid-derived polyimide was coated, showed spontaneous cracking initiated from damage during scribing the polyimide film to peel strips and propagated underneath the polyimide/silica interface, mainly due to the relaxation of residual tensile stresses in the polyimide film. On peel strips where substrate cracking was minimal, however, the peel strength increased further up to 107 g/mm with the failure of polyimide during peeling. Thus, a drop in the peel strength with thicker polyimide can be attributed more to weakening of the fracture resistance at or near the interface by residual stresses [22] than to difficulty in bending the thicker film [18, 23].

3.3. Locus of failure and reaction mechanisms

The locus of failure after peeling was studied by analysis of both the polyimide and the ceramic failure surfaces by XPS. The results are shown in Tables 2 and 3. Focusing first on the PMDA–ODA acid case (Table 2), the failure locus on Al$_2$O$_3$ and SiO$_2$ is within the apparent weak boundary layer of the polyimide, which is

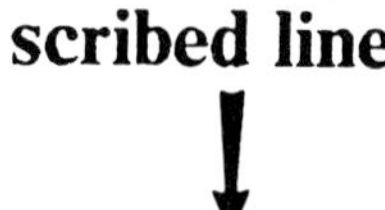

Figure 4. Scanning electron micrograph of a silica substrate showing cracking of the substrate during scribing 62 μm-thick acid-derived polyimide film to peel strips. Arrows indicate crack-initiation points.

Table 2.
XPS analysis of the failure locus of PMDA–ODA acid-derived polyimide/ceramic substrates

Substrate	Failure surface[a]		Composition (%)					
			C	O	N	Si	Al	Mg
SiO$_2$	SiO$_2$	1	29	49	2	20	—	—
	PI	1	75	18	7	—	—	—
	SiO$_2$	2	37	41	3	18	—	—
	PI	2	77	16	7	—	—	—
Al$_2$O$_3$	Al$_2$O$_3$	1	41	37	3	—	18	—
	PI	1	75	18	7	—	—	—
	Al$_2$O$_3$	2	54	29	5	—	12	—
	PI	2	76	17	7	—	—	—
MgO	MgO	1	27	39	2	—	—	31
	PI	1	73	19	7	—	—	2
	MgO	2	41	32	3	—	—	25
	PI	2	73	18	7	—	—	2

[a] PI 1 and 2 were peeled from substrate 1 and 2, respectively.

Table 3.
XPS analysis of the failure locus of PMDA–ODA ester-derived polyimide/ceramic substrates

Substrate	Failure surface[a]		Composition (%)					
			C	O	N	Si	Al	Mg
SiO_2	SiO_2	1	41	38	4	17	—	—
	PI	1	73	19	8	—	—	—
Al_2O_3	Al_2O_3	1	54	28	5	—	13	—
	PI	1	77	17	7	—	—	—
	Al_2O_3	2	46	33	4	—	17	—
	PI	2	74	18	8	—	—	—
MgO	MgO	1	37	33	3	—	—	27
	PI	1	75	18	7	—	—	—
	MgO	2	37	33	3	—	—	27
	PI	2	77	16	7	—	—	—

[a] PI 1 and 2 were peeled from substrate 1 and 2, respectively.

formed between the surface-bound polyimide chains and the bulk polyimide (Fig. 5). With a N/Si ratio of 0.10–0.17 for PMDA–ODA acid on SiO_2, the locus of failure was judged to be about 20 Å into the polyimide film [4]. The N/Al ratio is between 0.10–0.42 for Al_2O_3, which suggests a comparable failure locus to that on the SiO_2 substrate, which is further supported by the comparable peel strengths on these substrates (Fig. 2).

The PMDA–ODA acid on MgO, however, exhibits a different failure mode. As shown in the XPS data (Table 2 and Fig. 6), Mg was found also on the peeled polyimide failure surface. The fracture toughness of MgO is comparable to that of SiO_2 ($0.8\ \mathrm{MPa}\sqrt{m}$ and $0.7\ \mathrm{MPa}\sqrt{m}$, respectively); thus, the presence of Mg on the peeled polyimide surface cannot be due to the fracture toughness relation of these two materials. There is another plausible reason for this: the isoelectric point (IEPS) of the MgO surface is about 12, compared with about 8 for Al_2O_3 and about 2 for SiO_2 [6, 7]. The degree of acid–base interactions between the oxide surfaces and the organic polymers can be described using Δ values as shown in the literature [6, 7, 24]. A large positive value of Δ may indicate interactions strong enough to cause chemical attack or corrosion of the oxide surface, while negative values indicate weak or no interaction of the polymer with the oxide surfaces [6, 7, 24].

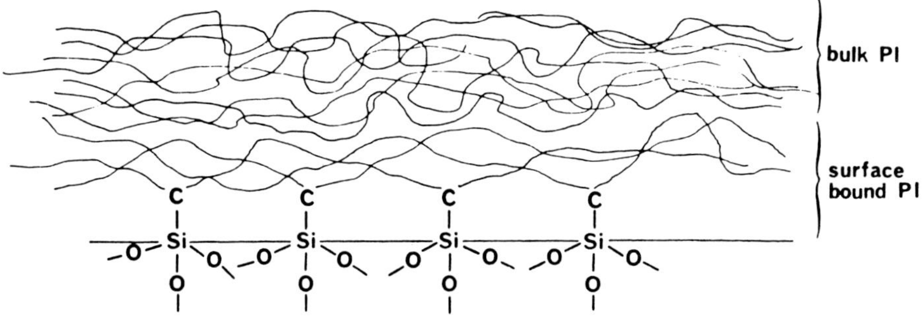

Figure 5. Schematic illustration showing the apparent weak boundary layer formed between the surface-bound polyimide chains and the bulk polyimide on SiO_2 substrate.

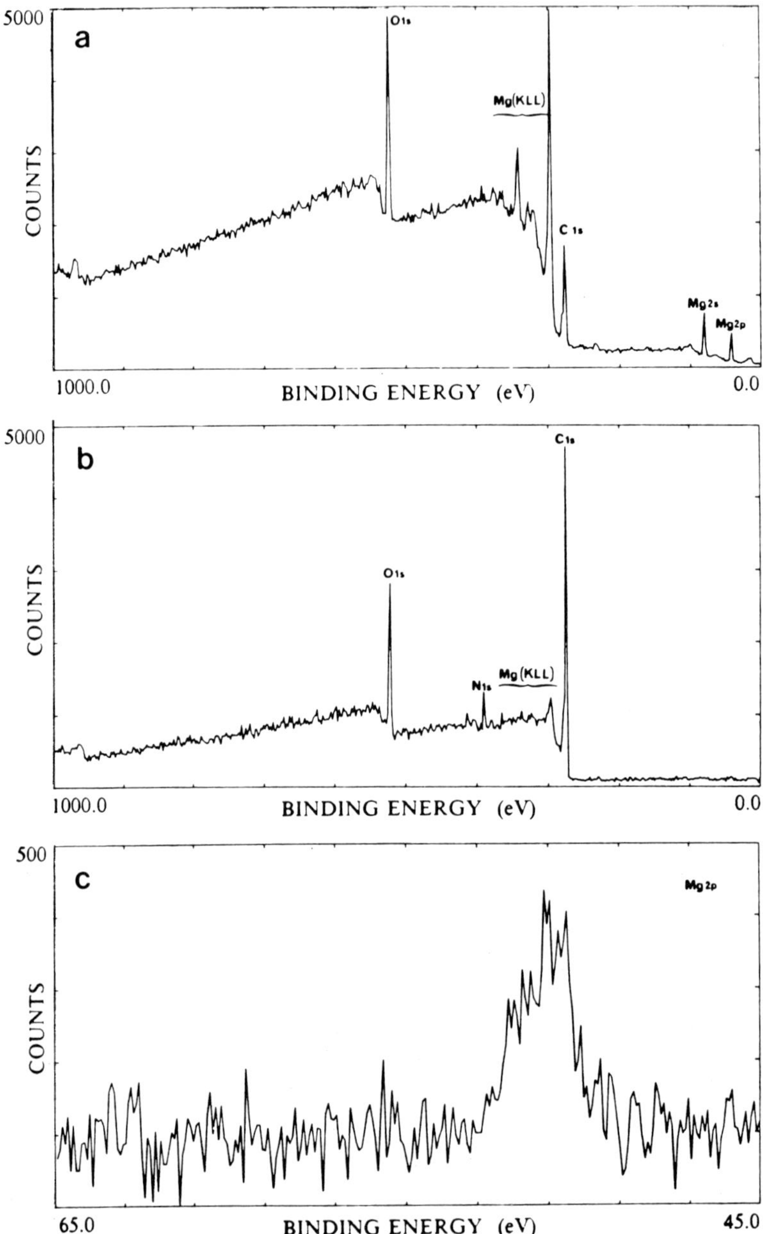

Figure 6. XPS analysis of the failure locus of PMDA–ODA acid-derived polyimide on MgO for (a) the MgO substrate failure surface, (b) the polyimide failure surface showing the presence of Mg, and (c) the Mg 2*p* peak on the polyimide.

For surface reaction with organic acids

$$\Delta_A = \text{IEPS} - pK_A,$$

where IEPS is the isoelectric point of the oxide surface and pK_A is the acid dissociation constant of the organic acid. Carboxylic acid ($pK_A = 4.5$) is the dominant reacting polar group in polyamic acid [3, 24]. Using the pK_A value of benzoic acid ($pK_A = 4.2$ [7]), Δ_A is -2.2 for SiO_2, 3.8 for Al_2O_3, and 7.8 for MgO, respectively. The acid–base interaction of the PMDA–ODA acid with the rather basic MgO surface is thus strong enough to cause salt formation and removal of the MgO from its crystalline structure, so that the weakest point is no longer the polyimide (although at some points this is still the failure mode), but the reacted or weakened MgO. Chemical reactions can result in good adhesion. However, the formation of a reaction product is ultimately detrimental to adhesion enhancement due to volume mismatch strains at the product/substrate interfaces [25]. Insoluble products can be formed with metal $2+$ ion salts of polymeric acid by crosslinking two carboxylic counter-ions located on different polymer chains and slow exchange of the carboxyl ligands [3]. Although the details of the interfacial reactions between PMDA–ODA acid and MgO require further investigation, it could be suggested that Mg carboxylate was formed by the reaction between PMDA–ODA acid with Mg^{2+} ions of the MgO surface, resulting in peel-crack propagation at the carboxylate/MgO interface and in the polyimide. The failure mode of PMDA–ODA ester on the ceramic substrates examined by XPS (Table 3 and Fig. 7) is comparable to that of the corresponding acid on Al_2O_3 and SiO_2 substrates. The main difference between the PMDA–ODA acid and ester is the interaction with MgO. Figure 7 shows no Mg on the polyimide failure surface. Thus the locus of failure is within the ester-derived polyimide, while it is a mixed-mode failure in the acid case. The ester is considered to be a neutral species, therefore having a milder reaction with the MgO surface than what is achieved with the polyamic acid, and causing no apparent 'corrosion' of the MgO.

3.4. Stick–slip behavior of peeling

Scanning electron micrographs of the failure surface of peeled polyimide films (Fig. 8) illustrate that, with strong adhesion as on Al_2O_3 and SiO_2 substrates, the peel crack propagates with a discontinuous stick–slip process. No such periodic striations on the ceramic failure surfaces indicate that the stick–slip peeling of polyimides occurred solely by non-uniform bending characteristics during the debonding process. Contrary to the stick–slip failure of pressure-sensitive tapes where the interspacing between transverse striations increased with the free length of the tape [17, 26], the interspacing on polyimides was constant with a small peel force fluctuation on the load–displacement curve.

Peel-crack propagation is a moment-controlled process with stress concentration at the peel–crack tip. Thus, during the peel test, the peel force increases as the radius of curvature of the peel strip at the peeling edge decreases until the stored strain energy exceeds the fracture resistance of the weakest bond at or near the interface [18, 23, 27]. During this stage, the peeling rate is less than the imposed machine rate, owing to the progressive bending of the peeled strip [26]. As shown in Fig. 9, the peel strength of acid-derived polyimide films on Al_2O_3 substrates increases with the peeling rate, mainly due to an increase of the yield strength of viscoelastic polymers with the elongation rate [18, 19, 26]. With strong interfacial adhesion, the cohesive failure occurs within the polyimide. Then a localized plastic hinge of near 90° bending can be formed, without further stress concentration at the peel-crack tip, by counterbalance of the increased bending

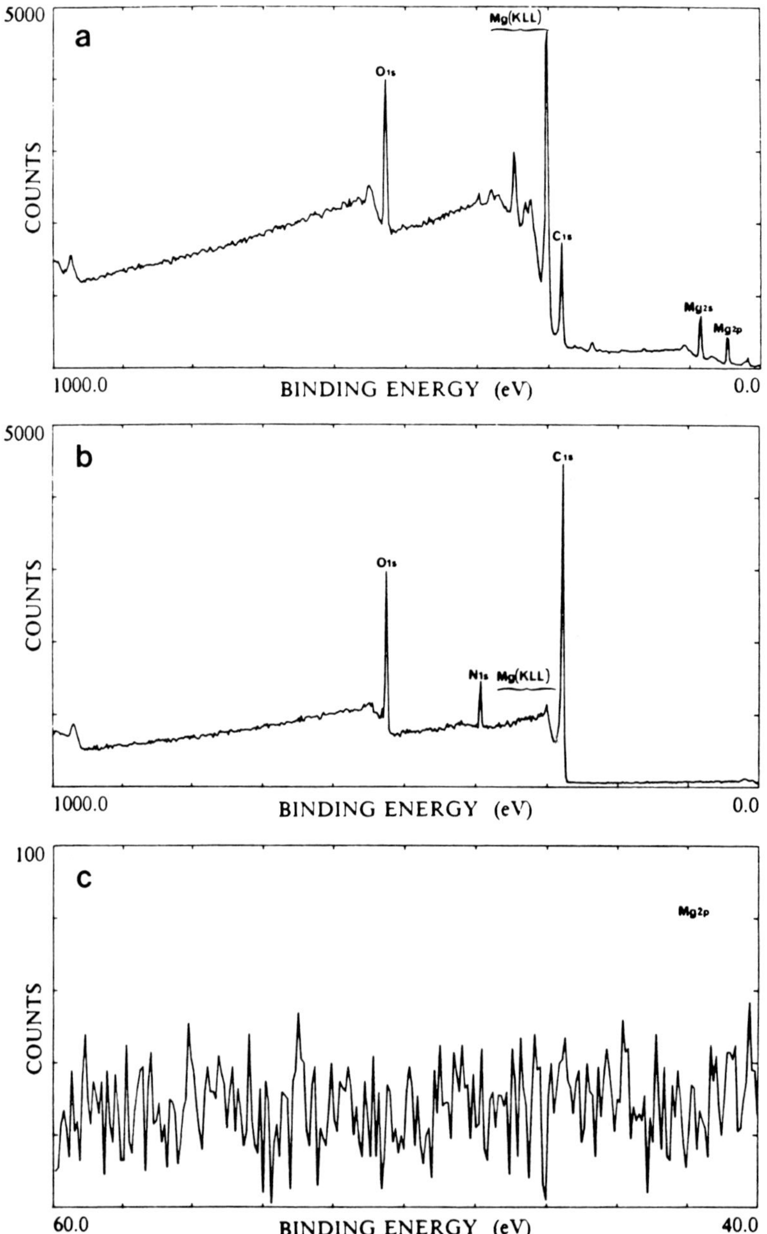

Figure 7. XPS analysis of the failure locus of PMDA–ODA ester-derived polyimide on MgO for (a) the MgO substrate failure surface, (b) the polyimide failure surface showing no magnesium, and (c) Mg $2p$ region on the polyimide. Note that the Mg $2p$ peak is different from the corresponding peak on PMDA–ODA acid-derived polyimide (Fig. 6).

moment with stretching polyimide chains behind the peel-crack tip. The peel strength increases further with increasing peeling rate [26]. With the breaking of the stretched polyimide chains behind the peel–crack tip at its rupture limit, catastrophic failure occurs with increasing radius of bending curvature, resulting in a

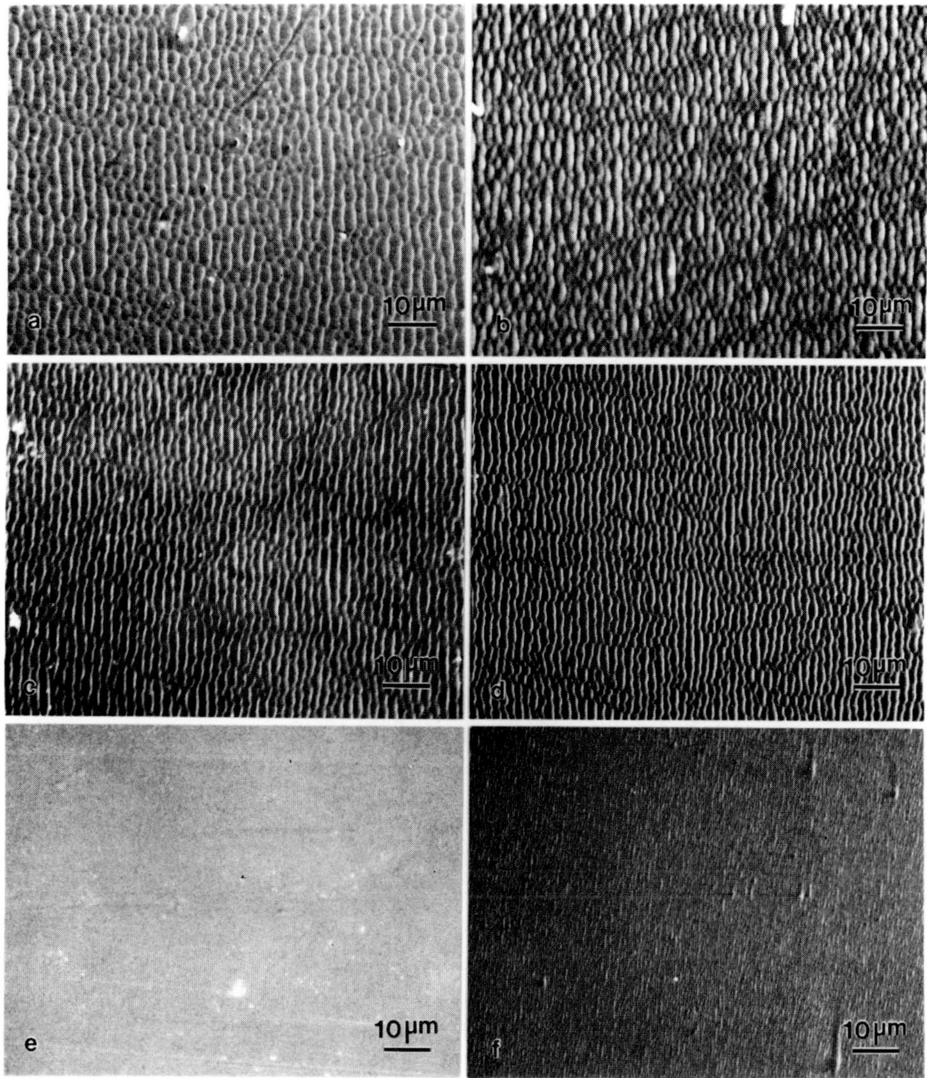

Figure 8. Scanning electron micrographs of the peeled surfaces of (a) acid-derived polyimide on SiO_2, (b) ester-derived polyimide on SiO_2, (c) acid-derived polyimide on Al_2O_3, (d) ester-derived polyimide on Al_2O_3, (e) acid-derived polyimide on MgO, and (f) ester-derived polyimide on MgO. Thickness of all polyimide films was 30 μm.

drop in the peel force. Peeling is then arrested at the point where the stress concentration at the tip, decreased by energy dissipation, is lower than the rupture strength of the weakest bonds at or near the interface. The peel force then rises again to form a new plastic hinge, producing periodic striations on the peeled strip.

The fracture strain of polyimides limits the amount of bending moment counterbalanced by the stretched polyimide chains and, thus, the tensile loading of the film. This results in constant interspacing of the transverse striations with the free length of peeled polyimide films. Thus, for peeling of polyimides on ceramic

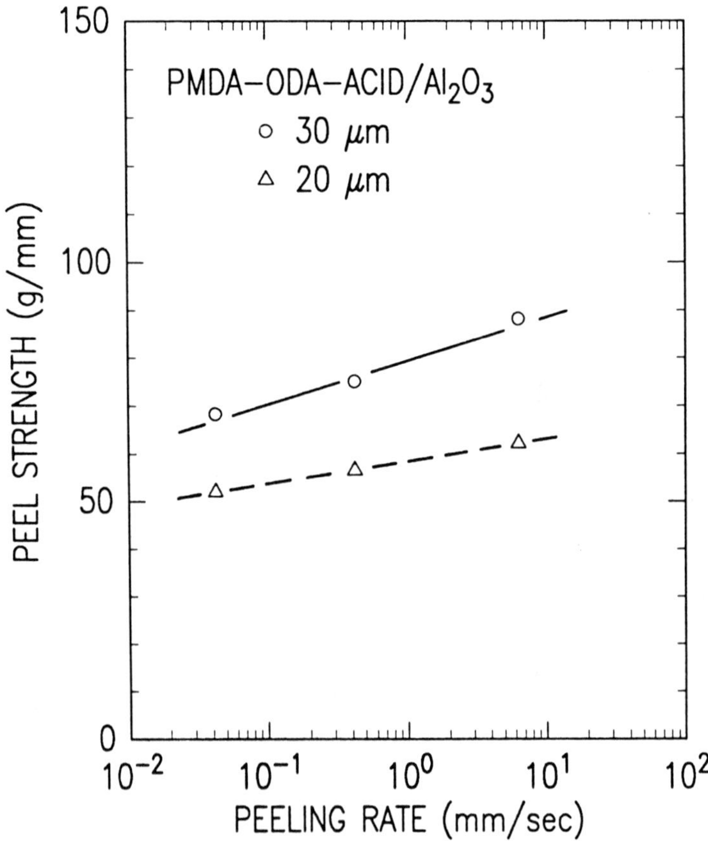

Figure 9. Peel strength of 20 μm and 30 μm-thick PMDA–ODA acid-derived polyimides on Al_2O_3 as a function of the peeling rate.

substrates, plastic deformation of the bending mode is the major energy dissipation process with a minimal contribution from tensile loading. The regions between striations on the peeled strips (Fig. 8) were not fully plastically bent due to the faster crack-propagation process [26]. As shown in Fig. 8, similar transverse striations were formed for acid- and ester-derived polyimides peeled on Al_2O_3 and SiO_2. With 45 g/mm of peel strength, however, stick–slip striations are significantly different on ester-derived polyimide peeled on MgO, compared with those peeled on Al_2O_3 and SiO_2 where 70 g/mm of peel strength was obtained. Although the details of stick–slip behavior require further investigation, the scanning electron micrographs of the peeled acid-derived polyimide surfaces on Si wafers (Fig. 10) clearly illustrate that the spacing between such plastic hinges is dependent on the interfacial adhesion strength and polyimide thickness. Figure 11 shows the peel strength as a function of the polyimide thickness on Si wafers with and without the coupling agent γ-aminopropyltriethoxysilane (γ-APS), which is widely used in microelectric packaging technology to improve adhesion [28–30]. 0.1 vol.% γ-APS solution in distilled water was spun on Si wafers and cured at 85°C for 30 min prior to spin-coating of the polyimide precursors. As shown in Fig. 10, the striation spacing becomes larger with the improved adhesion strength

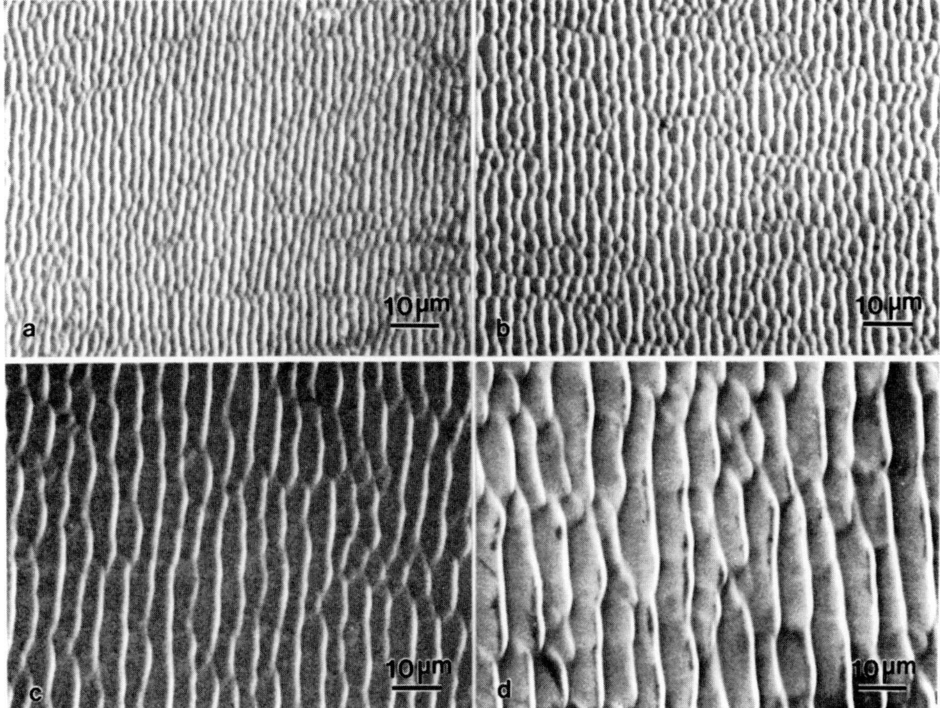

Figure 10. Scanning electron micrographs of the peeled surfaces, on Si wafers, of (a) 14 μm-thick and (b) 42 μm-thick acid-derived polyimide, and (c) 21 μm-thick and (d) 52 μm-thick acid-derived polyimide with γ-APS. For samples c and d, 0.1 vol.% γ-APS was applied to Si wafers prior to spin-coating of the polyamic acid. Note that the striation spacing increases with the film thickness and adhesion strength.

with γ-APS and larger deforming capacity with thicker polyimide films. With a low interfacial adhesion strength, as shown for acid-derived polyimide/MgO systems (Fig. 8e), bond breaking before development of local plastic hinges leads to rather continuous peel propagation without striation formation.

4. CONCLUSION

Based on the study of the adhesion strength and peeling behavior of acid- and ester-derived polyimides on Al_2O_3, SiO_2, and MgO substrates, the following conclusions can be made:

(1) The peel strengths of both PMDA–ODA acid- and ester-derived polyimides on SiO_2 and Al_2O_3 are almost identical and about two to three times higher than that on MgO at the same polyimide thickness. On MgO, the peel strength of ester-derived polyimide is about four times higher than that of acid-derived polyimide. Such a large improvement in the practical adhesion with ester-derived polyimide on MgO is not due to the mechanical properties of the polyimide, but due to the different interfacial reactions between each polyimide precursor and MgO surface.

(2) The failure locus of both PMDA–ODA acid- and ester-derived polyimides on SiO_2 and Al_2O_3 is within the polyimide. On MgO, which is rather basic with

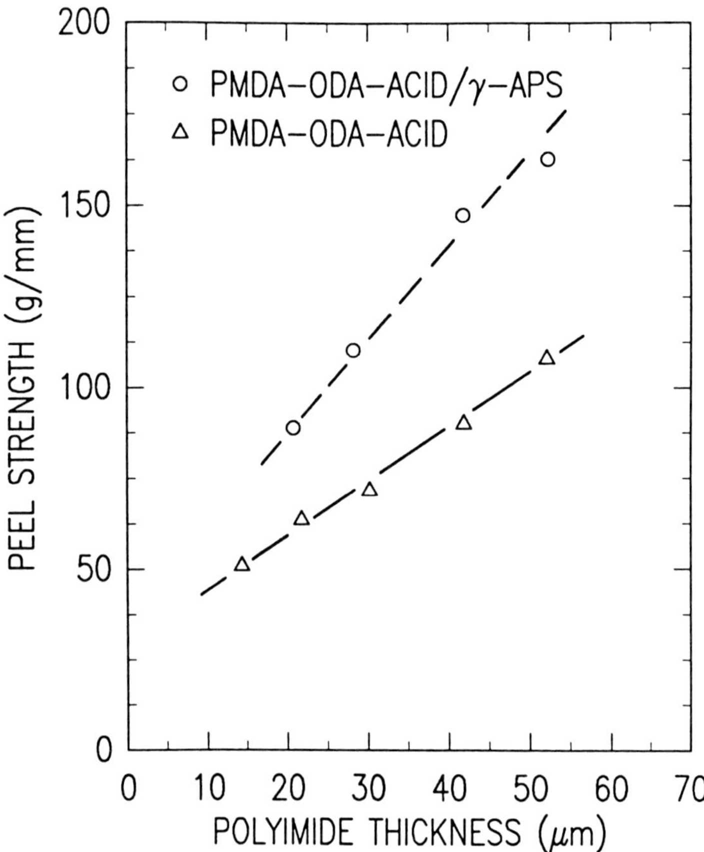

Figure 11. Peel strength of PMDA–ODA acid-derived polyimide on Si with and without 0.1 vol.% γ-APS coupling agent as a function of the film thickness.

an isoelectric point of about 12, XPS analysis implies a mixed-mode failure in the polyimide and MgO. Such a failure mode might be due to insoluble carboxylate salt formation and degradation of MgO by strong acid–base interactions between the polyamic acid and the basic MgO surface. With neutral polyamic ethyl ester, however, the locus of failure was changed to the apparent weak boundary layer of the ester-derived polyimide.

(3) With strong interfacial adhesion as on SiO_2 and Al_2O_3, peeling of polyimides shows a discontinuous stick–slip behavior due to non-uniform bending characteristics during the debonding process. As confirmed by the constant interspacing between transverse striations on the peeled polyimide surface, bending is the major deformation mode with little contribution from tensile loading. With a low adhesion strength like that of PMDA–ODA acid-derived polyimide/MgO interfaces, such transverse striations could not be observed, possibly due to bond breaking before the development of local plastic hinges.

Acknowledgements

Special thanks are due to Dr. K. W. Lee for helpful discussions, and to Mr. G. F. Walker and Mr. P. Lauro for experimental assistance.

REFERENCES

1. R. R. Tummala, R. W. Keyes, W. D. Grobman and S. Kapur, in: *Microelectronics Packaging Handbook*, R. R. Tummala and E. J. Rymaszewski (Eds), pp. 673–725. Van Nostrand Reinhold (1989).
2. Y.-H. Kim, J. Kim, G. F. Walker, C. Feger and S. P. Kowalczyk, *J. Adhesion Sci. Technol.* **2**, 95–105 (1988).
3. Y.-H. Kim, G. F. Walker, J. Kim and J. Park, *J. Adhesion Sci. Technol.* **1**, 331 (1987).
4. L. P. Buchwalter and J. Greenblatt, *J. Adhesion* **19**, 257 (1986).
5. S. P. Kowalczyk and J. L. Jordan-Sweet, *Chem. Mater.* (in press).
6. L. P. Buchwalter, *J. Adhesion Sci. Technol.* **1**, 341 (1987).
7. J. C. Bolger, in: *Adhesion Aspects of Polymeric Coatings*, K. L. Mittal (Ed.), pp. 3–18. Plenum Press, New York (1983).
8. F. M. Fowkes, D. C. Tischler, J. A. Wolfe, L. A. Lannigan, C. M. Ademu-John and M. J. Halliwell, *J. Polym. Sci., Chem. Ed.* **22**, 547 (1984).
9. D. L. Allara, F. M. Fowkes, J. Noolandi, G. W. Rubloff and M. V. Tirrell, *Mater. Sci. Eng.* **83**, 213–226 (1986).
10. L. P. Buchwalter, *J. Vac. Sci. Technol. A* **7**, 1772 (1989).
11. S. P. Kowalczyk, Y.-H. Kim, G. F. Walker and J. Kim, *Appl. Phys. Lett.* **52**, 375 (1988).
12. L. P. Buchwalter, B. D. Silverman, L. Witt and A. R. Rossi, *J. Vac. Sci. Technol. A* **5**, 226 (1987).
13. P. M. Cotts, in: *Polyimides: Synthesis, Characterization, and Applications*, K. L. Mittal (Ed.), vol. 1, pp. 223–236. Plenum Press, New York (1984).
14. K. R. Chen, private communication, IBM East Fishkill Site, Hopewell Jct., NY 12533 (1989).
15. L. P. Buchwalter and R. H. Lacombe, *J. Adhesion Sci. Technol.* **2**, 463 (1988).
16. D. C. Hofer, Organic packaging materials 'a customers viewpoint', presented at National Research Council, National Academy of Sciences, Washington, DC (1988).
17. D. W. Aubrey, G. N. Welding and T. Wong, *J. Appl. Polym. Sci.* **13**, 2193–2209 (1969).
18. A. N. Gent and G. R. Hamed, *J. Appl. Polym. Sci.* **21**, 2817–2831 (1977).
19. A. N. Gent and R. P. Petrich, *Proc. R. Soc. London, Ser. A.* **310**, 433–448 (1969).
20. K. L. Mittal, *Polym. Eng. Sci.* **17**, 467 (1977).
21. K. L. Mittal (Ed.) *Adhesion Measurement of Thin Films, Thick Films and Bulk Coatings*, STP 640, pp. 5–17. American Society for Testing and Materials, Philadelphia, PA (1978).
22. K. Kendall, *J. Phys. D, Appl. Phys.* **6**, 1782–1787 (1973).
23. K.-S. Kim and J. Kim, *Trans. ASME, J. Eng. Mater. Technol.* **110**, 266–273 (1988).
24. S. Wu, *Polymer Interfaces and Adhesion*, p. 600, Marcel Dekker, New York (1982).
25. M. G. Nicholas and D. A. Mortimer, *Mater. Sci. Technol.* **1**, 657–665 (1985).
26. A. J. Duke, *J. Appl. Polym. Sci.* **18**, 3019–3055 (1974).
27. W. T. Chen and T. F. Flavin, *IBM J. Res. Dev.* **16**, 203–213 (1972).
28. H. Ishida, in: *Adhesion Aspects of Polymeric Coatings*, K. L. Mittal (Ed.), pp. 45–105. Plenum Press, New York (1983).
29. H. G. Linde and R. T. Gleason, *J. Polym. Sci., Chem. Ed.* **22**, 3043–3062 (1984).
30. D. Suryanarayana and K. L. Mittal, *J. Appl. Polym. Sci.* **29**, 2039–2043 (1984).

Acid-Base Interactions, pp. 303-311
Eds. K.L. Mittal and H.R. Anderson, Jr.
©VSP 1991

The influence of surface acidity and basicity on adhesion of poly (ethylene-co-acrylic acid) to aluminum

M. F. FINLAYSON* and B. A. SHAH†

Dow Chemical Company, Polyolefins Research, B-3827, Freeport, TX 77541, USA

Revised version received 21 April 1990

Abstract—This work demonstrates the usefulness of flow microcalorimetry for surface characterization of metal foils (aluminum) and polymer [poly (ethylene-co-acrylic acid)] fibers. It shows that the polymer to aluminum adhesion is dominated by Lewis acid/Lewis base type interactions. These interactions are predictable from the measured heats of surface adsorption and desorption of probe molecules from dilute solution. The heats of interaction are a measure of the strengths of these sites. Adhesion between basic aluminum foil and acidic polymer resin increases with increasing numbers of either acidic sites on the polymer or basic sites on the foil. The calorimetry and adhesion results are in good agreement. This study supports recent observations *vide infra* that wettability of the aluminum is much less important for polymer/aluminum adhesion than chemical bonding.

Keywords: Flow microcalorimetry; aluminum; adhesion; poly (ethylene-co-acrylic acid); surface.

1. INTRODUCTION

The adhesion of polymeric materials to metals is important in a wide variety of technological areas. In this paper we are interested in the adhesion of poly (ethylene-co-acrylic acid) (EAA) to aluminum foil. This combination is often found in the food industry as part of a laminated beverage container consisting of paperboard, aluminum, EAA and polyethylene. The EAA functions as an adhesive between the aluminum foil and polyethylene (PE). We show here that both EAA and aluminum foil surfaces may be characterized as to their Lewis acidity and basicity and that optimum adhesion is obtained when the acid sites of the polymer interact with a maximum of basic sites on the foil. The technique of flow microcalorimetry (FMC) is used for surface characterization.

There are many techniques available to probe the surface atomic composition of materials, e.g. ESCA [1, 2], SIMS [3] and Auger [4] spectroscopies. These techniques undoubtedly yield precise and accurate information as to the presence and nature of surface species, but generally require analysis in UHV (ultra high vacuum) situations. Furthermore, these techniques analyze only very small samples and the samples are often subjected to cleaning steps prior to measurement. In an industrial situation, neither the cleanliness of the substrate nor its homogeneity are well controlled.

In order for two materials to adhere at the interface, they must develop intimate contact, which means one material must adsorb on the other. Since adsorption is

---

*To whom all correspondence should be addressed.

†Present address: Himont R&D Center, 800 Greenbank Rd, Wilmington, DE 19808, USA.

a spontaneous process, the free energy of adsorption, is negative, i.e. $\Delta G = \Delta H - T\Delta S$ is negative, where: ΔG is the change in free energy; ΔH is the change in enthalpy; ΔS is the change in entropy; and T is temperature (degrees Kelvin).

Adsorbed species are more ordered than solution species, therefore ΔS is always negative for adsorption, which makes the second term in the free energy expression positive. For adsorption to occur, ΔH must be sufficiently negative to overcome the entropic contribution. The negative ΔH results in heat being given up to the surrounding medium.

Macroscopic bond strength (adhesion) increases with increased bonding at the molecular level [5]. The bonding of polymers to inorganic surfaces involves both the general (London–Lifschitz dispersion force) and specific (acid–base) forces of attraction [6]. Acid–base interactions in organic media are exothermic [7, 8], and it has been shown [9] that the 'polar' interactions determining adsorption from non-aqueous media are purely of the Lewis acid–base type and not at all of the dipole interaction type. It is reasonable to assume, then, that increased adhesion results from increased adsorption, which is measurable as the exotherm associated with the adsorption.

The importance of Lewis type acid–base interactions has also been demonstrated for polymer–filler systems [6, 10] where it was observed that basic polymers only chemisorb on acidic fillers and acidic polymers only chemisorb on basic fillers. Such interactions should thus also be important in applications such as the adhesion of an acidic polymer (EAA) on a basic substrate (aluminum foil).

2. FLOW MICROCALORIMETRY

The FMC technique yields information as to the number, strength and types of Lewis surface sites available for interaction. Samples may be analyzed without special cleaning or ultra high vacuum requirements.

As the name implies, the technique measures the heat associated with adsorption and desorption of probe molecules onto a substrate from a flowing liquid stream. Figure 1 shows a schematic of the instrumental setup. Adsorption of the probe occurs from dilute (25 mM) solution in an inert solvent (in this case cyclohexane). The solvent (or solvent plus probe) flowrate is about 3 ml/h. Sample size employed is commonly 100–200 mg. The enthalpy decrease accompanying the adsorption means that heat is given up to the surrounding medium. This heat change in the flowing liquid stream is sensed by two thermistors placed directly into the sample. These are electrically connected in a Wheatstone bridge arrangement with another pair of thermistors in the sample chamber block. The voltage required to balance the bridge is proportional to the heat evolved. The relationship of the voltage to actual heat evolution is determined by a calibration procedure which involves injecting known amounts of heat into the sample with a resistive heating element. The heat evolution, combined with a knowledge of the sample weight and surface area, is used to obtain the total energy associated with adsorption onto—or desorption from—the sample surface sites (acidic or basic). Addition of the downstream UV detector further allows quantitative measurement of the actual number of molecules retained or released from the surface sites of each type.

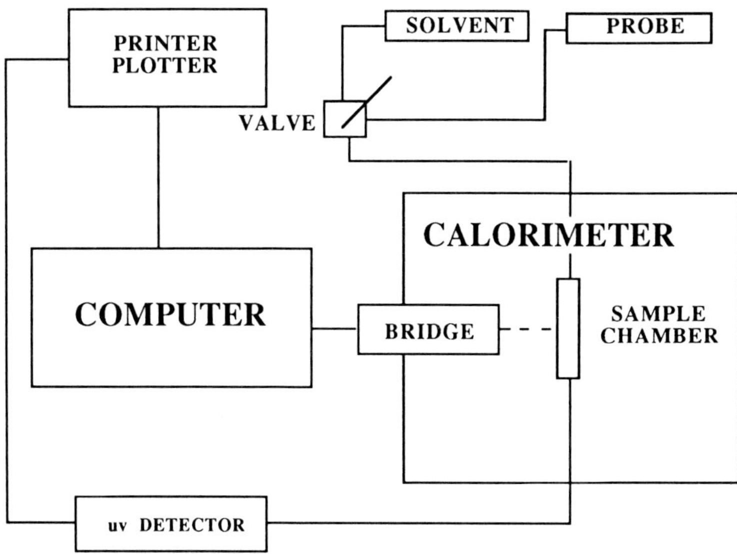

Figure 1. Experimental setup of flow mircocalorimeter. Delivery of solvent or probe solution is chosen via a valve arrangement. The solution flows over the sample in the sample chamber where the heat changes are recorded by the computer. Sample flows through to the UV detector and detector output is fed directly to a printer/plotter.

For example, pyridine is a Lewis base and will preferentially adsorb on Lewis acid sites of the sample substrate. The heat liberated is a measure of the total energy of surface acidic sites (physisorbable and chemisorbable). The number of pyridine molecules adsorbed is obtained from the time traces of the UV detector when the sample is present and when it is not. When the liquid flow is switched back from solvent containing probe to pure solvent, desorption occurs from those sites only capable of physisorbing the probe. Again, the heat changes associated with this process may be used to determine the energy of physisorption type acidic sites. In the same manner, we use phenol (a Lewis acid) as a probe molecule for determination of the strength, number and type of Lewis base sites on the sample surface. A profile of a typical run for phenol adsorption on aluminum foil is shown in Fig. 2.

3. EXPERIMENTAL

Heats of adsorption were measured on a Microscal Model 3 Flow Micro-Calorimeter equipped with two syringe pumps and a Perkin-Elmer LC-95 downstream UV absorption detector. The signal from the calorimeter is interfaced to a lab computer (IBM PC-AT) and the output from the UV flow cell detector is connected to a strip chart recorder. Probe solutions (pyridine and phenol) were prepared as 25 mM solutions in spectragrade cyclohexane. All chemicals were obtained from Aldrich, and once opened the chemicals and solutions were stored in a desiccator over Drierite™ prior to use. Moisture control is important in the FMC experiment since water is amphoteric and can react with both the acid and basic sites of the substrate. The water content of these solutions was measured to

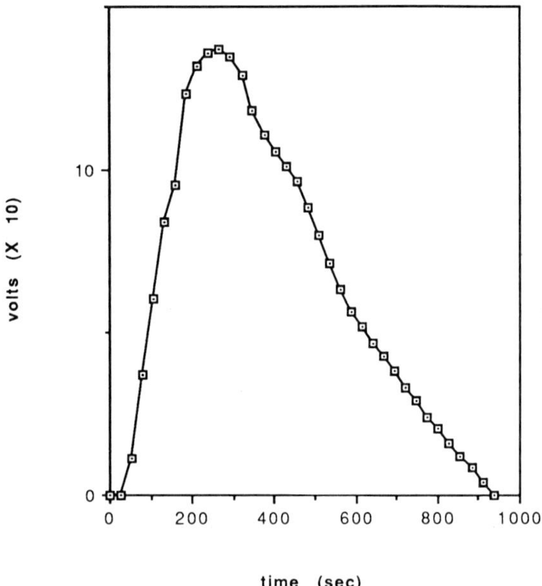

Figure 2. Typical output curve for Phenol adsorption on type 'C' aluminum. The area under the curve is related to the total adsorption energy. The conversion from area to energy is accomplished by a calibration procedure which involves injecting known amounts of heat with the sample in place.

be in the range of 0.1–0.12 mM, which should not appreciably affect the results [11] since probe concentration in the solutions is 200 times this concentration.

Aluminum foil (approx 25 μm thick) samples were obtained from Alcoa. According to the manufacturer, the surfaces of these foils are smooth (periodic hills and valleys 0.20–0.33 μm crest to trough). Three samples of aluminum foils of varying wettability (dryness codes A, B, C as measured by the manufacturer*) were analyzed. Wettability of these foils is difficult to interpret on an atomic level as the wettability is related to; (a) the extent to which residual rolling oil has been removed in the annealing process of the Al foil production; and (b) the amount of surface oxidation. Wettability is a standard quality control test at the point of manufacture, details of the test are reproduced here for completeness. Wettability is measured by placing a uniform and level stream of test liquid across the test width of a foil supported at 30° from the horizontal. Using distilled water, if the sheet is completely wet with no breakaway or drop formation, the wettability is termed 0 and the foil is 'A' type. This refers to 0% alcohol. If breakaway occurs, a 10% ethyl alcohol solution is used ('B' type foil). The alcohol content is increased until the minimum percent alcohol required for complete wetting is determined for each foil.

The aluminum foils were cut into narrow (approximately 1 mm) strips and wound around a special bobbin attachment for the FMC which was developed by Microscal for fiber measurements. A schematic of the bobbin is presented in Fig. 3. Aluminum foils were analyzed on the FMC unit with no cleaning step on our part other than drying in a vacuum oven for one hour at 50°C to remove gross

---

*Wettability is frequently referred to as dryness.

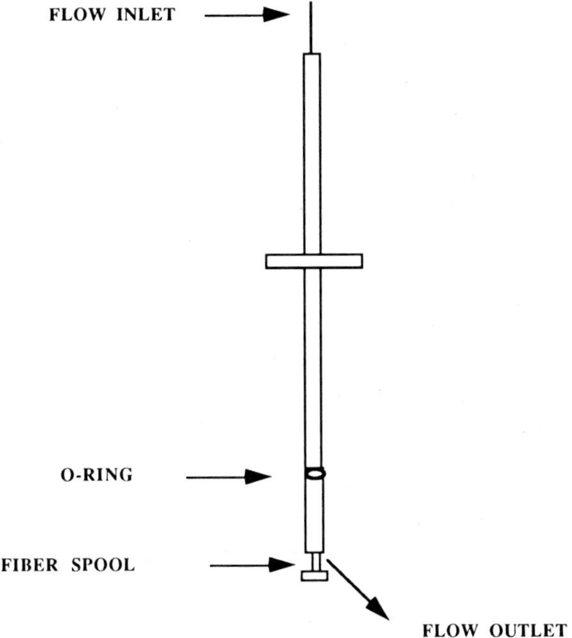

Figure 3. Bobbin attachment for fibers. Fibers are wound loosely around the fiber spool. Flow is delivered to the top of the fibers through a series of channels cut into the top of the fiber spool.

surface moisture. Further moisture evolution was obtained by vacuum treatment (0.1 torr) of the foils after mounting in the FMC unit. An endotherm, corresponding to residual surface water desorption, is commonly observed when samples are evacuated in the sample chamber. This endotherm is monitored and solvent flow is not started until the system has returned to equilibrium, indicating that the surface is substantially 'dry'. Approximately 0.02 g foil was used for each measurement, which translates into a surface area of approximately 2.9 cm^2.

Two Dow poly (ethylene-co-acrylic acid) copolymer (EAA) resins were employed in the FMC study; these are designated EAA1 and EAA2. The resins are characterized according to their melt flow rate and acid comonomer content;

Sample	MI	% Acid
EAA1	300	20
EAA2	600	08

The melt flow rate, most commonly referred to as the Melt Index (MI), is the amount of polymer in grams that is extruded through a given orifice at a particular temperature under a given applied pressure. It is inversely proportional to the melt viscosity and is proportional to the molecular weight up to the critical threshold molecular weight. Above this value it is related to the molecular weight raised to the 3.4th ($M_w^{3.4}$) power. In industry, the Melt Index is a specification commonly used for polyolefin type resins. The polymers in this study had MI measurements carried out according to ASTM 1238.

Both of these EAA resins were melt spun through a spinerette into thin (approximately 50 μm diameter) fibers. The fibers were tested on the FMC using the same procedure that was employed for the aluminum foils. An initial exotherm is usually observed when solvent is passed over the sample. This is a measure of the heat of wetting of the sample by the solvent. The FMC runs employing probe molecules were, therefore, only measured after the system had returned to equilibrium following this solvent wetting exotherm. The heat of solvent wetting is interesting in its own right as it yields information as to polymer/solvent interactions.

These copolymers are capable of forming sub-surface (bulk) complexes with zinc and other metals. It is, therefore, possible that the exotherms measured for the fibers include some sub-surface contribution. The data, however, support the notion that bulk phenomena are not a significant contribution to the exotherms measured here. The EAA1 fibers have 20% acrylic acid (M_w 72) content and thus have 0.2/72 = 2.77E-3 moles of acid sites/g fiber. If adsorption throughout the bulk is significant, then the total number of moles of adsorbed pyridine per gram of EAA fiber should approach this value. We measure only about 3E-5 moles pyridine/g fiber, i.e. only about 1% of the bulk acid content. When sodium or zinc are used to replace protons in this polymer, essentially 100% of the acid groups are neutralized. It is apparent, therefore, that at least under the conditions of these measurements, the polymer acid groups in question are of the surface and/or near surface type.

The evaluation of adhesion between EAA resins and the aluminum foil was done using extrusion coating grade EAA resins. These are;

Sample	MI	% Acid
EAA3	11.0	3.0
EAA4	11.0	6.5
EAA5	10.0	9.0

The adhesion samples were prepared by heat sealing aluminum foils between two layers of EAA in a hydraulic press at 177°C. The adhesion between the aluminum foil and EAA was measured as a 90° peel force on 2.54 cm wide strips using an Instron Tensile testing instrument. The peel speed was 5.0 cm/min and the fiber ends were pneumatically clamped in the Instron jaws.

4. RESULTS AND DISCUSSION

As stated earlier, FMC is a relatively new technique to industry. For this reason, we will first discuss reproducibility. To demonstrate reproducibility, we measured pyridine adsorption on high surface area SiO_2 particles. SiO_2 was obtained from Davison (SYLOID 245, 420 m^2/g, pore volume 1.7 cm^3/g, mean pore diameter 16.2 nm). The SiO_2 was dried here at 800°C in a nitrogen atmosphere for 6 h in order to remove all traces of water and reduce surface hydroxyl concentration [12]. Table 1 shows the calorimetry results. In run number 1 the heat of adsorption of pyridine was 22.4 J/g SiO_2. Upon washing with solvent (cyclohexane) the heat of desorption of physisorbed pyridine was 9.1 J/g SiO_2, indicating that 13.3 J/g is the heat associated with chemisorbed pyridine. This experiment was repeated on a

Table 1.

Heat of adsorption of pyridine on SiO_2 powder

Run no.	ΔH_{ads} ($-J/g$)	ΔH_{des} (J/g)
1	22.4	9.1
2	21.1	7.1

fresh sample of SiO_2 and the results shown as run number 2. The values from these runs agree within 5%.

Due to the small sample volume available in the FMC sample cell, it is desirable to have samples with the maximum amount of surface area. This is provided by fibers. The reason for using fiber grade (EAA1 and EAA2) for the FMC studies rather than the extrusion coating grade (EAA3, EAA4 and EAA5) resins is that the extrusion coating grade materials could not be melt spun into fibers because of their very low MI. Currently, we are developing methods to generate acceptable (for the FMC technique) powders from the extrusion coating grade materials. The lower MI resins were used for adhesion studies because they are extrusion coating grade resins and in an industrial situation the food containers are prepared by extrusion coating. In any case, the main intent of the work here is to show that both aluminum foils and polymers may be characterized by FMC and that the principles of acid–base interactions are applicable to the adhesion of these coatings.

The adsorption exotherm is also shown in terms of surface area in this table. This is easily calculated from a knowledge of the surface area of the sample. Surface areas of the foils were calculated from weights of 10×10 cm squares cut from the sample sheets. The heat of chemisorption is easily obtained as the sum of the heats of adsorption and desorption. It is the chemisorbed species which are important for adhesion. The FMC data on the three grades of aluminum foil presented in Table 2 show that the heat of chemisorption of phenol is highest on grade C foil. This means that grade C foil has the highest affinity for resins with Lewis acid functionality. Foil grade B shows the least affinity for Lewis acids.

The characterization of the Lewis acid surface sites of the two EAA resins is shown in Table 3. The heat of adsorption of pyridine that is associated with

Table 2.

Flow microcalorimetric surface characterization of aluminum foils

Foil	ΔH_{ads} ($-mJ/g$)	$\Delta H_{ads}{}^a$ ($-mJ/m^2$)	ΔH_{des} ($+mJ/m^2$)	$\Delta H_{chem}{}^b$ ($-mJ/m^2$)	Probe
A	507	7.32	0.89	6.43	phenol
B	743	6.83	5.17	1.66	phenol
C	988	13.3	6.26	7.04	phenol
A	565	8.16	—	—	pyridine
B	506	4.65	0.92	3.73	pyridine
C	307	4.14	0.05	4.09	pyridine

[a] The conversion from heat per unit weight (J/g) to heat per unit surface area (J/m^2) is obtained by multiplying the heat per unit weight by weight per unit surface area (g/m^2).

[b] The heat of chemisorption is obtained as the sum of the heats of adsorption and desorption.

Table 3.
Flow microcalorimetric surface characterization of EAA fibers

Sample	ΔH_{ads} ($-$ mJ/g)	ΔH_{ads} ($-$ mJ/m^2)	ΔH_{des} ($+$ mJ/m^2)	ΔH_{chem} ($-$ mJ/m^2)	Probe
EAA1	278	66.1	38.7	27.8	pyridine
EAA2	87	20.7	9.67	11.0	pyridine

chemisorption increased by a factor of $[(-)66.1 + 38.7]/[(-)20.7 + 9.67] = 2.48$ which is in excellent agreement with the increase in acid content $(20/8) = 2.5$.

The high affinity of aluminum foil grade C for resins with higher Lewis acidic surface site content is reflected in the adhesion values shown in Fig. 4. Grade C foil has the highest adhesion to all three EAA resins. Grade B foil has the lowest heat of chemisorption of phenol. This foil also exhibits the lowest adhesion with all three EAA resins.

The plateauing effect of adhesion strength to foil type 'C' with increasing acid content of the polymer is interesting. It represents a change in fracture mechanism from interfacial at the lower levels to cohesive at the plateau level. The phenomenon results from the adhesion strength becoming greater than the tear strength of the films.

With respect to the apparent correlation of adhesion with chemisorption found in this study, we note that the importance of chemical bonding in similar systems has already been established. For example, IR spectra have shown the presence of an aluminum carboxylate species for adhesion of polyethylene grafted with acrylic acid [13]. Also, an Al oxide–carbide species is formed when Al is sputter deposited onto polyacrylic acid, whereas only an Al carbide-like species is formed

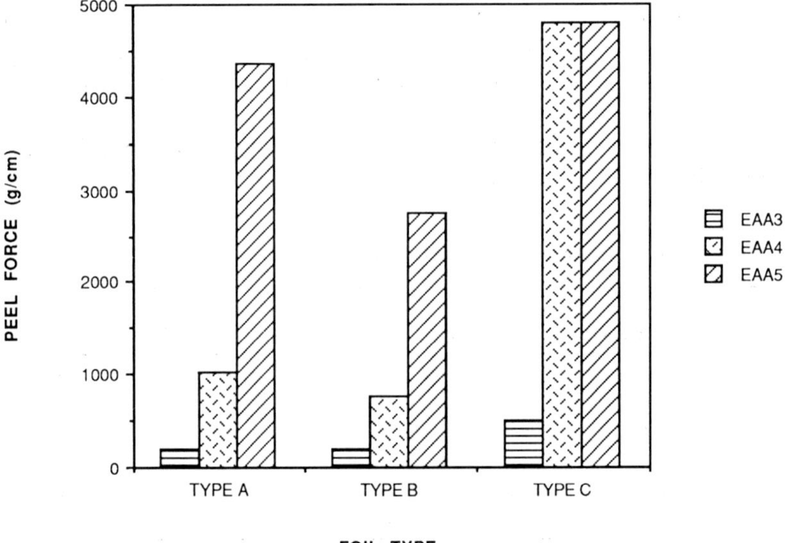

Figure 4. Effect of acid content and aluminum foil type on adhesion of EAA resins to aluminum foils. EAA 3, 4 and 5 contain 3, 6.5 and 9% acrylic acid, respectively. Increasing acid content of the polymer results in increased adhesion. FMC shows the order of foil basicity to be C>A>B. Increased basicity of the foil also results in increased adhesion.

when Al is sputter deposited onto low density polyethylene [14]. Finally, it has recently been concluded from a comparison of surface properties assessed by wettability data that surface polarity of the aluminum is not the main determining parameter for adhesion [15]. In that report it was concluded that polyolefin adhesion to aluminum occurs through chemical bonding of acid groups grafted in the polymer bulk with basic groups present on the aluminum.

5. SUMMARY

The FMC technique offers a route to directly and quantitatively determine the numbers and strengths of acidic and basic surface sites on polymers and on thin metal foils. In the case of poly (ethylene-co-acrylic acid) adhesion to aluminum, the relative orders of adhesion strengths of three different types of foil with three different EAA resins are correctly predicted. The acid–base contribution to adhesion is clearly demonstrated for this system. FMC also allows quantitation of the physisorbable sites. Physisorption contributes further to adhesion and in the absence of chemical bonding it is the main contribution. Direct information as to the kinetics of adsorption from solution is also provided by the FMC technique. In short, FMC can yield a plethora of useful information for quantitative surface characterization.

Acknowledgements

The authors with to thank Mary Vaughn for carefully performing the calorimetric measurements.

REFERENCES

1. D. W. Dwight and W. M. Riggs, *J. Colloid Interface Sci.* **47**, 650 (1974).
2. D. Briggs, D. M. Brewis and M. B. Konieczko, *J. Mater. Sci.* **11**, 1270 (1976).
3. G. H. Morrison, in: *Characterization of Metal and Polymer Surfaces*, L. H. Lee (Ed.), vol. 1. Academic Press, New York (1977).
4. J. A. Kelber, R. R. Rye, D. R. Jennison and J. E. Houston, in: *Physicochemical Aspects of Polymer Surfaces*, K. L. Mittal (Ed.). Plenum Press, New York (1983).
5. K. L. Mittal, in: *Adhesion Science and Technology*, L. H. Lee (Ed.), pp. 129–168. Plenum Press, New York (1975).
6. F. M. Fowkes, in: *Surface and Colloid Science in Computer Technology*, K. L. Mittal (Ed.), pp. 3–25. Plenum Press, New York (1987).
7. R. S. Drago, G. C. Vogel and T. E. Needham, *J. Am. Chem. Soc.* **93**, 6014 (1971).
8. R. S. Drago, L. B. Parr and C. S. Chamberlain, *J. Am. Chem. Soc.* **99**, 3203 (1977).
9. F. M. Fowkes, in: *Physicochemical Aspects of Polymer Surfaces*, K. L. Mittal (Ed.), vol. 2, pp. 583–603. Plenum Press, New York (1983).
10. F. M. Fowkes and M. A. Mostafa, *I&EC Prod. R&D* **17**, 3 (1978).
11. S. T. Joslin and F. M. Fowkes, *I&EC Prod. R&D* **24**, 369 (1985).
12. R. K. Iler, *The Chemistry of Silica*, p. 645. Wiley, New York (1979), and references therein.
13. J. Schultz, A. Carre and C. Mazeau, *Int. J. Adhesion Adhesives* **4**, 4 (1984).
14. B. M. DeKoven and P. L. Hagans, *Appl. Surface Sci.* **27**, 199–213 (1986).
15. J. Schultz and L. Lavielle, *Makromol. Chem., Macromol. Symp.* **23**, 343 (1989).

Acid-Base Interactions, pp. 313-328
Eds. K.L. Mittal and H.R. Anderson, Jr.
©VSP 1991

The effects of filler/polymer acid–base interactions in model coating systems

A. C. TIBURCIO[1],* and J. A. MANSON[2],†

[1] *General Electric Co., Electromaterials Group, 1350 S. Second St., Coshocton, OH 43812, USA*
[2] *Materials Research Center, Lehigh University, Bethlehem, PA 18015, USA*

Revised version received 5 June 1990

Abstract—The water vapor permeability of glass-bead-filled phenoxy films was shown to depend strongly on the degree of interfacial interaction between the filler and matrix; the greater the adhesion, the lower the permeability. Scanning electron microscopy (SEM) was used to characterize the glass surface and the corresponding degree of adhesion between the filler and polymer matrix. Maximum interaction between the acidic phenoxy and glass filler was obtained when the glass had been treated with an aminopropyltriethoxysilane, which yielded a basic surface overall. Retention of the cellosolve acetate solvent was also reduced by the glass filler, especially by the more basic glass. The dynamic mechanical properties were affected primarily by the presence of residual solvent.

Keywords: Acid–base interactions; interfacial interactions; filled phenoxy coatings; glass surface; aminosilane.

1. INTRODUCTION

The interaction between a pigment or filler with a polymer binder affects the permeability and other properties of composite materials. This fact is important in protective coatings because the permeation of water and oxygen, and the mechanical properties play important roles in corrosion [1, 2] and durability [3]. Thus, as part of a major program to understand the fundamentals of corrosion protection, a study on the effects of interfacial interaction in actual and model coatings was begun, with emphasis on permeability and mechanical behavior.

Within the past few years, it has become apparent that Lewis acid–Lewis base (or electron donor–acceptor) interactions often dominate such phenomena as miscibility, adsorption, and adhesion [4–20]. Therefore, it was of interest to investigate various combinations of pigments or fillers with binders having similar or opposite acid–base characteristics. Indeed, preliminary studies showed that the permeability behavior of zinc-rich epoxy coatings could be interpreted in terms of the acid–base concept as adapted by Fowkes and Mostafa for specific interactions of filler–matrix interfaces [6]. Maximum adhesion was obtained when the pigment–binder system constituted an acid–base pair and solvent interactions with either component were negligible [13, 20].

In this investigation, glass beads (both untreated and treated with an amino-silane coupling agent) and a phenoxy binder were used to study the water vapor permeability, dynamic mechanical behavior, and solvent retention of a coating as a

---

*To whom correspondence should be addressed.
†Deceased.

function of the filler concentration, and filler/binder acid–base characteristics. Qualitative differences between the surfaces of the treated and untreated fillers were observed by scanning electron microscopy (SEM) and by electron spectroscopy for chemical analysis (ESCA). The acid–base concept was demonstrated to be useful in understanding filler–binder interfacial effects.

2. LEWIS ACID–LEWIS BASE THEORY

According to the Lewis acid–Lewis base theories of Drago, as adapted by Fowkes for specific filler–matrix interactions [6, 21–27], the enthalpy of adsorption may be divided into three main components: dispersion, acid–base, and dipole–dipole interactions [6, 22]. However, the dipole–dipole interactions are negligible in comparison with the dispersion and acid–base interactions [6, 22]. Moreover, for composite systems, the contribution of dispersion interactions is small compared with donor–acceptor interactions. Therefore, maximum filler/polymer adsorption results when the heats of acid–base interactions are maximized.

All polymers with the exception of saturated hydrocarbons (which are neutral materials) behave as electron donors, acceptors, or both, and their interactions with other species are governed by acid–base interactions. In the presence of a solvent, the adsorption of polymers onto the surface of inorganic fillers involves competitive interactions between the polymer, filler, and solvent. Adsorption results only when acid–base interactions between the polymer and filler dominate. Correspondingly, polymer adsorption onto the filler may be enhanced by modifying the filler surface with an appropriate coupling agent.

3. PERMEABILITY

Various permeability equations have been derived which are useful in interpreting experimental data by showing trends associated with some effects and setting limits to be expected of some parameters [28–31]. In the limiting case of good adhesion, the relative permeability coefficient, P_r, for *spherical* fillers is given by [28]:

$$P_r = P_c/P_{PL} = V_p/(1 + 0.5 V_f), \tag{1}$$

where P_{PL} is the permeability of the permeant through the bulk polymer; V is the volume fraction of an isotropic inclusion; and the subscripts c, p, and f correspond to the composite, polymer, and filler, respectively. However, this simple model fails to consider the contributions of interfacial interactions and the solubility of vapors or liquids in the system, which may vary from the interface to the bulk polymer. To account for such cases, more general expressions are needed.

Two such equations, developed by Nielsen, are considered below [29]. Equation (2) expresses the permeability of the limiting case where the interfaces in a composite form channels all the way through the composite:

$$P_r = P_c/P_{PL} = [P_i V_{Li}/P_{PL}\tau^*] + [(V_P + V_{LP})/\tau] \tag{2}$$

where P_i is the permeability of the permeant through the interfacial part; V_{Li} and V_{LP} correspond to the volume fraction of permeant dissolved in the interface and polymer, respectively; τ^* is the tortuosity factor for the interfacial region; and τ represents the tortuosity factor through the bulk polymer. Equation (2) divides the total permeability into two parts (the interface and bulk polymer) and takes into account the contribution of dissolved permeant. In this case, one important parameter to consider is the ratio P_i/P_{PL}, the relative permeability of the interfacial region and the bulk polymer. Thus, equation (2) should model a composite that exhibits a low degree of filler–matrix interaction.

Another factor to consider is filler distribution heterogeneity. In this case, the void volume in the aggregates increases the permeability of the material. A system where the filler adheres strongly to the polymer matrix produces a more uniform specimen, whereas one that exhibits low filler–matrix adhesion enhances the formation of filler aggregates [32]. Equation (3) represents this type of behavior, where P_a is the permeability of the diffusing molecule through the porous aggregate. For equation (3), the value of P_a/P_{PL} has a significant effect on the overall permeability.

$$P_r = [AV_f/(V_f^{1/3} + AB)] + (V_P + V_{LP})$$ (3)

$$A = P_a/P_{PL}$$ (4)

$$B = 1 - V_f^{1/3}.$$ (5)

4. EXPERIMENTAL

4.1. The model system

A model system was selected, consisting of glass beads as fillers, a high-molecular-weight linear phenoxy binder, and cellosolve acetate solvent. A linear binder eliminates the complication associated with crosslinked systems, namely, the problem of residual curing; while a single solvent formulation minimizes the number of components in the system. The selection of glass beads provides fillers whose surfaces can be readily modified. Untreated and treated glass beads were used. Also, a polyamide-cured epoxy was used as another binder, using glass beads and zinc particles as fillers. Furthermore, compression-molded specimens of unfilled and glass-bead-filled phenoxy samples were made to study the effects of residual solvent and interfacial interaction separately.

4.2. Materials

The phenoxy resin from Union Carbide (PKHH) constituted one binder. An epoxy based on Epon 1001F (Shell) cured with Polyamide 1511 (Emerez) served as the other. Glass spheres, type 3000 (untreated) and type 3000-CP03 (treated by the manufacturer with an aminopropyltriethoxysilane coupling agent), with a mean particle size of less than 25 μm were supplied by Potters Industries.

4.3. Sample preparation

Solvent-cast phenoxy films were prepared by dissolving 20 parts-by-weight phenoxy in 80 parts-by-weight cellosolve acetate and adding the desired amount of glass spheres: 0.1, 0.2, and 0.4 volume fractions. The specimens were cast in small aluminum pans and dried in two stages: 1-day drying at ambient room temperature followed by vacuum-oven drying at 40°C to a constant film weight. All films were 0.1 mm thick. Other films were cast: poly(styrene) in xylene and chlorinated poly(vinyl chloride) in THF.

Extruded samples (unfilled phenoxy polymer and phenoxy containing 0.1 volume fraction glass) were compression-molded at approximately 150°C and 3.4 MPa pressure for 10 min. The specimens were then quickly quenched in warm water.

4.4. Permeability testing procedures

Moisture-vapor permeability, P, was determined using a permeation cup method (ASTM D-1653) with calcium chloride as the desiccant. Values of P were calculated over a time range from t_{i-1} to t_i from the values of the weight loss of water ($G_{i-1} - G_i$) using the following equation [33]:

$$P = \frac{(G_i - G_{i-1})\, d}{A\,(t_i - t_{i-1})\, \Delta p} \tag{6}$$

where P is the permeability coefficient (in g cm/cm^2 Pa s), G is the water loss (g), d is the film thickness (cm), $\Delta p = (p_1 - p_2)[p_1$ and p_2 being the partial pressures of water vapor on opposite sides of the film (Pa)], A is the sample area (cm^2), and t is the time (s).

4.5. Solvent retention

After the phenoxy films were cast in aluminum pans, solvent loss was measured gravimetrically at intervals throughout the two stages of drying.

4.6. Differential scanning calorimetry (DSC)

Specimens were scanned in a DSC unit (Perkin-Elmer, model DSC-1B) at a rate of 20°C/min to obtain the glass transition temperature, T_g.

4.7. Scanning electron microscopy (SEM)

Fracture surfaces were coated with an Au/Pd layer and examined using an ETEC SEM unit.

4.8. Dynamic mechanical spectroscopy

An Autovibron unit (Toyo Baldwin Co., Model DDV III) was used to measure dynamic mechanical properties, including the storage and loss moduli (E' and E'', respectively), tan δ (the loss tangent, E''/E'), and T_g.

4.9. Spectroscopy

A Mattson Sirius 100 and a JEOL FT-NMR FX900 were used for infrared and NMR measurements, respectively. ESCA measurements were performed on a Physical Electronics Model 548 ESCA/Auger Spectrometer.

5. RESULTS AND DISCUSSION

5.1. Filler characterization

The nature of the glass bead surface is of importance because it controls the filler–matrix interface. As mentioned above, one type of glass bead was treated with an aminopropyltriethoxysilane coupling agent while the other was left untreated. SEM micrographs (Fig. 1a and 1b) show the treated glass bead surfaces to be rougher in appearance than those of the untreated ones, while ESCA studies indicate the presence of nitrogen on the treated glass beads but not on the untreated one, confirming the presence of the amino groups on the surface of the treated filler surface. These differences in surface properties manifest themselves in the adhesion behavior of each filler type.

Fractured surfaces of the various solvent-cast films revealed that the untreated glass spheres do not interact to a significant degree with any of the following polymeric binders: acidic epoxy and chlorinated poly(vinyl chloride) (Cl-PVC), and basic polystyrene (PS) [14–16] and poly(methyl methacrylate) (PMMA) [34]. In contrast, the treated filler adheres well with the acidic epoxy (Fig. 1c and 1d) but not with the basic PS or PMMA, nor with the acidic Cl-PVC (Fig. 2).

According to the acid–base concept, either the untreated glass is effectively neutral or else the solvent interacts more strongly with the glass or the polymer, inhibiting filler–matrix interaction. On the other hand, the treated filler whose surface contains an amino group acts as an electron donor, and interacts with the epoxy matrix but not with the PS or PMMA. The low degree of filler–matrix interaction between the treated filler and Cl-PVC further suggests the effect of solvent on the polymer–filler interaction. When dioxane, chloroform, and THF were used as solvents, the Cl-PVC showed little or no bonding with the treated filler, an

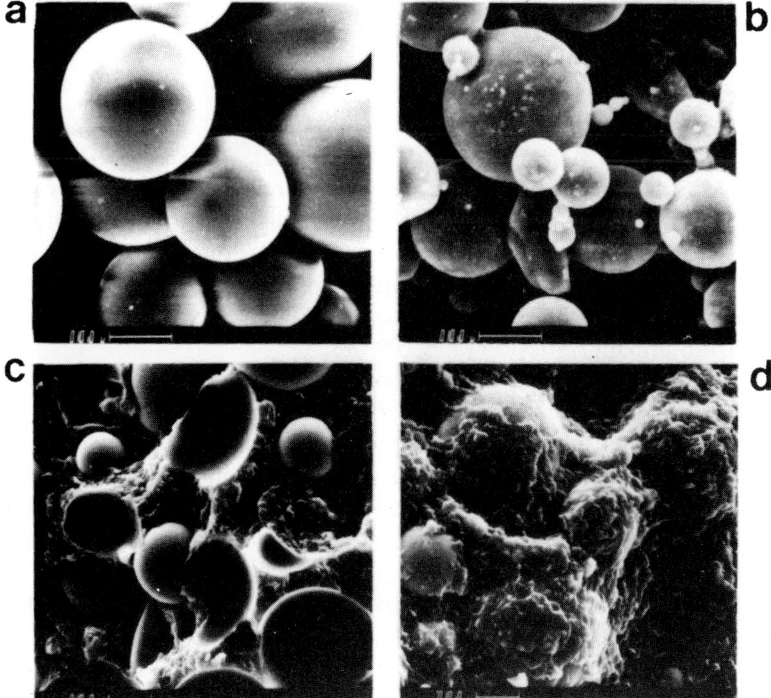

Figure 1. SEM micrographs of glass spheres: (a) untreated; (b) aminosilane-treated. SEM micrographs of fractured glass-sphere-filled epoxy films: (c) untreated; (d) aminosilane-treated.

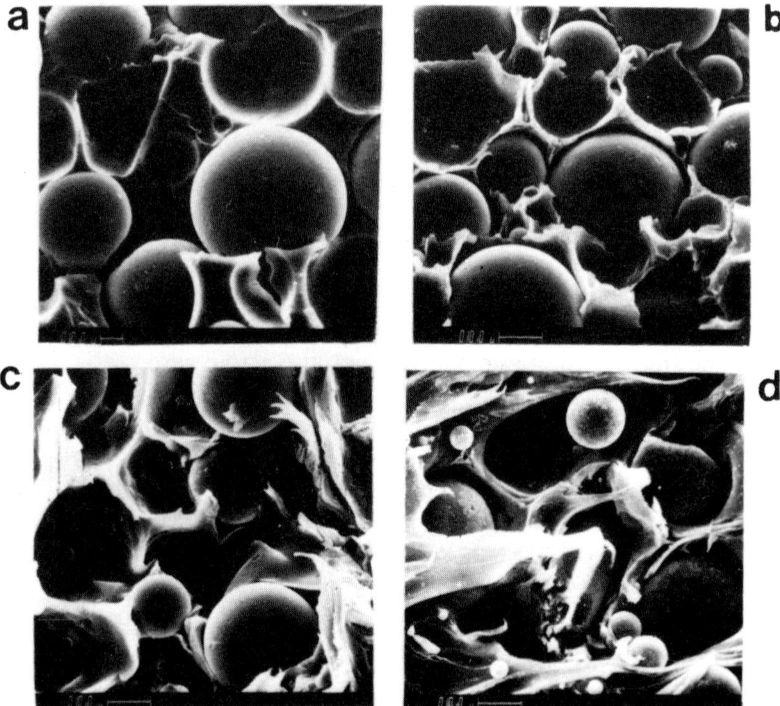

Figure 2. SEM micrographs of fractured glass-sphere-filled films. Polystyrene cast in xylene: (a) untreated; (b) aminosilane-treated. Chlorinated poly(vinyl chloride) cast in THF: (c) untreated; (d) aminosilane-treated.

observation made by Marmo with Cl-PVC (acidic) and $CaCO_3$ (basic) in THF [35]. The basic solvents (dioxane and THF) prevent Cl-PVC from interacting with the treated glass, while the acidic chloroform adsorbed preferentially onto the treated glass, inhibiting polymer–filler interaction. Thus, even when the filler–binder matching with respect to H-bonding is appropriate, the use of a solvent that is more basic or acidic than the polymer or filler competes with the polymer–filler adsorption, minimizing adhesion [5, 6, 35].

Figure 3 exhibits the fracture surface of solvent-cast and compression-molded glass-bead-filled phenoxy samples. In both cases, the untreated filler does not adhere to the polymer matrix while the phenoxy binder adsorbs strongly onto the treated filler. Since the phenoxy behaves as an electron acceptor [14, 36], these results are consistent with the findings above. The solvent does not interact strongly enough with the binder or the glass spheres to reduce the degree of filler–matrix adhesion between the phenoxy binder and the treated filler.

The adsorption behavior of glass beads has also been confirmed by inverse gas chromatography (IGC) studies [18]. The IGC analyses, which measure the acid–base character of the glass bead surface, have shown the treated glass to be more basic than the untreated glass bead. The heat of adsorption of basic probes (dimethyl ketone and diethyl ether) with the two glass types differed by at least 9.4 kJ/mol, a magnitude which is sufficient to promote or inhibit adsorption.

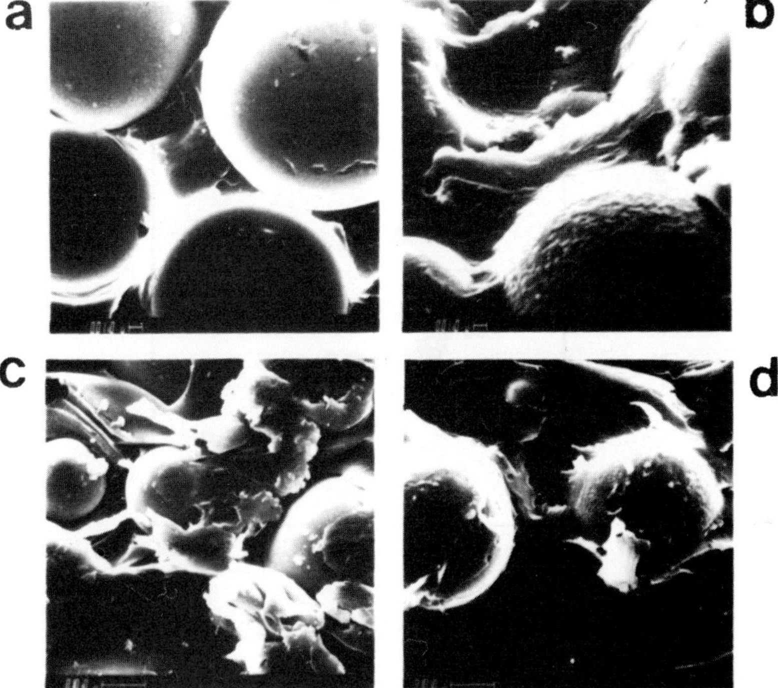

Figure 3. SEM micrographs of fractured glass-sphere-filled phenoxy films. Solvent-cast films: (a) untreated; (b) aminosilane-treated. Compression-molded: (c) untreated; (d) aminosilane-treated.

5.2. *Drying behavior*

Figure 4 shows the drying behavior of the two glass types at different filler concentrations. The y-axis corresponds to the amount of solvent retained in the phenoxy binder.

Several trends are evident. It can be seen that both systems exhibit the two stages of solvent release proposed by Hansen [37, 38]: an initial rapid period of solvent loss due to surface evaporation, followed by a slower rate of solvent loss arising from bulk diffusion. After an initial 24 h when the rate of solvent loss is high, the rate of solvent release decreases gradually, reaching equilibrium some time after $3\frac{1}{2}$ days.

5.3. *Bound solvent*

The solvent content steady-state values (determined gravimetrically) are shown in Table 1. Clearly, significant amounts of residual solvent are retained in the binder. As expected, the basic character of the carbonyl and ether groups of the cellosolve acetate results in a strong degree of hydrogen bonding with the acidic hydroxyls of the phenoxy resin. Indeed, the FT-IR spectrum of an unfilled solvent-cast phenoxy film exhibited a strong absorption band in the region corresponding to the carboxyl group of the cellosolve acetate ($1720–1760$ cm^{-1}). Since the phenoxy has no carbonyl groups of its own, the presence of a carbonyl peak can only be attributed to residual solvent.

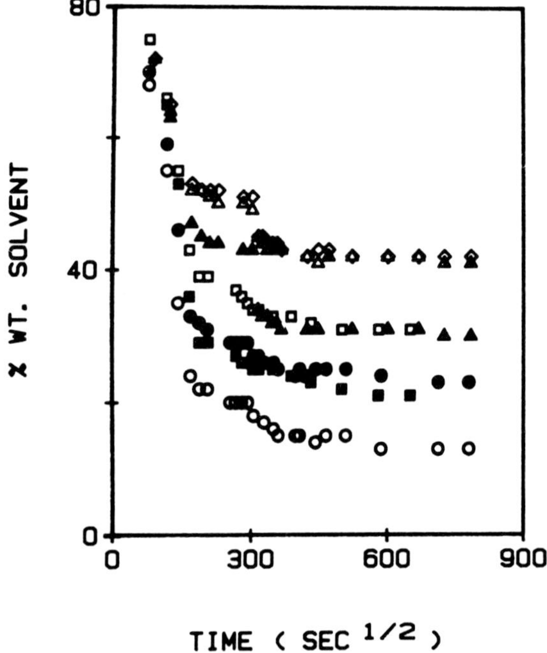

Figure 4. Solvent retained in the glass-sphere-filled phenoxy films as a function of the (drying time)$^{1/2}$. ($\diamond$) Unfilled control; ($\triangle$, $\blacktriangle$) $v_f = 0.1$; ($\bigcirc$, $\bullet$) $v_f = 0.2$; ($\square$, $\blacksquare$) $v_f = 0.4$. Open and closed symbols for filled systems correspond to untreated and aminosilane-treated beads, respectively. The discontinuities reflect the change from air to vacuum-oven drying.

Table 1.
Retained solvent^a in glass/phenoxy films as a function of the filler content

Vol. fraction of glass beads	% retained solvent at equilibrium		
	Untreated beads	Treated beads	Theoretical^b
0	(42.1)^c	(42.1)^c	—
0.1	41.0	30.0	37.9
0.2	13.0	23.3	33.7
0.4	31.1	21.0	25.3

^{a} Determined gravimetrically.
^{b} Estimation based on the actual volume fraction of the binder.
^{c} Phenoxy only.

The solvent content of the pure matrix can also be estimated from the lowering of T_g (see Fig. 5 and Table 2) using one of several equations [39, 40]. Closest agreement with the gravimetric results was obtained using the limiting form of the Kelly and Bueche equation [41]. In this calculation, the cellosolve acetate T_g was taken to be 143 K, based on values of the ratios of T_g to the boiling and melting points given by Bondi and Tobolsky for low-molecular-weight esters and other polar liquids [42]. However, the calculated values (24 and 22%-by-wt, from the Autovibron and DSC data, respectively) are significantly lower than the gravimetric calculations. Such a deviation from linearity in the T_g-composition curves in

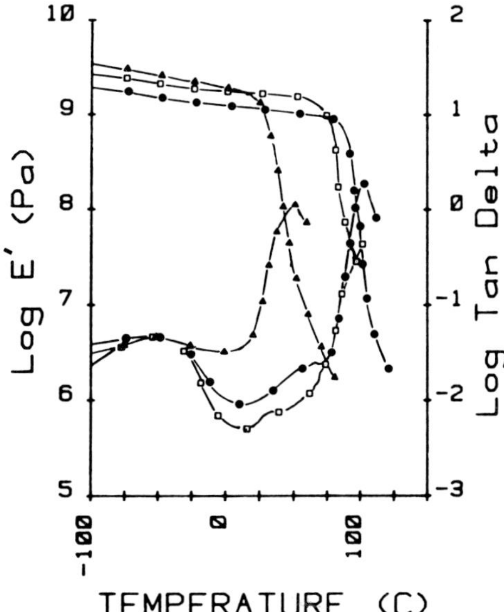

Figure 5. Dynamic mechanical spectra of phenoxy specimens: (▲), solvent-cast; (●) compression-molded dry resin; (□) compression-molded resin containing 0.7 wt% water. Upper and lower sets of curves correspond to log E' and log tan δ, respectively.

Table 2.
Effect of residual water and solvent on the $T_g{}^a$ of the phenoxy binder

Resin	Vol. fraction of filler	Autovibron tests[b]		DSC tests	
		T_g (°C)	% wt loss[c]	T_g (°C)	% wt loss[d]
Solvent-cast	0.0	33	1	26	1
Containing 0.7 wt% water	0.0	85	0.4	74	0.6
Dry resin	0.0	93	—	83	—
Dry resin (treated beads)	0.1	90–95	—	83	—
Dry resin (untreated beads)	0.1	90–95	—	83	—

[a] The difference between the two sets of T_g values is attributable to the time scale of the experiment (Autovibron at 110 Hz vs. DSC).

[b] T_g defined as E'' maximum. Note that Figs 5 and 6 show values of log tan δ vs. T (tan $\delta = E''/E'$).

[c] After completion of testing.

[d] After second DSC scan.

polymer–diluent systems has been interpreted in terms of specific interactions [43].

In general, for each glass bead type, the sample with the highest amount of binder retains the most solvent, while for the same volume fraction of filler, the untreated glass spheres contain more residual solvent (about 10%). The reason for the anomalous behavior of the material with 0.2 volume fraction filler is not known, but its steady-state value seems low in comparison with the other samples.

The significant difference between the solvent retention behavior of the treated and untreated specimens implies that filler–matrix interaction plays a role. As shown in Fig. 3, the phenoxy binder adheres rather well onto the treated filler but not to the untreated one. If this is the case, the formation of a 'tight' interfacial region, which is believed to reduce permeant transport through a composite material, might also be expected to inhibit solvent release. At the same time, the increased porosity of the untreated glass-bead-filled samples resulting from low filler–matrix interaction would be expected to provide an easier path for bulk diffusion. However, the experimental results show just the opposite trend. In fact, the *untreated* filler samples retain more solvent than the *treated* glass-sphere-filled specimens, while with treated beads, less solvent is retained than in the pure matrix. Thus, when adhesion is high, the effective fraction of phenoxy available to interact with the solvent is reduced.

Therefore, we proposed that the basic solvent (cellosolve acetate) must interact with the acidic phenoxy binder strongly enough so that any residual solvent is tightly bound to the binder. Consequently, the strong solvent and polymer matrix interaction can overcome the presumably easier diffusion due to the weak interface of the untreated beads. In contrast, with the treated glass, the amount of solvent retained must be limited by the lower entropy of the polymer fraction involved in the tight and strong interface.

5.4. Viscoelastic behavior

Log tan δ and log E' are plotted in Fig. 5 as a function of temperature for three unfilled phenoxy specimens: solvent-cast film, compression-molded phenoxy containing 0.7% water by weight, and a compression-molded dry specimen. It can be seen that residual solvent or water acts as a plasticizer. Thus, the T_g, tan δ, and Young's modulus vary with the amount of retained solvent or water.

The magnitude of the T_g is of interest because it reflects the degree of plasticization. Table 2 shows that the T_g value follows the order pure polymer $>$ 0.7% water $>$ solvent-cast. Other plasticization effects manifest themselves in the dynamic mechanical behavior of the samples. The increase in modulus in the glassy region parallels the increase in the amount of residual solvent present (an anti-plasticization effect). Similarly, tan δ of the solvent-cast film rises in comparison with that of the pure polymer, and the sample containing 0.7% water reflects the higher damping capability of the more plasticized polymer material.

The effects of the filler are also of interest (Fig. 6). As expected, the modulus of the phenoxy in the glassy state increases with the incorporation of filler particles. Apparently, the magnitude of the increase does not depend on the type of glass spheres, at least not at this filler content. Similarly, the tan δ behavior of the two glass fillers exhibits no significant difference, but the presence of a rigid second phase lowers the damping behavior of the phenoxy matrix relative to the unfilled matrix. Also, the results shown in Table 2 indicate that the filler particles do not significantly change the segmental mobility of a large enough fraction of phenoxy to affect the T_g, at least not for the 0.1 volume fraction glass.

5.5. Permeability of phenoxy systems

The permeability of water vapor through phenoxy films containing various volume fractions of glass spheres was measured. The transport behavior exhibited an initial dynamic period of rapid permeation followed by a more gradual decrease in permeability until the steady state was reached after 200 h. Equilibrium values of P_c along with calculated P_r values are reported in Table 3. It is seen that the P_r

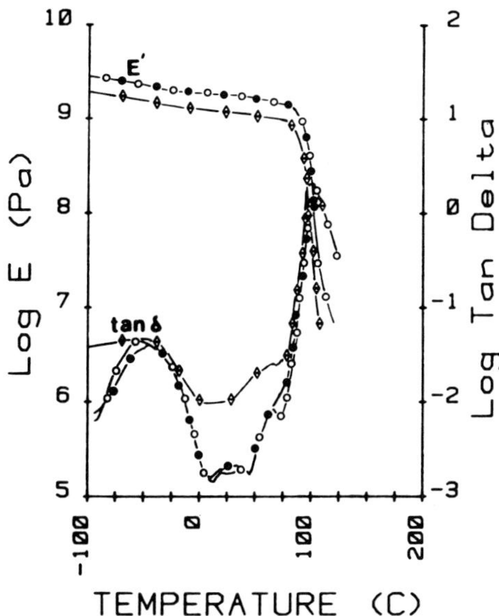

Figure 6. Dynamic mechanical spectra of compression-molded phenoxy specimens: ⬦, dry resin; (○) 0.1 volume fraction untreated glass spheres; (●) 0.1 volume fraction aminosilane-treated glass spheres.

Table 3.
Experimental[a] and theoretical[b] permeabilities of glass bead/phenoxy films

Vol. fraction of filler	Experimental permeabilities, $P_c{}^a$ $(P_r)^b$		Theoretical relative permeabilities, P_r		
	Untreated glass	Treated glass	Eqn. (1)[c]	Eqn. (2)[d]	Eqn. (3)[e]
0	3.04 (1.0)	3.04 (1.0)	1.0	1.0	1.0
0.10	7.51 (2.47)	6.53 (2.15)	0.86	2.0	1.08
0.20	8.71 (2.87)	5.68 (1.87)	0.73	3.0	1.28
0.40	28.0 (9.21)	6.40 (2.11)	0.50	5.0	2.12

[a] Units: g cm/cm^2 s Pa $\times\,10^{14}$; under ambient conditions and 0% relative humidity.
[b] $P_r = P_c/P_{PL}$.
[c] Calculated assuming ideal case of good adhesion [equation (1)].
[d] Calculated from equation (2) representing interfacial channeling; $P_{PL} = 1$, $V_{Li} = 0.1\ V_f$, $\tau^* = 1$, $\tau = 1 + 0.5\ V_f$, and $P_i = 100$.
[e] Calculated from equation (3) which predicts filler aggregation.

values are all greater than unity, indicating that the presence of the filler results in an increase in permeability rather than the decrease predicted for a simple, well-bonded composite system [28, 31]. Thus, it is likely that solvent casting yields a system that contains pathways for diffusion that would not exist otherwise. However, a strong difference exists between systems containing untreated and treated glass beads, suggesting filler–matrix effects.

For the untreated glass-bead system, weak acid–base interaction increases the permeability monotonically with the filler content, while for the treated glass-bead system (strong acid–base interaction), the permeability tends to level off with the filler content, rather than decrease as predicted by theory. Decreases have been noted in some pigmented paint films and filled epoxy resins [44–48]. At the same time, exceptions to the predicted decrease are well known in cases where the filler is not well bonded or wetted by the resin, or where the filler or interface absorbs the penetrant [44–47, 49–51]. Comparison in more detail with the Nielsen models distinguishes which factors may be important. Three cases are of interest: the limiting case of good adhesion [equation (1)], interfacial channeling [equation (2)], and filler aggregation [equation (3)].

Figure 7 exhibits the transport behavior corresponding to the various models for the phenoxy system. As mentioned above, the filled specimens have permeabilities higher than the values predicted for a well-bonded, impermeable second phase (curve 'Ideal'). Since the polymer binder adsorbs strongly onto the treated glass beads, the difference between the 'Ideal' curve and the experimental curve for the treated glass suggests a matrix which has a high free volume, a finding discussed in more detail below. As expected, the 'Ideal' curve dramatically fails to fit the experimental curve of the untreated filler system.

Another model to consider is the one representing incomplete filler dispersion or aggregation, curve 'P_a'. While curve 'P_a' roughly approximates the transport behavior of the treated glass-bead-filled phenoxy systems, the effects of filler aggregation should be less evident for the treated filler samples because good filler–matrix interaction promotes a uniform filler distribution [32]. More importantly, curve 'P_a' rises at a higher volume fraction of filler, a trend uncharacteristic of the treated glass-bead-filled phenoxy system. Filler aggregation may contribute to the increase in the permeation rate of the untreated system but curve 'P_a' yields values much lower than the untreated glass/phenoxy films, even when P_a/P_{PL} equals infinity. Thus, incomplete filler dispersion does not strongly influence the behavior of these polymer systems.

Interfacial channeling more appropriately models both composite systems, particularly the untreated filler. In the case of the treated glass system, the

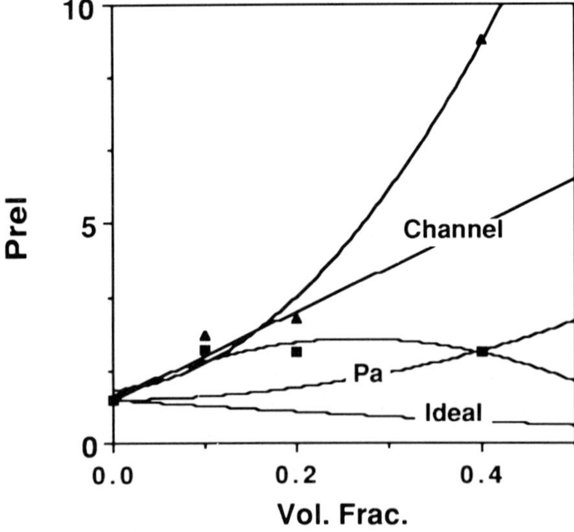

Figure 7. Comparison of the permeability behavior of the glass-bead-filled phenoxy films with the Nielsen model predictions from Table 3. The symbols ▲ and ■ correspond to the untreated and aminosilane-treated glass beads, respectively. The curves designated as 'Channel', 'P_a' and 'Ideal' correspond to equations (3), (2), and (1), respectively.

selection of a lower P_i/P_{PL} (with respect to the values chosen for the 'Channel' curve in Fig. 7) will produce a curve approximating its permeability behavior. Thus, a filler–binder matching which produces a low degree of adsorption tends to increase the permeability in comparison with a well-bonded one.

The higher permeability values of the treated glass system relative to those of the 'Ideal' curve may also be due, in part, to an increase in P_{PL} (the permeability through the polymer portion). Since the filler and matrix form a tight interface, the increased permeability may be attributable to the higher effective free volume in the binder portion. In effect, channeling may occur in the polymer matrix fraction instead of in the interfacial region. Consequently, the overall permeability of the composite will increase. Of course, this type of relaxation mechanism is also present for the untreated filler system, but its contribution will be small in comparison with interfacial channeling.

An interesting factor to consider is the residual solvent. Judging from the amount of retained solvent present in the composite, the solvent could well play a role in the permeability. Interestingly, the unfilled phenoxy which retained the highest amount of solvent is the least permeable of all (Table 3). In view of the strength of the interaction between the solvent and binder, the solvent may act to restrict water vapor transport. Recalling the antiplasticizing effect noted above, anomalies in densities and permeabilities due to specific interaction are known to exist [39].

For the treated glass spheres, at lower filler loadings, the decrease in the amount of residual solvent in the binder (in comparison with the unfilled sample) leads to a slight increase in P_r. As the volume fraction of the filler increases, the formation of a tight interfacial region more than compensates for the reduction of solvent in the polymer matrix, the net effect being a lower P_r value. In the case of the untreated filler, the higher amount of residual solvent in the binder, as compared with the treated specimens at the same filler loading, cannot overcome the higher fraction of porous interfacial region, resulting in a much higher P_r value than that of the treated systems. Thus, interfacial channeling is the most dominant factor.

5.6. *Comparison of permeability with other systems*

The results may also be compared with those obtained for a typical corrosion-resisting zinc/epoxy [13, 16] and a glass bead/epoxy systems [13]; the volume fractions of glass and zinc were 0.61 and 0.36, respectively. The degree of interfacial adhesion was consistent with the predictions of the acid–base concept. The untreated glass, treated glass, and zinc behaved as an acceptor, donor, and acceptor species, respectively, while the epoxy behaved as an electron acceptor. Only the aminosilane-treated glass adhered strongly to the epoxy. The polyamide curing agent interacted sufficiently with the epoxy to inhibit adsorption onto the acidic zinc surface.

Inspection of Tables 3 and 4 reveals several interesting comparisons. First, the permeability constant of the linear phenoxy resin is higher than that of the cross-linked epoxy. This may be partly due to an effect of crosslinking coupled with the lower concentration of hydroxyl groups in the polyamide (Versamide)-cured epoxy. In contrast, as shown in Table 4, the zinc-filled epoxy behaved almost as predicted for a matrix containing a well-bonded, impermeable second phase. This

Table 4.
Effect of zinc and glass beads on the steady-state permeability[a] of epoxy films

Material	Experimental P^a ($P_r^{\,b}$)	Theoretical predictions (P_r)	
		Eqn. (1)	Eqn. (2) ($P_i/P_{PL} = 10$)
Epoxy matrix	3.1 (1.0)	1.0	1.0
Zinc-filled ($v_f = 0.36$)	2.0 (0.65)	0.54	1.3
Glass-filled (3000, $v_f = 0.61$)	3.9 (1.3)	0.30	1.6
Glass-filled (3000CP03, $v_f = 0.61$)	1.7 (0.55)	0.30	1.6

[a] Units: g cm/cm^2 Pa s $\times 10^{14}$, $T = 20$–$25°C$.
[b] $P_r = P_c/P_{PL}$.

behavior is caused by the production of zinc corrosion products at the relatively open interface between the zinc particles and epoxy matrix. Evidence for such a phenomenon has been cited elsewhere [13]. Indeed, with zinc-rich coatings, *weak* interfacial adhesion is desired, in order to permit a high degree of pore sealing by the zinc compounds, which are also believed to serve as corrosion inhibitors [52–54]. Although there are insufficient data to compare the values of P_r for the epoxy and phenoxy resins over the entire range of filler content, it seems likely that the epoxy systems will generally have lower permeabilities. Second, as with the epoxy resin, use of treated rather than untreated beads invariably yields lower P values, though the observed permeabilities are higher than expected theoretically.

6. CONCLUSIONS

It was shown that treatment of glass spheres with an aminopropyltriethoxysilane coupling agent changes the filler–matrix interaction with phenoxy films: the higher the interfacial adhesion, the lower the permeability to water vapor. However, for these solvent-cast films, the interface was not as well bonded as expected for perfect adhesion and complete wetting. Moreover, for solvent-dried glass bead/phenoxy films, the residual solvent content was high enough to greatly lower the T_g and change the dynamic mechanical response. It was also shown that the stronger the interfacial adhesion and the higher the filler content, the lower the solvent retention. At the same time, the dynamic mechanical response, including T_g, is affected more by retained solvent than by interfacial adhesion. More importantly, the acid–base concept is useful in interpreting solvent retention, interfacial wetting and adhesion, and permeability behavior.

Acknowledgements

We acknowledge the support and funding of the Materials Research Center, Lehigh University, Bethlehem, PA and the Office of Naval Research under contract No. N000-14-79-C-0731.

REFERENCES

1. W. Funke, in: *Corrosion Control by Coatings*, H. Leidheiser, Jr. (Ed.), p. 1. Science Press, Princeton, NJ (1979).
2. Z. W. Wicks, Jr., in: *Corrosion Control by Coatings*, H. Leidheiser, Jr. (Ed.), p. 29. Science Press, Princeton, NJ (1979).

3. P. Nylen and E. Sunderland, *Modern Surface Coatings*, Chap. 15. Interscience, New York (1965).

4. P. Sorensen, *J. Coatings Tech.* **7**, 431 (1975).

5. M. J. Marmo, M. A. Mostafa, M. A. Jinnai, F. M. Fowkes and J. A. Manson, *Ind. Eng. Chem. Prod. Res. Dev.* **15**, 206 (1976).

6. F. M. Fowkes and M. Mostafa, *Ind. Eng. Chem. Prod. Res. Dev.* **17**, 3 (1978).

7. J. Williams and J. A. Manson, *Org. Coatings Plast. Chem. Preprints* **42**, 175 (1980).

8. F. M. Fowkes, in: *Adhesion and Adsorption of Polymers*, Part 1, L. H. Lee (Ed.), p. 43. Plenum Press, New York (1980).

9. F. M. Fowkes, in: *Surface Treatments for Improved Performance and Properties*, J. J. Burke and V. Weiss (Eds), p. 75. Plenum Press, New York (1982).

10. J. A. Manson, in: *Proc. Int. Symposium on the Interface/Interphase*, p. 1. SPE Benelux, Liège, Belgium (October 1983).

11. D. C. McCarthy, Ph.D. Dissertation, Lehigh University, Bethlehem, PA (1984).

12. S. T. Joslin, Ph.D. Dissertation, Lehigh University, Bethlehem, PA (1984).

13. A. C. Tiburcio and J. A. Manson, Annual Report, Pigment/Binder Interactions and Corrosion Protection, Office of Naval Research Contract No. N00014-79-C-0731, Lehigh University, Bethlehem, PA (1982).

14. A. C. Tiburcio and J. A. Manson. Annual Report, Effect of Acid–Base Interaction on the Permeability and Mechanical Behavior of Model Protective Coatings, Office of Naval Research Contract No. N00014-79-C-0731, Lehigh University, Bethlehem, PA (January 1984).

15. A. C. Tiburcio and J. A. Manson, Annual Report, 'Acid–Base Effects in Glass Bead/Phenoxy Model Coating Systems', Office of Naval Research Contract No. N00014-79-C-0731, Lehigh University, Bethleham, PA (December 1984).

16. A. C. Tiburcio, M.S. Report, Lehigh University, Bethlehem, PA (1983).

17. A. C. Tiburcio, Ph.D. Dissertation, Lehigh University, Bethlehem, PA (1988).

18. A. C. Tiburcio and J. A. Manson, *J. Appl. Polym. Sci.* (in press).

19. J. A. Manson, J. Lin and A. C. Tiburcio, *Org. Coatings Polym. Sci. Proc.* **46**, 121 (1982).

20. F. M. Fowkes, *J. Adhesion Sci. Technol.* **7**, (1987).

21. R. S. Drago and B. B. Wayland, *J. Am. Chem. Soc.* **87**, 3572 (1965).

22. R. S. Drago, G. C. Vogel and T. E. Needham, *J. Am. Chem. Soc.* **93**, 6014 (1971).

23. R. S. Drago, M. S. Nozari and G. C. Vogel, *J. Am. Chem. Soc.* **94**, 90 (1972).

24. R. M. Martin and R. S. Drago, *J. Am. Chem. Soc.* **95**, 759 (1973).

25. F. M. Fowkes, *J. Phys. Chem.* **66**, 382 (1962).

26. F. M. Fowkes, *Ind. Eng. Chem.* **56**, 40 (1964).

27. F. M. Fowkes, D. O. Tischler, J. A. Wolfe, L. A. Lannigan, C. M. Ademu-John and M. J. Halliwell, *J. Polym. Sci., Polym. Chem. Ed.* **22**, 547 (1984).

28. J. C. Maxwell, *Electricity and Magnetism*, Vol. 1. Clarendon Press, New York (1904).

29. L. E. Nielsen, *J. Macromol. Sci., Chem.* **1**, 929 (1967).

30. G. S. Crank and W. R. Park, *Diffusion in Polymers*. Academic Press, London (1969).

31. J. A. Manson and L. H. Sperling, *Polymer Blends and Composites*. Plenum Press, New York (1976).

32. H. P. Schreiber, W. R. Wertheimer and M. Lambla, *J. Appl. Polym. Sci.* **27**, 2269 (1982).

33. E. Nilsson and C. M. Hansen, *J. Coatings Tech.* **53**, 61 (1981).

34. J. S. Lin, Unpublished data, Lehigh University, Bethlehem, PA (1983).

35. M. J. Marmo, M.S. Report, Lehigh University, Bethlehem, PA (1976).

36. C. Saint Flour and E. Papirer, *J. Colloid Interface Sci.* **91**, 69 (1983).

37. C. M. Hansen, *Off. Dig.* **37**, 57 (1965).

38. C. M. Hansen, *J. Oil Colour Chem. Assoc.* **51**, 27 (1968).

39. O. Olabisi, L. M. Robeson and M. T. Shaw, *Polymer–Polymer Miscibility*. Academic Press, New York (1979).

40. P. R. Couchman, *Macromolecules* **11**, 1156 (1978).

41. F. N. Kelly and F. Bueche, *J. Polym. Sci.* **50**, 549 (1961).

42. A. Bondi and A. V. Tobolsky, in: *Polymer Science and Materials*, A. V. Tobolsky and H. Mark (Eds), p. 95. Wiley–Interscience, New York (1971).

43. J. R. Fried *et al.*, Preprints, 3rd Int. Symp. on PVC, IUPAC, p. 186. Cleveland, OH (1980).

44. J. Bardeleben, *Kunststoffe* **53**, 162 (1963).

45. W. Funke, U. Zorll and B. K. G. Murthy, *J. Paint Tech.* **41**, 210 (1969).

46. A. S. Michaels, *Off. Dig.* **37**, 638 (1965).

47. D. Y. Perera and P. M. Heertjes, *J. Oil Colour Chem. Assoc.* **54**, 774 (1971).

48. J. A. Manson and E. H. Chiu, *J. Polym. Sci. Symp. Ser.* **41**, 95 (1973).

49. W. I. Higuchi and T. Higuchi, *J. Am. Pharm. Assoc. Sci. Ed.* **49**, 598 (1960).

50. D. R. Paul and D. R. Kemp, *Am. Chem. Soc., Div. Org. Coat. Plast. Preprints* **32**, 156 (1972).

51. R. L. Stallings, H. B. Hopfenberg and V. Stannett, *Am. Chem. Soc., Div. Org. Coat. Plast. Preprints* **32**, 131 (1972).

52. K. S. Rajagopalan, S. Guruviah and V. Chandrasekharan, *Paintindia* **25**, 13 (1975).

53. F. Theiler, *Corros. Sci.* **14**, 405 (1974).

54. T. K. Ross and J. Wolstenholme, *Corros. Sci.* **17**, 341 (1977).

Acid-Base Interactions, pp. 329-342
Eds. K.L. Mittal and H.R. Anderson, Jr.
©VSP 1991

Effects of oxygen and ammonia plasma treatment on polyphenylene sulfide thin films and their interaction with epoxy adhesive

H. F. WEBSTER and J. P. WIGHTMAN*
Chemistry Department, Center for Adhesive and Sealant Science, Virginia Polytechnic Institute and State University, Blacksburg, VA 24061, USA

Revised version received 20 September 1990

Abstract—X-Ray photoelectron spectroscopy (XPS) and infrared reflection absorption spectroscopy (IRRAS) were used to study the chemical modification of polyphenylene sulfide (PPS) thin films on exposure to both oxygen and ammonia plasmas. The XPS results for the oxygen plasma treatment indicated a large oxygen increase with the incorporation of various oxidized carbon species as well as oxidized sulfur. For the ammonia plasma, both nitrogen and oxygen were incorporated. IRRAS proved to complement the XPS results, showing a wide range of C—O and C=O functionalities incorporated on oxygen plasma exposure. For the ammonia plasma treatment, an increase in hydrocarbon, alkene-type fragments, and possibly amine groups was detected. Both the XPS and IRRAS results indicated that exposure of plasma-treated surfaces to epoxy with subsequent carbon tetrachloride washes removed most of the modification originally present after plasma treatment. IRRAS analysis showed that a thin layer of epoxy remained after repeated solvent washes and that the film seemed to be cured. For untreated PPS films, a non-cured epoxy film adsorbed. This work suggests that the plasma-modified layer plays a role in the formation of a covalent interphase region between PPS and epoxy.

Keywords: XPS; reflection absorption spectroscopy; polyphenylene sulfide; plasma treatment; adhesion; interphase; epoxy.

1. INTRODUCTION

Adhesion of materials involves interactions occurring within the 'interphase' between two distinct materials. Methods to alter the physicochemical nature of the adherends such as acid etches, abrasion, and the use of coupling agents are often used to promote adhesion; however, the specific mechanisms involved in adhesion are usually not well understood. Changes in surface texture, wettability, and acid–base character are but some of the mechanisms proposed. One commonly used method to increase adhesion strength is the surface modification by low-pressure gas plasma treatment. The advantage of such a treatment is that the modification is restricted to the uppermost layers of the substrate, thus not affecting the overall desirable bulk properties of the substrate adherend. The energetic environment of the plasma creates surface free radicals that can lead to the possible cleaning, ablation, crosslinking, or chemical modification of the substrate surface [1].

---

*To whom correspondence should be addressed.

To analyze the chemical modification of polymer substrates, many studies have used X-ray photoelectron spectroscopy (XPS), while comparatively few other spectroscopic techniques have been utilized. These have included the use of secondary ion mass spectrometry (SIMS) and neutron activation analysis (NAA) to study nitrogen and ammonia plasma treatment of polystyrene and polycarbonate [2]. The use of infrared techniques has also proven useful in the study of plasma modification. Attenuated total reflectance has been used to study the ammonia plasma treatment of polytetrafluoroethylene [3], while reflection absorption spectroscopy has been used to study the surface modification by various plasma gases on thin polymer films [4, 5]. Ellipsometry has also been used in a number of these studies to monitor the thickness of thin films before and after plasma treatment. Contact angle analysis is another technique that has also been utilized to investigate the effect of initial plasma treatment on wettability and to monitor the change in contact angle as a function of time after treatment [6].

The increase in adhesion strength on plasma pretreament has been well documented and the excellent review by Liston outlines a number of studies in this area [1]. In many cases, an increase in adhesion strength is seen with plasma treatment, but the chemical nature of the particular polymer used, as well as the specific plasma treatment conditions, seems to play a vital role in the resultant adhesion strength.

Polyphenylene sulfide (PPS) is a high performance, semicrystalline thermoplastic resin exhibiting a wide range of desirable properties including high strength, resistance to chemical attack, and high thermal stability. For these reasons, it is an extremely suitable matrix resin for use in many composites. Surface modification by plasma treatment can be an important step for proper composite-to-composite bonding [7]. Further, a composite surface is often composed of almost exclusively matrix resin. Thus, knowledge of the interaction between residual surface species left after plasma treatment and the adhesive used could be valuable in the understanding and prediction of adhesion strength and durability.

In this study, the surface chemical modification of polyphenylene sulfide by both oxygen and ammonia plasmas was investigated. Spectroscopic analysis was performed using XPS, infrared reflection absorption specroscopy (IRRAS), and ellipsometry. These techniques were also used to probe the nature of the interaction between plasma-modified PPS and a model epoxy adhesive.

2. EXPERIMENTAL

Ryton™ polyphenylene sulfide (PPS) was obtained from Phillips Petroleum. A high-temperature custom spin coater was used to coat PPS from benzophenone solutions. The solvent (Fisher, certified grade) was degassed with nitrogen at 70°C for 1 h prior to use. The metal substrate used was ferrotype plate obtained from Thompson Photo Products (Bedford Park, IL) consisting of a 0.51 mm steel base plated with an approximately 80 nm surface layer of chromium. The plates were cleaned by two 15 min oxygen plasma etches followed by water and acetone rinses prior to use. Plate and solution temperatures were maintained at 270–290°C and this temperature range was critical for uniform polymer coating.

The coating procedure was carried out under a stream of nitrogen to prevent excessive oxidation of the chromium surface. Solution concentrations ranged from 0.4 to 1.5% (wt/wt) and after coating, the plates were immediately brought to room temperature. The coated plates were then used as is, or reheated to 340°C for 2 min, quenched in liquid nitrogen, and annealed at 150°C for 5–60 min to crystallize the thin PPS film.

For the plasma treatment of thin films of PPS, a Tegal Plasmod™ was used to produce a radio frequency (13.56 MHz) generated 50 W plasma of oxygen and ammonia. The pressure inside the treatment chamber was typically 1.2 Torr.

Ellipsometry was used to determine the film thicknesses of the spin-coated samples. The instrument used was a Gaertner L116A dual mode automatic ellipsometer. Both linear and circular polarizations were used and a continuous rotating analyzer was employed with automatic data sampling at 72 points for each state of polarization. A 1 mW helium–neon laser (632.8 nm) was used as the incident light source and the spot size analyzed was 1 mm. Only amorphous films were measured (after quenching), as the state of crystallinity in extremely thin PPS films and its variation with thickness are not known.

XPS analysis was performed using a Perkin-Elmer PHI 5300 spectrometer employing a Mg K_α (1253.6 eV) achromatic X-ray source operated at 15 keV with a total power of 400 W. Typical operating pressures were $<1\times10^{-7}$ Torr and the surface area analyzed was either a 1 mm circular spot or a 1×3 mm rectangle. The spectrometer was calibrated to the $4f_{7/2}$ photopeak of gold at 83.8 eV and the $2p_{3/2}$ photopeak of copper at 932.4 eV, and all binding energies were referenced to the main C—H photopeak at 285 eV.

IRRAS analysis was performed using a Nicolet 510 spectrophotometer employing 4 cm^{-1} resolution. Reflectance measurements were taken using a retro mirror design, and careful alignment of all components allowed use of an 85° angle of incidence. In all cases, p-polarized light was used, and typically 500–1000 scans were taken to achieve the desired signal-to-noise levels.

For epoxy interaction studies, a standard bisphenol A epoxy, Epon 828, from Shell Co. was used. The resin was used as received, and no crosslinking agent was used. Both untreated and plasma-treated PPS samples were placed in an epoxy bath held at 95–100°C for 40 min prior to analysis. Plates were removed from the bath and the excess epoxy was drained from the surface. Samples were then washed in 50 ml of carbon tetrachloride (Fisher, certified grade) at 100 rpm for at least 5 min, removed and blown dry with a dry stream of nitrogen, and reheated to 150°C for 2 min to remove residual solvent. This procedure was repeated until all excess epoxy was removed from the substrate.

3. RESULTS AND DISCUSSION

3.1. *Ellipsometric analysis of oxygen- and ammonia-treated PPS*

Figure 1 shows the results of the thickness removed by etching vs. time of exposure for both oxygen- and ammonia-treated PPS. The original thickness values for these films ranged from 30 to 50 nm. The results for the oxygen plasma treatment show a limited and almost linear ablation rate. By contrast, the ammonia plasma treatment shows a relatively rapid etch rate and the plot seems to level off after 2–3 min. Owing to the thermal treatment of the film following

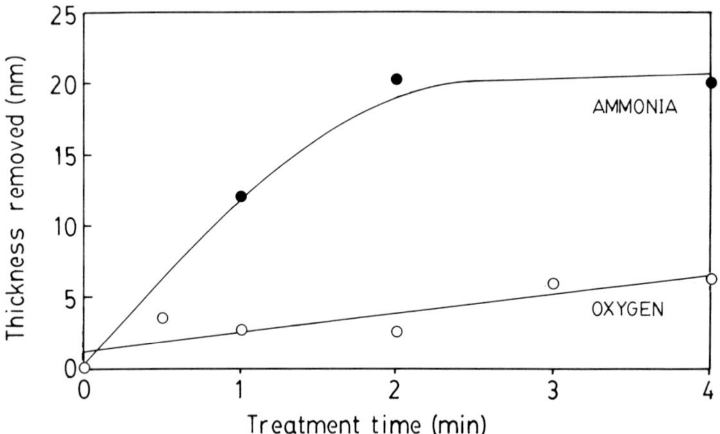

Figure 1. Thickness removed for PPS thin films as a function of the oxygen and ammonia plasma etch time.

plasma treatment and prior to the thickness determination (re-melt and quench), the results can only be viewed as approximate, although the relative rates of etching for ammonia and oxygen plasmas are certainly valid.

3.2. XPS analysis of oxygen- and ammonia-treated PPS

Figure 2 shows the atomic concentrations of major elements found after various oxygen plasma treatment times of PPS. The plot shows a rapid incorporation of oxygen, reaching almost 30% after only 1 min exposure time. In comparison with the ellipsometry results, the relatively unchanging atomic concentration levels with time after the first measurement (30 s) indicate that steady-state conditions must exist for the plasma conditions used.

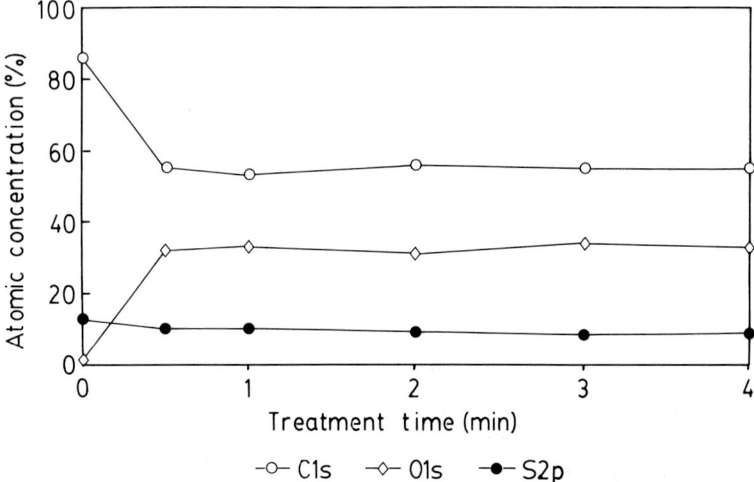

Figure 2. XPS analysis of atomic concentration changes for PPS thin films vs. the oxygen plasma exposure time.

Figure 3 shows the C 1*s* narrow scan spectrum of PPS after 1 min oxygen plasma treatment. The C 1*s* spectrum for the control sample with no treatment is also given and shows no oxidized carbon species, indicating that little oxidation occurred during the elevated temperature coating procedure. The shake-up satellite at about 291.5, due to the aromatic nature of the polymer, is also evident. The spectrum for the plasma-modified PPS shows the presence of many oxidized carbon species ranging from 0.8 to almost 5 eV higher than the main C—H photopeak position at 285 eV. Curve resolution of the carbon spectrum indicates four peaks at 285.8, 287.5, 288.5, and 289.6 eV. Although all individual binding energies are somewhat lower than that reported for oxygen plasma treatment of polystyrene [8], the photopeaks may be tentatively assigned to C—O, C=O, O—C=O, and carbonate species. By lowering the exit angle from 90° to 10°, the mean sampling depth can be decreased and the technique made more surface-sensitive. If the mean free path (λ) for the C 1*s* electron is taken as approximately 1.4 nm [9], the sampling depth can be calculated as approximately $3\,\lambda\,\sin\theta$, where θ is the take-off angle. Therefore, the sampling depths for 90° and 10° are found to be 4.2 and 0.7 nm, respectively. Figure 3 also shows the narrow scan spectrum for a 10° 'take-off' angle. Here, the relative intensity of the oxidized carbon species to the main C—H photopeak has increased, indicating that the modifications due to plasma treatment are restricted to the topmost layers of the PPS substrate.

Analysis of the S 2*p* photopeak in Fig. 4 shows that oxygen plasma treatment results in a broad higher binding energy photopeak at 169.4 eV, clearly separated from the sulfide doublet at 163.8 and 164.8 eV. This high binding energy can be attributed to highly oxidized forms of sulfur, and the wide peak indicates more than one oxidized form. Similar results were obtained for PPS composites exposed to atomic oxygen in low earth orbit [10]. This may show the usefulness of plasma treatment in simulating at least some of the effects expected for candidate materials for space applications. The dramatic increase in the oxidized

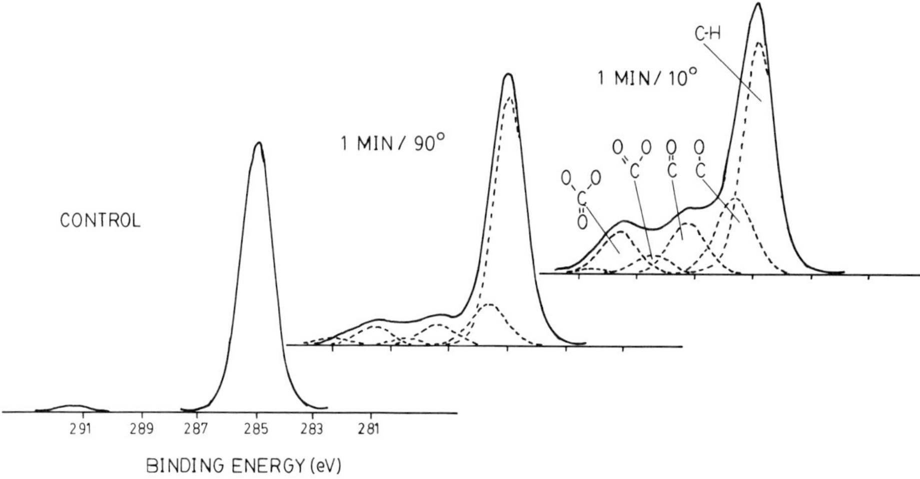

Figure 3. C 1*s* line shape for untreated and oxygen-plasma-treated PPS films at both 90° and 10° exit angles.

sulfur species compared with non-oxidized substrate sulfur when a 10° exit angle is used again that the modification is restricted to top surface layers of PPS.

Analysis of the atomic concentration found by XPS for various exposure times of PPS to ammonia plasma discharge is shown in Fig. 5. An increase in the nitrogen content from 0 to 6–7% is oberved, and the N $1s$ binding energy was approximately 400 eV, indicating an amine-type functionality. There was also a

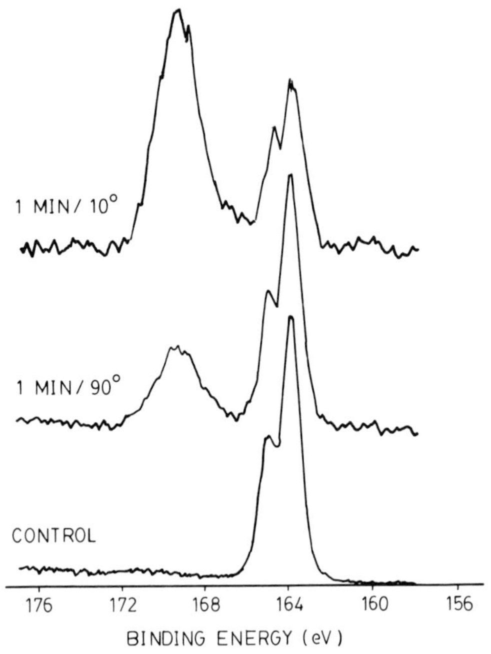

Figure 4. S $2p$ line shape for untreated and oxygen-plasma-treated PPS films at both 90° and 10° exit angles.

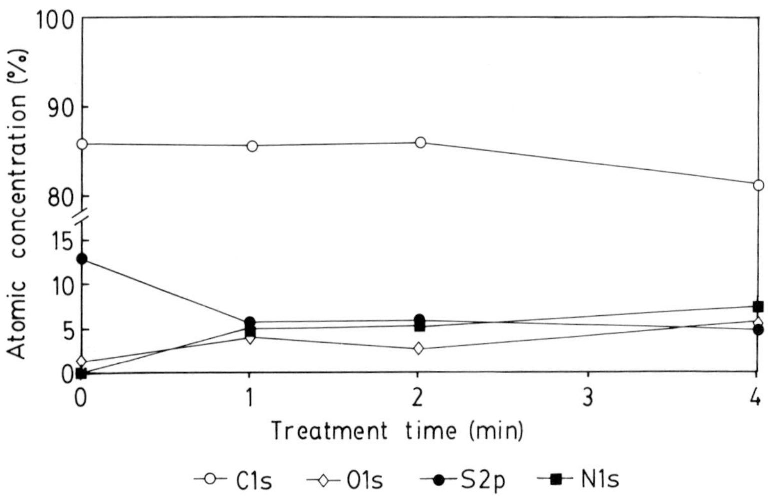

Figure 5. XPS analysis of atomic concentration changes for PPS thin films vs. the ammonia plasma exposure time.

similar small increase in the oxygen content. The increase in oxygen content with non-oxygen-containing plasmas is a common phenomenon found in many studies [3, 11], and is possibly due to the oxidation of residual free radicals when the sample is exposed to air. The increase in nitrogen and oxygen contents was accompanied with a 50% reduction in the sulfur content and little change in carbon. The sulfur narrow scan showed little difference when compared with control samples, indicating that the reduction in sulfur concentration most probably is attributed to an overlayer of modified material containing little sulfur. As with the oxygen plasma treatment, a steady state was achieved within 1 min even though ellipsometry indicated significant ablation of the surface film by the ammonia plasma. The C 1*s* narrow scan spectrum in Fig. 6 shows a new photopeak curve fitted at approximately 0.8–1 eV higher than that of the main C—H peak and this peak can be assigned to a C—N or C—O functionality. However, even with the incorporation of oxygen, no evidence for higher oxidation states of carbon is seen.

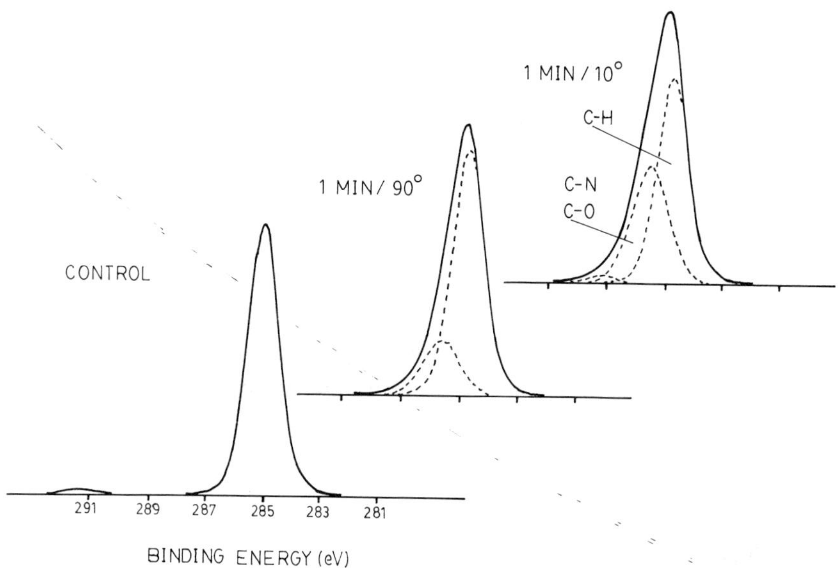

Figure 6. C 1*s* line shape for untreated and ammonia-plasma-treated PPS films at both 90° and 10° exit angles.

3.3. IRRAS analysis of oxygen- and ammonia-plasma-treated PPS

The IRRAS spectrum of a 30 nm thick polyphenylene sulfide film before and after oxygen plasma treatment is shown in Fig. 7. The plasma-induced changes are evident in the C—O and C=O absorbance regions. Figure 8 shows the difference spectrum where the untreated PPS film is used as a reference and its intensity reduced by a factor to account for the amount of material removed during plasma treatment. Two relatively intense and broad absorption bands are located at 1760 and 1230 cm^{-1}, again indicating mainly C=O and C—O functionalities. The breadth of the carbonyl peak indicates a wide range of functional groups created during treatment including possibly ketone, aldehyde,

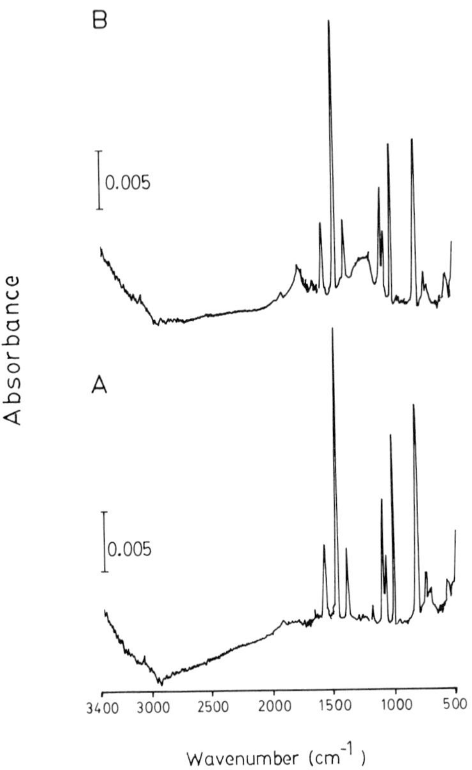

Figure 7. IRRAS spectra of a 30 nm PPS film before (A) and after (B) a 1 min oxygen plasma exposure.

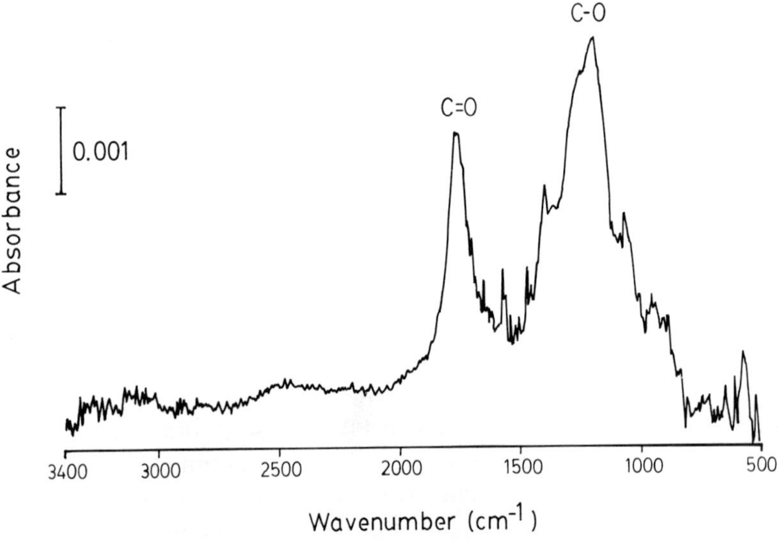

Figure 8. IRRAS difference spectrum of a 1 min oxygen-plasma-treated PPS film. The reference was multiplied by a factor to account for the amount of polymer removed.

ester, and carboxylic acid moieties. Relatively little absorbance is seen, however, in the OH stretch region, indicating little hydroxyl or carboxylic acid formation. The large absorption band due to primarily C—O functional groups is also in agreement with XPS data that show these to be in the largest relative concentration compared with other forms of oxidized carbon. These functionalities may be associated with various ether linkages as well as C—O stretches associated with ester or carboxylate groups. Oxidized forms of sulfur should also be detected in this region and the absorbance bands associated with these groups probably overlap with various C—O absorbances. For example, absorbance bands for the asymmetric and symmetric stretches of sulfonic acids fall in the ranges 1340–1350 and 1150–1165 cm^{-1}, respectively [12], which are well within the envelope observed.

The difference spectrum of ammonia-plasma-treated PPS is shown in Fig. 9. All the bands are relatively weak although the absorbance band at 2940 cm^{-1} can be assigned to a C—H stretch associated with increased hydrocarbon functionality after ammonia plasma treatment. The absorbance bands at 1690 and 1450 cm^{-1} are most probably due to the C=C stretch and C—H deformation bend in CH$_2$ groups, respectively. Similar spectra have been reported for the ammonia treatment of polytetrafluoroethylene [3]. The shoulder at 1600–1650 cm^{-1} may also indicate a N—H deformation absorbance band although the poor signal-to-noise level does not allow assignment of a N—H stretch absorbance in the region above 3000 cm^{-1}. The apparent absence of C—O absorbances indicates that the XPS curve-fit C 1s peak at 285.8–286 eV can most probably be assigned to a C—N functionality. The absence of carbonyl absorbance bands is in agreement with the XPS curve-fitted results.

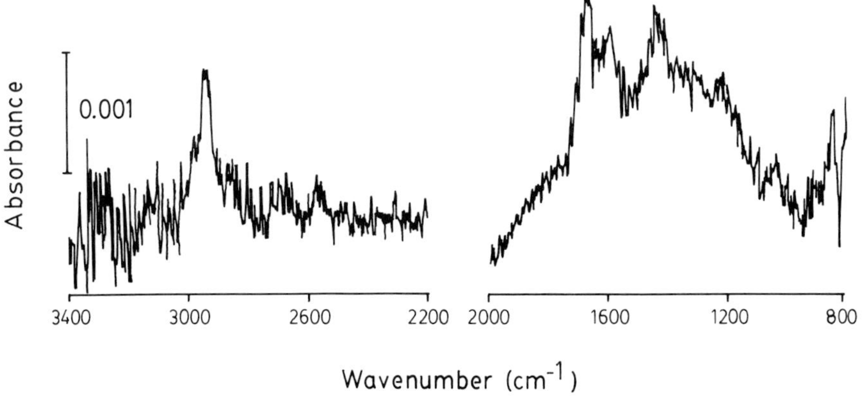

Figure 9. IRRAS difference spectrum of a 2 min ammonia-plasma-treated PPS film. The reference was multiplied by a factor to account for the amount of polymer removed.

3.4. XPS analysis of epoxy interaction with oxygen- and ammonia-plasma-treated surfaces

Table 1 shows the atomic concentrations obtained for 1 min oxygen- and ammonia-plasma-treated surfaces before and after exposure to epoxy with subsequent CCl$_4$ washes. For the oxygen-plasma-treated PPS surfaces, the oxygen

Table 1.

XPS percent atomic concentrations of untreated PPS film, initial 1 min oxygen- and ammonia-plasma-treated samples, and films exposed to epoxy followed by three CCl$_4$ rinses

		Oxygen		Ammonia	
	Control	Initial	W/epoxy	Initial	W/epoxy
C 1s	86.5	61.8	82.2	85.0	85.1
O 1s	0–1	28.4	10.0	2.4	3.3
N 1s	0	0	—	6.8	0.8
S 2p	12.5	6.7	7.8	5.9	10.8
S 2p (ox)	0	3.1	0	0	0

content is dramatically reduced after epoxy treatment while the carbon and sulfur contents increased. The sulfur fraction has been divided into oxidized (ox) and non-oxidized species, and the epoxy treatment resulted in complete removal of all surface oxidized sulfur species. For the ammonia plasma treatment, the most noticeable change is the almost complete loss of nitrogen functionality with a doubling of the sulfur content.

For both cases, the results seem consistent with the loss of much of the surface functionality induced by plasma treatment. To support this assumption further, the XPS C 1s narrow scan spectrum is shown in Fig. 10. For the oxygen treatment, only a C—O functionality remains with no higher oxidation states of carbon present. For the ammonia plasma treatment, the curve-fitted peak previously assigned to either a C—N or a C—O functionality is greatly reduced.

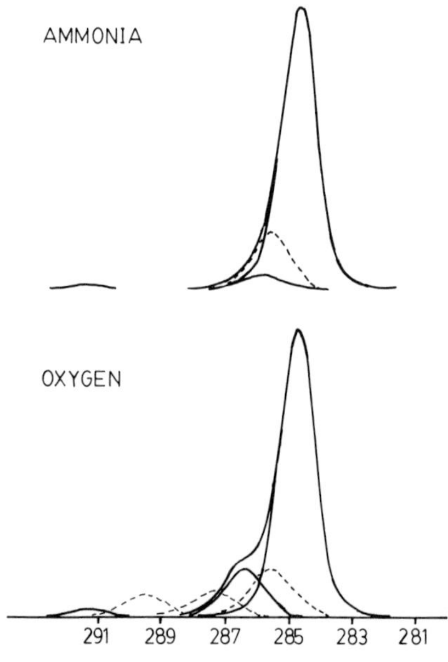

Figure 10. C 1s line shape for 1 min oxygen- and ammonia-plasma-treated PPS films before (– – – –) and after (———) exposure to epoxy followed by three CCl$_4$ rinses.

Also, for both cases, the shake-up satellite is significantly increased, probably indicating the exposure of an unmodified PPS surface. These results are consistent with the loss of oxygen and nitrogen functionalities seen for oxygen and ammonia plasma treatment, respectively. The XPS results therefore indicate that only a C—O functionality for oxygen plasma and C—O or C—N for ammonia plasma are present after exposure to epoxy followed by CCl_4 rinses.

3.5. IRRAS analysis of oxygen- and ammonia-plasma-treated PPS with epoxy

Infrared analysis can be used to identify the chemical nature of the residual functionality seen above in the XPS analysis after epoxy treatment of plasma-modified PPS. IRRAS analysis of plasma-treated PPS exposed to epoxy is shown in Fig. 11. The spectra represent the difference spectra between plasma-modified PPS after exposure to epoxy with a varying number of CCl_4 washes, and control

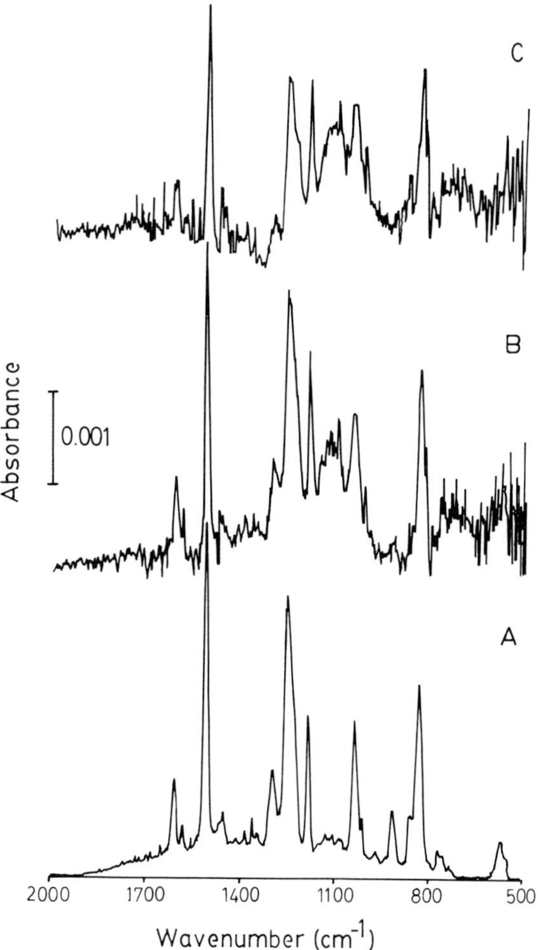

Figure 11. IRRAS difference spectra of a 1 min oxygen-plasma-treated PPS film exposed to epoxy at 100°C after (A) one CCl_4 rinse, (B) two CCl_4 rinses, and (C) three CCl_4 rinses. Spectrum (A) was multiplied by a factor of 0.16 for scaling purposes.

samples receiving no treatment. Therefore, the spectra represent both modification due to the plasma treatment and any spectral changes due to the epoxy/ CCl_4 treatment. The first solvent wash after removal from the epoxy bath shows a relatively thick layer of residual epoxy remaining on the surface. The thickness of this film, as measured by ellipsometry, was in the range 50–70 nm. All bands can be assigned to the epoxy structure and assignments can be found elsewhere [13]. Clearly, however, the epoxy ring vibration is present at $916 \, cm^{-1}$, indicating no crosslinking in this film. After the second wash, however, several interesting spectral features can be noted. The most evident is the complete lack of a carbonyl functionality previously seen in Fig. 8 for oxygen plasma treatment of PPS. This is consistent with the XPS results and must again indicate that much of the surface modification can be removed after epoxy treatment, therefore possibly contributing little to bonding. A second feature is the relative absence of the epoxy ring vibration compared with the other absorbance bands present. Also, an absorbance band at approximately $1120 \, cm^{-1}$ can be detected which may be assigned to C—O stretching vibration of either an alcohol or an alkyl ether structure. This absorbance is much weaker in intensity than that seen for the plasma-modified surface before epoxy treatment and has shifted from the peak maximum at $1230 \, cm^{-1}$ seen previously. The loss of the epoxy vibration with a relatively intense C—O absorbance band is consistent with the spectrum of a cured epoxy film, indicating that crosslinking has occurred in the surface film. Similar spectral features have been seen for the transmission spectra of epoxy films crosslinked with a variety of curing agents [14]. The particular curing agent involved here is not known although the lack of carbonyl absorbance bands indicates that it does not contain this moiety. Reaction with a number of functional groups including ketones, alcohols, or acetals could possibly lead to C—O absorbance bands, although the conditions required for reaction can be varied [14]. The overall low concentration of reactive groups at the surface may be below the detection limits of the IRRAS technique, and the poor signal-to-noise levels in the hydroxyl region unfortunately do not allow detailed analysis in this region of the spectrum.

The IRRAS difference spectrum for an ammonia-plasma-treated PPS surface after treatment with epoxy and three CCl_4 washes is shown in Fig. 12. A spectrum similar to that found with the oxygen plasma treatment is obtained although the amount present seems to be much smaller. The absence of the epoxy ring vibration again indicates crosslinking, and the C—O absorbance is also evident at $1150 \, cm^{-1}$ although the band is not as intense.

To test for epoxy interaction with PPS samples without plasma treatment, control samples were heat-treated with epoxy under the same conditions and the difference spectrum is shown in Fig. 13. The spectrum clearly shows that epoxy will adsorb to untreated PPS surfaces, although the presence of the epoxy ring vibration at $915 \, cm^{-1}$ shows that the polymer is adsorbed in a non-crosslinked form. Also, no C—O absorption bands are detected in the region of 1100–$1200 \, cm^{-1}$. The mechanism of adsorption is not known, although it most probably consists of a combination of dispersion forces and, more importantly, acid–base contributions. Since carbon tetrachloride is a neutral solvent, even weak acid–base interactions would not be disrupted by this solvent, and maximum adsorption would occur. In work reported earlier by Fowkes and

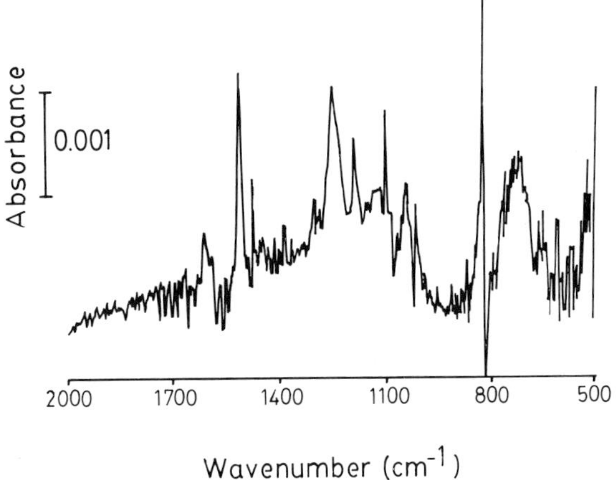

Figure 12. IRRAS difference spectrum of a 1 min ammonia-plasma-treated PPS film exposed to epoxy at 100°C followed by three CCl$_4$ rinses.

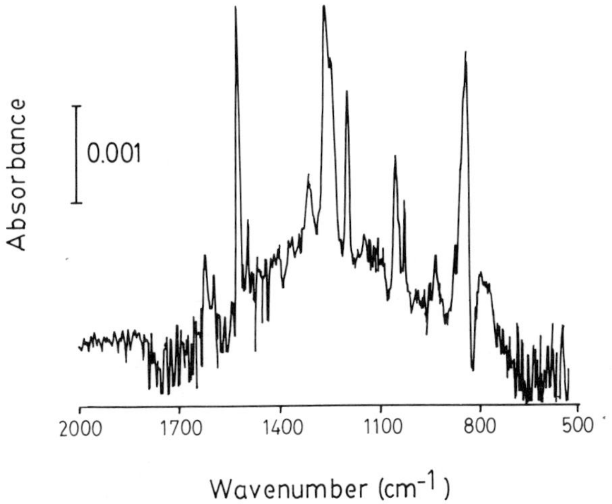

Figure 13. IRRAS difference spectrum of untreated PPS films exposed to epoxy at 100°C followed by three CCl$_4$ rinses.

Mostafa [15], the use of this solvent was shown to give the maximum adsorption of basic PMMA on acidic silica.

4. SUMMARY

XPS has shown that the oxygen plasma treatment of PPS results in an overall increase in the oxygen content of the surface region. This oxygen is associated with both oxidized forms of carbon and sulfur. Ammonia plasma treatment resulted in the incorporation of nitrogen and oxygen as either a C—N or a C—O functionality. Treatment with epoxy followed by CCl$_4$ rinses, however, seemed to remove most surface functionalities, leaving only C—O or C—N moieties, as determined by the C 1s narrow scan XPS spectra. IRRAS showed that this was, in part, due to C—O functional groups from a thin epoxy layer at the surface. The absence of the epoxy ring vibration indicated a crosslinked surface film for both oxygen- and ammonia-plasma-treated PPS surfaces, although the surface density or thickness was much less for the ammonia treatment. Although the specific curing site could not be identified, it certainly resulted from the plasma surface treatment. Epoxy treatment of non-plasma-treated PPS films showed that an uncured film could also adsorb to the PPS surface.

The use of IRRAS as a complementary tool to XPS has shown great promise in the determination of the mechanisms involved between plasma-modified surfaces and a model adhesive. The indication of a direct bond between a plasma-induced surface species and epoxy shows some of the first evidence for the formation of covalent bonds in the interphase for such a system.

Acknowledgements

Financial support of this work by Phillips Petroleum and the Adhesive and Sealant Council is gratefully acknowledged. The technical advice of Dr. T. W. Johnson was appreciated.

REFERENCES

1. E. M. Liston, *J. Adhesion* **30**, 199 (1989).
2. J. Lub, F. C. B. M. van Vroonhoven, E. Bruninx and A. Benninghoven, *Polymer* **30**, 40 (1989).
3. N. Inagaki, S. Tasaka and H. Kawai, *J. Adhesion Sci. Technol.* **3**, 637 (1989).
4. D. S. Dunn and D. J. McClure, *J. Vac. Sci. Technol.* **A5**, 1327 (1987).
5. J. L. Grant, D. S. Dunn and D. J. McClure, *J. Vac. Sci. Technol.* **A6**, 2213 (1988).
6. A. D. Katnani, A. Knoll and M. A. Mycek, *J. Adhesion Sci. Technol.* **3**, 441 (1989).
7. G. K. A. Kodokian and A. J. Kinloch, *J. Mater. Sci. Lett.* **7**, 625 (1988).
8. D. T. Clark and A. Dilks, *J. Polym. Sci., Polym. Chem. Ed.* **17**, 957 (1979).
9. D. T. Clark and H. R. Thomas, *J. Polym. Sci., Polym. Chem. Ed.* **15**, 2843 (1977).
10. C. Evans & Associates, *Application Note* **5**(2) (1990).
11. D. T. Clark and R. Wilson, *J. Polym. Sci., Polym. Chem. Ed.* **21**, 837 (1983).
12. L. J. Bellamy, *The Infrared Spectra of Complex Molecules.* John Wiley, New York (1964).
13. J. Filbey, Ph.D. dissertation, Virginia Polytechnic Institute and State University, Blacksburg, VA (1987).
14. H. Lee and K. Neville, *Handbook of Epoxy Resins.* McGraw-Hill, New York (1967).
15. F. M. Fowkes and M. A. Mostafa, *Ind. Eng. Chem. Prod. Res. Dev.* **17**, 3 (1978).

Acid-Base Interactions, pp. 343-361
Eds. K.L. Mittal and H.R. Anderson, Jr
©VSP 1991

The effect of donor–acceptor interactions on the mechanical properties of wood

DAWN MA,[1] WILLIAM E. JOHNS,[1]* A. KEITH DUNKER[2] and
ABDEL E. BAYOUMI[1]

[1] *Department of Mechanical and Materials Engineering, Washington State University, Pullman,
WA 99164-2920, USA*

[2] *Department of Chemistry Biochemistry/Biophysics Program, Washington State University, Pullman,
WA 99164-4660, USA*

Revised version received 7 March 1990

Abstract—This paper explores the empirical correlations between the mechanical properties of
solvent-treated wood and solvent parameters. Wood beams (10 inch × 0.7 inch × 0.4 inch) of lauan,
birch, and Douglas fir were soaked in five solvents (benzene, dioxane, methanol, dimethyl sulfoxide,
and distilled water) for 4 months at room temperature (23°C) and then mechanically tested in bending.
The acoustic emission (AE) energy (related to failure energy), the modulus of rupture (MOR), and the
modulus of elasticity (MOE) of the specimens were determined. It was found that all of these
properties of the specimens decreased with the solvent strength parameter which was related to
acceptor number (AN), donor number (DN) and molecular volume (MV). The linear correlation of
the AE energy of the specimens with the solvent parameter $(AN + DN)^{1/2}$ was verified. Alternative
relationships between the AE energy, MOR or MOE of the specimens and the modified solvent
strength parameters, $2.5\ DN + AN$ or $DN/MV^{1/3}$, were explored. The microstructures of fracture
surfaces of lauan specimens were examined by scanning electron microscopy (SEM) and showed
dramatic changes when the solvent strength increased, such as microfibril pull-out and a decrease in
roughness of the fracture surfaces of cell walls. All these observations reflected the changes in
mechanical properties and the mean AE energy qualities of the specimens.

Keywords: Lewis acid; Lewis base; donor; acceptor; wood swelling; acoustic emission; solvent strength
parameter.

1. INTRODUCTION

Donor–acceptor interactions, based on donor numbers (DN) and acceptor
numbers (AN), may be key factors in understanding certain mechanical properties, especially when applied to polar materials. The concept of donor–acceptor
interactions is an extension of Lewis acid–base reactions, dealing with coordinate
bonds which are formed by sharing a pair of electrons between donor and
acceptor species. Efforts have been made to establish a correlation between the
material properties and donor–acceptor interaction parameters. Fowkes [1, 2],
using the four-parameter system, related either basicity or acidity to the adhesion
between polymers and inorganic surfaces. While the data showed a systematic
response to DN and AN, no simple, clear mathematical model was identified.
Schleicher [3], using the two-parameter system, noted that the square sum of
donor and acceptor values $(AN + DN)^2$ was weakly correlated with the amount of

---

*To whom correspondence should be addressed.

liquid retained in pulp pads subjected to high gravitational loads during centrifugation. Recently, Larsson and Johns [4] analyzed the data involving tensile energy absorption (TEA) of paper sheets soaked in various organic liquids with the donor and acceptor values for these liquids and found a simple mathematical relationship between the TEA of paper and the square root of the sum of AN and DN of these liquids, i.e. the TEA of the paper decreased linearly with increasing $(AN + DN)^{1/2}$. They proposed that the values of $(AN + DN)^{1/2}$ were analogous to an energy spectrum, showing the range of secondary bond strengths noted by Schleicher [3]. In other words, the values of $(AN + DN)^{1/2}$ would be a relative measure of the energy required to break the bonds within paper.

2. OBJECTIVES

The work in this paper is intended to amplify and develop further the approach of Larsson and Johns [4]. Specifically, the objectives are:

(1) to verify the implication of the Larsson and Johns model when applied to solid wood;
(2) to show how factors other than DN and AN of the solvent might be involved in modeling the behavior of the wood–solvent system;
(3) to explore the behavior of the modulus of rupture (MOR), the modulus of elasticity (MOE), and the acoustic emission (AE) energy of solvent-treated wood; and
(4) to provide visual information about the fracture surfaces of wood after treatment in solvents and bending to failure.

3. BACKGROUND

3.1. Basic aspects of donor–acceptor interactions

The theory of donor–acceptor interactions is not new and is based on the Lewis acid–base reaction which was first formulated in 1923 [5]. This theory deals with the formation of coordinate bonds which are formed by sharing a pair of electrons between acid (acceptor) and base (donor) species. Gur'yanova *et al.* [6] pointed out that the concept of intermolecular donor–acceptor reaction is associated with the interaction of molecular orbitals. It is assumed that the formation of a donor–acceptor bond between two molecules involves the transfer of an electron from the highest filled donor orbitals to the lowest vacant molecular orbital of the acceptor. Gutmann [7] further proposes that in donor–acceptor interactions, the charge density rearrangement caused by charge transfer and the polarization effect leads to the changes in bond length, bond properties, and chemical behavior of the system. The extended donor–acceptor concept is based on a comparison of equilibrium structures in different environments irrespective of the nature of the binding forces [7].

Gutmann's two-parameter system is important because the empirical values of AN and DN show their fundamental roots in the bond lengths and bond strength of shared electron pairs, and the evaluated AN and DN do not treat each species as being either an acid or a base, thus making allowance for amphoteric behavior. Using Gutmann's measured DN and AN values, a solvent can be qualitatively classified as being primarily donor (large *DN*, small *AN*), acceptor (small *DN*, large *AN*), amphoteric (large *DN* and *AN*), or 'inert' (small *DN* and *AN*) [5].

The basic features of donor–acceptor interactions based on Gutmann's model are:

(1) The approach of donor–acceptor interaction is "... the visualization, rationalization and extension of previously expressed Lewis acid–base ideas" [8] and is based on the observable changes in bond length and properties such as spectroscopic data. These changes are caused by the charge density rearrangement as a result of donor–acceptor interaction.

(2) The formation of a coordinate bond beween donor and acceptor molecules influences the adjacent bonds within each molecule, leading to changes in these bond strengths and, in turn, in the mechanical properties of the material.

(3) Two solvent parameters, *AN* and *DN*, can characterize most kinds of solvents and provide an opportunity to develop an understanding of solvent systems and solvent strengths.

(4) Donor–acceptor interactions can be regarded as universal interactions in that charge density rearrangement can range from very weak (physical) to very strong (chemical) interaction [7] and it is extremely useful in "... understanding and explaining a vast amount of hitherto unrelated phenomena in solution chemistry as well as other types of molecular interactions such as interface and crystallization phenomena" [8].

In a 1983 review, Schleicher [3] noted that in the previous decade different authors, including himself, verified that the connections between the extent of cellulose swelling and any characteristic data of the liquids used, such as dielectric constant, solubility parameter, and donor number (*DN*), were not successful. He attempted to relate the swollen celluloses with both the acceptor number (*AN*) and the donor number (*DN*) of the swelling solvents. He plotted his swelling factor 'Q' against the square sum of *DN* and *AN* $[(AN + DN)^2]$ of the swelling solvents used and found that the correlation was weak. Schleicher [3] explained that because the values of *DN* and *AN* of the swelling solvents used were not fully available at that time, he could not form any conclusions.

Recently, Larsson and Johns [4] reviewed the literature data on interactions between liquids and cellulose or lignocellulose. They suggested that the acid–base properties of a solid might be measured via the interactions of that solid with a series of liquids of known acidity or basicity. For a porous, swelling polymeric material such as paper, the interaction can be determined by measurements of the tensile energy absorption (TEA) of paper sheets soaked in liquids [9]. They correlated the data obtained from the literature with several solvent parameters and found rather loose correlations with traditional solvent properties, such as dielectric constant, cohesive energy density, molar volume, and qualitative hydrogen bonding properties, but a fair correlation with *DN* and *AN*, as shown in Table 1.

The results shown in Table 1 are consistent with the statement of Fowkes [10], i.e. in solution theory, hydrogen bonds and acid–base interactions, i.e. donor–acceptor interactions, are important and the minor role of dipole–dipole and dipole-induced–dipole interaction can be ignored. Larsson and Johns successively removed certain independent variables and recalculated the residual variances. The results, summarized in Table 2, show that removing the donor/acceptor

Table 1.

Correlation coefficients (r) from linear regression analysis [4]

Independent variable	Correlation coefficient of the TEA with the variable
$D_{\mathrm{II.I}}$	-0.86
DN	-0.86
AN	-0.75
MV	0.78
e	-0.71
δ	-0.75
δ_{d}	0.20
δ_{p}	0.81
δ_{h}	-0.97

DN: donor number; AN: acceptor number; MV: molar volume; e: dielectric constant; δ: solubility parameter; δ_{d}: dispersive component of δ; δ_{p}: polar component of δ; δ_{h}: hydrogen bonding component of δ; TEA: tensile energy absorption.

Table 2.

Partial least squares (PLS) modeling with latent variables analysis [14]

Independent variables used in the PLS model	Percentage of the variance in the independent variables accounted for ($100\ r^2$) by the PLS model
$DN_{\mathrm{II.I}}$, DN, AN, MV, e, δ, δ_{d}, δ_{p}, δ_{h}	87.2
MV, e, δ, δ_{d}, δ_{p}, δ_{h}	64.6
$DN_{\mathrm{II.I}}$, DN, AN, MV	84.5
$DN_{\mathrm{II.I}}$, DN, AN	82.9
$DN_{\mathrm{II.I}}$, DN, MV	85.6
$DN_{\mathrm{II.I}}$, DN	78.5

For symbols see Table 1.

parameters is disastrous for the model, whereas removing the solubility parameters and the dielectric constant only slightly deteriorates the fit. Further removal of either the acceptor number or the molecular volume only affects the fit slightly, while the removal of both deteriorates the fit.

Larsson and Johns concluded that the basicity and acidity of the solvent along with the involvement of molecular volume are the most important factors in the interactions. They also explored the possibility that simple relationships may exist between these factors and the physical properties of the adhering system. Their mathematical model showed that the TEA of the paper sheets decreases linearly as the value of $(AN + DN)^{1/2}$ of the solvent increases. About 84% of the variability of TEA of soaked paper sheets could be accounted for by this model. No fundamental reason was offered for why the relationship takes the form of the square root of DN plus AN. The correlation shown by Larsson and Johns, however, improves upon that given by Schleicher. Larsson and Johns have presumed that since the empirical values of AN and DN have their fundamental roots in the bond length and bond strength of shared electron pairs, they would be associated with the energy that relates to the bond strength in all bonded systems [4].

4. METHOD

4.1. Materials

Lauan (known commercially as Philippine mahogany), birch, and Douglas fir were used in this research. Selection of these species was based on anatomical characteristics. Lauan has a coarse, uniform texture; birch is dense with a uniform texture; and Douglas fir is straight-grained, with clear abrupt boundaries between the early and late wood, forming a nonuniform texture. For this study textural differences were considered vital. We had no idea how the wood texture might influence acoustic emission data, which were, in turn, influenced by the various solvents. Having a range of texture types allowed for a range of possibilities in the data.

4.2. Solvents

The solvents used in this research were deionized/distilled water, benzene, dioxane, methanol, dimethyl sulfoxide (DMSO), dimethyl formamide (DMF), and pyridine. All organic solvents were produced by J. T. Baker, Inc., Phillipsburg, NJ. The donor and acceptor numbers for these solvents were taken from the literature [5, 7, 8] and are given in Table 3.

4.3. Specimen preparation

The wood was obtained as 1-inch lumber and cut into small beams 10 inches long, 0.7 inch wide, and 0.4 inch thick. The lengths of the beams were parallel to the grain of the wood. The beams were randomly assigned to the first five solvent treatments in Table 3, or to no treatment after being oven-dried at 102°C for 24 h. Specimens were soaked for 4 months at 23°C.

With the completion of the original batch of data, it was felt necessary to add two new solvents. Time did not permit the above treatment, so an accelerated vacuum-pressure method was used instead. The specimens were placed in a 0.01 Torr vacuum for 24 h, after which the solvent was drawn into the container by the vacuum. Lauan specimens were then soaked for 1 week at 25°C to reach the equilibrium swelling condition. The alternate solvents were DMF and pyridine.

Table 3.

Values of the donor and acceptor numbers of solvents used

Solvent	AN	DN
Benzene	8.2	0.1
Dioxane	10.8	14.8
DMSO	19.0	29.8
Methanol	41.3	19.0
Water	54.8	18.0
DMF	16.0	26.6
Pyridine	14.2	33.1

Dioxane was also used in this treatment as a control to test this alternate treatment effectiveness.

4.4. Testing equipment

Two devices were used, an Instron universal tension machine model TTD and an acoustic emission device. Acoustic emission (AE) refers to a class of phenomena wherein transient elastic waves are generated by the rapid release of energy from localized sources within a material [11]. Thus, the acoustic signal is an elastic wave generated by deforming material. An acoustic signal is usually called an 'event' or a 'hit'. The AE characteristics of interest in this research were the number of hits and the energy for total failure. The number of hits relates to the mode of failure. Generally, a large number of hits imply plastic deformation and ductile fracture; a small number of hits identify elastic deformation and brittle fracture. The AE energy is a relative value and may relate to the energy required for deforming the bonds within the material.

4.5. Testing procedure

Before testing, the cross-sectional area of the soaked specimen was measured. Then each specimen was tested by a three-point static bending method on the tension machine. The loading rate was 0.5 inch/min and the testing temperature was 23°C. The transducer of acoustic emission was placed on the surface of the solvent-wet specimen. After testing, the fracture surfaces of lauan specimens were examined in a scanning electron microscope (SEM).

5. RESULTS AND DISCUSSION

5.1. The change in acoustic emission energy of solvent-treated wood specimens

The AE energy value was calculated as the AE energy per hit, which indicated the mean energy for each deformation to total failure, and is called the mean AE energy. Initially, the same mathematical model as that given by Larsson and Johns [4] was used. The correlations between the mean AE energy of the specimen and the solvent parameters $(AN + DN)^{1/2}$ are shown in Figs 1, 2, and 3.

In general, the mean AE energy of solvent-treated birch and lauan decreases linearly with an increase in the solvent parameter $(AN + DN)^{1/2}$ (Figs 1 and 2). This is similar to the trend shown by Larsson and Johns. However, the mean AE energy of solvent-treated Douglas fir has little correlation with the solvent parameter, and shows a scatter pattern (Fig. 3). The three species were selected to provide for a range of anatomical qualities. Douglas fir is a wood featuring an abrupt transition from early wood to late wood; birch has a uniform texture; and lauan has a uniform texture but is much coarser than birch. Prior to the actual experiment, it was not known how these qualities would be reflected in the AE data. This study has shown that the abrupt nature of the transitions in Douglas fir wood provided a failure mode that overwhelmed any effect of the solvent in the AE data. The data from this phase of the experiment found no correlation between the solvent properties and AE energy. It is interesting to note that the overall response of all three species to strain, as measured in the bending strength and stiffness, correlated extremely well to solvent properties, as will be discussed

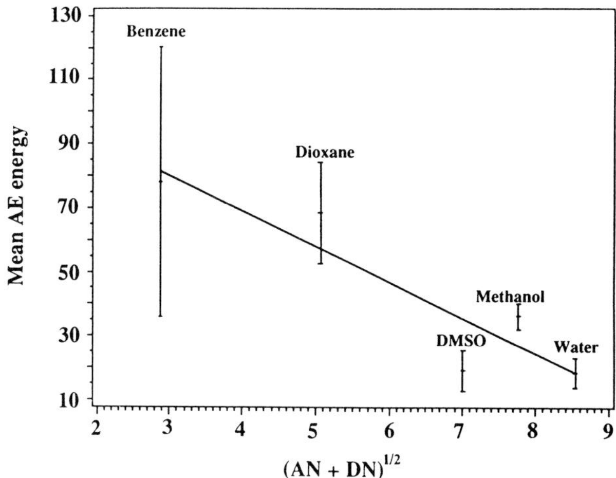

Figure 1. The mean AE energy of solvent-treated birch vs. the solvent parameter $(AN + DN)^{1/2}$.

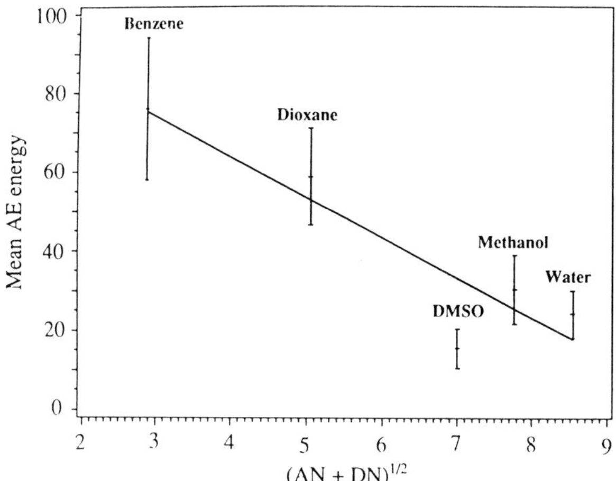

Figure 2. The mean AE energy of solvent treated lauan vs the solvent parameter $(AN + DN)^{1/2}$.

later in this paper. For the rest of this section, the discussion will be limited to birch and lauan species, where the anatomical features did not have an adverse effect on the relationships between the solvent properties and AE data.

DMSO is substantially off the linear relationship generated by the other solvents (Figs 1 and 2), reducing the correlation coefficient. To explore this further, specimens were prepared with DMF and pyridine, using the accelerated soaking technique mentioned earlier. These two solvents were selected because of their similarity in *DN* and *AN* with DMSO. The results for this phase of the research are shown in Fig. 4. Two conclusions can be drawn:

(1) the DMSO solvent does not behave unusually and just reflects the nature of this solvent; and
(2) the deviation of the model points to the possibility that the form of the solvent parameter $(AN + DN)^{1/2}$ might be inappropriate or that some other factors

                                    Dawn Ma et al.

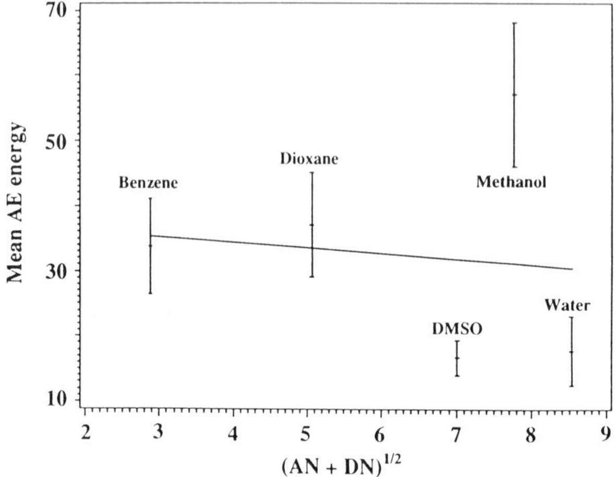

Figure 3. The mean AE energy of solvent-treated Douglas fir vs. the solvent parameter $(AN + DN)^{1/2}$.

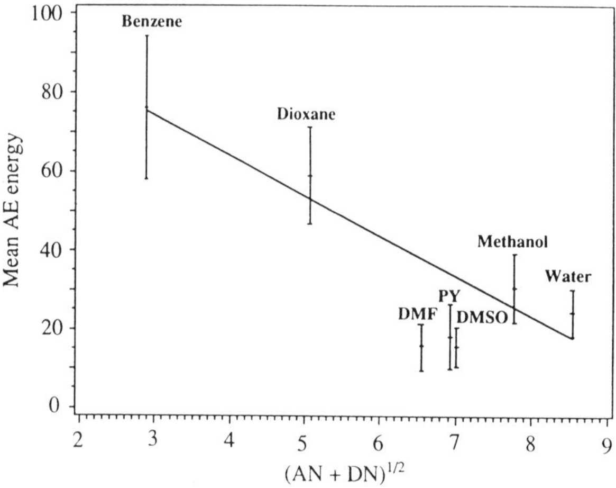

Figure 4. The mean AE energy results of lauan treated with the new solvents compared with the initial results.

involved in the donor–acceptor interaction of wood–solvent systems should be considered.

First, the form of the solvent parameter $(AN + DN)^{1/2}$ should be reconsidered. According to the definition of AN and DN, AN is a dimensionless number and DN has the energy units of kcal/mol. These two values cannot be added directly. Further, the values of AN and DN given by Gutmann [7] show that for the same pair of solvents, $SbCL_5$ and $(Et)_3PO$, there are different scales. $(Et)_3PO$ reacts with the reference acid $SbCl_5$, yielding a DN value of 40 kcal/mol; $SbCl_5$ reacts with the reference base $(Et)_3PO$ using the ^{31}P NMR shift, defining the AN value of 100. To make these two scales equal, we defined the normalizing factor as

$$(DN \text{ kcal/mol})(100)/40 \text{ kcal/mol} = 2.5\, DN, \tag{1}$$

where the units on the left-hand side of equation (1) cancel out and 2.5 *DN* is now a dimensionless number. Our scaling factor is based on the DN units of kcal/mol. Appropriate adjustments must be used when values are calculated in kJ/mol. Thus redefined, *AN* and 2.5 *DN* can be handled in a mathematical manner. The data were again correlated with the mean AE energy of birch and lauan specimens by a linear model. The best combination is the sum of 2.5 *DN* and *AN*, which gives a better correlation than the older model $(AN + DN)^{1/2}$. Approximately 90–94% of the variation between AE energy and solvent strength for lauan and birch specimens could be accounted for with this new model. The results of the mean AE energy of the lauan and birch specimens with the modified solvent parameter 2.5 *DN* + *AN* are shown in Fig. 5.

Second, the swelling effect should be considered because it is an obvious factor affecting the wood properties. It was found that DMSO, DMF, and pyridine produced a larger swelling effect than the other solvents. Swelling in wood is controlled by two factors. One is the attraction between the wood material and solvent molecules. The second factor is the size of the solvent molecule. Thus, the

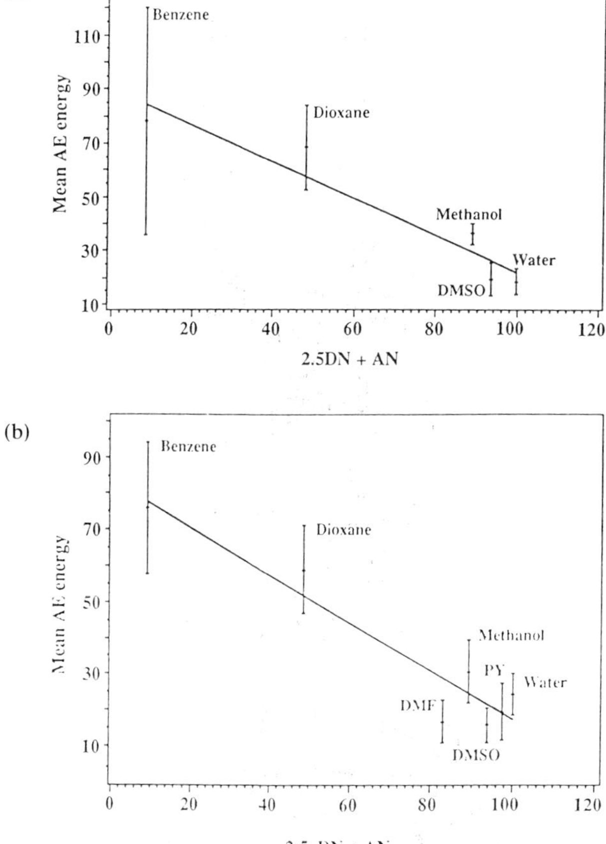

Figure 5. The mean energy of solvent-treated wood vs. the solvent strength parameter 2.5 *DN* + *AN*. (a) Birch; (b) lauan.

possible interaction of both attractive forces, as characterized by the donor–acceptor interaction and the molecular volume (MV) of the solvent, should be considered. This was suggested in the work of Larsson and Johns [4], where the relative importance of MV was initially seen via statistical analysis (Table 2). This avenue was explored.

The MV values of the solvents used were calculated with MACROMODEL [12], a molecular modeling program. The swelling effect of all specimens, and the values of AN, DN, and MV of the solvents are given in Table 4. A correlation between the mean AE energy of the specimen and the solvent parameter that included the MV factor was attempted. Several models were developed and correlated linearly with the mean AE energy of the specimen. It was found that the best model was $DN/MV^{1/3}$. This model is as accurate as the previous modified solvent paramenter, i.e. about 90–94% of the variation between the AE energy and the solvent parameter for birch and lauan could be accounted for, as shown in Fig. 6. It should be noted that the units of $DN/MV^{1/3}$ are *force*; when DN is in J/mol this force is measured in N/mol. Thus, by analogy, the solvent parameter $DN/MV^{1/3}$ could physically represent a numerical measure for the concept of the solvent strength.

It is not surprising that only the DN is seen to be important in reactions with wood. Wood is inherently acid. Acid functionality is found in the hydroxyl-rich cellulose and hemicellulose cell-wall material. Further, wood has a nonstructural component, referred to as extractive material, which is also quite acid. In general, the pH of wood is in the 5.5–3.8 range.

The correlation of the mean AE energy of solvent-treated lauan and birch with the modified solvent parameters lends credence to the observation of Larsson and Johns, namely, either AN or DN or DN along with MV of the solvent are the most important factors characterizing the donor–acceptor interactions in cellulose materials. The linear correlation indicates that there is indeed a range of secondary bond strengths in wood. Direct breakage of these bonds depends on the solvent strength parameter represented as either 2.5 $DN + AN$ or $DN/MV^{1/3}$; the higher the solvent strength parameter, the stronger the donor–acceptor interactions taking place within wood, the more the original acid–base bonds of wood being broken, and, finally, the weaker the AE energy of wood. However, from the

Table 4.

The swelling effect of solvent-treated wood with the solvent parameters AN, DN, and MV

Solvent	MV (Å^3)	Swelling (%)		
		Lauan	Birch	Douglas fir
Benzene	84.3	1.1	1.8	1.1
Dioxane	87.5	9.5	12.1	7.0
DMSO	70.8	15.2	23.8	29.0
DMF	75.8	14.9	ND	ND
Pyridine	79.6	15.1	ND	ND
Methanol	37.4	11.7	16.3	13.4
Water	20.1	12.4	17.7	13.4

ND = not determined. Swelling(%) = percentage of difference between dry and wet cross-sectional areas of the specimen.

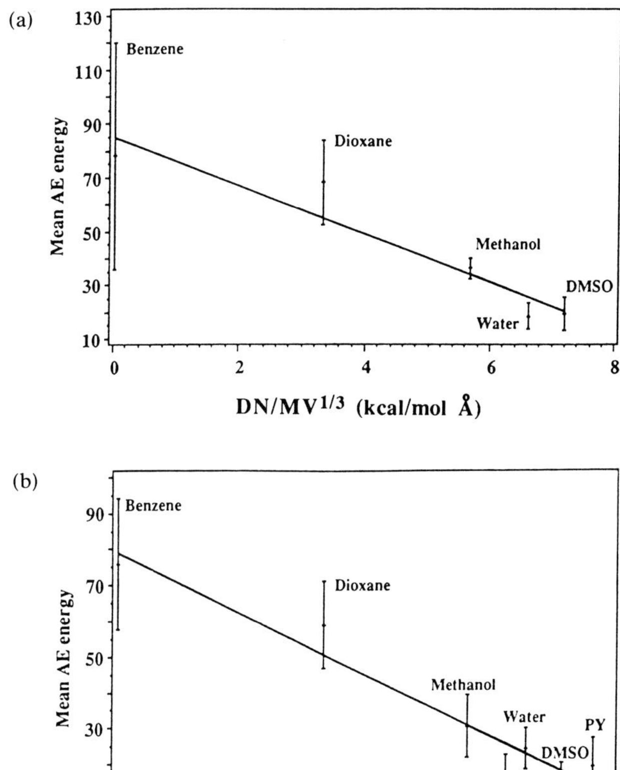

Figure 6. The mean AE energy of solvent-trated wood vs, the solvent strength parameter $DN/MV^{1/3}$. (a) Birch; (b) lauan.

figures above, the decrease in the mean AE energy of specimens tends to be constant when the solvent strength is sufficiently high. This might show that the donor–acceptor interactions within wood have reached a state in which most secondary bonds have been broken. The mean AE energy of wood specimens tends to a minimum value.

An interesting observation is seen in the figures that show AE values plotted against various DN/AN models. The errors bars tend to decrease in size as the solvent parameter (x-axis) increases, i.e. the variability within the data decreases with increasing solvent strength. This phenomenon might be related to the acoustic emission approach used, the properties of the wood, and the degree of donor–acceptor interactions within the wood. The energy calculated is the mean AE energy per hit. Thus, the number of hits obtained influences the final mean value. It was found that the variability within the data was directly related to the number of AE hits, ie. greater variability was associated with relatively few hits. For example, some benzene-treated birch failed with less than 50 hits. On average, the number of hits was less than 500 for benzene, while some water-treated birch failed with over 9000 hits. On average, the number of hits for water-treated birch was more than 2000.

The number of hits collected by the transducer of an acoustic emission device is related to how the material fails. If the material breaks suddenly, i.e. via brittle fracture, the failure may be caused by a few major events within the wood and few acoustic signals are generated. In contrast, if the wood breaks gradually, i.e. ductile failure, the failure results from numerous smaller events within the wood and more acoustic signals are generated. Wood tends to experience brittle fracture when dry. When soaked in benzene, the benzene does not penetrate the cell walls [13]. Thus, the benzene-treated wood still behaves as though it were dry. With increasing solvent strength, the accessibility of cellulose/lignin material to solvent, the swelling effect, and the donor–acceptor interation increase. The wood becomes flexible and softer. The fracture becomes gradual and ductile, resulting in failures with large numbers of hits. The accessibility accompanied by the swelling effect seems to be related to the reduced variability, or spread in data. In other words, the variability or confidence interval somehow indirectly probes the behavior of the donor–acceptor interaction within wood material. That this change in the size of error bars is not seen in the two bending tests described below is, in all likelihood, related to the nature of the bending test when compared with the AE tests. Each bending test provides one data point which assumes that the beam cross-section is homogeneous and anisotropic. AE testing can provide thousands of discrete data points per specimen.

5.2. The change in MOR and MOE of solvent-treated wood specimens

The results of *MOR* and *MOE* of solvent-treated birch, lauan, and Douglas fir were determined. According to the definition of *MOR* and *MOE*, the results obtained from the swelling effect, and the change in mean AE energy, it is reasonable to suggest that the stronger the solvent, the larger the swelling effect; the weaker the specimen, the lower the *MOR* and the *MOE*. Based on this, the modified solvent parameter $DN/MV^{1/3}$ is useful in describing the behavior of the mean *MOR*, *MOE* of the specimen with solvents. The results of this modified solvent parameter are shown in Figs 7 and 8. The *MOR* and *MOE* of solvent-treated birch, lauan, and Douglas fir decrease as the solvent strength parameter increases. The reduction in *MOR* and *MOE* is related to the weakness or breakage of the secondary bonds through donor–acceptor interactions. This result agrees with that of the mean AE energy discussed earlier, i.e. the lower the *MOR* and *MOE*, the more the secondary bonds of wood are weakened or broken.

The correlation of *MOR* and *MOE* of all the solvent-treated specimens with the solvent strength parameter shows an excellent quadratic model. About 95–99% of the variation in *MOR* and *MOE*, when correlated with solvent strength, is accounted for by this model. A complete summary of the models developed and the correlation coefficients for all the systems is given in Table 5.

The *MOR* curves show a critical value for the solvent strength parameter. When the solvent strength parameter is lower than this value, the *MOR* does not change significantly, which implies that few donor–acceptor interactions take place; when the solvent strength parameter is close to this value, the *MOR* decreases considerably. Strong donor–acceptor interactions occur, with an increasing solvent

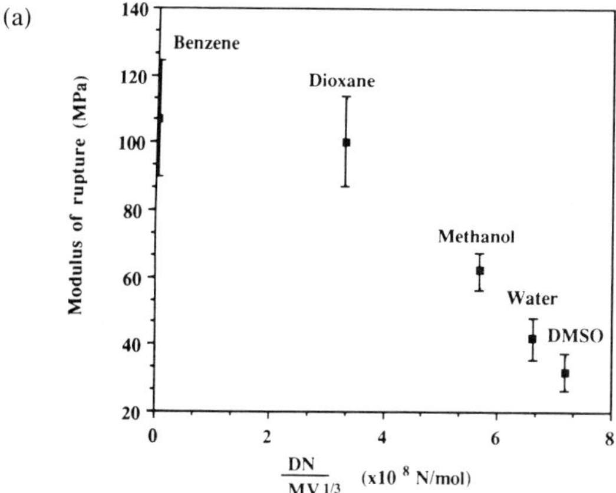

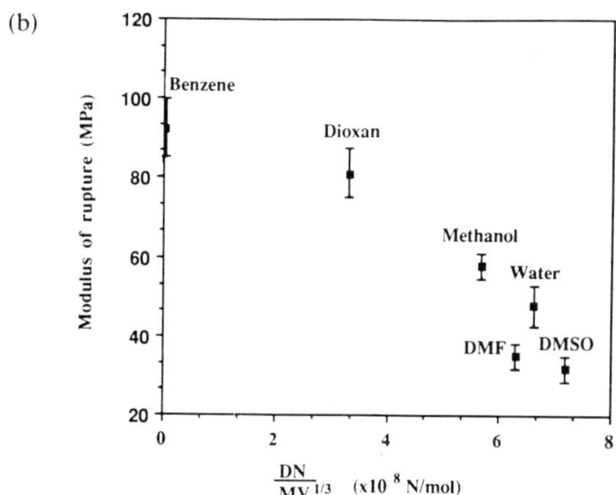

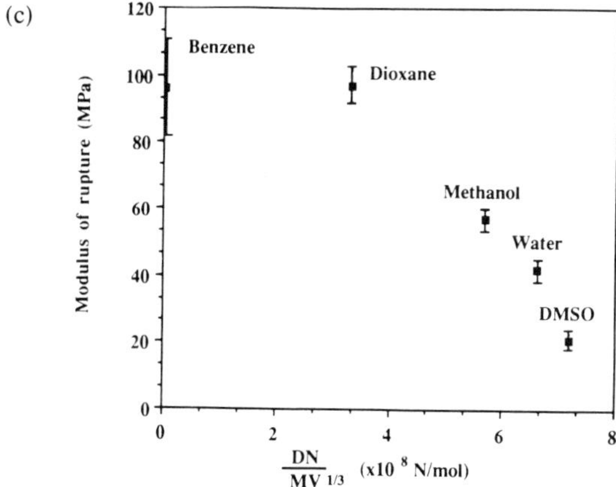

Figure 7. *MOR* values of solent-treated wood vs. the solvent strength parameter *DN/MV*$^{1/3}$. (a) Birch; (b) lauan; (c) Douglas fir.

 Dawn Ma et al.

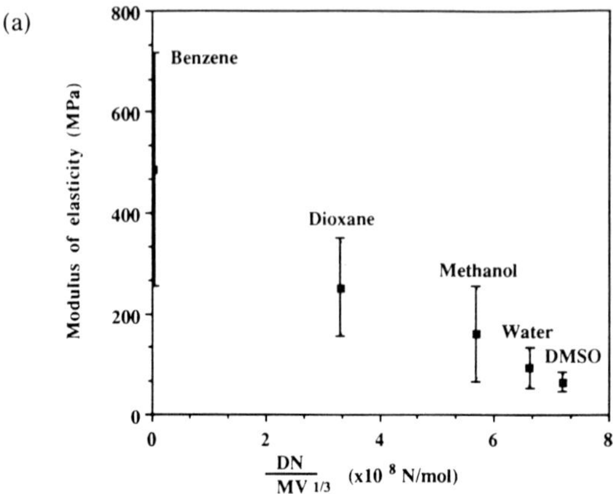

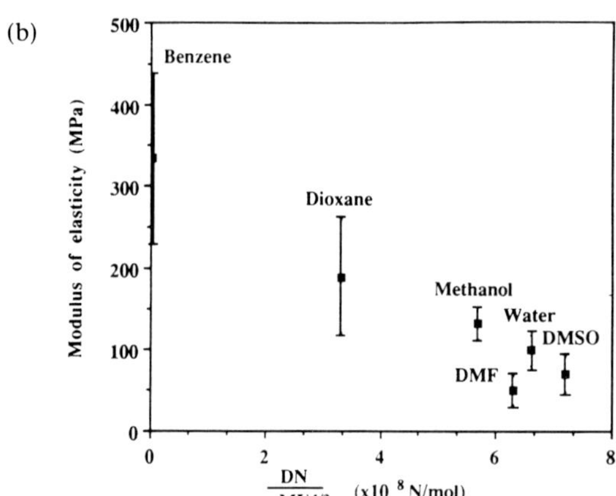

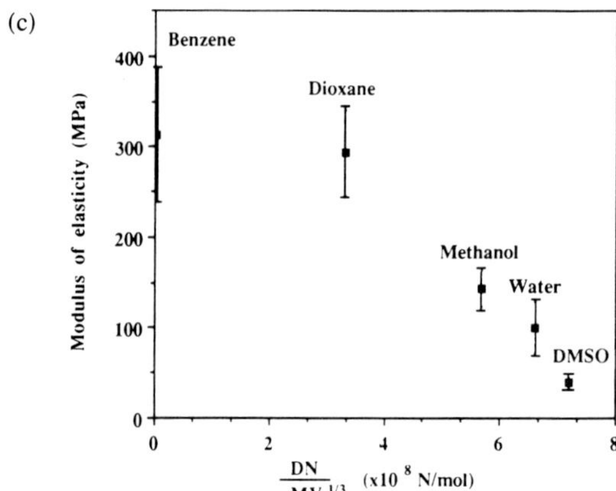

Figure 8. *MOE* values of solvent-treated wood vs. the solvent strength parameter $DN/MV^{1/3}$. (a) Birch; (b) lauan; (c) Douglas fir.

Table 5.

The correlation coefficients of the properties of solvent-treated cellulose materials with solvent parameter models

Properties of wood	Solvent parameter model	Mathematical relation	r^2
TEA	$(AN + DN)^{1/2}$	Linear	0.828
$E_{AE,\,lauan}$	$2.5\,DN + AN$	Linear	0.904
$E_{AE,\,birch}$	$2.5\,DN + AN$	Linear	0.941
$E_{AE,\,lauan}$	$DN/MV^{1/3}$	Linear	0.941
$E_{AE,\,birch}$	$DN/MV^{1/3}$	Linear	0.910
MOR_{lauan}	$DN/MV^{1/3}$	Quadratic	0.943
MOR_{birch}	$DN/MV^{1/3}$	Quadratic	0.990
$MOR_{Douglas\ fir}$	$DN/MV^{1/3}$	Quadratic	0.994
MOE_{lauan}	$DN/MV^{1/3}$	Quadratic	0.955
MOE_{birch}	$DN/MV^{1/3}$	Quadratic	0.994
$MOE_{Douglas\ fir}$	$DN/MV^{1/3}$	Quadratic	0.990

TEA: tensile energy absorption of solvent-treated paper sheet; E_{AE}: mean acoustic emission energy of solvent-treated wood; *MOR*: modulus of rupture of solvent-treated wood; *MOE*: modulus of elasticity of solvent-treated wood.

strength parameter, more molecular bonds are affected or broken and the *MOR* drops continuously. Finally, when the solvent strength parameter is sufficiently high, the donor–acceptor interactions within the wood reach an equilibrium state, after which the *MOR* flattens out and maintains a minimum value. Similar behavior is found in the mean AE energy of the specimens discussed earlier.

5.3. Microstructure

The changes in the mechanical properties of materials are commonly associated with changes in the microstructures. The fracture surfaces of lauan specimens were examined with scanning electron microscopy (SEM) and are shown in Figs 9–11. The microstructure of lauan treated by a low strength solvent, such as benzene, is similar to that of dry lauan. The cellulose microfibrils are bonded very well and are not pulled out. The fracture surface of the cell wall is rough (Figs 9 and 10). The microstructure of lauan treated with methanol, however, shows a significant change compared with the microstructures of dry lauan and benzene-treated lauan, where large wood fibers were pulled out and the fracture surface of the cell wall was smooth (Fig. 11). These photographs show the disruption of the structural integrity of the wood when treated with methanol. In other words, the solvent strength of methanol critically affects the mechanical properties of the wood. The microstructure results are consistent with the numerical results discussed earlier.

5.4. Additional results

Larsson and Johns [4] established the solvent parameter as $(AN+DN)^{1/2}$ and found a negative, linear correlation between this solvent parameter and the tensile energy absorption (TEA) of the paper sheets treated by various solvents. This work used the modified solvent parameter $2.5\,DN + AN$ on the TEA data and found that the correlation is improved. About 96% of the variation in TEA of the

                    Dawn Ma et al.

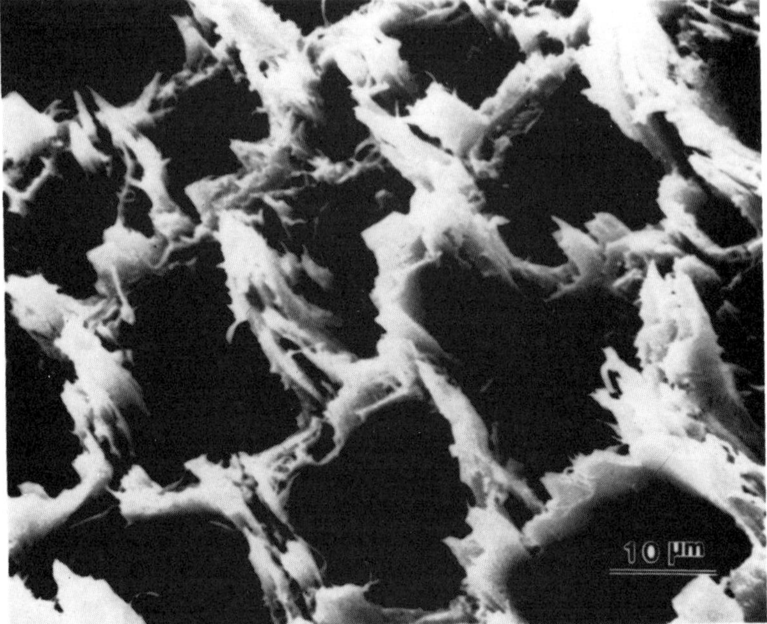

Figure 9. The fracture surface of dry lauan (× 1600).

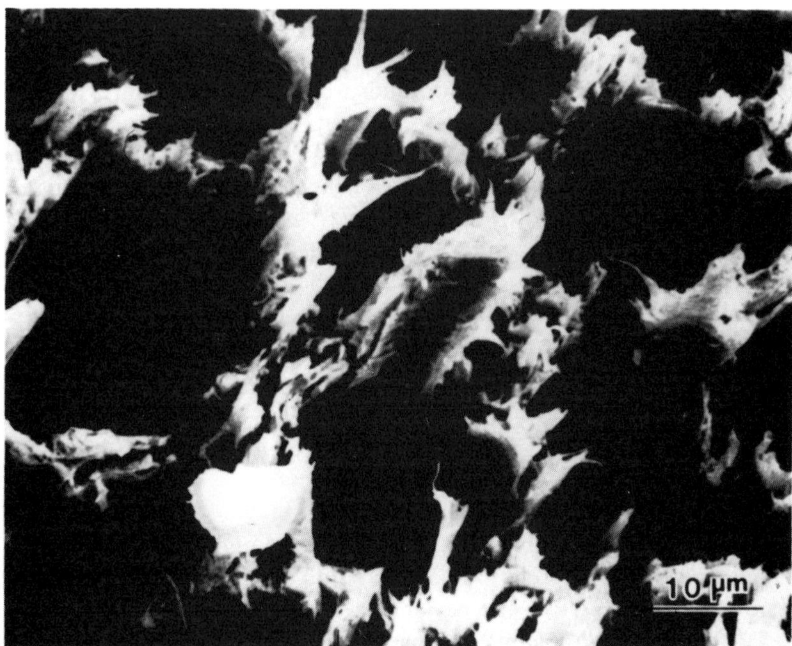

Figure 10. The fracture surface of lauan soaked in benzene (× 1545).

Figure 11. The fracture surface of lauan soaked in methanol ($\times 822$).

paper sheets can be accounted for by the modified solvent strength parameter 2.5 $DN + AN$, only about 84% by the old solvent parameter $(AN + DN)^{1/2}$, as shown in Table 5. For comparison, the TEA data against the old solvent parameter $(AN + DN)^{1/2}$ and the modified solvent parameter 2.5 $DN + AN$ are respectively listed in Table 6 and plotted in Fig. 12.

Table 6.
TEA data of solvent-treated paper sheets with the solvent parameters $(AN + DN)^{1/2}$ and 2.5 $DN + AN$

Solvent	$(AN + DN)^{1/2}$	2.5 $DN + AN$	TEA (g/cm)
Bz	2.88	8.45	20.5
CCl$_4$	3.41	16.1	20.3
NB	4.38	25.8	16.5
CHCl$_3$	5.49	40.6	15.9
Dox	5.06	47.8	14.2
Ac	5.43	55.0	9.5
EtOH	7.65	87.1	6.7
MeOH	7.77	88.8	3.75
Py	6.88	96.95	2.9
DME	6.53	82.5	2.11
FA	7.99	99.8	1.43
W	8.53	99.8	0.90

Bz = benzene; CCl$_4$ = carbon tetrachloride; NB = nitrobenzene; CHCl$_3$ = chloroform; Dox = dioxane; Ac = acetone; EtOH = ethanol; MeOH = methanol; Py = pyridine; DMF = dimethyl formamide; FA = formamide; W = water.

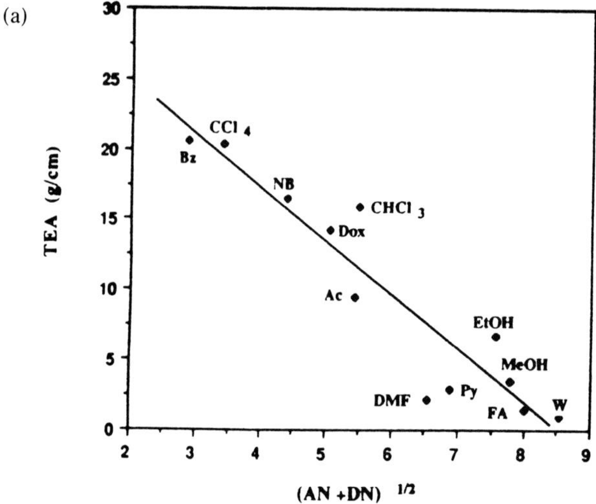

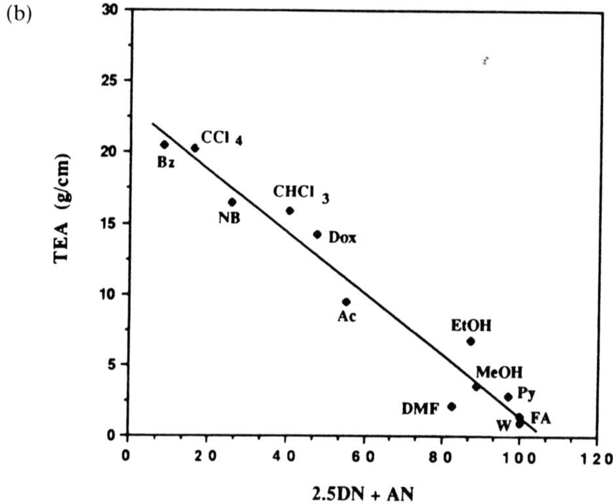

Figure 12. Plots of the tensile absorption energy (TEA) of solvent-treated paper sheets with the old solvent parameter $(AN + DN)^{1/2}$ (a) and with the modified solvent strength parameter $2.5\ DN + AN$ (b).

6. CONCLUSIONS

(1) Through the model of Larsson and Johns and the acoustic emission approach, the mean AE energy (related to failure energy) of solvent-treated *uniform texture wood* has a reasonably good negative linear correlation with the solvent parameter $(AN + DN)^{1/2}$.

(2) The modified solvent strength parameters $2.5\ DN + AN$ and $DN/MV^{1/3}$ are a significant improvement over the linear correlation based on $(AN + DN)^{1/2}$. This verifies that *DN* and *AN*, and *DN* along with the *MV* of a solvent, are the most important factors in characterizing the behavior of donor–acceptor interactions in

cellulose material, and that there is a measurable range of secondary bond strengths in wood.

(3) The *MOR* and *MOE* of solvent-treated birch, lauan, and Douglas fir decrease quadratically as the modified solvent parameter $DN/MV^{1/3}$ increases.

(4) The microstructure changes in the fracture surface of solvent-treated lauan specimens, such as the wood fiber pull-out and the decrease in roughness of the cell-wall fracture surface, parallel the numerical results.

Acknowledgement

We would like to thank the VADMS Center, Washington State University for the use of the Macromodel program and related computer and visualization facilities.

REFERENCES

1. F. M. Fowkes, *Rubber Chem. Technol.* **57**, 328 (1984).
2. F. M. Fowkes, *Ind. Eng. Chem. Prod. Res. Dev.* **17**, 3–7 (1978).
3. H. Schleicher, *Acta Polymerica* **34**, 63 (1983).
4. A. Larsson and W. E. Johns, *J. Adhesion* **25**, 121–131 (1988).
5. W. B. Jensen, *Chemtech* **12**, 755 (1982).
6. E. N. Gur'yanova, I. P. Gol'dshein and I. P. Romm, *Donor–Acceptor Bond*, p. 1. Keter Publishing House Jerusalem Ltd (1975).
7. V. Gutmann, *The Donor–Acceptor Approach to Molecular Interactions*, Chaps 1 and 2. Plenum Press, New York (1978).
8. V. Gutmann, *Electrochim. Acta.* **21**, 661–670 (1976).
9. A. A. Robertson, *TAPPI* **53**, 1331 (1970).
10. F. M. Fowkes, *J. Adhesion* **4**, 155–159 (1972).
11. J. R. Mathews and D. P. Hay, in: *Acoustic Emission*, J. R. Mathews (Ed.), pp. 1–13. Gordon & Breach, New York (1983).
12. W. C. Still, F. Mohamadi, N. G. J. Richards, W. C. Guida, M. Lipton, R. Liskamp, G. Chang, T. Hendrickson, F. Degunst and W. Hasel, *MACROMODEL* V2.5, Department of Chemistry, Columbia University, New York, NY 10027.
13. A. J. Stamm, *Wood and Cellulose Science*, pp. 196–199. Ronald Press, New York (1964).

Acid-Base Interactions, pp. 363-380
Eds. K.L. Mittal and H.R. Anderson, Jr.
©VSP 1991

Acidic and basic nature of ferric oxide surfaces. Adsorption, adhesion, zeta potentials and dispersibility in magnetic inks for hard disks

YAN CHU HUANG,* FREDERICK M. FOWKES and THOMAS B. LLOYD

Department of Chemistry, Lehigh University, Bethlehem, PA 18015, USA

Revised version received 8 August 1990

Abstract—The surface FeOH sites of ferric oxides are amphoteric in that they form interfacial acid–base complexes with either acids or bases. Adsorption of acidic or basic test molecules from neutral organic solvents was used to characterize the acid–base properties of γ-Fe$_2$O$_3$, a magnetic ferric oxide of importance in magnetic tapes and disks for computer memories.

Gafac RE-610, a dispersant commonly used in the dispersion of ferric oxide powders for fabrication of the magnetic polymer composites used in computer tapes and disks, is an amphoteric diester of phosphoric acid with a polyoxyethylene adduct of an alkylphenol. The POH group of RE-610 adsorbs as a strong acid onto basic ferric oxide sites, and the 10–12 ether groups adsorb as bases onto acidic ferric oxide sites. When RE-610 adsorbs onto ferric oxide, the first monolayer is spread out broadly over the surface, occupying 2.2 nm^2 per molecule, because several ether groups as well as the POH group are complexed to surface FeOH sites. A second monolayer adsorbs, a rare phenomenon in adsorption from non-aqueous solutions, and this second layer can be attached only through its POH acid group, as evidenced by an area per molecule of only 0.72 nm^2.

RE-610 is used as a dispersant for γ-Fe$_2$O$_3$. Such dispersions have a strong tendency to flocculate, for the magnetic inter-particle attractions add appreciably to the normal London–Lifshitz attractions, illustrated by comparing magnetic with non-magnetic ferric oxide dispersions. Such dispersions are much more deflocculated after adsorption of acidic polymers.

Keywords: Electrostatic repulsion; steric stabilization; multilayer adsorption; deflocculation of magnetic oxides; amphoteric adsorption.

1. INTRODUCTION

The adsorption of surfactants and of polymers onto inorganic solids from organic solvents has been shown to require acid–base complexation between acidic (or basic) functional sites of the surfactant or polymer and basic (or acidic) surface sites of the inorganic substrate [1]. Initial studies were done with an acidic powder (silica) and a basic powder (calcium carbonate) in carbon tetrachloride solutions of an acidic polymer (post-chlorinated polyvinyl chloride, CPVC) or of a basic polymer (polymethylmethacrylate, PMMA). It was found that the basic PMMA adsorbed strongly onto the acidic silica, that the acidic CPVC adsorbed strongly onto the basic calcium carbonate, but that there was no adsorption of the acidic polymer onto the acidic powder nor of the basic polymer onto the basic powder. This study also found that if acidic or basic solvents were used instead of the relatively neutral carbon tetrachloride, these solvents competed with either the polymer or the inorganic solids for complexation of basic or acidic sites, with a consequent reduction of the extent of polymer adsorption onto the particles.

---

*To whom correspondence should be addressed.

1.1. The amphoteric nature of ferric oxide surfaces

This paper is a report of a continuation of the above studies with amphoteric inorganic surfaces, having both surface acidity and surface basicity, and with adsorbates which are acidic, basic, or amphoteric in nature. Bolger and Michaels made a study in 1968 [2] of the acid and basic nature of metal oxides which pointed out that in the presence of moisture, metal oxides develop appreciable surface concentrations of M—OH sites which once formed are quite stable, even at elevated temperatures. Some M—OH sites are acidic (Si—OH), some are basic (Al—OH), and some are amphoteric (Fe—OH). Bolger and Michaels' studies were with oxides in aqueous media, but similar conclusions had been drawn by Fowkes from studies of the acidity or basicity of metal oxides in hydrocarbon solutions of test acids or bases [3]. In these studies the native oxide of aluminum was found to adsorb test organic acids much more strongly than organic bases; the oxide layer on 18-8 stainless steel was found to adsorb organic bases much more strongly than organic acids; but the oxide layers on carbon steel were found to adsorb both acids and bases strongly. The amphoteric nature of the FeOH surface sites of ferric oxides has also been demonstrated by the pH dependence of their ion-exchange capacity in a novel XPS study by Simmons and Beard [4]. In aqueous media at high pH the surface sites become negatively charged (FeO$^-$), and at low pH the surface sites become positively charged (FeOH$_2^+$); charge neutrality at about pH 7 indicates equal concentrations of both charged species.

1.2. Acid–base nature of adsorption in organic media

The acid strength or base strength of surface sites can be characterized by the heats of acid–base interaction of surface sites with test acids or bases of known acidic or basic strength. The heats of adsorption are actually heats of acid–base complexation, and these have been determined calorimetrically or by spectrometric shifts in infra-red (IR) spectra [5]. Calorimetric titrations of the acidic sites of silica and of rutile powders were done with a stirred suspension in cyclohexane as either ethyl acetate, pyridine, or triethylamine was added. After the adsorption was completed and all of the heat registered, the equilibrium concentration of titrant was determined by UV absorption and the molar heat of adsorption determined. Molar heats of adsorption were also determined from IR spectral shifts [5], using the previously calibrated shift of the carbonyl stretch frequency $\Delta \nu_{C=O}^{ab}$ with heats of acid–base complexation ΔH^{ab} of carbonyl groups with various acids [6].

1.3. Modern theories of hard and soft Lewis acids and bases

The strength of acids or bases in non-aqueous media is best correlated with the Drago E and C equation [7, 8]:

$$-\Delta H^{ab} = C_A C_B + E_A E_B, \tag{1}$$

where the E and C constants relate to the electrostatic and covalent contributions to acid–base complexation, and the subscripts A and B refer to the acid and base. Hydrogen bonds are examples of hard acid–base interactions dominated by electrostatic interactions, so that the second term of equation (1) predominates. Interactions of soft acids and bases, such as in the sulfur–gold interactions of self-assembled monolayers, are dominated by covalent interactions, and the first term of equation (1) predominates. The E and E constants for about 40 acids and 40 bases were determined by Drago and co-workers from heats of acid–base interaction determined in dilute solution in neutral solvents such as cyclohexane, using calorimetry as the basic tool, and also using spectral shifts (IR and NMR) calibrated by calorimetry. E and C constants for additional acids or bases could then be determined from their heats of complexation with test acids or bases of known E and C constants. This procedure even allows determination of the E and C constants for the acidic or basic sites of inorganic oxides, using the heats of adsorption of test bases or acids absorbing onto stirred suspensions of silica or rutile powders.

Another popular correlation of acid and base strength was developed by Mayer and Gutmann in Vienna, involving the determination of a measure of base strength (the donor number, DN) and a measure of acid strength (the acceptor number, AN) [9]. The donor number is the molar heat of complexation (in kcal/mol) of $SbCl_5$ interacting with a basic liquid in dilute solution in 1,2-dichloroethane. Since $SbCl_5$ is a very soft acid, the donor number emphasizes the covalent contribution to basicity and tends to be proportional to Drago's C_B. The acceptor number (AN) is determined by the ^{31}P-NMR shift for triethylphosphine oxide (Et_3PO) when added to an acidic liquid; the shift for hexane is arbitrarily assigned an AN of zero, and the shift for $SbCl_5$ in 1,2-dichloroethane is arbitrarily assigned an AN of 100. AN values were intended to be a measure of acid strength, but because the Et_3PO test probe is added directly to the liquid under test rather than to a solution of the liquid, the AN value has an appreciable van der Waals (London dispersion force) contribution. For instance, pyridine has an AN of 14.2, which suggests that this basic liquid also has appreciable acid character; however, subtraction of the van der Waals contribution reduces the AN for pyridine to 0.5, demonstrating that pyridine has negligible acid character [10]. It should be noted that Et_3PO is a hard oxygen base, and so acceptor numbers (after correction) tend to be proportional to Drago's E_A constants.

Gutmann has taught that all compounds have some acidity and basicity and therefore have some degree of acid–base self-association. Current calorimetric research on acid–base self-association in organic liquids tends to bear this out [11, 12]. For instance, at 25°C, the heat of vaporization of acetone is 28.7 kJ/mol, of which 8.8 kJ/mol is required to dissociate the acid–base bonds of acetone. In Drago's studies the basic strength of acetone is characterized, but not the acidity; in fact, in all of Drago's studies only the acidity or only the basicity of a tested compound has been addressed, and never both. In this study, both the acidity and the basicity of ferric oxide surface sites will be considered and evaluated.

1.4. Determination of the E and C constants of FeOH sites

In earlier studies we used flow microcalorimetry to determine the heats of adsorption of pyridine from cyclohexane solutions onto α- and γ-ferric oxides [13, 14]. Although each commercial batch of ferric oxide powders tends to differ in distribution of strengths of acidic surface sites, the range of heats of adsorption of pyridine from cyclohexane was from about -6 to -12 kcal/mol, with the most uniform batch (a γ-Fe_2O_3) having a measured heat of adsorption which increased with coverage from -9.0 kcal/mol (at zero coverage) to -11.3 ± 0.2 kcal/mol (at 84%

coverage) [14]. The increase in heat of adsorption with coverage for pyridine on γ-ferric oxides is believed to result from increased van der Waals interactions as the adsorbing pyridine molecules enter monolayers increasingly richer in pyridine. In another study of pyridine adsorption on γ-Fe$_2$O$_3$ the initial heat of adsorption was also -9 kcal/mol, but it decreased with surface coverage [15]. In general, the acid–base interaction of pyridine with γ-ferric oxide is nearly as strong as that observed for pyridine interaction with phenol, silica, or rutile.

In studies with α-Fe$_2$O$_3$ powders, heats of adsorption of triethylamine, pyridine, and ethyl acetate allowed estimation of the Drago E_A and C_A constants for the FeOH surface sites [13]: $E_A = 4.5 \pm 1.1$ (kcal/mol)$^{1/2}$ and $C_A = 0.8 \pm 0.2$ (kcal/mol)$^{1/2}$. E_B and C_B constants for ferric oxides have yet to be determined.

1.5. Acid–base nature of electrostatic charging in organic media

The classical electrophoretic measurements in aqueous media were used by Bolger and Michaels [2] to determine the acidity or basicity of metal oxides. Non-aqueous electrophoresis has been so much more difficult that it is only in the last decade that modern instrumentation has allowed such measurements to be made easily and accurately. The low dielectric constants of organic solvents make it quite unlikely that ionic emulsifiers, dispersants, or other surfactants can dissociate as they do in aqueous media where adsorbed ions tend to determine the measured zeta-potential. The mechanism of charging of particles in organic media involves acid–base interaction between the surface sites of the particles and the dispersant molecules or micelles. In the initial studies by van der Minne and Hermanie [16], it was found that carbon black particles in benzene developed appreciable zeta-potentials in the presence of commercial sludge dispersants: that some dispersants provided negative charging and others positive charging; that highly charged suspensions were very well stabilized against flocculation; and that mixing these negative and positive charging dispersants minimized charging and flocculated the dispersions. In subsequent studies, Fowkes *et al.* [17] concluded that the mechanism of charging of such carbon black particles depended on their acidic or basic surface sites interacting with the basic or acidic sites of the micellar or polymeric dispersants. It was shown that basic dispersants always provided negatively charged carbon black particles, and that acidic dispersants always provided positively charged carbon black particles. These rules suggested a dynamic process in which basic dispersants adsorb on the acidic sites of carbon black, pick up protons from the acidic sites, and desorb as positive ions into the diffuse counter-ion layer, leaving the particle negatively charged because protons had been taken from their acidic sites. This mechanism was proven in detail by using ^{14}C-tagged dispersants and tritium-tagged acidic sites on acidic pigments; electrodeposition of charged particles on the anode and of ^{14}C-tagged dispersant counter-ions with tritium on the cathode showed that the dispersant had picked up tritium from the acidic surface sites of the pigments, producing positively charged counter-ions [18].

In these studies with γ-ferric oxide particles in organic media, electrophoresis has been used extensively to determine acid–base interactions between the surface sites of the oxide particles, the basic solvent mixture, the acidic epoxy resin, and the basic PVME, using the principles described above for carbon black systems.

The findings have been useful for determining whether acid–base interactions have taken place, for determining the relative basicity or acidity of competing interactants, and for predicting electrostatic stabilization of charged particles.

2. EXPERIMENTAL

2.1. Materials

Several batches of Pfizer γ-ferric oxide 2300 (Charles Pfizer and Co.) were used in this study, and are identified by lot number. Lot number 214023 was used for the adsorption and dispersion studies with Gafac RE-610; this lot had a specific surface area of 23.0 m^2/g (by multi-point BET), and after heating at 450°C for 8 h for conversion to the α-form, the specific surface area was 21.6 m^2/g. Prior to any measurements, the oxides were tumbled overnight in each solvent and then dried at 110°C overnight under vacuum. The first wash was in ethanol; the second in acetone; and the third in distilled water. Washings were monitored by UV and IR spectroscopy and by conductivity of aqueous washings. Lot number 214023 of γ-Fe$_2$O$_3$ was washed with distilled water until the conductivity of the washings was minimized; this led to a 0.5% loss of weight.

The Gafac RE-610 dispersant (from General Aniline and Film Corp.) is a mixture of a 76% diester and 24% monoester of phosphoric acid with a nonyl-phenoxy heptaethyleneoxy adduct, with an average molecular weight of 1120, an equivalent weight of 783, and a density of 1.10–1.12 g/ml [19, 20]. The EPON 2004 epoxy resin (from Shell Chemical Co.) has an equivalent weight of 875–975 per epoxy group. The phenolic crosslinker R108 is an allyl-capped phenol from BTL Specialty Resins of Rochester, New York. The polyvinylmethylether (PVME), molecular weight 55 000, is a product of BASF.

Phenol (loose crystals, 99 + %) and pyridine (anhydrous, 99 + %) were from Aldrich Chemical. All solvents were from Aldrich Chemical and were of HPLC grade except for EAK (5-methyl-3-heptanone), which was of spectrophotometric grade (99 + %), and isophorone (97%). The solvents were dried over freshly activated 3A molecular sieves (Fisher Scientific), and filtered through 0.2 μm Acrodisc chemical-resistant syringe filters (Gelman). A mixture of isophorone, EAK, and p-xylene was used for many experiments and will be referred to as the 'mixed solvent'; it had a density of 0.8765 ± 0.0008 g/ml, a viscosity of 1.1324 ± 0.0005 cp, and a refractive index of 1.4648 ± 0.0002 at 23°C.

2.2. Experimental procedures

The adsorption isotherms for pyridine and phenol adsorbing out of cyclohexane onto γ-Fe$_2$O$_3$ were determined after 5 days of tumbling at 23°C in screw-capped bottles, each with about 0.5 g of powder in 20 ml aliquots of pyridine or phenol solutions. Four or five replicates were determined for each of six different initial concentrations (10–70 mM). The equilibrium concentrations of the supernatant solutions (after overnight magnetic separation) were determined from the UV absorbance measured with a Perkin-Elmer Lambda 5 UV–visible spectro-photometer in a wavelength and concentration range where Beer's law holds.

Adsorption isotherms for Gafac RE-160 adsorbing on lot 214023 of the Pfizer 2300 γ-Fe$_2$O$_3$ were determined in a similar manner in the mixed-solvent system (isophorone, EAK, and *p*-xylene), using atomic emission with a SpectraSpan VI spectrometer (by Y.C.H. in the IBM San Jose Laboratories) to measure the atomic emission of phosphorus at 213.618 nm. The samples were digested in sulfuric and nitric acids to remove organic matter before atomization from an aqueous medium upon injection into the DC plasma of the spectrometer. The intensity of the emission fitted a polynomial equation of second degree for the phosphorus concentration, with a correlation coefficient of 0.999989. The adsorption isotherms were determined at 23°C with quadruplicate samples of 16 different concentrations of RE-610 over a very wide range, from 4.6 mmol/m^3 to 22.3 mol/m^3. In each sample, 0.5 g of dry γ-Fe$_2$O$_3$ powder was equilibrated with 20 ml of RE-610 solution as in the above adsorption studies with pyridine and phenol.

Gafac RE-610 was found to dissolve some of the γ-Fe$_2$O$_3$ powder; so after adsorption equilibrium was attained, the oxide was washed with the mixed solvent, then with acetonitrile and with acetone before vacuum-drying at 110°C until a constant weight was attained.

Zeta potentials of γ-Fe$_2$O$_3$ powders in about 100 ppm suspensions in the mixed solvent (with various solutes present) were measured with the Pen Kem 3000 electrophoresis apparatus. The reported zeta potentials (ζ) were calculated from the measured electrophoretic mobilities (μ) with the Hückel equation

$$\zeta = 3\eta\mu/2\varepsilon_r\varepsilon_o. \tag{1}$$

The exact compositions used in these measurements are listed in Table 1.

The dispersibility and dispersion stability of α- and γ-Fe$_2$O$_3$ (lot 214023) powders in solutions of RE-610 in the mixed-solvent system, and of the same lot of γ-Fe$_2$O$_3$ in solutions containing the three polymeric components (PVME, Epon 2004, and the R108 crosslinker) were tested after vigorous milling. Four grams of the oxide in 7 ml of mixed solvent (plus specified amounts of the above additives) and 20 g of 80 mesh Ottawa sand were milled in a vibratory energy mill (Model 2601, Pitchford Scientific Instruments), where it was vibrated with about 2 cm strokes at 3500 cycles/s. The samples were in screw-topped polyethylene bottles fitted tightly into a steel tubular milling chamber. After 10–15 min of milling (with forced air cooling), the bottles were tumbled overnight at about 100 rpm to ensure equilibrium adsorption, followed by another 15 min of vigorous milling.

Table 1.

Zeta potentials of γ-Fe$_2$O$_3$ dispersions in the mixed-solvent system as a function of the other components

% γ-Fe$_2$O$_3$	% Epoxy	% PVME	% RE-610	% R-108	ζ (mV)
0.0039					$-37, -32$
0.0030	2.9				$+10, +7$
0.0039		1.1			$-23, -24$
0.0035			0.1		$+10, +10$
0.0038				1.2	$-6, -7$
0.0034	2.8	1.1			$-14, -17$
0.0031	2.9		0.1		$+14, +17$
0.0026		1.1	0.09		$+18, +20$
0.0026			0.09	1.1	$+31, +32$
0.0026	2.8	1.1	0.08		$+35, +39$
0.0020	2.8	1.1	0.08	1.1	$+42, +44$

Particle size distributions of the freshly milled ferric oxide dispersions were determined with a Coulter N4 photon correlation spectrometer on suspensions micropipetted from the milling concentrates and sonicated into enough mixed-solvent system to reduce the ferric oxide concentration to about 20 ppm, so that the spectrometer was operating in its optimum photon flux (preferably 10^5–10^6 counts/s). After a 30 s auto-ranging adjustment by the instrument to determine an optimum sample time and pre-scale factor, particle size distributions were determined sequentially over five 30 s periods, twenty 200 s periods, and five 500 s periods. Thus, the flocculation rates were determined over a period of about 2 h, and in this relatively short period differences in the formulation of the dispersions showed large differences in the flocculation rates.

3. RESULTS AND DISCUSSION

3.1. Adsorption isotherms and test acids and bases

Figures 1 and 2 are adsorption isotherms for pyridine and phenol, respectively, adsorbing onto γ-Fe$_2$O$_3$ powders from cyclohexane at 23°C. In each graph there are two plots: one with a dashed curve in which the equilibrium interfacial concentration Γ (right ordinate, in μmol/m^2) is plotted against the equilibrium bulk concentration C_{eq} (in mol/m^3 or mmol/l), and one with a solid straight-line Langmuir plot of C_{eq}/Γ vs. C_{eq}, where

$$C_{eq}/\Gamma = 1/K_{eq}\Gamma_m + C_{eq}/\Gamma_m. \tag{2}$$

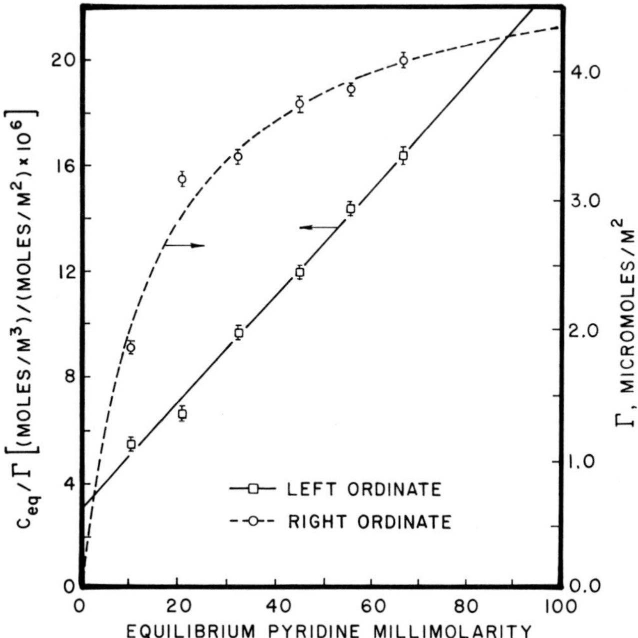

Figure 1. Langmuir isotherm for the adsorption of pyridine from cyclohexane onto γ-ferric oxide for the determination of the acidic surface sites of this oxide. The dashed line is a plot of the surface concentration vs. the equilibrium bulk concentration, and the solid line is plotted according to equation (2).

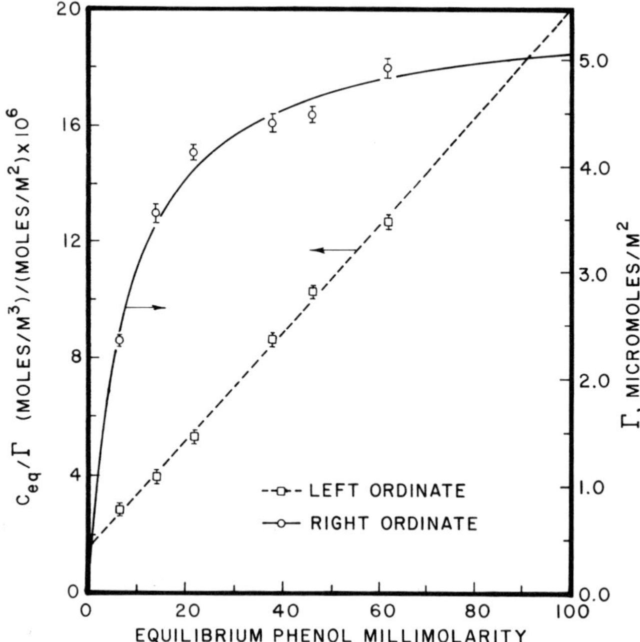

Figure 2. Langmuir isotherm for the adsorption of phenol out of cyclohexane onto γ-ferric oxide for the determination of the basic surface sites of this oxide, plotted as in Fig. 1.

The slope of the solid lines in Figs 1 and 2 determines Γ_m, the monolayer capacity of the interface when saturated with the adsorbate, 4.99 μmol/m^2 for pyridine and 5.48 μmol/m^2 for phenol. An interface with such saturation concentrations would average 3.0 sites per nm^2 for pyridine and 3.3 sites per nm^2 for phenol, values close to the 2.6 FeOH surface sites per nm^2 determined by Simmons and Beard by ion exchange in XPS measurements of aqueous ferric oxide interfaces at various pH values [4]. The surface FeOH sites are really quite similar in aqueous or in non-aqueous media, for they do not dissociate upon heating in air to several hundreds of °C; nor should they disappear upon immersion in an organic solvent. It, therefore, appears reasonable to believe that the acidic sites that adsorb pyridine and the basic sites that adsorb phenol are the same FeOH sites that provide the observed ion-exchange capacity in aqueous ferric oxide interfaces.

3.2. *Adsorption of RE-610 on* γ-*Fe$_2$O$_3$ powders*

The experimental data for the adsorption of RE-610 onto γ-Fe$_2$O$_3$ powder at 23°C are plotted in Figs 3 and 4, plots similar to those of Figs 1 and 2. The difference between Figs 3 and 4 is the concentration range; Fig. 3 covers adsorption at very low concentrations, up to C_{eq} values of 1.54×10^{-2} mol/m^3, while Fig. 4 covers adsorption at concentrations which are ten to a thousand times greater. Figure 3 is concerned with the adsorption at very low concentrations of the first monolayer of RE-610, in which the molecules are spread out flat enough on the surface to occupy 2.2 nm^2 per molecule, as determined from slope of the solid line in Fig. 3.

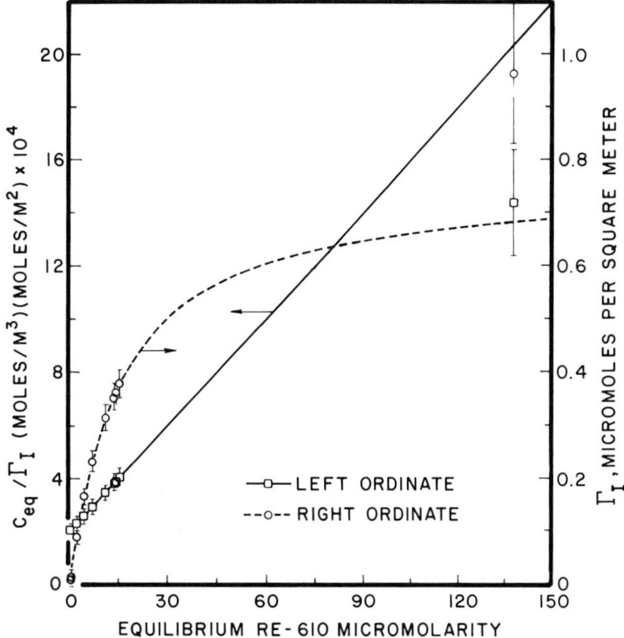

Figure 3. Langmuir isotherm for the adsorption of low concentrations of Gafac RE-610 out of the mixed solvent onto γ-ferric oxide, plotted as in Fig. 1.

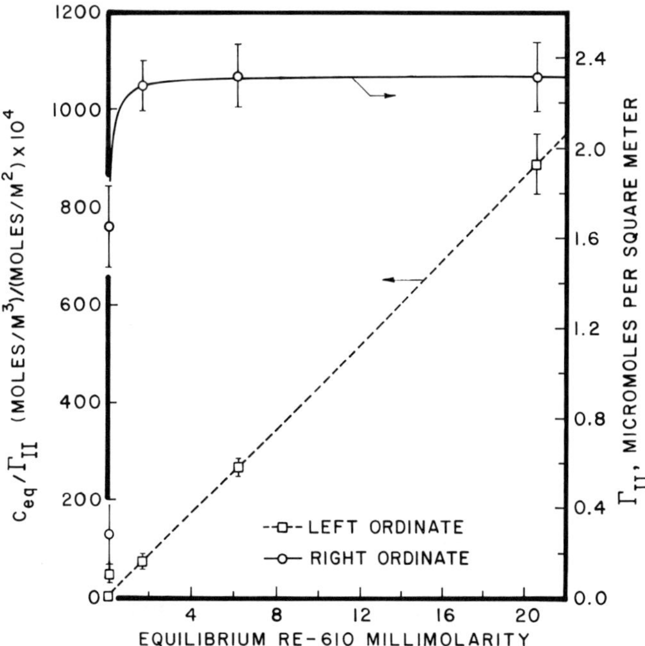

Figure 4. Langmuir isotherm for the adsorption of high concentrations of Gafac RE-610 out of the mixed solvent onto γ-ferric oxide, plotted as in Fig. 1.

At much higher concentrations a second monolayer adsorbs, but the molecules are packed more tightly; they occupy only 0.72 nm^2 per molecule, as determined from the slope of the solid line in Fig. 4.

Multilayer adsorption from non-aqueous solution is a rare phenomenon. This is because the adsorption of surfactants from an organic solvent requires the formation of acid–base complexes between acidic or basic sites of the adsorbate with basic or acidic surface sites of the substrate. Thus, pyridine adsorbs from cyclohexane onto the acidic surface sites of ferric oxide (Fig. 1), and no second monolayer can adsorb, because pyridine molecules of a second layer cannot form acid–base bonds with the adsorbed pyridine molecules of the first layer. However, in the case of RE-610, a surfactant with a strongly acidic P—OH group and with 12 basic ether sites, the first layer most probably adsorbs with several of the ether sites bonded to acidic FeOH sites, because it is found to occupy a very much larger area per molecule (2.2 nm^2) than would be required if it were to adsorb only by the phosphoric acid site. Of course, the first adsorbed layer of RE-610 shields the ferric oxide surface itself from further adsorption, but each molecule has several free ether sites to which the phosphoric acid site of molecules in a second layer of RE-610 can adsorb. Figure 5 is a plot of the amount adsorbed as a function of the logarithm of the equilibrium bulk concentration of RE-610; this plot combines in one diagram the two-step adsorption process which occurs in such widely separated concentration ranges. Figure 6 is a schematic diagram to illustrate the bonding and orientation of RE-610 molecules in the first and second adsorbed monolayers. It is aimed at showing that the first adsorbed monolayer shields the ferric oxide, so that the second monolayer can adsorb only on the basic ether sites of the first monolayer. This two-step bilayer adsorption process for RE-610 on γ-ferric oxides has also been observed by DasGupta (private communication).

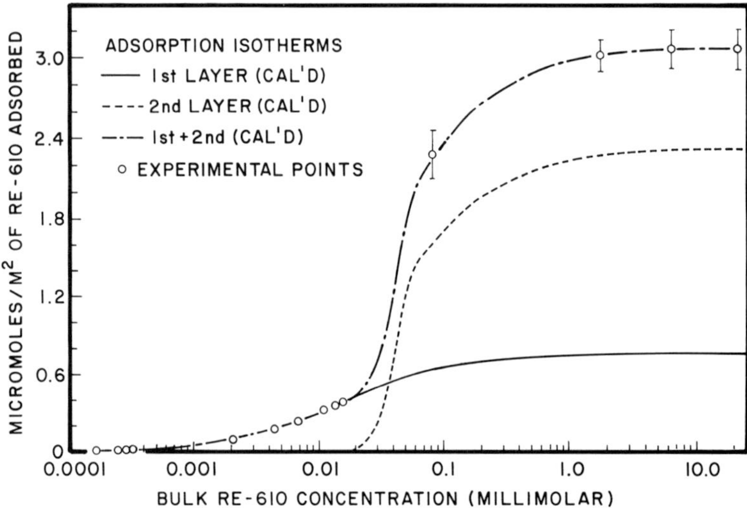

Figure 5. RE-610 adsorption out of the mixed solvent onto γ-ferric oxide plotted against the logarithm of the equilibrium concentration so as to illustrate the adsorption of the first monolayer at low concentrations, and of the second monlayer at high concentrations.

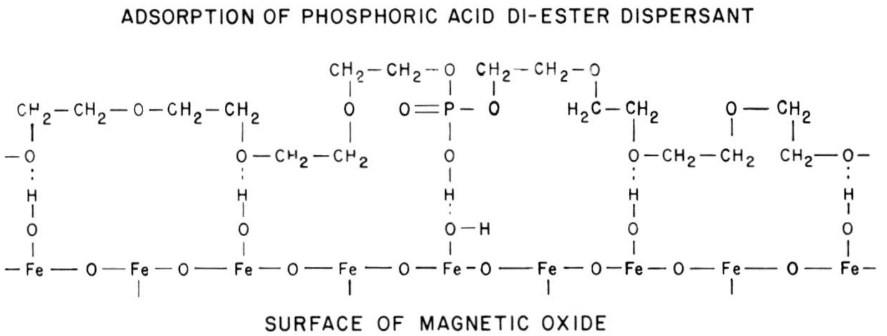

ADSORPTION OF PHOSPHORIC ACID DI-ESTER DISPERSANT

SURFACE OF MAGNETIC OXIDE

Figure 6. Schematic diagram of the first adsorbed monolayer of RE-610 on γ-ferric oxide, indicating the exposed and unbonded ether groups which can interact with the acidic POH groups of the second adsorbed monolayer of RE-610.

3.3. *Adsorption of acidic and basic polymers on γ-iron oxides*

In this investigation of the acid–base interactions of γ-ferric oxide, the study included interactions of the oxide with acidic and basic polymer components, and their interaction with RE-610, because these interactions are very important to the performance of the γ-ferric oxide dispersions. The adsorption onto γ-ferric oxides of an epoxy resin (Epon 2004) and of polyvinylmethylether (PVME) was determined by photoacoustic FTIR spectroscopy of the oxide powders filtered from the polymer solutions in a mixed solvent after adsorption was complete [14].

The epoxy resin is acidic, as has been determined from the 3 ppm shift of the ^{31}P-NMR peak of triethylphosphine oxide (Et_3PO) upon complexation with secondary alcohol groups of Epon 2004 in carbon tetrachloride [11]. Et_3PO is a basic probe for measuring the strength of acidic sites; the 3 ppm shift corresponds to a heat of acid–base complexation of about 20 kJ/mol with Epon 2004. The acidity of Epon 2004 was also demonstrated in electrophoresis measurements of γ-ferric oxide particles suspended in Epon 2004 solutions in the mixed solvent (Table 1). In this basic (electron-donating) solvent, the γ-ferric oxide particles had a zeta potential of -35 mV, but when 2.8 wt% of Epon 2004 was added, this acidic (electron-accepting or proton-donating) resin changed the zeta potential to $+9$ mV.

The phenolic R-108 crosslinker was also observed to be acidic, as demonstrated by the ^{31}P-NMR chemical shifts in complexes with Et_3PO [11], and by the positive increases in the zeta potential of the iron oxide particles in each of the three formulations in which it was tested (Table 1).

The PVME polymer is basic, for aliphatic ethers are rather strong oxygen bases. The basicity of PVME was also demonstrated by electrophoretic measurements with γ-ferric oxides as shown in Table 1. To the above-described suspension of γ-ferric oxide in a solution of 2.8 wt% of Epon 2004 (with a zeta potential of $+9$ mV), 1.1 wt% of PVME was added, and its basicity (electron-donating or proton-accepting ability) was large enough to overcome the acidity of the epoxy resin and change the zeta potential to -16 mV.

Because the ferric oxide surface sites are both acidic and basic, the acidic epoxy and the basic PVME were both adsorbed by γ-ferric oxide particles from a solution of 2.8 wt% of Epon 2004 and 1.1 wt% of PVME in the mixed solvent, as shown in Fig. 7, a photoacoustic FTIR spectrum of adsorbed polymer films on

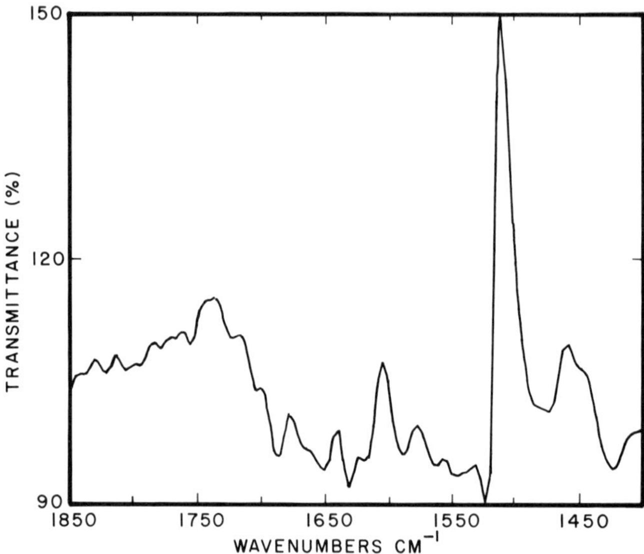

Figure 7. IR absorption spectrum of the adsorbed mixed film of epoxy resin (1510 and 1605 cm^{-1}) and of polyvinylmethylether (1460 and 1740 cm^{-1}) obtained by photoacoustic FTIR or ferric oxide powders with such adsorbed films.

γ-ferric oxide powders. The absorption peaks for the epoxy resin are at 1510 and 1605 cm^{-1}, and for the PVME at 1740 and 1460 cm^{-1}. Adsorption is a three-way competitive process between (1) the solutes, (2) the solvent, and (3) the surface sites of the oxide powder [1]. In this case, the adsorbing basic PVME polymer competed successfully against the basic solvent for the acidic surface sites of the oxide, and the basic surface sites of the oxide competed successfully against the basic solvent for acid–base interaction with (and adsorption of) the acidic epoxy resin. However, when the powder with its adsorbed film of acidic and basic polymers were rinsed in an ultrasonic bath of the basic mixed solvent, the basic solvent eventually won out against the basic sites of the oxide for acid–base interaction with the acidic epoxy resin, which was dissolved off [14].

3.4. Colloidal attributes of adsorbed RE-610 bilayers

Adsorbed films of RE-610 should provide some colloidal stability to ferric oxide particles, and the contribution of such films to steric and electrostatic stabilization against flocculation can be predicted from their properties. The average thickness of the first monolayer, calculated from its Γ_m of 0.75 μmol/m^2 (corresponding to 2.2 nm^2 per molecule) and from the bulk density of RE-610 (1.11 g/ml), should be only 0.8 nm if solvent-free, and probably is never thicker than about 1.0 nm. The second monolayer is appreciably thicker, as demonstrated by its denser packing of 0.72 nm^2 per molecule. If solvent-free, this second monolayer should have a film thickness averaging 2.3 nm, but since the polyether chains are not tied down, they are free to extend up to the full extension of the polyether and nonylphenol groups (about 3.7 nm). Thus, when two oxide particles collide and each has two monolayers of RE-610, the thickness of each film should be of the order of 4.0 nm upon first contact and these can be compressed in the collision to no thinner than 3.0 nm. Thus, the ferric oxide surfaces of colliding particles with adsorbed bilayers of RE-610 can approach to within 6–8 nm of each other, close enough that the van der Waals attractive forces (plus magnetic forces) can exceed several kT and the particles will tend to flocculate into a deep 'secondary minimum' unless sufficient interparticle electrostatic repulsion is present, as is illustrated in Fig. 8, taken from Shaw [21]. The left-hand diagram shows the potential energy vs. interparticle distance when the interparticle forces consist of the van der Waals attraction (V_A) and a steric barrier (V_S); a deep (negative) secondary minimum exists, indicating the interparticle distance at which flocculation occurs. The right-hand diagram shows the same system to which electrostatic repulsion (V_R) has been added; the secondary minimum is gone, and no flocculation occurs, even though the electrostatic repulsion by itself would be insufficient to deflocculate the dispersion.

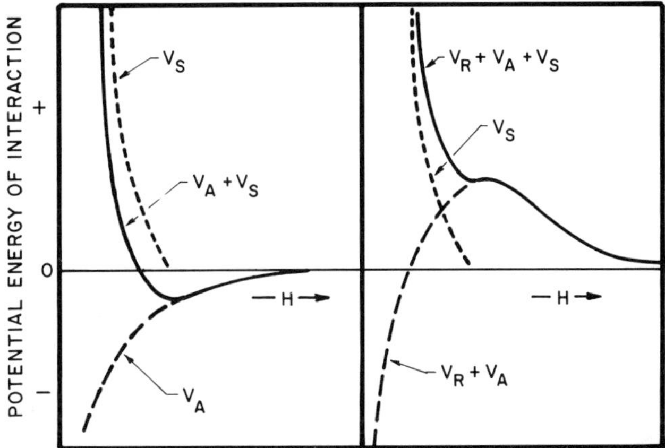

Figure 8. Potential energy diagrams for interparticle energy (V) vs. interparticle separation distance (H), summing van der Waals attractions (V_A), steric repulsion (V_S), and electrostatic repulsion (V_R). After Shaw, [21].

The electrostatic charging of particles in organic media results from the interaction of organic-soluble electron-acceptors (acids) with electron-donor (basic) surface sites of the particles, or from the interaction of soluble electron-donors (bases) with electron-accepting surface sites of the particles. We therefore expect γ-Fe$_2$O$_3$ particles in the mixed solvent (all three components of which are basic) to be negative because the basic solvent tends to accept protons from (or give up electrons to) the acidic FeOH surface sites; the experimental result (Table 1) was -29 ± 5 mV. Upon adding 0.1% of RE-610 to a 100 ppm suspension the zeta potential was reversed to $+10 \pm 2$ mV, indicating that the easily desorbed outer layer of RE-610 is attached through its acidic site, as expected for 1 mM RE-610.

The magnitude of the electrostatic repulsion U^e_{121} between particles with an effective radius a, and a surface potential of ψ_0, in a medium of dieletric constant ε_r, and at an interparticle separation H of 8 nm is given by the Deryagin equation [22]:

$$U^e_{121} = 2\pi\varepsilon_r\varepsilon_0 a\psi_0^2 \ln(1 + e^{-\kappa H}), \qquad (3)$$

where ε_0 is the permittivity of free space and $1/\kappa$ is the Debye length, which is predictable for non-aqueous media from the conductivity σ and the diffusion constant D for the charge-carrying species [23]:

$$1/\kappa = (\varepsilon_r \varepsilon_0 D/2\sigma)^{1/2}. \tag{4}$$

Earlier studies in non-aqueous media had concluded that the Debye lengths $(1/\kappa)$ were enormous, and that equation (3) was not applicable because of the lack of a counter-ion atmosphere; but when equation (4) was used and the conductivity σ was measured with the disperse phase present, it was found for many systems that the Debye length in organic media were often in the range of 10–100 nm, where equation (3) applies [24, 25]. In the earlier studies which concluded that the Debye lengths were enormous in liquids of low dielectric constant, the concentration of counter-ions was estimated from the conductivity of the dispersant solution, rather than from the conductivity of the disperson. Since the counter-ions are formed by acid–base interaction of the dispersant with the surface sites of the particles, there should be no counter-ions in the dispersant solution unless the particles are present. Johnson has made a strong point of this principle in his studies, and has measured the increase in conductivity of a non-aqueous suspension as the dispersant is titrated up to the concentration range where electrostatic stabilization occurs [25].

Equation (3) may be used for calculating the energy of electrostatic repulsion between flocculating ferric oxide particles, where the effective radius for the clumps of acicular particles (a) may be a poor approximation for the primary particles, but is a much better approximation for the clumps, which have the form of a 'hand' of bananas. Using equation (3) for particles with a Debye length of 50 nm and a particle radius of 100 nm, the electrostatic repulsion in the depth of the secondary minimum (at 8 nm separation) is 1 kT for a ψ_0 (or zeta potential) of 20 mV, 4 kT at 40 mV, and 10 kT at 60 mV. As the particles flocculate and the effective radius of the clumps doubles, the electrostatic repulsion also doubles, until the electrostatic repulsion is large enough to overcome continued flocculation. This process is illustrated in Figs 9–12, which are plots of the time dependence of average particle diameters, determined by photon correlation spectroscopy with the N4 photon correlation spectrometer. In all of these plots, the rate of floc growth tends to taper off as the floc size increases, often reaching a size where the rate of floc growth becomes negligible, a characteristic of electrostatic repulsion or of the combination of steric and electrostatic repulsion, but which is not observed with pure steric stabilization. With pure steric stabilization the smaller particles are the more stable, for these have the shallowest secondary minima; while with electrostatic repulsion, the larger particles have the stronger repulsion and have a shallow secondary minimum or none at all.

3.5 Dissolution of γ-Fe$_2$O$_3$ in RE-610 solutions

From weight loss and atomic emission studies of γ-Fe$_2$O$_3$ in solutions of RE-610 it was found that the amount of ferric oxide dissolved into the solutions varied from 0.07 to 0.3 wt%, depending on the RE-610 concentration. This function of RE-610 may be of importance in deflocculating the clumps of γ-Fe$_2$O$_3$ during the milling process, for dissolution of the attachment points between particles should be favored during milling because the enhanced mechanical stress can increase the Gibbs free energy of such attachments so that they are preferentially dissolved and the clumps broken up.

3.6. Dispersibility and dispersion stability of α- and γ-ferric oxides in RE-610 solutions

Suspensions of 4 g of α-Fe$_2$O$_3$ or γ-Fe$_2$O$_3$ were vigorously sand-milled into 20 ml RE-610 solutions of various concentrations, and small aliquots were immediately diluted down to about 20 ppm, sonicated, and then subjected to particle size analyses with the Coulter N4 photon correlation spectrometer for the next 2 h. The results are shown in Figs 9 and 10. Figure 9 shows that RE-610 alone at concentrations of 4.5 wt% and above has appreciably slowed down the flocculation of the non-magnetic α-oxides at Stokes diameters of about 160 nm, but in Fig. 10 the magnetic γ-oxides continually flocculate as the Stokes diameters climb from about 400–500 nm to 600–900 nm during the 2 h run. RE-610 has slowed down the flocculation rate of the magnetic γ-oxides appreciably, but not to any useful extent. Clearly, the magnetic attractive forces between γ-oxide particles have made dispersibility and dispersion stability much more difficult.

3.7. Colloidal stability with combined RE-610 and polymeric components

Figures 11 and 12 illustrate the increased dispersibility and dispersion stability attained with α- and γ-oxides when the polymeric components (Epon 2004, PVME, and the R-108 crosslinker) are added to the formulation. The polymeric

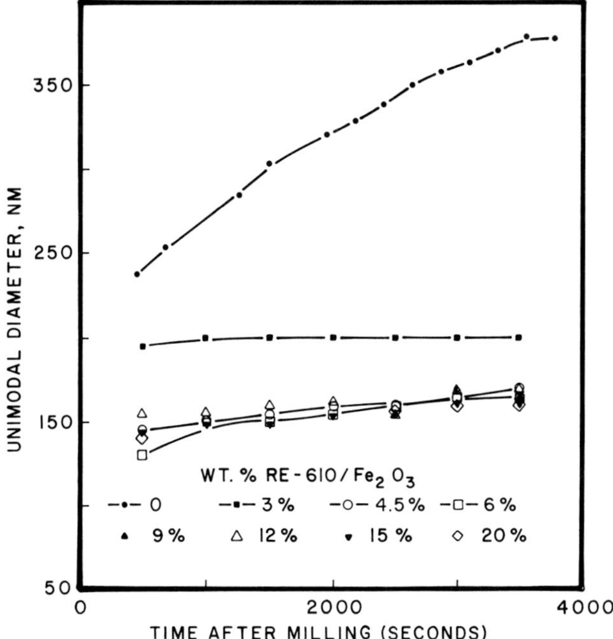

Figure 9. Flocculation rates of dispersions of α-Fe$_2$O$_3$ in RE-610 solutions in the mixed solvent; unimodal Stokes diameter distributions determined by Coulter N4 photon correlation spectrometry.

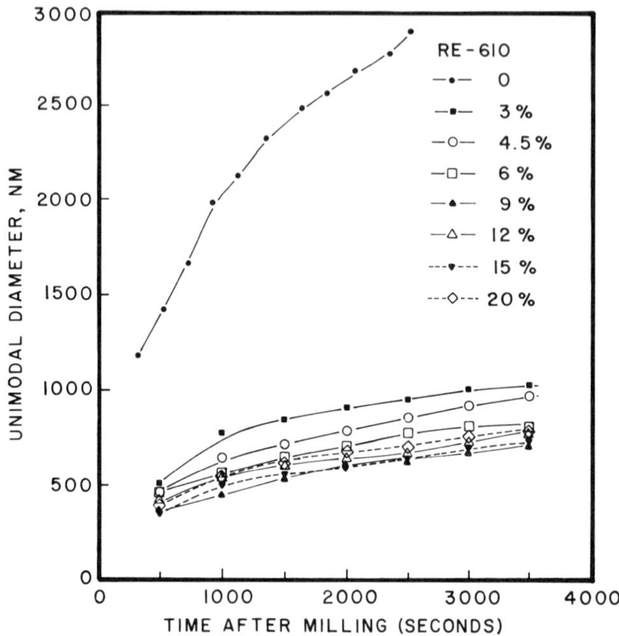

Figure 10. Flocculation rates of dispersions of γ-Fe$_2$O$_3$ in RE-610 solutions in the mixed solvent; unimodal Stokes diameter distributions determined by Coulter N4 photon correlation spectrometry.

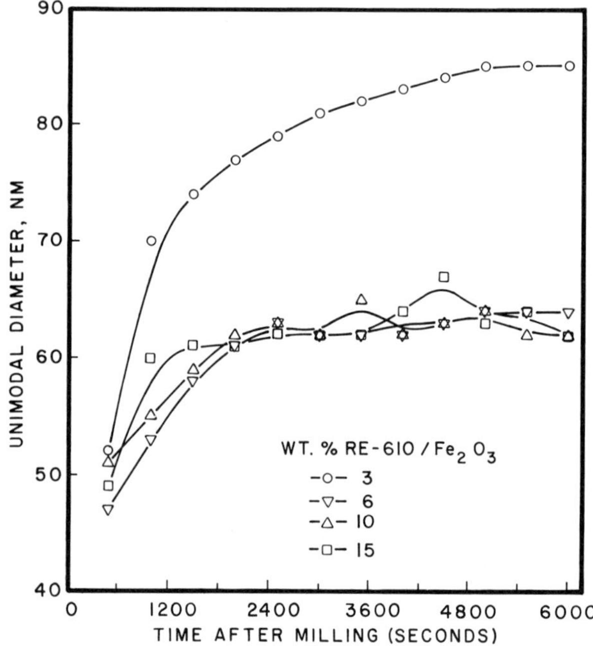

Figure 11. Flocculation rates of dispersions of α-Fe$_2$O$_3$ in solutions of RE-610 with 2.8 wt% of epoxy resin, 1.1 wt% of polyvinylmethylether, and R-108 phenolic crosslinker; unimodal Stokes diameter distributions determined by Coulter N4 photon correlation spectrometry.

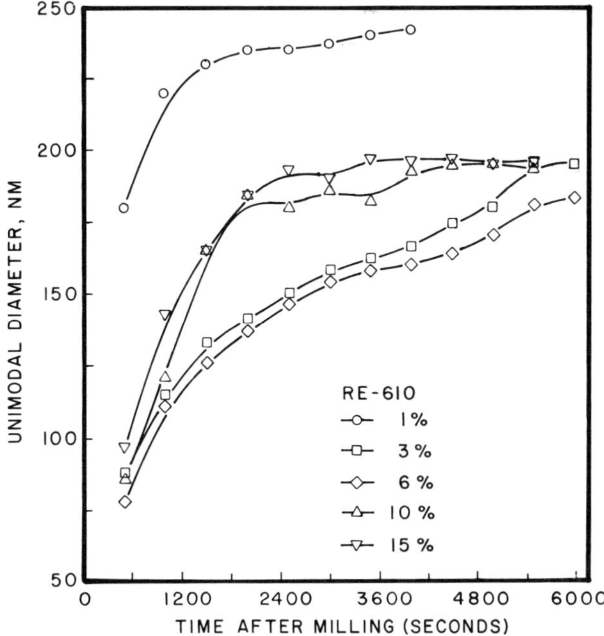

Figure 12. Flocculation rates of dispersions of γ-Fe_2O_3 in solutions of RE-610 with 2.8 wt% of epoxy resin, 1.1 wt% of polyvinylmethylether, and R-108 phenolic crosslinker, unimodal Stokes diameter distributions determined by Coulter N4 photon correlation spectrometry.

components undoubtedly provide a much thicker steric barrier, but there is evidence in Table 1 which suggests that magnetic γ-oxide particles with adsorbed PVME are more basic and that RE-610 tends to donate protons to the adsorbed PVME and thereby provide quite high zeta potentials ($+43 \pm 1$ mV).

Figure 11 shows that the α-oxides are now dispersed at a much smaller Stokes diameter (60–70 nm) than without the polymeric components (160 nm), but only when the RE-610 concentration is 6 wt% or more.

Figure 12 shows that the magnetic γ-oxides in the formulation with polymeric components now have a much greater stability, and the particle size is now much smaller than before. Stokes diameters of 170–190 nm are found with the γ-oxides in the polymeric formulation (when the RE-610 concentration is 3 wt% or higher), as compared with 600–900 nm Stokes diameters when the polymeric components are absent, and the rate of flocculation is quite low, thanks to the combination of steric and electrostatic interparticle repulsion.

4. JUSTIFICATION AND RECAPITULATION

The dispersion of γ-ferric oxides in a solution of epoxy resin, phenolic crosslinker, and PVME polymer in a mixture of basic solvents is used to cast films of magnetic polymer composites which are cured at elevated temperatures. The mechanical properties of the composite, especially toughness and abrasion resistance, as well as its magnetic properties, tend to depend to a large extent on the degree of dispersion of the magnetic particles, and on the magnitude of the work of adhesion of the matrix polymer to the surface sites of the particles [26]. Both the work of adhesion and the degree of dispersion of the magnetic particles depend on the acid–base interactions of the surface sites of the particles with the matrix polymer, for the adsorption of the polymer components onto the oxide by acid–base interaction made possible the steric stabilization required for this system, and also provided some degree of electrostatic stabilization. The recognition that the surface sites of ferric oxides have both acidic and basic character, and that both acidic and basic polymers adsorb on such surfaces, is an important step in the understanding of these systems and in the adhesion of polymers in general to the surface of ferric oxides, including steel surfaces.

Acknowledgements

The support of this project by IBM, San Jose, California is gratefully acknowledged, especially the active participation of Dr. Randall Simmons and Dr. Manfred Cantow. Michael Herman's expert assistance in the zeta potential measurements is much appreciated.

REFERENCES

1. F. M. Fowkes and M. A. Mostafa, *Ind. Eng. Chem. Prod. Res. Dev.* **17**, 3–7 (1978).
2. J. C. Bolger and A. S. Michaels, in: *Interface Conversion for Polymer Coatings*, P. Weiss and G. D. Cheever (Eds), pp. 3–60. Elsevier, New York (1968).
3. F. M. Fowkes, *J. Phys. Chem.* **64**, 726–728 (1960).
4. G. W. Simmons and B. C. Beard, *J. Phys. Chem.* **91**, 1143–1148 (1987).
5. F. M. Fowkes, D. C. McCarthy and D. O. Tischler, in: *Molecular Characterization of Composite Interfaces*, H. Ishida and G. Kumar (Eds), pp. 401–411. Plenum Press, New York (1985).
6. F. M. Fowkes, D. O. Tischler, J. A. Wolfe, L. A. Lannigan, C. M. Ademu-John and M. J. Halliwell, *J. Polym. Sci., Polym. Chem. Ed.* **22**, 547–566 (1984).
7. R. S. Drago, G. C. Vogel and T. E. Needham, *J. Am. Chem. Soc.* **93**, 6014 (1971).
8. R. S. Drago, L. B. Parr and C. S. Chamberlain, *J. Am. Chem. Soc.* **99**, 3203 (1977).
9. V. Gutmann, *The Donor–Acceptor Approach to Molecular Interactions*. Plenum Press, New York (1978).
10. F. L. Riddle, Jr. and F. M. Fowkes, *J. Am. Chem. Soc.* **112**, 3259–3264 (1990).
11. F. M. Fowkes, *J. Adhesion Sci. Technol.* in press.
12. F. M. Fowkes and D. A. Cole, manuscript in preparation.
13. S. T. Joslin and F. M. Fowkes *Ind. Eng. Chem. Prod. Res. Dev.* **24**, 369–375 (1985).
14. F. M. Fowkes, Y. C. Huang, B. A. Shah, M. J. Kulp and T. B. Lloyd, *Colloids Surf.* **29**, 243–261 (1988).
15. G. F. Hudson and S. Raghavan, *Colloids Surf.* **29**, 263–272 (1988).
16. J. L. van der Minne and P. H. J. Hermanie, *J. Colloid Sci.* **7**, 600 (1952); **8**, 38 (1953).
17. F. M. Fowkes, H. Jinnai, M. A. Mostafa, F. W. Anderson and R. J. Moore, in: *Colloids and Surfaces in Reprographic Technology*, ACS Symposium Series 200, pp. 307–324. Am. Chem. Soc. (1982).
18. K. Tamaribuchi and M. L. Smith, *J. Colloid Interface Sci.* **22**, 404 (1966).
19. Personal communication with R. L. Bradshaw, IBM General Products Division, Tucson, Arizona (1987).
20. *Gafac Surfactants, Technical Bulletin.* GAF Corporation, New York (1987).
21. D. J. Shaw, *Introduction to Colloid and Surface Chemistry*, p. 211. Butterworths, London (1966).
22. F. M. Fowkes and R. J. Pugh, in: *Polymer Adsorption and Dispersion Stability*, E. D. Goddard and B. Vincent (Eds), ACS Symposium Series 240, pp. 331–354. Am. Chem. Soc. (1984).
23. A. Klinkenburg and J. L. van der Minne, *Electrostatics in the Petroleum Industry*, p. 40. Elsevier, New York (1958).
24. F. M. Fowkes, in: *Ceramic Powder Science and Technology; Advances in Ceramics*, vol. 21, pp. 411–421 (1987).
25. R. E. Johnson, Jr. and W. H. Morrison, Jr., in: *Ceramic Powder Science and Technology; Advances in Ceramics*, vol. 21 p. 323 (1987).
26. F. M. Fowkes, D. W. Dwight, J. A. Manson, T. B. Lloyd, D. O. Tischler and B. A. Shah, *Mater. Res. Soc. Symp.* **119**, 223–234 (1988).